HARMONY, THE HEARTBEAT OF CREATION

HARMONY,
THE HEARTBEAT OF CREATION

*The Convergence of
Ancient Wisdom and Quantum Physics
in the Triune Pulse of Nature's Forms*

MONIQUE POMMIER

Lindisfarne Books | 2020

© 2020 by Monique Pommier

Published by Lindisfarne Books
an imprint of SteinerBooks/Anthroposophic Press, Inc.
402 Union Street #58, Hudson, NY 12534
www.steinerbooks.org

Cover paining © by Marie-Odile Pommier

Library of Congress Control Number: 2019944472

ISBN: 978-1-58420-925-6 (paperback)
ISBN: 978-1-58420-926-3 (eBook)

All rights reserved. No part of this publication may be reproduced, stored in a retrieval system, or transmitted, in any form or by any means, electronic, mechanical, photocopying, recording, or otherwise, without the prior written permission of the publisher.

CONTENTS

Foreword	xi
I. Prelude and Premises: The Enigma of Twelve	1
Two sides of reality: world and self and the dual genesis of this book	1
Zodiac and soul: A two-leaved portal into the mystery of twelve	9
The gap between cosmology and psychology—and the bridge of astrology	20
II. Overture: The Mandorla of Mediation	25
The inner reality of twelvefold structures: mediation	25
Symbols of mediation: Vesica and mandorla	27
The substance of mediation: Harmony	31
Harmonics, musical scale, standing waves	34
Third archetype of twelvefold mediation: The musical scale	38
The genesis of the twelvefold and its twin, the sevenfold	44
The circle of the fifth principle	46
III. The Harmonic Composition of the Universe	55
The Platonic cosmogony:	
Creation of the world soul by means of proportion and harmony	55
Soul and selfhood—a primordial twelvefold unit	68
Corporealization of the world soul into a soulful world	72
Physicalization of the world soul into five Platonic elements-solids	76
First groundings of Plato's harmonic universe:	
Kepler's celestial physics and Bode's law	80
Modern insights into the harmonic structure of	
Platonic solids and atomic elements	90
The golden fifth, the living bond of creation	104
The secret physics of Life: bonding elements	
into 12fold mediating molecules—Chlorophyll and DNA	108
Similar twelve-toned matrices in Indian, Chinese and Celtic cultures	113
The dodecahedron—the quintessential form of the universe	118
The zodiac: Etheric matrix of the solar system and time frame of the soul	121
Zodiac and soul: "Means" of harmony making, and milieus of evolution	125

The gradual silencing of the harmonic borderland between Heaven and Earth	126
The 12fold/7fold harmonic composition of the human body	134
The threefold ground of the archetypal sevenfold	144
The golden mean—the geometric vessel of love, the soul of life	148

IV. The Spiraling Dynamics of the Universe 163

From harmonic structures in space to spiraling motions in time	163
Logarithmicity: The "rhythm of the Logos"	172
Two operations woven into one: Addition and multiplication	174
The logarithmic pulse of forms of nature and processes of perception	176
The dynamics of logarithmicity: A threefold bundling process	178
The threefold operation of perception	182
The logic of world dimensions and levels of being	184
The logarithmic chain of Being	187
Sense perception sheds light on the inner reality of logarithmicity	192
Cosmos and Nature's common sense	195
The fundamental dynamics of Creation and its reflection in the human mind	199
Threefold, twelvefold and sevenfold:	
The archetypal bundle of spirit, soul, and body	205
The archetypal logarithmic spiral: The golden spiral	211
The Fibonacci series: The golden spiral adapted to finite ends	220
Phi, the "figure" and profile of the Logos	228
The golden dynamics of the Logos: The pulse of the world	230
The spiraling universe: Physical facts of nature, body and cosmos	235
The grand self-sameness of the forms of nature, and cosmos:	
The spherical vortex	259
The dual vortex nature of the human experience	277

V. A Once and Future Cosmogony 289

Fundamental oneness	293
Polarization and triangulation of the One: The ancient view	300
Modern science rediscovers the tri-unity of the world	307
Inklings about dualization and threefolding in contemporary cosmology	319
The triune ground of matter: Field, mass, energy	321
The quantum paradigm: Emergence of a logarithmic order	329
Quantum order of matter: Quantum dynamics of evolution	338
Spin, whirling holomovement, and double helix	343
String theory and the logarithmic model	354

The fourth cosmological step:
    The deployment of the triad into a 3–7–12fold form 358
From threefold field to 7/12fold form: From resonance to consonance . . 365
The mystery of sevenfoldness:
    Law of the octave and principle of self-replication 372
The twelvefold inwardness of the sevenfold form 384

VI. Facets of a Unitive Science 389

Toward a harmonic science of the forms and systems of nature . . . 389
Toward a science of nature's formative process:
    A logarithmic bundling of cosmic quanta 392
Toward a science of the evolution of consciousness as a process
    of harmonization: From Yoga to alchemy and Jung's
    depth psychology—from Four back to Three, Two, and One . 397
Astrology: A cosmo-psychological science 424

VII. Universality of the Harmonic Model of Creation 442

From the Dogons to Indian musicology to the Sumerian Pantheon . 442
Mysterious numbers: The numbers of the mystery of creation . . 445
Indian musicology and the Sumerian Pantheon:
    A radical light on cosmology 449
The ancient significance of numbers 464
A mystery-number in quantum electrodynamics—137 . . . 470
Numbers and the fundamental form of life 487

Epilogue 495

Illustration Credits 502

Bibliography 503

Index of Topics, Names, and Numbers 509

To Marie-Odile
who lighted this journey
from the other side

Human beings, vegetables, or cosmic dust, we all dance
to a mysterious tune intoned in the distance by an invisible piper.

Everyone who is seriously involved in the pursuit of science
becomes convinced that a spirit is manifest in the laws of
the Universe—a Spirit vastly superior to that of man.

I have deep faith that the principle of the
universe will be beautiful and simple.
—Albert Einstein

Just as we place images, Imaginations, before the soul, so
on still higher levels the inner power of numbers is
placed before human beings. Human beings have
to learn to experience the inner proportions
of numbers as spiritual music.
Of particular importance
is the proportion
1 : 3 : 7 : 12.
—Rudolf Steiner

The sought-after borderland between physics and
psychology lies in the secret of the number.
— Carl Jung

FOREWORD

Not until the expedition that resulted in the present book was well on its way did I become aware of its actual destination—and my unrealized intent—to understand how the world is made. Not a scientist by training or vocation, I never entertained the aspiration to climb my way to a point of view on creation—let alone the thought I could ever set eyes on what was to loom as a deeply heartening ground of cosmological understanding.

Not until the trails started to assemble into a coherent landscape did a majestic unity begin to rustle around its inner folds and a unison gather from within the whirling—a single tone that continued to grow in volume and depth as it reverberated around the peripheries of world physicality and consciousness.

However satisfying facts may be, whether mathematical or geometric, biological or astronomical, the understanding they grant is relatively local and contextual. One aspect of reality is visited, elucidated and marveled at. It may brush up with the reality of the world as a whole, it may deliver applications that revolutionize human experience. In and of themselves however, isolated facts cannot embrace, let alone explain the living phenomena of nature and cosmos. Abstracted from nature's formative processes and cycles of becoming, facts never seem to catch up with the intelligence of life.

The fractal character and symmetric arrangement of the forms of nature, the recurrence of the golden mean and Fibonacci series in its geometries, the dynamics of the electromagnetic field, the quantum mechanics of matter, the periodic order of the elements stand out as some of the most eloquent facts of nature. Each is an expression of the phenomenon of phenomena: creation. Each also represents a seal on its mystery.

This book shares the adventure that took me to a panoramic viewpoint on the beginnings of the universe and the generation of the forms of nature. As the journey unfolded from intrigued broodings over cosmic order to lucid marveling at its underpinnings, trails were laid for future travelers to find their way from admiring nature's wonders to understanding the vibrant principles at work in their formation and composition—an understanding at once scientific and spiritual. What emerged in the end was a unitive view of nature and being, cosmos and consciousness, physics and psyche.

Who has not been awed by the precise geometries of crystals and flowers, the constructs of shells and fruit, the wondrous coincidences among cosmic measurements? Why it might be so, however, and what operation could govern these enigmatic orderings of nature, is the great mystery veiled in its exuberant foliage and myriads of stars. To lift a corner of this veil proved to be the essential purpose of this cosmo–spiritual journey.

> If you have entertained such questions as: What is the relation between consciousness and matter? If there is unity in the universe, on what does it rest? And how might it coexist with its obvious diversity? Could there be a universal pattern at work in all forms of life? Why is the golden mean so prevalent in nature's forms? What is the secret behind the rhythmic intervals between successive branches and leaves? If such questions have crossed your mind, this journey is for you.
>
> If you wonder what quantum physics might be saying about the world, and how it could possibly fit with our ordinary perception of reality,
>
>> If you are intrigued by key numbers (such as seven and twelve) in the age-old partitioning of time, space and consciousness,
>>
>> If you once probed into the mystery of such recurrent numbers and rhythms in nature or cosmos, but felt dismayed by the shoddy approximations this apparent cosmic order is fraught with,
>>
>> If love of music draws you to explore the Pythagorean pathways to cosmology hidden in notes and intervals,
>>
>> If you believe astrology rests on deeply logical grounds yet to be fathomed,
>>
>> This cosmological journey is definitely for you.
>
> If an interest in psychospiritual development is leading you to seek an integrative understanding of human nature and becoming,
>
>> If, a Buddhist or a Christian, Jewish or Moslem, you sense the great constant behind the many figures and symbols of God,
>>
>> If you ponder the grounds of a unitive approach to objective and subjective worlds, cosmos and consciousness,
>>
>> If esoteric traditions have brought you to see the underlying unity of creeds and cosmogonies, and you anticipate that science will eventually join hands with the world's ageless intuitions about Creation to elaborate a unified cosmogony,
>>
>> The trails ahead will take you to such a unitive viewpoint.

I. PRELUDE AND PREMISES:
THE ENIGMA OF TWELVE

Two sides of reality: world and self and the dual genesis of this book

> *Without the slightest doubt there is something through which material and spiritual energy hold together and are complementary. In the last analysis, somehow or other, there must be a single energy operating in the world.*
> —Teilhard de Chardin[1]

> *The measuring device has been constructed by the observer, and we have to remember that what we observe is not nature in itself but nature exposed to our method of questioning.* —Werner Heisenberg[2]

Scientific questions about the universe and spiritual questions about self surge from one and the same source: the mind. They arise from the encounter between self and world, which is the pulse of the human experience and the mystery of humankind. How these two sides of reality—outside world and inner self—actually connect and relate, however, eludes the unified understanding that science aspires to reach and religion means to provide. Science has been traditionally occupied with the question of matter, and psychospiritual quests with the question of consciousness. Beyond these two questions lies the more central one, which quantum physics has brought to the attention of science and spirituality has forever aimed at mastering: What is the relation between matter and mind, subjectivity and world physicality?

The only sources to offer a unitary view of self and universe are ancient cosmologies and world theosophies. Contemporary science remains largely disconnected from these inclusive accounts of the creation of the universe and the evolution of beings. Admittedly, however charged such teachings may be with knowledge that illumines the multidimensional grounds of human experience, they do not provide, for the nonphysical realities they include, the physical experiments, data or

---

1 De Chardin, *The Phenomenon of Man*, p. 63.
2 Heisenberg, *Physics and philosophy*.

equations that modern science requires to give serious attention to a paradigm, let alone a seal of "accepted theory." On the other hand, contemporary science does not account for the majority of human questions and experiences. While casting consciousness and mind outside of its field may have been a necessary step along the way, and a fruitful one, this very field has now disabused science of the illusion of pure objectivity. The discovery that elementary particles are not independent of the human intervention that isolates and measures them has rendered purely materialistic premises obsolete and propelled the question of the relation between world and mind to the foreground of scientific pondering, prompting scientists to reach for more integrative approaches to the universe. If the "object" of observation shifts under the observing subject, no longer can objective and subjective worlds be regarded as absolutely distinct. What their relation is, however, remains a mystery.

To approach this mystery is to engage the controversial question of what came first, matter or spirit. Are life, and later mind, chance epiphenomena of matter's increasing complexity? Or does consciousness precede matter and generate the physical world and its creatures as ancient cosmogonies and world Scriptures describe? Scientific progress itself has brought these questions, traditionally regarded as "metaphysical," to the center stage of physics. To see the tangibility of matter dissolve into an immaterial field of waves and quasi-particles is to witness that an invisible reality (field) precedes and underlies the visible (things). To the confounding discovery that the condition of matter (particle or wave) is entangled with the act of observation, the Big Bang theory has added a reformulation of world beginnings. Both have made the issue of what came first a vital scientific question.

So two great mysteries stand before us: the relation between psyche and world, and the question of which is primary: consciousness or matter. Although the simple fact that mind creates science would argue for the primacy of mind, modern science focuses exclusively on the empirical investigation of the physical world, trusting that further dissection of physical components and analysis of biophysiological functions will eventually explain how life and mind arise from matter. By contrast, ancient wisdom regards matter as the densest expression of spirit and the physical universe as the deployment of an intelligent living spirit Being. While scientific knowledge builds itself "up" from an intricate ground of quantitative data and experiments, ancient cosmologies lay out the descent and diversification of the world from a primeval oneness. If science offers sophisticated penetrations into the details and laws of the physical world, ancient teachings offer comprehensive vistas of the organization and evolution of the universe.

Since physics has not been able, thus far, to shed light on the relation between matter and mind, it would seem simply logical to envision exploring the reverse perspective, and see if valuable insights might not arise from the premise that matter and forms *derive* from an intelligent life, and that the physical world is a manifestation of a mindful source, be it referred to as spirit or "God." The materialistic premise of modern science is no less a "theory" than the "idealistic" premise of a "living One" informing and pervading all material forms. Whitehead pointed out the importance of distinguishing "between the authority of science in the determination of its own methodology and the authority of science in the determination of the ultimate categories of explanation." If the Newtonian laws provided an invaluable understanding of the physical world, which supported the development of the analytical clarity and ingenious instrumentation of modernity, this framework is now turning into a creative hurdle for the scientific mind seeking its way to the transdual paradigm of quantum physics. In fact, confronted by their revolutionary findings, a number of quantum physicists felt the necessity to step beyond the materialistic mindset and ponder the new facts in the light of ancient sources of knowledge.

The cosmological descent of a transcendent source of life into matter and forms is the less traveled road of inquiry this book will follow, in an attempt to ascertain whether ancient descriptions of world creation can offer a reasonable, "acceptable" account of such a process, one apt to meet the factual requirements of the modern mind, and so shed light on the relation between the two sides of our reality. After all, "accepted theory" is science's seal of approval. Only the inevitable institutionalization of science leads us to believe that its latest findings and shifting paradigms are absolute truths. Scientists would agree that science is and will always be transitional.

In this case, acceptability cannot rest solely on physical criteria, since such criteria exclude nonphysical realities such as mind or consciousness. Empirical research in the field of mind and body connection has demonstrated the possibility to register variations in qualitative states of consciousness, thus revealing the existence of a link between subjective experience and objective data. However, the nature of this bridge remains an enigma. On the other hand, the subjective intimation of truth that may seize the reader of religious and theosophical cosmologies does not have the universal character of a sense perceptible fact. If science has been eviscerated of subjectivity, ancient wisdom lacks the objective grounds that the modern intellect requires to have a fully satisfying experience of plausibility.

Beyond the seemingly radical dichotomy between empirical verification and intuition of truth, which maintains an abyss between modern science and ageless wisdom, there exists a middle way of noble logic, which honors the reasoning requirements of the intellect yet does not bury its living, wholistic spirit in fragmented quantitative data. Such a "noble logic" emulates the logic of life. Rather than pull apart the fabric of nature to analyze its smallest components, it seeks to grasp its form building operations by blending matter-of-fact observation with attentiveness to the wholistic "spirit" of life's productions. It focuses less on separate parts and their features than on the dynamic phenomenon that results in a physical form. To ponder the logic of a particle, for instance, would mean to ponder the phenomenon of its formation. How does the universe produce a particle? How do plants, shells, bodies, atoms unfold from the invisible? Such logic is not based on a purely external approach to the object, aimed at isolating its smallest components, but on embracing, from within, the dynamics that shaped it. Rather than cumulative experiments and quantitative measurements, this approach to knowing—in other words, to "science"—rests on the qualitative ground of analogies and patterns. It gains insights from whole perspectives rather than partial data. When the intellect allows its logic to make space for the noble, nature-making wisdom of life, human logic is ennobled by cosmic logic. Being inevitably the logic of *our* nature, the logic of nature does more than satisfy the curiosity of our brains, it heartens our being.

Since the approach to universe and consciousness proposed in this book reverses the perspective of contemporary science, willingness to suspend philosophical beliefs and lay aside scientific biases may be needed on the part of the reader to allow for a genuine dialogue with the various sciences engaged in this work and the unitive ground summoned up by their convergence. The greatest stretch to the current mindset is the premise of this approach that the universe is a living system—a live entity. If the ancients believed or rather "knew" this as experiential evidence orchestrated by their cosmologies, the modern intellect can also find scientific grounds for this view in sources such as James Lovelock's Gaia hypothesis. In the 1960s, Lovelock made a convincing case for his proposal that the Earth is a self-regulating system that can be thought of as a single live organism, a view that more recent information about the cosmos invites to extrapolate to the universe as a whole and to its systems, from plants to cells, and from planets to stars.

The Big Bang theory, which proposes that everything in the cosmos emanates from a single point of time–space–energy, implies that a singular source precedes and encompasses the multiplicity of the universe—which could be described, in the beginning, as a resounding and expanding cosmic whole, oneness, or "entity." Similarly, the embryonic development of a single cellular unit into trillions of cells and complex organs with vastly different functions points to singular origin and wholeness of being for the human "entity." The notion of a continuous electromagnetic, then quantum field subtending the universe and connecting particles via instantaneous means that transcend mechanistic laws has also contributed to an emerging conception of the universe as a highly integrative, coherent system—born as a singularity and evolving as a coherent whole.[3] So, even though mainstream science might consider it irrelevant to its domain to ponder the implication of such oneness of beginnings—which is that the universe is one great system entity—this notion can no longer be reduced to an Eastern "mystical" view.

To embrace the hypothesis of an intelligent living universe is clearly necessary if one is to explore the connection of mind and matter from the ancient perspective that the physical world is the densest manifestation of a world being. If the world is alive and one, as such perspectives richly explain and majestically proclaim, the core scientific question is: How does the universe actually propagate and maintain its primordial oneness throughout its parts? Where is such oneness to be found in the infinitely diverse formations of matter, nature, and cosmos? Is there a common pattern inherent to systems of cosmos and forms of nature that might point to a universal identity—an *identity* of the universe? Should such a signature exist, it would signal that a one-minded "subjectivity" presides over the formation of world parts and particles. It would also shed light on the connection between observing mind and observed matter, which physics has witnessed but not explained. It would provide insights into the elusive rapport between subjective consciousness and objective world.

A few decades prior to Lovelock's Gaia hypothesis, the Austrian biologist Ludwig von Bertalanffy proposed a "general systems theory," which introduced system thinking and laid the ground for the emergence of a systemic approach to domains as diverse as biology, ecology, sociology and organizational psychology. The notion that the whole has "emergent" properties that cannot be derived from the properties of its constituents, defined a new paradigm in philosophy and science. Challenging the familiar mechanistic framework, which views the world as an aggregate of parts, system thinking introduced a "whole-centered" approach to organisms and organizations, focused on internal patterns and functional dynamics. While making use of the specialized knowledge that continues to be the gift of modern science, this new way to apprehend reality has birthed a developing field of unifying research and cross-pollinating insights between relatively separate domains—such as the biological and the sociological.

The core principle of the new paradigm—"self-organizing" and self-regulating—introduced a seemingly benign yet radical shift in the perception of "external" realities. The idea that systems self-organize and self-correct implies that they harbor a certain inwardness. Recognizing *self*-organizing properties in socioeconomic structures or biological arrangements surreptitiously grants them interiority and subjectivity. Imbued with internal feedback dynamics and modulations, seemingly

---

3   Laszlo, *Science and the Akashic Field*, pp. 144–145.

random external assemblages turn out to be quasi-"entities," relatively autonomous wholes, complete with a defining identity (self), a definite form or structure (system), and formative–transformative processes (organizing). A significant moment in cultural history, system theories point to a realignment of the compass of "science" to its pole star, which, rather than calling for a maximum dissecting of the world into objective parts, points to the subjective and objective essence of cognition reflected in the notion of "self-organizing system," and calls for a unifying grasp apt to satisfy the requirements of the intellect for objective facts, as well as the subjective claim of the soul for meaning and coherence.

When Einstein stated, around the same period, *it is the theory that decides what one can observe,* he placed his finger on the fact that science itself is a self-organizing system, a sobering truth for a domain based on the "theory" of objectivity. The blow had already been dealt, however, to the old premise, and the notion of objective particles forever compromised by the incontrovertible presence of the observing subject.

Science is a self-organizing system, and, from the perspective of ancient wisdom, so is the world. The world being the object of science, it would seem logical for the self-organizing of science—its theory—to aim at coinciding with the *self-organizing* mysteries of the universe.

This self-organizing logic of nature and cosmos was to be the golden trail this book would follow—inside out and outside in.

The less traveled road of a cosmological expedition from spirit to matter was not a road I intentionally stepped on, nor is this work a book I set out to write. The adventure that took me, somewhat inadvertently, to the "world-making" heart of the universe caught me by surprise, although to call it an entirely spontaneous adventure would be to ignore the power of long standing inquiries to bring about the convergence of currents that can pull us out of a familiar stream and drop us, blindfolded, on the way to a Garden of the Hesperides. While I had no intention to tackle cosmological mysteries, I can also see, having returned from the unforeseen ends of this journey, that what brought it about were years of pondering the mystery of the number twelve. This number, along with its complementary, seven, would turn out to epitomize the noble logic of cosmos and harbor a cosmological nucleus.

A fascinating character of the adventures of the Grail knights is how they simply "ride into the forest," with no plan or intent other than questing for the mysterious Grail, and how magical encounters or places *find them*—a castle, a damsel in distress, a hermit—and numinous happenings unfold. At the end of the day, an

I. Prelude and Premises: The enigma of twelve

undefinable charge of meaning has accrued from this simple trusting forth. What leads them is not a thoughtful intention, but the sure compass of their inquiring heart, ever preceding them from the future.

Not unlike the medieval seekers, trusting forth was to be this journey's mode of transportation. Not unlike a Grail, the enigmatic "attractor" of twelveness is what led me to the intertwined realities of universe and self, spirit and matter, lodged in Plato's cosmology. It deposited clues under trees, left hints hanging between branches. The compelling resonances between Plato's harmonic scheme of world beginnings and spiritual–scientific tenets about "twelve" prompted me to want to test its factual accuracy by walking, step by step, down the twists and turns of supportive scientific findings, to the bottom line of facts. Physical corroborations of Plato's "metaphysical" mathematics in turn paved a bridge to the "trans-physical," immaterial field of contemporary physics, opening the way to a potentially integrative view of the formative matrix of the world.

Consequently, this book is not organized by themes, nor is it structured by a priori reflections. Instead, it follows the geo-cosmographical pathways presented by the unraveling of its own questioning-finding momentum. Ideas and insights arise from the viewpoints offered by the progression of the journey of discovery itself. Observations on history or ponderings on such notions as logos or reason may thus reappear at different times, as if brought to view by a different round of climbing up and down, in and out, the world land.

A lay inquirer and not a physicist, I found myself rather ill equipped for the scientific nature of the terrain. However, serendipitous openings and unexpected findings steadily invited further penetration into the unfamiliar surroundings. The gradual unfolding of a landscape of world beginnings, which displayed no less than the common ground of world and mind, became rapidly compelling, even breathtaking. The insights provided by lineages of seekers cleared pathways through the luxuriant veils of nature. Ultimately, the vivifying wholeness and panoramic coherence that gathered around this steadfast venturing forth generated the ground of the work. Whether the profound beauty of a unitive dynamics is a convincing enough counterpoint to the quantitative solidity of intricate data, empirical verifications and elaborate equations, will be for the reader to decide. I can only let the symphonic vitality of the universal grasp of reality it lays out vouch for itself. If the cosmic organicity to emerge from the entire journey offers, admittedly, but a general outline of the body of the world, this body template has a heartbeat and a breath that may be found to "verify" its life—which in turn verifies its "truth," if we embrace the Goethean view that the veritable criterion of truth is its power to bear fruit, that is to generate life. What bears fruit is by essence: alive, which points to life and life giving as the secret core of truth—and of the bearer of truth: the mind. "It is not always needful," says Goethe, for truth to take a definite shape; it is enough if it hovers about us like a spirit and produces harmony."[4]

To approach the world as a living entity is to approach a "Creature" that is, as we are, multidimensional and self-contained. Any attempt to penetrate its mysteries is bound to turn into a spherical and multilayered venture, rather than a sequential one. Such is the character of this book—spherical and layered.

The first part unravels the harmonic underpinnings of fundamental structures of our world and consciousness. The second follows the vital course of life around the forms of nature and cosmos, tracing its unique flow and discovering its secret rhythm. The third part swirls ancient wisdoms and current sciences into a unitive cosmological understanding. The fourth part gathers into one science of harmonization the formative processes of nature and the evolutionary processes of the soul, up the psychospiritual lining of the world spheres. The fifth part orchestrates with worldwide cultural resonances the harmonic pattern of deployment of the world-being and essentializes it to an ultimate, archetypal "round."

Each round reflects the two-faceted character of its world model, as it moves in and out of subjective and objective realms, physical facts and nonphysical experiences. Treading the philosophical and the scientific, the historical and the geometric, each clears a path that is at once cosmological and psychological—around the enigmatic hub that bonds world and self and hosts their common core.

The itinerary is carefully mapped and light gears (scientific and esoteric) supplied, so anyone intrigued by the intelligent physics of the universe and willing to examine a few intricacies of nature's laws, cross a mathematical bridge or two, brave a gentle climb through the cutting edges of science, can make his/her way to panoramic sightings of the cascading unfolding of the universe and pause in luminous intimations of the vibrant intelligence that runs through the minutest particles and vastest configurations of nature and cosmos. Dense thickets along the way break into "viewpoints" that are signaled—✪—as places of special significance to take in the latest vista and contemplate the "perspective" gained.

In the end, my hope is that the patient self-educating made necessary by my lack of scientific equipment will translate into clarity and simplicity of access for the reader unversed in mathematics and physics. Credit must be given to the same for the bold innocence of stepping into the forbidding field of quantum physics and seeing its large strokes expose the same vibrant core of reality. Limited scientific training can have the value of preserving a spaciousness of perspective, free from the sophisticated knowledge that can obscure simpler modes of understanding.

---

[4] Goethe, *Maxims and Reflections*.

I. Prelude and Premises: The enigma of twelve

It is also my hope that the glorious sight of the world soul's profile at rest in the shapes of nature, the tempo of life pulsing through the branching of trees, the harmonic modulations toning plants and beings, will compensate for the incompleteness of the view proposed. To unravel the secrets of world foundations held in the inconspicuous numbers of clocks and calendars will have been the unanticipated purpose of the adventure. To discover the rhythmic grace that shapes nature and moves cosmos out of an overflowing fount of love-filled harmony will have been its gradually self-unraveling gift.

Moreover, the waves of joy that surge from rediscovering the world through its center and seeing how its forms swell out of one mysterious formative hub will have been what fueled and sustained the entire labor; riding them, what carried the journey onward. Certainly, waves can plunge one deep under before they take us onward, conjure up dread before raising us to exhilarating crests, and pause to an ominous stillness before momentum gathers again, but waves of the heart of life they are, and trusting their suprahuman course brings us inevitably to the very source of our being—not only the awe inspiring sight of the whole, but the bliss of partaking, if only in the leap of the heart of life within us, in the ongoing act of creation.

ZODIAC AND SOUL: A TWO-LEAVED PORTAL INTO THE MYSTERY OF TWELVE

> *Socrates asked Xenophon from whence we have conceived the soul, if there is none in the world. And I ask, whence speech, whence the regular harmony of speech, whence song?* —CICERO

If understanding how a primordial "one" becomes and pervades a manifold universe is, in the end, the main thrust of this book, it was not its starting point. What prompted this journey was yet another walk around the more modest question I had been contemplating for a number of years: the mystery of "twelve."

Why preoccupy yourself with the number twelve, anyone may ask? And yet, the intriguing recurrence of the number twelve in the most daily and sacred contexts of human experience might as well elicit the reverse question: why hasn't more attention been given to twelve?

First and foremost, twelve presides over the grand partition of space and time: 12 hours to a day and a night, 12 months to a year, 12 zodiacal ages to the 25,800 years precessional cycle referred to as the Platonic year. Twelve organizes time by dividing in equal sectors the sphere of cosmic space in which the Earth rotates and experiences its collective rhythms of night/day, months and seasons. Replicated on all the clocks and calendars of the globe, this 12fold partitioning of space and time is ordained cosmically by the cyclical relation between Earth, Sun and Moon—revolution cycle of the Earth around the Sun and cycle of the Moon around the Earth—which results in approximately 12 lunar cycles within the yearly cycle of the Sun.

Twelve is often found in the company of another number: seven. Where there is a twelvefold, a sevenfold is generally not far.[5] The organization of time, for instance—12 hours, 12 months, and 7 days of the week—mirrors the cosmic fundamentals of 12fold zodiac and 7 planets, which constitute the parameters of astrology: 12 zodiacal constellations and 7 traditional planets. The musical scale refers alternatively to a diatonic set of 7 notes and a chromatic set of 12 notes. 12 and 7 are present side by side in our physical and energetic constitution: 7 cervical and 12 thoracic vertebrae of the spine; 7 energy centers (chakras) and 12-petal lotus of the heart center.

Twelve and seven intertwine as well in our spiritual history since ancient times: The Rig Veda, the most ancient book in Indo–European language,[6] lays it out as the grand cosmic enigma:

> Twelve spokes, one wheel, navels three
> Who can comprehend this?
> On it are placed together three hundred and sixty like pegs;
> They shake not in the least. (Dirghatamas, *Rig Veda*, Hymn 1.164.48)
>
> A seven-named horse does draw this
>   three-naved wheel
> Ageless and irresistible as well,
> Which props all worlds.
> Seven steeds draw the seven-wheeled chariot...
> Wherein are placed the sacred notes seven...(Ibid., 1.164.15)

---

[5] The alternate use throughout the book of *12* vs. *twelve*, *7* vs. *seven*, *12fold* vs. *twelvefold*, *7fold* vs. *sevenfold* is intended to highlight two facets of each number or structure. Depending on the context, *12* and *7* emphasize the *particulate* facet of fundamental structures from the quantitative point of view. *Twelvefold* and *sevenfold*, on the other hand, emphasize the qualitative tenor of such structures and support a feeling for the unique forms of *wholeness* and cosmic organisms they represent. In general, numbers have been intentionally kept as figures rather than words throughout the text to highlight the mathematical key-operations they speak to in the organization of cosmos. On the other hand, spelled out words felt most apt, when appropriate, to emphasize the inner reality and cosmological significance of numbers.

[6] Feuerstein, Kak, and Frawley note in their book *In Search of the Cradle of Civilization*: "The Rig-Veda is the oldest book in the Sanskrit language, indeed in any Indo-European language. In addition, if we are correct, it is the oldest book in the world.... The fact that the Rig-Veda mentions a stellar configuration that corresponds to a date from 6000 to 7000 BC suggests that it is of that age, rather than the generally assumed dating of 1500 and 1200 BC."

I. Prelude and Premises: The enigma of twelve

The Bible opens with the 7 days of Genesis and ends with the 12fold heavenly city of the Apocalypse, sealing its "heavenly" knowledge along the way in such iconic groupings as the 7 pillars of wisdom, the 12 tribes of Israel, the 12 disciples of Jesus, the 7 horses and seals of the Book of Revelation. Tibetan wisdom refers to 12 creative hierarchies and 7 levels of being; Islam, to 12 imams and 7 cosmographic regions or "climates." The emperor Julian spoke enigmatically of the god with seven rays.[7] Theosophy distinguishes 7 evolutionary rounds, globes, and root races and 12 ages. Central in the Rig Veda are 7 primary seers or Rishis. Associated, not only in India but also in Persia and China, with the 7 stars of the Big Dipper, they are regarded as the guiding lights of the 7 chakras.

As inert and inconsequential as it may appear around the worldwide clocks of time and the age old wheels of astrology, the number 12 hints again and again, along with its twin, 7, at fundamentals of our solar systemic universe and meaningful structures of humankind.

The maps of the constitution of the human being and the solar system presented in theosophical teachings also emphasize these two numbers. The picture laid out by Alice Bailey (fig. 1) displays 7 levels of being, ranging from physical to spiritual. In their midst is the central structure of "selfhood": the 12fold Egoic Lotus.

So the question arises, what is twelveness about, that it accompanies such great orderings of time, space, and consciousness? Is there a fundamental law hidden in this number, which might shed light on the underpinnings of our space–time reality and its connection with subjectivity?

A small event resurfaced in my memory a few years ago, as thoughts were gathering once again around the question of twelve. It may have been the winter of 1993. I was waiting for the subway on the cold windy platform. A middle-aged bearded man was sitting next to me on the bench. On his lap was a plastic bag filled with what looked like large blue stars made of strong glossy paper. With a smile, he pulled one out and showed it to me, explaining in a soft-spoken voice that they were twelve-pointed stars he made and was selling for a few dollars. Charmed by the star, his maker, the magic of the approaching solstice, I reached for my wallet, at which point he handed one to me and said it was a gift. I never saw him again. I realized much later that this was the time when my interest in the mystery of twelve was germinating. I wonder who this Magi-like passer-by was who seemed to know my way better than I did and gently pointed to it with an epiphanic gleam in his eyes?

At the time, I had been practicing as an astrologer for quite a few years, while continuing the esoteric studies that had led me to astrology. A few months earlier, reading Alice Bailey's depiction of the 12fold Egoic lotus had intensified my

---

7   D'Olivet, quoted in Godwin, *The Harmony of the Spheres*, p. 346.

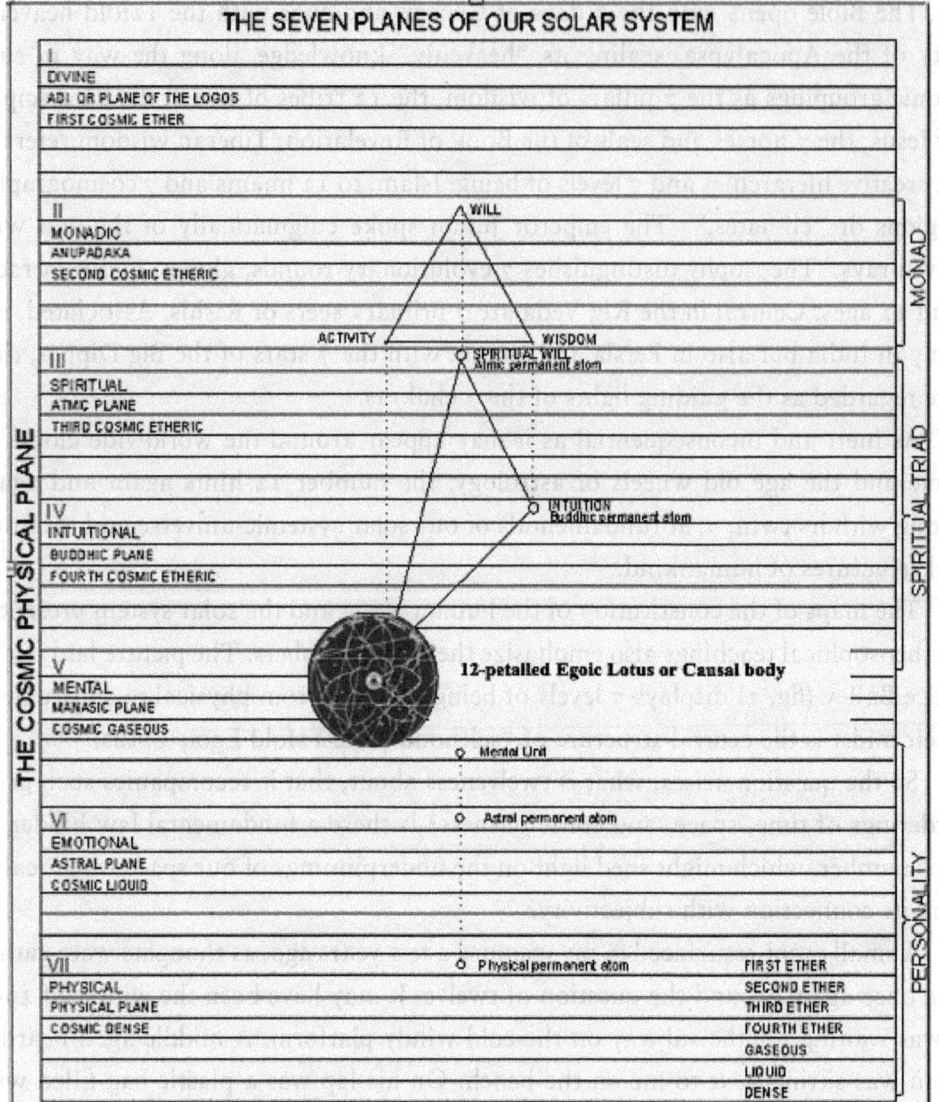

Fig. 1. Constitution of the solar system and the human being.
(Source: Alice Bailey, Initiation Human and Solar)

awareness of the significance of "twelve." Not only did this number bundle hours, months and ages into whole rounds of days, years, and "Great years," but it also bundled the forces of the human soul into a whole "self" structure. In other words, it did no less than delineate the very framework of human experience—from the most expansive cosmic sphere around us to the most intimate subjective one within. Echoing my probing looks, the static set of numbers staring vacantly from the kitchen clock began to shimmer with an archetypal gaze and outgrow its frame

I. Prelude and Premises: The enigma of twelve

until, one day, the impression widened to a circle of twelve presences above the pond I frequently walked around. I set out to inquire.

A decade of pondering the reverberations of the twelvefold zodiac in soul-inspired literary and philosophical creations led to a book, for which I was in the process of writing an article, when a third twelvefold circle landed most unexpectedly on the stage, bringing with it a new set of questions and insights, and prompting the adventure of the present book. This third twelvefold held in it what I now regard as the core mystery of twelve, the secret of its cosmological function. Inadvertently, the inquiry about twelve had pulled at the fringe of the cosmological question, and had begun to tease out its golden thread.

This third twelvefold reality, the awareness of which prompted the present expedition and opened up its unanticipated cosmological landscape, was the "circle of fifths" of music theory. We will be exploring this key notion in due time. For now, suffice to say that the revelation of a fundamental twelvefold circle of tones at the basis of the laws of harmony and music struck a meaningful resonance with the twelvefold structuring of the utmost periphery and inmost center of human experience—zodiac and soul—whose kinship had been the focus of my previous book. Penetrating into this new, harmonic version of twelveness was to be the first step in tracing the significance of this number to its operative role in the formation of the universe.

A brief evocation of the ground of the first book will serve to introduce the first two twelvefold formations—zodiac and soul—and provide the spiritual scientific premises necessary to grasp their reality from the perspective of this work. It should be added that the two books are entirely independent and there is no sequentiality that would necessitate being acquainted with one before reading the second.

The Twelvefold Archetype of the Human Soul was inspired by two statements of Alice Bailey, the author, in collaboration with the Tibetan master Djwhal Kuhl, of extensive works on the inner reality of the human being and the universe. The first one states, "The structure of selfhood—the 'human soul'—is twelvefold." The second, that "the zodiacal wheel is essentially a cosmic heart center...a twelve-petal lotus." These two pronouncements beamed up the pillars of a portal through which I stepped into the mystery of twelve.

The first statement defines what the term *soul* will refer to in the present context. Rather than the emotional life and the more or less conscious movements and moods of the psyche, what it designates here is the specific, albeit subtle *organism* of transpersonal, non-egocentric selfhood at the core of the human self. The term *soul* and its equivalents—*Ego, Egoic lotus, higher self*, and *causal body*—refer to the unique "individuality" we build over many life times out of the wisdom, love, and strength distilled from daily experiences. The term *lotus* is simply an alternate designation for the chakras, evocative of their dynamic flower-like appearance to clairvoyant sight.

According to Alice Bailey, the twelvefold flower like structure of the soul—displayed at the center of the map above—is a nonphysical, "mental substance" formation. As generic to humanity as the physical form of our being, it is the vehicle of higher selfhood at the heart of the mental nature, a "lotus" destined to blossom into a unique center of universal intelligence, love, and strength, out of the gradual maturation and transcending of ordinary thinking, personal feeling and egotistic will. The developmental necessities of this "mindful body" of selfhood are what guides or "causes" the selection of incarnational circumstances and the major brushstrokes of a destiny—hence the term *causal* body. Experientially, we recognize the soul as the quiet voice of moral intelligence (wisdom), empathic feeling, and selfless motive of our true center. In this context, the term *moral* refers to the soul's inherent attunement to the truth and good of the whole, in contrast with the self-centered response of the small "e" ego and its self-serving choices.

Careful observation confirms what theosophical sciences and perennial philosophy teach, which is that our being is composed of a physical body, a physiological, vital or "etheric" body, an emotional or "astral" body, a mental body, the higher levels of which host the causal body where the monad—our "spirit"—dwells. The first four vehicles constitute what is often referred to as "personality" or "ego." The causal body, which is made of substance from the three highest subplanes of the mental plane (see map), vehiculates the soul or Ego, our true "individuality," charged with the monadic spark of spirit.

The first four have their correspondence in the vehicles of the world being: the kingdoms of nature. Indeed, said Lao Tsu, "The universe is a man on a large scale." The physical body has its correspondence in the mineral kingdom. The "etheric" or energy body has its correspondence in the plant kingdom and ensures all vital functions such as breathing (air), blood circulation (liquid), digestion, growth and reproduction (earth), and body temperature (fire). As the solar energy in nature organizes and vitalizes the formation of plants from mineral matter, etheric energy shapes and vitalizes our physical organism through the agency of seven major energy centers called "chakras." The emotional body, the sentient organism associated with desires and fears, has its correspondence in the animal kingdom. The mental body, the field of personal self-awareness, is specific to the human kingdom.

I. Prelude and Premises: The enigma of twelve

The creative source of these structures lies beyond the human sphere of intelligence and creative capacity. No human can create a fish, engineer a plant or a human body, let alone a cell or a particle of matter. The *Timaeus* says it lucidly: "The difference between the human and the divine nature [is] that god both adequately knows and is able to mix many things into one and, again, to dissolve one into many, while no human being is adequate to either of these tasks, now or ever."[8] Oneness points to the "divine" order of the mystery of *life*, the wholeness of living "beings."

However, while the physical body is a "God-given" vehicle, made of the same elements as nature and cosmos, and the twelvefold structure of the causal body is equally divine in origin, the *content* of the causal body is, so to speak, "human-made." This "higher nature" of ours gradually unfolds out of the forces and qualities of wise understanding, selfless love, and steadfast purposefulness distilled from experience. We cannot create living organisms, but we "grow" *consciousness*. Soul strengths and individual qualities develop from the kind of choices we make. While the secretions of our body pertain to the wisdom of the universe in us, it behooves our "self" to secrete the wisdom, love and spiritual strength that bring our individuality to maturity. The soul is akin to the Sun in our "nature," and the distillation of solar/soul forces, day after day, life after life, out of the "raw material" of existential matters, is what brings about the unfolding of the causal body and its eventual blossoming into a unique fullness of Egoity or individuality.

This process of inner growth is the central work of human evolution—the grand Opus of Alchemy—the "Great Work" of transmutation of existential turmoils into the solar gold of the soul. The causal, Egoic body is what world Scriptures refer to as the "chariot of the Sun" or the "temple made not by hand." Made not by hand, it owes its edification to the conscious striving to grow wiser, more loving and spiritually strong. The causal self is the individuality in-progress Jung referred to as the individuating self. Jung did indeed view the human psyche as an *organism*, one that is "part of the inmost mystery of life and *has its own structure and form like every other organism.*"[9]

Alice Bailey describes this organism of the soul on the upper mental plane as a "twelve-petal lotus," a flower like structure comprising four sets of three petals: three petals of knowledge, three petals of love and three petals of will sacrifice.[10] The fourth set of three forms a nucleus, which hides and eventually reveals the vibrant gem of monadic (universal) selfhood referred to as the "jewel in the lotus."

---

8   Plato, *The Timaeus*, 68c-d.
9   Jung, Collected Works (CW) 9i, Para 187.
10 Bailey, *Treatise on Cosmic Fire*, pp. 536–544.

Paradoxical as this may seem initially, the more individuated the human being, the more universal s/he becomes.

The developing individuality, soul or Ego, is an intermediary reality between the personality driven, self-protecting and self-seeking ego (with a small e) and the universal spirit alive, but unconscious, in the depths of our being. Whereas the term *personality* designates a set of character traits, behavioral patterns and emotional dynamics associated with a particular lifetime, the term *soul* refers to the central "selfhood"-in-progress whose mindfulness transcends our instinctual, emotional, and mental automatisms, whose journey of growth transcends incarnational cycles, and whose character is vibrant presence, wise, loving and strong. The complete unfolding of the Egoic lotus marks the fulfillment of human being-ness as we know it, at which point the form referred to as causal body is relinquished, destroyed by greater vibrancy of "being." The individuality-based consciousness is superseded by the universal consciousness of the "monad"—identified with world needs and world good. The blossoming of the twelve-petal lotus of the soul heralds the evolutionary stage of universalized personality and individualized universality characteristic of the accomplished, self-realized human being. "Great" human beings across history and cultures exemplify such unique universality of individuality. The term *initiate*, in its most general acceptation, speaks to this stage of consciousness and being.

The psycho-cosmic map (fig. 1) displays the seven planes of the inner geography of cosmos and consciousness: Physical–etheric (energetic), astral (emotional), mental (concrete to higher mind), Buddhic (universal love), Atmic (radiatory will), monadic (spirit being) and Logoic (divine being). As shown in the illustration, the Egoic lotus is located at the center-top of the mental plane. This twelvefold center has its denser correspondence on the VIIth, physical-etheric plane, in the heart center. The heart chakra is indeed the outpost of the soul. Tracing the chain of heart centers and following it imaginatively, opens out a contemplation of the path of expanding identification that leads from personal to universal identity: from physical–etheric to astral–emotional heart center, to mindful individuality (the Egoic, causal self or soul), to the heart of humanity (the Christos–Buddhi principle, surrounded by the twelve), to the 12fold Sun at the heart of the cosmic being in whom the solar system is a center, to the 12fold heart in the head center of this being—the zodiac. Alice Bailey states that the Sun "is literally, from the higher cosmic planes, seen as a vast blue twelve-petal lotus, each petal being formed of forty-nine lesser petals..." and that it is the embodiment of a cosmic heart center: "The center in the body of this greater life of which our solar system is the embodied force is the heart center."

✸

I. Prelude and Premises: The enigma of twelve

The second statement that prompted my venturing into the mysteries of number 12 is a momentous pronouncement, also by Alice Bailey, which fleshes out the connection between zodiac and soul intimated by their twelve-based similarity of structure, and unveils the live, organic content of this similarity: its *chakric* nature. Causal body and zodiac are chakras. In other words, they represent life centers in a specific entity, individual or cosmic:

> The zodiacal wheel is essentially a cosmic heart center; it is a twelve-petalled lotus...it is a twelve-petalled lotus within the thousand-petalled lotus of an unknown Entity, the One referred to in my earlier books as the One About Whom Naught May Be Said.[11]

This remarkable statement affirms the objective reality of the zodiac. It opens an understanding of the zodiac as an etheric center—the chakra of a cosmic being in whom the entire solar system has its being. This view echoes the reference found in old Indian scriptures to the wheel or "cakra" of the zodiac. Since "chakra," in Sanskrit, means "wheel," a more eloquent designation for the zodiacal wheel might be indeed "zodiacal cakra."[12] Alice Bailey's statement at once "solidifies" and vitalizes the zodiac. It grants an organic reality, albeit of a nonphysical, "etheric" order, to the twelvefold circle of constellations that populates the outermost expanse of our universe. It suggests that behind what is sometimes regarded as a fanciful frame of figures projected on the skies by our ancestors' mythological imaginations, is a subtle yet *objective* twelvefold structure. It lends ontological grounds to our time space organization, and cosmological foundations to the field of astrology. The astrological wheel, the framework of a cosmic logic of self and evolution, becomes grounded in an objective etheric structure of a cosmic order.

The ancient, now widely accepted, yogic notion of the chakras as subtle energy centers in the human system eases access to the idea of a similar "chakric" constitution of macrocosmic systems, such as the solar system and its seven planetary centers, and the much vaster system whose heart-in-the-head center is the zodiac. Thus twelve speaks to a central structure not only in the human microcosm and the "local" interplay of Earth–Moon–Sun, but in the grand being in whom the solar system itself is a chakra. This notion implies a *subjectifying* of cosmos—a view of cosmos as a great living being, whose constitution mirrors our own chakric constitution on a macrocosmic scale—a greater version of our selves, not only alive but endowed with subjectivity. This conception of the universe as a living being in whom we have our being was the worldview and experience of ancient civilizations. While it is still, at least theoretically, the ground of religions, it has become a rather

---

11 Bailey, *Rays and the Initiations*, p. 339.
12 According to Dirghatamas, *Rig Veda* I.155.6, "With four-times-ninety names [360], he [Vishnu] sets in motion moving forces like a turning wheel [*cakra*]."

unfamiliar and alien "belief," dismissed if not outright rejected by the tacit cultural and scientific mindset that we live in a world of things and objects, domestic and cosmic, cleanly separate from our "selves"—from self. The "god-ness" or being-ness of the world is left to poets and mystics.

Yet, the findings of physics have been pressing upon the Newtonian and Cartesian framework of a neatly divided reality of mind and matter, and submerging it with the confused germinations of a new paradigm. The necessity to make sense of a world that is not as solid, "particulate" and separate from consciousness as was assumed led a number of scientists to reach to the transphysical—"metaphysical"—grounds of ancient wisdoms and their "subjective" order. The view derived from system thinking that the cosmos is not a purely material "thing," but a multileveled living system, with discernable gradients of self-organizing inwardness—from the association of atoms into molecules to the photo sensitive response of plants, from the sentiency of animals to the self-awareness of humans, to the rhythmic order and light-giving potency of star systems—this view points to an "emergent property" of subjectivity in our representation of the system-universe.[13] The emergence of world-awareness in the human being is its counterpart.

"World-self" is a *relative* designation of identity. For a cell in our body, the "world-self" is the human self. For humankind, it is the intelligent Life of our solar system—what ageless wisdom refers to as the solar "Logos." For the solar system as a whole, the world-self is the being Alice Bailey refers to as the "One about Whom Naught May Be Said," the vast entity in whom the solar system is a chakra. What the term *world-self* will primarily refer to in this work is the central consciousness of "our system"—the living system of the Sun, the organism of the "solar Logos." The term *universe* will refer to our solar system and its context, which is defined physically by the Orion arm of the Milky Way Galaxy and ontologically by the cosmic entity in whom it constitutes a chakra and whose heart-in-the-head chakra is the zodiac.

The glimpses into the kinship between zodiac and soul sparked off by Alice Bailey's statements open new perspectives on astrology, intimating the possibility of a cosmological basis to its claims that the *cosmic* wheels and their motions provide insights into *self* and self-development. The heart-in-the-head nature of the zodiac and the zodiacal structure of the soul offer inklings of a potential cosmic logic in support of astrology's view of the zodiac as template of the soul, and of self-development as the "soul" of the zodiac.

The similarity of nature and structure between zodiac and soul invites us to envision the zodiacal signs as twelve facets of the soul—twelve forms of soul

---

13 An *emergent property* is one that pertains to a complex system and appears with it. None of the parts of the system has in and of themselves such a property.

I. Prelude and Premises: The enigma of twelve

consciousness, twelve signatures of the soul. This perspective inspired what was to be the central theme of *The twelvefold archetype of the human soul*: the relationship between a zodiacal sign emphasized in a person's natal chart and his/her philosophical worldview. Focusing an imaginative, archetypal eye on philosophical landscapes revealed in their design and "texture" zodiacal like configurations, motives, and dynamics. Constellational outlines shimmered in the fabric of philosophical interpretations of reality. Rich correspondences emerged between the structural depths of a particular worldview and the zodiacal signature emphasized in its author's birth sky—intimating that the zodiac is the collective sky of the human soul, and constellations configure universal "forms" of consciousness. What became apparent is, on one hand, the cosmic "logic" of individual worldviews and, on the other hand, the soul logic of the zodiac.

Once the book was completed, there seemed to be no further to go in my inquiry about "twelve." However, not unlike the vast castle gardens judiciously created for the visitor's esthetic pleasure and optical marveling, in which a whole ground of architectural creation remains invisible until one walks down one or two levels of terraces and water basins,[14] the twelvefold rotunda of time and space, clock and calendar, which astrology turns into a psychic map and spiritual sciences identify as the *heart* of reality and the garden of self-*growth*, this twelvefold castle garden of humankind had one more depth to reveal, which holds the key to the essence of the whole architectural mystery.

This deeper level of the twelvefold garden came into view, momentous with serendipity, in the midst of a casual conversation. Crossing the lips of a musician friend I was conversing with, the expression "circle of fifths" stood in mid air, vibrant in the simple garb of its own wording. The combination of twelve, circle and fifth, for reasons that will become clear, split the moment with the vibrant clash of a cymbal. Ushered into this next circle of twelve, I watched a new momentum pull me onward from insights into facts, down an uncharted road from ancient philosophical worldviews to modern physics. The present book retraces this journey into the inner reality of twelve.

Before walking toward this third ground, where the adventure of the present work truly begins, two more circumambulations within the twelvefold rotunda will allow us to probe the implications of the structural mirroring of soul and zodiac for psychology and astrology.

---

14 The castle of Vaux le Vicomte near Paris, designed by André le Nôtre for the finance minister of Louis XIV, is a good example of such landscape architecture.

THE GAP BETWEEN COSMOLOGY AND PSYCHOLOGY—
AND THE BRIDGE OF ASTROLOGY

Although astrology per se is not the focus of this work, nor was clarifying its ground the motivation of my inquiring into "twelve," its 12/7fold framework implicitly includes it in the adventure of this book and gives it a special presence in the background. Moreover, the essence of astrology as a cosmos based psychology is immediately relevant to the central question introduced at the beginning of the book—the relation between world and self. What astrology proposes is precisely a perspective and a language for this relation, which is what its 12/7 based charting of cosmic cycles purports to observe and articulate. Even though the lack of a rational ground to justify astrology's claims leaves it outside the perimeter of modern science, and makes the validity of the world-self philosophy it practices questionable, the relevance quickly noted here will prove to gain significance in ways that will not only serve the core purpose of this work but also shed light on the grounds of astrology. For all these reasons, it is vital to draw a quick profile of astrology on the cultural backdrop of science (world) and psychology (self).

The inventor of the first commercially licensed helicopter, Arthur Young, an engineer by training and a cosmologist by vocation, once remarked about astrology:

> Science would insist that the facts of astrology have not been proved; but since I have, to the best of my ability, verified many of the claims of astrology, I would have to answer science as Newton answered Halley when the latter criticized him for his study of astrology. "I have studied the matter; you have not." In any case, the judgment of science against astrology is based on a belief of science, and science has progressed by revising its beliefs, not by insisting on them.[15]

Astrology bases its "cosmological psychology" on the symbolic interpretation of what it regards as meaningful correspondences between the planetary configurations of a moment of time and the subjective experience of this moment, collective or individual. For the individual, the moment of birth is viewed as critical, in that the first imprint of celestial configurations on the "crowning" head of the newborn child, and the first inbreath of its unique composition of energies become the seed-signature of his/her psychic identity for an entire life time—a signature referred to as horoscope or natal chart. With the passage of time, the changing relations, resonant or dissonant, between current planetary positions and "natal horoscope," chart new maps of symbolic access to the psychospiritual processes of growth at work in subjective experiences as well as external events.

---

15 See http://www.arthuryoung.com.

I. Prelude and Premises: The enigma of twelve

Any serious form of relation to astrology, whether as consultant, student or client, tacitly acknowledges that the logic of the cosmos informs the logic of the psyche, and that the cosmic wheels and their motions shed symbolic light on our psychological experiences. Many have felt the benefits of receiving and dialoguing with this "cosmic" information and accessing a guidance that integrates subjective and objective windows into their lives. The question is, what supports such "reading" of our selves in the arrangement of cosmic wheels? What justifies inquiring about our evolution in their cycles? To embrace the zodiacal circle of constellations as the meaningful backdrop and fabric of our soul life (or an extension of our soul in space) and consider the 7 planets as the main functions of our psyche is hardly a proposition we can afford to acknowledge, let alone explore, in the physicalist context of current science, since its variables have no basis in physical reality—no tangible fact to show for the soul, and none for the fantastic figures of the zodiac. However, the likelihood that this remarkable source of perspective on our lives does not rest on physical grounds need not be a truth to recoil from, since we use astrological "data" to "figure out" the workings of this most intangible yet most real dimension of our being—the soul and its journey. Nor should it negate the possibility of its pertaining to another order of science. Such an acknowledgment frees us in fact to consider what these nonphysical grounds might be, and so shift our "scientific" position, if necessary, off the exclusively physical framework of mainstream science.

What we can surmise is that this subtle ground of astrology as a "science" of the psyche was a definite reality for the consciousness from which it originated—a consciousness that experienced its connection with the surrounding cosmos, was in touch with its laws as the laws of its own being, and charted planetary cycles to make observations and divinations about human experience. The consciousness persists in a practice, but the practice is cut off from an intelligence of its roots.

Ancient teachings offer a conceptual access to this "cosmically literate" consciousness of old. The two statements by Alice Bailey introduced above, for instance, refer to a science of subtle planes of reality: etheric and mental. While they are "esoteric"—in other words, indemonstrable on physically tangible grounds—they participate of a profound logic of the universe's constitution. We note how the unparalleled intimations they offer about the vital reality of the zodiac and the *similar chakric nature* and structure of zodiac and causal body—outermost cosmic matrix and innermost psyche—expose a potential common ground for "self" and "world," hence the first inkling of an intelligible foundation for the *psycho*logical interpretation of *cosmo*logical wheels at the heart of astrology. While this correspondence is clearly far from granting astrology rational status as a soul science, it makes for a compelling start to explore the main feature of this common ground: the number 12.

Seven, its companion number, brings up a similar order of correspondence between chakras and planets.

If the structural mirroring of zodiac and soul hints at a potential ground for astrology's claim to psychological relevance, it highlights by contrast the absence of an objective, cosmological background in the mainstream science of the psyche—psychology. The lack of a cosmological perspective on the nature of self and psyche limits the understanding of "health," the interpretation of disorders, and the capacity to optimize healing and growth. The disconnection between subjective experience and cosmological framework characteristic of psychology prior to Jung parallels in fact the radical separation between object and subject characteristic of modern science prior to quantum physics. They illustrate the discontinuous reality in which we still tend to operate, while the seeds of quantum physics and Jungian depth psychology, among others, continue to germinate into the inklings of a new paradigm.

Paradoxically, to note the absence of a defined cosmological context informing psychology leads us to realize that an implicit cosmology is at the center of every psychology. A "religious" cosmological framework such as prevailed in centuries of Judeo–Christianity, for instance, rests on intuitive or inculcated faith in a transcendent ground of divinity and afterlife, which induces a moral psychology of self-improvement and societal altruism. In the cosmological framework of a mechanistic universe, in which man appears by way of evolutionary survival as a finite unit in empty space, psychology studies personality development and pathologies, aims at improving emotional health and personality integration, and seeks to strengthen and optimize societal adaptation. Freud's, then Jung's pioneering explorations of the unconscious significantly expanded this framework. Jung's insights in particular took psychology beyond the divide between subjective and objective reality, self and world, self and god, even self and other. He explored disciplines such as astrology, alchemy, the *I Ching*—all of which address internal experiences and external events as *one* process. He inquired into synchronicities, the acausal phenomena of meaningful coincidences between inner preoccupations and external happenings. His notion of the Self as the supraconscious center of the totality of the psyche, drawing psychic components such as ego and shadow, masculine and feminine, subjectivity and objective circumstances, conscious and unconscious, to new levels of integration, restructured the cultural understanding of the psyche. The distinction he made, having observed it in himself, between two levels of self redefined the notions of self-knowing, self-healing, and self-development. Self-actualization became the road to health as a road to *wholeness*—facilitated by tending the subtle conversation

I. Prelude and Premises: The enigma of twelve

between self and "Self" presented by dreams and life events, and activated by the therapeutic process. The psychological focus shifted from identifying pathologies to fostering integration and integrality of being by bringing conscious and unconscious, self and circumstances—psyche and world—to bear on each other, allowing the tension between such polarities to elicit the emergence of a more integrative ground of identity—a process Jung referred to as the process of individuation.

Jung's notion of individuation raises the question of the nature of "individuality" and its developmental arc. What is "individuality" and where does the process of individuation end? Arguably, both questions are cosmological, and outside the traditional purview of psychology. Yet they are fundamental to a full understanding of individuation. Despite Jung's remarkable insights into the weaving of outside and inside, world and self, constitutive of this generic process of human unfoldment, the absence of a cosmological framework leaves its ultimate "objective" context in the dark. Without a cosmological grasp of selfhood as a "unit of being" in the making, and individuation as a finite cycle—the growth cycle of this unit—with a beginning *and an end*, depth psychology lacks a full grasp of the scope of the process.

Admittedly, the evolutionary maps provided by spiritual teachings involve the notion of successive lives, a significant hurdle for the collective mindset, yet a fact that a logic of individuation simply necessitates, since the maturing of human Egos into universalized human beings cannot be accomplished in one lifetime.

If Jung stopped short of defining a cosmology that would ground his unitive (personal–universal) and comprehensive grasp of the human psyche, he himself clearly pointed out that such was not the focus of his work. Jung viewed himself as an empiricist of the psyche and related to metaphysics and religion as productions of the psyche—as *psycho*logical—which indeed they are, though no more than psyche is a transphysical, "meta"-physical, production. Nonetheless, his study of alchemy and kundalini yoga, along with his interest in astrology brought him within the cosmological context of the human constitution and evolution.

We will return to Jung's position in the sixth part of this book and look more closely into how and where he stopped short of fully embracing the cosmological picture offered by the alchemical and yogic wisdom he studied in depth—hence how his tremendously insightful psychology did not step into its full spiritual–scientific potential.

The unitive psycho-cosmology implicit in Jung's approach resonates fully with the premises of astrology, whose cosmic factors are regarded to signal specific features and timings of the psyche. The understanding of self that astrology "reads" in planetary alignments involves the same transcending of the split between subjective experience and objective facts as Jung's notion of the Self. Both see and address self and circumstances—whether personal, collective or cosmic—as *one "Self."* However, while they represent two unitive fields of approach to world and self, Jungian psychology keeps objective, cosmological considerations at bay, and astrology does not have a rational basis to explain how the cosmic spheres could have anything to do with human psychology. As satisfying and illuminating as such wholistic cosmo-psychologies may be in a looming search for a paradigm of body–mind unity, matter–consciousness interplay, and universe–self oneness, the question remains: On what do they rest?

II. OVERTURE: THE MANDORLA OF MEDIATION

> *Matter is the vehicle for the manifestation of soul on this plane of existence, and soul is the vehicle on a higher plane for the manifestation of spirit, and these three are a trinity synthesized by life, which pervades them all.*—H. P. BLAVATSKY

THE INNER REALITY OF TWELVEFOLD STRUCTURES: MEDIATION

What twelvefold centers have in common is that they represent middle grounds. The heart chakra is the fourth (of seven) chakra in the etheric body, and the causal body is the intermediary realm of our body–*soul*–spirit being—the soul vessel—"the middle point between spirit and matter."[1] Mediating structures between the "above and below," source and periphery of our being, they embody the intercourse of spirit life with bio-instinctual patterns. They fulfill this intercourse and they bear its fruit.

While the physical heart carries out the circulation of the blood between the oxygenating loop of the lungs (outside air) and the deoxygenating, oxidizing loop of the organism (inner earth)—the intercourse of inside and outside in the body—the soul hosts the intercourse of spirit and body, subjective life and objective world (see fig. 64). The existential interplay between instinctual reactions and spiritual impulses, emotional patterns and core values, opinions and universal truths, which the encounter with the outside world generates, prompts soul growth and builds individuality. It is this intercourse of "Heaven and Earth" in us that generates consciousness—the mindful consciousness that gives substance to the causal body of the soul.

Contrary to the mainstream understanding that the heart "pumps" the blood and generates circulation, the spiritual scientific view proposed by Rudolf Steiner is that the heart is generated and "organized"—in other words, *made into an organ*—by the dynamic currents of exchange between inside and outside at work in the

---

1   Bailey, *Treatise on Cosmic Fire*, p. 248.

systemic and pulmonary loops of the blood. In the same way that individuality unfolds out of the intercourse of spirit and matter, the physical heart *results* from the dynamics of opposite currents at work in the blood. Such a reversal of the habitual perspective raises again the question of what came first: does the organ generate the energy and motion of life or does the dynamics of life shape the organ?

As middle grounds and mediating structures, heart and causal body have as their main function to foster the creative interplay of spirit and body, inner being and outside world. In fact it is this conjugation of polarities that produces an "organ" in their midst: heart (in the body) and Egoic lotus (in the soul)—in the same way that the interplay of Earth and Sun forces produces plants, flowers, and fruit.

Physical heart, etheric heart, and Egoic lotus epitomize this mediating function at the physical, vital, and mental levels of being. The *physical–etheric heart* regulates the exchanges of outside oxygen and internal carbon dioxide in the blood. The live, beating heart is of necessity a physical *and* energetic, "vital" reality. The corpse alone is a purely physical reality, one that rapidly disintegrates, precisely, when the organizing (etheric) forces of life withdraw at the moment of death.

The *etheric–astral heart* is the fourth and *middle* chakra of the human energy system. Halfway between the base and crown chakras, it balances the "earthbound" and "heavenbound" within us. It is also, Rudolf Steiner describes, thereby revealing the cosmic dimension of this mediating function, a unique hub of exchange *between cosmos and self*—the only such place in the etheric body. "On our descent into the earthly world," Steiner explains, "we take with us in our etheric body an image of the cosmos." As the etheric body is about to unite with the physical body, it is like "a sphere complete with stars and zodiac and Sun and Moon."[2] The period leading up to puberty sees a gradual drawing inward of the radiant-star ether body into a distinct structure—the etheric heart—in the midst of which the physical heart is suspended. While the cosmic radiance concentrates into an etheric heart, a concomitant process begins to take place, whereby all that a human being does, all the "gestures" and features of his personal (astral) rapport to the world begin to insert themselves into his etheric heart. Having gathered the cosmos to a central organ in early life, the heart now gathers psychospiritual patterns from within the person. When the time comes for the soul to release its vehicles after death, the heart pours out its qualitative content into the cosmos. In collecting cosmos into self at birth, and pouring out its self-patterned subtle formations into the cosmos at death, the heart plays a central role in commingling microcosm and macrocosm. What Steiner's insight illumines is the co-creative intercourse of personal and universal being at work in the human heart. The heart is the subtle organ by means of which

---

2   Steiner, *The Human Heart*, p. 4.

our feeling–thought patterns, actions, and creations play their part in the process of world becoming. It is the mediating hub through which the cosmos lives in us and each of us participates in evolution:

> The entire universe, which is there within him as an essence, receives all that he does and permeates itself with it. By this constant coming together—this mutual permeation—the opportunity is given throughout human life for human actions to be instilled into the essence of the images of the cosmos (configurations)...when the human being expands back into the cosmos.[3]

As for the *Egoic lotus* of the soul, it is the true "heart" of the human being, the middle ground of his body–*soul*–spirit constitution. The potential "flower" of human nature, this "organ" of fully conscious selfhood on the *mental plane* unfolds out of the dynamic interplay between the supraconscious pole of monadic spirit and the subconscious pole of instinctual, bodily forces.

Symbols of mediation: Vesica and mandorla

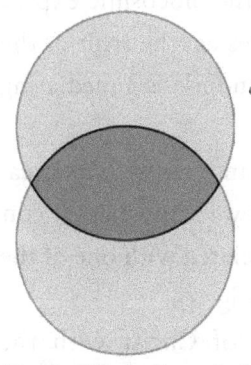

Fig. 2. "Vesica" symbol of the soul as mediating milieu between spirit and matter

Ancient iconographic and geometric symbolizations illustrate the association of the soul with mediation and with number twelve. Pythagorean philosophy likened the soul to the geometric "intersection" of two circles—a circle of spirit and a circle of matter—into an almond shaped figure called vesica piscis or mandorla (fig. 2). A symbol of interpenetration and communion of opposites, the vesica piscis suggests a polarity transcending milieu. The shape evokes the geometry of the human eye, drawing attention to the unique character of this organ—at once physical and soulful. Objective organ of flesh and epiphany of a "self," the eye transcends the separate categories of "inner" and "outer."

The solar God Mithras or Phanes (fig. 3), whose pre-Christian cult spread from Babylonia to Persia, then to Greece and later to Rome, is represented in a second-century marble relief standing within an almond-shaped, vesica piscis-like form, etched with the 12 signs of the zodiac. The mandorla replicates the shape of the egg of the world from the flaming ends of which an androgynous youth is shown emerging—one half of the egg above the figure's head, one half under the feet. Figure 3 shows a sixteenth-century rendering of this ancient representation of the god.

---

3   Ibid., p. 12.

Figure 3: (left) The Sun God Mithras–Phanes hatched from the cosmic egg in flames, in the world egg—inlaid with the zodiacal signs (sixteenth-century drawing by Francesco de Rossi); Figure 4: (right) Christ amid the four evangelists (Chartres Cathedral)

A central image of ancient cosmogonies, the world egg is the biocosmic expression of the intersecting circles of spirit and matter—the live vesica, the fruit of the "intersection–intercourse" of the world parents. The 12fold mandorla is a mediating space of birth—the birth of self-awareness.

Christian iconography filled this archetypal space of the mandorla with its great *mediator*—son of God and "sun of the world"—surrounded by 12 apostles. A condensed version of the 12 depicts the four evangelists, each associated with one of the fixed signs of the zodiac, occupying the four corners of space (fig. 4).

The mandorla also mirrors the traditional association of Christ with the "fish"[4]—a symbol of choice for the archetypal representative, in this Age, of the vesica *piscis* (literally, "fish bladder") of worlds. The Christ unites in his being the sphere of the human and the sphere of the divine, the spirit self within ("I") and the physical world without ("this is my body"). Resurrected body and "incarnation of god," he embodies the "intermediating" reality that the mandorla symbolizes. Redeemer of Earth and embodiment of the Sun being, he epitomizes the function of the soul, which is to solarize nature and bring divine consciousness into the depths of physicality. As appropriate for a "Sun God," Jesus Christ is born with the rebirth of the Sun at the winter "*sol*stice." He turns water into wine—a feat the Sun alone can accomplish. The Sun grows dark, eclipsed, at the time of his death: the Sun is passing through the Earth.

---

4   The Greek word for "fish," *ichtus*, is an acronym for *iēsous Christos, Theou Huios, sōtēr*, or Jesus Christ, Son of God, Savior.

II. Overture: The Mandorla of Mediation

The zodiacal signs surrounding the Mithraic cosmic egg display the gestating womb of the Sun God: the twelvefold wheel of time and space. This wheel is the womb of the "solar gods" we humans are, in potential. The zodiac is at once the spatial sphere in which we live, and the wheel of time in which we cultivate soul growth—the wheel of the time needed for nature to grow flowers, for the human self to grow radiant, for the world to become self-aware, and for humanity to become the brotherhood of "holy men" that twelvefold circles such as the apostles and the Grail knights symbolize—the soul of humankind.

If divine–human beings and sons of God are set amidst zodiacal twelvefolds, it simply heralds the fact that their aura (their auric "egg") radiates this universal fullness of being and is organized accordingly. In the light of this iconography, the zodiac itself, the heart-in-the-head chakra of the cosmic being we are, stands revealed as the great mediating structure between the spirit of our universe and its manifestation, between eternity and time, infinity and space—the grand Vesica of time-and-space consciousness.

Aion, the god of time, is another name of the god Mithras. Aion is the divinity of time, not as chronological sequence, the realm of Chronos, but as portal to timelessness—the vesica-like intersection of time and timelessness where the soul is being born, *in the midst of existence.* A special moment in the opera Parsifal dramatizes the experience of this numinous quality of time as "soul space." Invited by Gurnemanz to witness the mystery ritual of the Grail, Parsifal follows him to the sacred place.

As they walk, he notices how he seems to scarcely move, yet is traveling far. Gurnemanz explains that, in this realm, "time becomes space" (*Zum Raum wird hier die Zeit*). After an orchestral interlude, they arrive in the castle hall where the knights are gathered in procession to celebrate their communing rite. The Aionic dimension of time introduces the "foolish" self (Parsifal, Perzi fal, means "holy fool") to the castle of the soul and initiates him to timeless communion with the heart of being.

The iconography of the Sun God involves also, albeit implicitly, number seven. Wrapped around the body of Mithras–Aion is a serpent, whose coils actualize in the spiraling movement of evolution the theme of time displayed in the zodiacal constellations surrounding it. The spiraling up of the serpent of time underlies the *evolving* (Latin: *volvere* = "to roll") of selfhood from the early self-protectiveness of the egg to the

Fig. 5. *The World, XXI arcana, Tarot de Marseilles*

cosmic self-awareness of the god. The gradual arising, through the chakras, of the Kundalini serpent initially coiled at the base of the spine marks seven evolutionary stages of consciousness toward the eventual apotheosis: the birth of a "solar god."

The twenty-first major arcana of the tarot—"The World" (fig. 5)—depicts the same archetype of a god/dess like figure inscribed in a mandorla framed by the four fixed signs of the zodiac. A loincloth across his/her body evokes the Aionic serpent of time coiled around Mithras' body. The fully blossomed human being is a *world* conscious being.

Mithras is also known as Phanes—the epi-"phany" (Greek: *phanein* = to appear) of the divine at the beginning and at the end of evolution. From the perspective of world beginnings, he portrays the surging into existence of a primordial oneness (egg), the appearing of the world *"phaino"*-menon out of the great noumenon, the birth of timeless Being in time and space. From the perspective of the end of human evolution, he represents the human being giving birth to his solar self through the cycles of time. As such, he embodies universalized selfhood. This recurrent iconographic theme suggests how the twelvefold vesica matrix bears the world's Sun God at the beginning of time, as well as the human apotheosis at the end of time.

The winged youth of Phanes has the glow of an apparition—the glow of the mystery of apparition itself. Phanes–Kosmocrator, as he is often designated, is the divinity of the "appearing" of Being into world—the divinity of cosmogony. As an Orphic god of initiation mysteries, he also heralds the essentialization of man out of earthly existence. He transcends the duality of divine essence and human existence, and speaks at once to the divine advent of creation and the initiatory advent of resurrection. A god of cosmogenesis and human theosis, Phanes is the revelation of the twelvefold mystery of time and space, which presides over the descent of Being in spatial forms and the solarizing ascent of consciousness in time.

❂ This simple and profound icon of the solar god is a rich source of contemplation of the polarity-transcending milieu of the Vesica and the inner reality of twelve. Suspended in this figure of the god is nothing less than the passage of Being in and out of existence, its appearing in the human form through the universal womb of time and space—as well as its breaking free from existential limitations.

The solar being who appears within the *zodiacal* mandorla is an archetypal image of the *soul*. The zodiac is fundamentally connected with the birth, growth and "solarization" of the human being. In this symbolic space of the twelvefold mandorla, soul and zodiac lose their separateness; individuality and world being blend their orders of magnitude to divinize the twelvefold gate of world manifestation, which is also the gate of the man reborn as a god at the end of time, when the primordial Outside (the zodiac) has become the ultimate Inside (the Egoic lotus). ✪

Tracking the enigma of "twelve" in the mediating function associated with the twelvefold structures of heart and soul has led us to a figure charged with the solar radiance of the soul. In time, we shall discover in its midst the key to the twelvefold mystery of mediation—a musical key, the "clef of sol"—the G-clef (*clef* is archaic French for *cle*, or key). The affinity between the words soul and solar, Sun and Son, soul and sol is more than a fortuitous happening of language. Linguistic kinship often points to similar realities captured by similar patterns of vibration. The Sun stands at the center of our universe in the same warming, illuminating, vitalizing way as the soul does within our being. The note "sol" or G, the musical interval of the fifth, will emerge as a third kin to soul and Sun (Latin: Sun = *sol*)—charged with the same radiance.

The substance of mediation: Harmony

> *The number 12, applied to the universe and all that represented it, was always the harmonic manifestation of the natural principles 1 and 2.*—Fabre d'Olivet[5]

The function of mediating between spirit and "nature," self and world, which is the essence of heart, self, and soul, speaks word for word to the core principle of Pythagorean teachings: "harmony." Harmos, in Greek, means "joining or fitting together." Pythagoras defined harmony as a "fitting together" of ideal extremes.[6] He was also the first to use "kosmos," the Greek term for order and beauty, to designate the universe. Kosmos, he says, is a harmonia, a fitting together of the ideal extremes of Unity and diversity, Monad and Dyad.[7] The Monad is the creative source, the beginning and end of all things; the Dyad—the duality and multiplicity of manifestation. These fundamental poles of manifestation are represented by number 1 and 2—an obvious, yet also highly meaningful, designation of Monad and Dyad (spirit and matter) to which we shall return. Harmony is the *substance* of the mediating function of the soul.

For the Pythagoreans, this cosmic harmonia or "joining together" was most exquisitely expressed by musical harmony, and they regarded the musical proportion as its key formula. While the principles of harmony and the musical scale are

---

5   D'Olivet, quoted in Godwin, *The Harmony of the Spheres*, p. 345.
6   Fideler, *Jesus-Christ, Sun of God*, p. 58.
7   Ibid., p. 214.

generally attributed to Pythagoras, most scholars now agree that his works elaborated upon a fount of knowledge he received from the Babylonians during his travels in Egypt, then Babylon. Pythagoras was born around 570 BC in Samos, Greece, and went to Egypt in about 535 BC to study with the temple priests. He was sent to Babylon as a prisoner after the invasion of Egypt by the Persians and was initiated in the Babylonian sciences and Chaldean mysteries.[8] According to his biographer Iamblichus, "the Magi instructed him in their venerable knowledge and he arrived at the summit of arithmetic, music and other disciplines." He learned the theory of music from the priests of the mysteries into which he was received, and then pondered for several years the laws governing consonance and dissonance.[9]

After returning to his homeland, about fifty-six years of age, he established a philosophical and religious school at Crotona, a Greek colony in Southern Italy, where his many followers came to live and work in accordance with his philosophy. He taught occult mathematics, music, and astronomy.[10] He died in 495 BC. The school of the Essenes, where Jesus would receive his early education a few centuries later, was based on Pythagorean principles.

As we penetrate into the musicological grounds laid out by Pythagoras, patience will be needed to proceed through notions that may initially feel obscure. It seems to be part of the nature of this journey that the traveler, be it the author or the reader, is presented along the way with laws, realities, and phenomena that only gradually deploy their content. The sighting happens first, and only as a few more steps are taken do the realities concerned gather "definition." Like rocks on a well-marked trek, no seemingly opaque notion should deter the reader from proceeding onward, as a deliberately slow pace and trusting persistence will assuredly clear the way.

The most fundamental principle that Pythagoras "reputedly and plausibly brought home from Babylon," says musicological scholar Ernest McClain, is the musical proportion expressed in the formula: 6:8::9:12. It reads: six is to eight as nine is to twelve; and six is to nine as eight is to 12. (fig. 6). This supreme formula of harmony is the basis of the diatonic, heptatonic scale. The "miraculous" significance of the musical proportion, which is also, David Fideler points out, the most simple, perfectly symmetrical whole number proportion in the universe, is that it captures the ratios that define the fundamental intervals of music: Octave, fifth and fourth—the only absolute consonances of the Pythagorean system.[11] Folded in

---

8   Hall, *The Secret Teachings of All Ages*, pp. 191–192.
9   Ibid., p. 250.
10 Iamblichus, *The life of Pythagoras*.
11 Fideler, *Jesus-Christ, Sun of God*, p. 88.

II. Overture: The Mandorla of Mediation

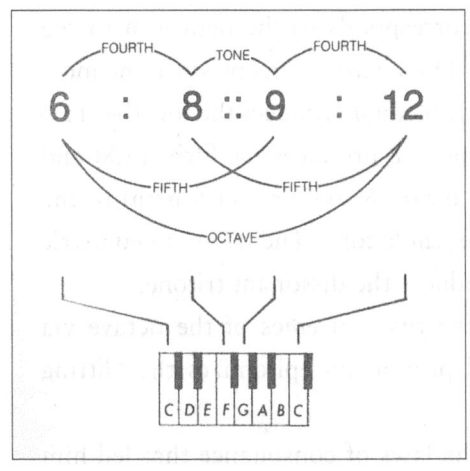
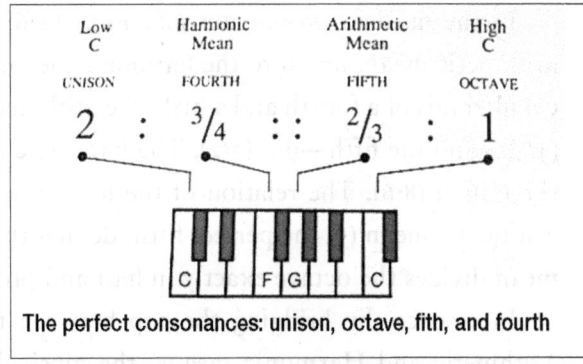
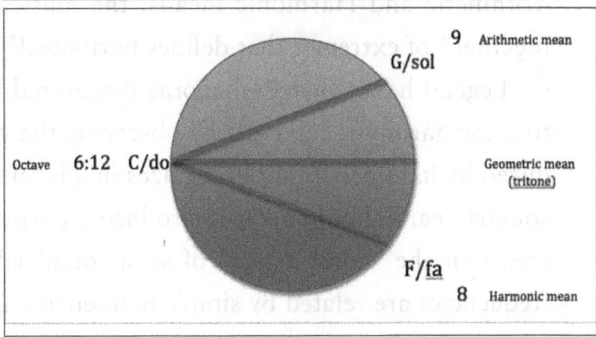

Fig. 6 (left & top images): The Musical Proportion (Fideler, Jesus-Christ, Sun of God);
Fig. 7 (below): The musical proportion on a circular graph of the scale

one compact formula are the foundations of harmony and the musical scale, not only in Western music, but as we shall see, in most musical systems.

(6:12) defines the 1:2 ratio of the octave.
(6:9 and 8:12) defines the interval of the perfect fifth (2/3 or 3/2)
(6:8 and 9:12) defines the perfect fourth (3/4 or 4/3)
(8:9), the ratio between perfect fourth and perfect fifth, defines the whole tone.

In other words, 6, 8, 9, and 12 are the smallest integers (i.e., whole numbers) that capture the system of interlocking ratios constitutive of the diatonic scale.

The musical proportion was also regarded as an expression of perfect proportionality because it encapsulates the principles, cherished by the Pythagoreans, of Arithmetic, Harmonic and Geometric *mediation*.[12] For the purpose of the present exploration, it is sufficient to grasp that the three "means" define key *intermediary* values and tones between the extremes of the octave, based on different types of mediation. Each mean is defined by a mathematical formula, which yields one of the major harmonic ratios (fig. 7).

---

12 In his *Introduction to Arithmetic*, Iamblichus confirms that Pythagoras learned about the "golden proportion" from the Babylonians—A : H = R : B, in which H and R are, respectively, the harmonic and the arithmetic mean of A and B.

In the musical proportion 6:8::9:12, 12:6 corresponds to the octave, 9 to the arithmetic mean, and 8 to the harmonic mean. These two means produce the musical intervals of a fourth and a fifth: the arithmetic mean 9 produces the fourth—12:9 (4/3)—and the fifth—9:6 (3/2). The harmonic mean 8 produces the fifth (12:8) and the fourth (8:6). The relation of the harmonic mean (8, the perfect fourth) to the arithmetic mean (9, the perfect fifth) defines the whole tone. The third or geometric mean divides the octave exactly in half and produces the dissonant tritone.

In essence, by bridging the gap between the two extremes of the octave via Arithmetic and Harmonic means, the musical proportion epitomizes the "fitting together" of extremes that defines harmony.[13]

Legend has it that Pythagoras discovered the laws of consonance that led him to these harmonic intervals by observing the variance in pitch between sounds produced by hammers of different sizes in a blacksmith's forge. He noticed that certain sounds heard simultaneously produce a particularly pleasant sensation and discovered that the "consonance" of such sound vibrations is due to the fact that their frequencies are related by simple mathematical ratios. For example, any two sound vibrations whose frequencies are in a 2:1 ratio—which is to say that one sound vibration is twice the frequency of the other—result in a particularly pleasing sensation. Their relation is called an octave.

It will prove significant in our context to realize that the perfect fifth and the perfect fourth, the pillars of musical harmony, are intimately related to one another. As shown on the first diagram above, moving down a perfect fifth from higher C (C to F) is the "same" as moving up a perfect fourth from lower C (C to F), while moving down a perfect fourth (C to G) is the same as moving up a perfect fifth (C to G). The interval of the whole tone is defined by the ratio of the perfect fourth to the perfect fifth.

HARMONICS, MUSICAL SCALE, STANDING WAVES

Before going further into the fundamentals of harmony, let the reader be assured that facts will be kept to the bare essentials necessary to unravel the mystery of twelve. Prompted by the quest for meaning, every step will stay closely engaged with meaning. At the same time, it must be emphasized how vital a patient walk through these basic physics and mathematics of harmony will be for the entire journey.

Pythagoras taught his theory of harmony on the monochord—a single string stretched between two pegs and supplied with movable frets. He showed that harmonious sounds result from plucking a string at specific intervals defined by

---

13 Fideler, *Jesus-Christ, sun of God*, p. 88.

simple mathematical ratios. When a string is stroked, it vibrates at its fundamental rate. When a finger or a bridge is positioned at the halfway point, and either half-length is stroked, a higher note is produced, which is twice the frequency of the fundamental note—resulting in the interval known as the octave. Divided in three, the string produces triple the frequency—resulting in the interval of the fifth, so called because it corresponds to the fifth note in the scale: C–G. Divided in four, four times the frequency—the interval of the fourth: C–F. Frequency and wavelengths are inversely proportional. Dividing a wavelength in half results in twice the frequency. A significant principle to emerge from this experiment is that the production of harmonious, musical tones depends on definite mathematical relationships between the string lengths involved: they must be related by whole number ratios and whole number ratios only.

A second fundamental fact of harmony is the phenomenon of harmonics (fig. 8). When a string vibrates "musically," which is what happens when one plucks a guitar string (simply because guitar strings are set to "musical intervals"), one hears not only the fundamental tone but also a whole series of fainter tones. A musical "note" is in fact a composite sound. The main tone has its source in the back and forth vibration of the whole string—the fundamental, also called "first harmonic." The secondary tones, called overtones, partials, or harmonics, result from secondary vibrations of the string induced or "entrained"[14] by the main vibration. When the string vibrates as a whole, it spontaneously vibrates as well in halves, thirds, fourths, and so on. These secondary sound waves produce subtle tones, called harmonics, which, though hardly audible, are responsible for the rich, musical quality of the resulting sound. We don't hear the harmonics as separate notes (or very faintly as "overtones"), but they give a sound or a voice its musical "timbre." The sound of a single frequency is a relatively mechanical, uninteresting, and unmusical sound.

Harmonics are whole number ratios of the fundamental—*whole fractions* of the fundamental string length and *whole multiples* of the associated frequency. They are based on the same principle of whole number division/multiplication as the consonant tones involving musical intervals. Vibrating in halves, the string produces the second harmonic (the octave interval, twice the fundamental frequency). Vibrating in thirds, it produces the third harmonic (the interval of the fifth, three times the fundamental frequency), and so on. Whole number multiples

---

14 *Entrainment* is the principle whereby a powerful force or frequency influences lesser forces in its vicinity. The term was coined by physicist Christian Huygens, who noticed that two pendulum clocks placed next to each other on a wall came to swing in unison. The vibration of the Sun entrains the vibration of the Earth. This is known as the "Schumann Resonance," or the Earth's heartbeat. The Sun itself is entrained by the galactic frequency.

of the fundamental frequency, harmonics are naturally related to each other by whole number ratios. All of these vibrations happen at the same time, resulting in a deeply harmonious sound.

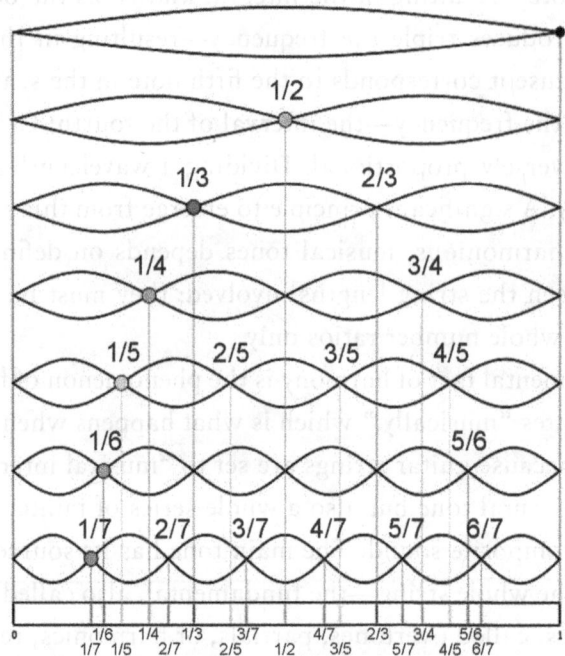

Fig. 8. Harmonics are produced by the nodes of a vibrating string. Nodes are whole number fractions of the string

The harmonics of a tone form a natural, predictable series of invariant intervals, called the harmonic series. No matter what the fundamental, the intervals between successive harmonics are always the same. Every term of the harmonic series after the first is "the harmonic mean" of its neighboring terms.

The question is, why are harmonious "musical" sounds tied to whole number ratios, or why is it that only whole number ratios produce harmonious sound? Why is it that no harmonic will be found to vibrate at the rate of 2.23 or 5.48 times the frequency of the fundamental?

Quite simply, such ratios reflect the periodic character of sound waves and the limit imposed on their combination by the whole cycles of space and time they represent. Sound waves are defined by their frequency, which is measured in cycles per second. It is the synchronous, constructive interference of sound waves that results in the experience of "harmonious" sound. If waves are to interfere harmoniously with themselves and with other waves, their half-cycles have to be whole units of division–multiplication of each other and of the entire space, string, or instrument. For instance, for two notes to be an octave apart, there must be exactly two waves

II. Overture: The Mandorla of Mediation

of one note for each wave of the other note. The interval of the fifth is "musical" because every third wave of the fifth synchronizes with every second wave of the base frequency. Musical sound is an experience of mutually reinforcing waves and sub waves of vibration, supported by physical instruments conceived to allow this constructive, "holistic" interplay. Musical intervals or notes are tied to whole number or "integer" ratios simply because their production depends on an *integrality* of wave cycles wholistically combining with each other (fig. 9).

From another perspective, one can say that musical sound indicates the presence of harmonics. What the presence of harmonics signals, physics wise, is that the wavelength of the vibration produced by the plucking of a string (or the setting in resonance of an object) matches the length of the string or the internal proportions of the object, so that the end nodes of the fundamental and its harmonics coincide with the physical bounds of the string or object. In which case, the outgoing wave of the fundamental bounces back and *conjugates* harmoniously with itself, while also entraining harmonic sub waves perfectly embedded within each other.

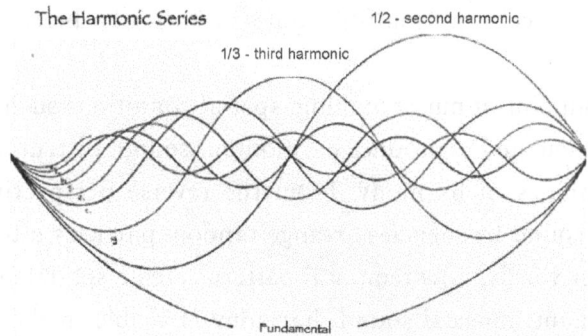

Fig. 9. Overtones created when any string or column of air, of fixed length, is sounded (thestarseedscompass.com/wp-content/uploads/2016/08/STA-Harmonic-Series.jpg)

Harmonic sub waves are whole-number divisions of the fundamental wave length, which is equivalent to saying that they are multiples of the fundamental frequency. Their half-periods fit perfectly in the same space. In other words, they are contained as whole half-wavelengths in the fundamental and move in harmonious, reinforcing interplay. A pattern of constructive interferences is produced, called a "standing wave pattern."

A standing wave is both a dynamic and stable formation generated by recurrent, self-reinforcing waves. Standing waves appear whenever waves are confined within a finite space, and one of them, an "incident" wave with a half wavelength that coincides with the bounds of the medium, begins to bounce back into a "reflective" wave that interferes constructively with the first, entraining harmonic sub waves and resulting in a wave formation that oscillates in place and does not appear to move. All objects have a "natural" frequency or set of frequencies at which they naturally vibrate when struck or plucked. Every such natural frequency is associated with a standing wave pattern. In other words, when an object resonates at one of its natural frequencies, it means that a standing wave is activated in the object.

Standing waves and their harmonic sub waves—waves whose half wavelengths represent whole fractions of the length of a string—are what musical sound is made of. What we perceive as musical tones are standing waves. Musical instruments are conceived to fit the mathematics of harmony, as does the adaptable shape of the larynx.

In summary, harmonious sound is the expression of a propitious rapport between a particular wavelength of vibration and the spatial dimensions of an object, string, or instrument. Harmony results from a perfect "fit," defined by simple mathematical ratios, between the periodicity (or *timing*) of sound waves and the physical boundaries of the *space* within which they are "resounding." Harmony is a perfect fit—hence, for the Greeks, a "soulful" combination—of vibrational frequency and spatial structure.

From the standpoint of music making, spatial conditions such as variable string lengths must be calculated to produce harmonious sound. Instruments are conceived to fit the mathematics of harmony. From the reverse perspective, simple experiments show how sound frequencies arrange random particles into orderly, periodic patterns—in other words, how temporal patterns create spatial geometric patterns. Audible in consonant, musical sound, harmony is visible in the rhythmical geometries created by self-reinforcing waves of vibration.

Third archetype of twelvefold mediation: The musical scale

As regards our inquiry into the mystery of twelve, the question is: how were the 12 notes of the chromatic scale obtained? Technically, they can be derived from the Pythagorean intervals: An ascending fifth (C to G), followed by a descending fourth (G to D) leads to a whole tone (C to D); adding a tone to D (E) establishes the semitone (between E and F). These basic intervals suffice to create the scale.

The most elegant way to arrive at the scale however, and the most illuminating for our context, is through the "ladder of fifths"—which may well have been the method used by Pythagoras himself. The "fifth," so called for its fifth position in the diatonic scale—G or sol being the fifth note or interval in reference to the tonic C

II. Overture: The Mandorla of Mediation

or Do—is the third harmonic. The third term in the harmonic series, it is produced by plucking a third of the string length. The first harmonic is the fundamental, the second harmonic is the octave (1/2), and the third is the musical interval of the fifth: 2/3. Similarly, the fifth harmonic, the note E, is called a major "third" because it is the third note of the major scale. Its half-wavelength, however, is one fifth of the string length.

Dividing the string at 2/3 of its length produces the fifth interval. By continuing to divide each 2/3 section into an ever smaller sequence of 2/3 ratios, a ladder of higher notes is produced. On the thirteenth step, the original note is sounded again—although slightly flat, as the last fifth oversteps the octave by a minute interval called the comma.

Fifths	Series of Fifths Harmonics	Note	Series of Octaves Harmonics	Octaves
			1	C_1 — first
first	2	C	2	C_2
	3	G		— second
second			4	C_3
third	4.5	D		— third
	6.75	A		
fourth			8	C_4
fifth	10.12	E		— fourth
	15.19	B		
sixth			16	C_5
	22.78	F sharp		— fifth
seventh			32	C_6
eighth	34.17	C sharp		— sixth
	51.25	G sharp		
ninth			64	C_7
tenth	76.88	D sharp		— seventh
	115.32	A sharp		
eleventh			128	C_8
twelfth	172.98	E sharp		— eighth
	259.48	B sharp	256	C_9

Fig. 10. Parallel tabulation of octave and fifth series
(Dane Rudhyar, The Magic of Tone and the Art of Music, p. 71)

The table (fig. 10) shows how the series of octaves and fifths come to quasi coincidence, but by a comma, at the 12th fifth and 7th octave: the 12th fifth is almost the same note as the 7th octave (counting as first the second octave, which contains the first fifth, what appears as the 8th octave on the figure is in fact the 7th in terms of the parallel series).

Climbing from fifth to fifth from fundamental to octave brings forth 12 pitches, which, transposed within the span of a single octave, give the 12 successive notes of the chromatic scale. Archetypally, touching the higher octave is equivalent to reaching the fundamental again, hence completing a "fundamental" cycle of 2/3 intervals. The musical scale is thus based on the combination of two cycles—octave cycle and cycle of fifths—a point that will prove of great significance.

The beauty of this background of the scale is that it shows the twelve notes emerging as a sequence cycle from *a single source: the third harmonic* (3/2 or 2/3). The 12 notes represent the unfolding of the third harmonic within the bounds of the octave. Mathematically and meaningfully, the third harmonic, which defines the major intervals of the fifth and the fourth, is the great "mean" (at once harmonic and arithmetic) between fundamental and octave—the harmonizing "sol" or "G" between lower and higher "C." The third harmonic is the acoustic expression of the mediating principle, the "son" of the primary two—the musical version of the vesica piscis.

We feel this mediating character of the fifth when we follow how the unfolding scale moves us from the pull of the tonic to the pull of the octave, through the perfect equilibrium of the fifth. To experience the scale means to shift our attention from the physical notes we hear to the intervals they generate: a note does not exist in a musical vacuum, but in relation to a fundamental (by convention—C). A note is an expression of relationship. It is an "interval." A "G" is a fifth, an "A" a sixth. Musical experience lives in the intervals. Armin Huseman offers this illuminating insight about the musical scale:

> The intervals from the prime to the fourth are, each in their own way, related to the tonic like to a physical gravitational point. On the other hand, the intervals of the sixth and the seventh are related to the octave such that they move toward it. The octave affects the sixth and the seventh as reference note without sounding physically...it happens in a way that makes us feel drawn to this (nonsounding) tone. The octave affects the fifth [just] sufficiently to balance the "pull of gravity" of the prime. That is the essential polarity of the scale: *motion held back from below changes in the fifth into intensifying motion unrestricted from above.*[15]

The cycle of the third harmonic, which chimes the twelve notes of the scale, is the cycle of the "child of the two"—fundamental and octave—the musical equivalent of the soul. Musician and cosmologist Dane Rudhyar elaborates,

> The octave symbolizes a whole of sound...The first octave symbolizes the purely subjective relationship of the one and the other that is its image, the love of Shiva and Shakti...the first act of self-duplication...The power release acting through number two produces number three, the symbolic child.[16]

☉ The twelvefold scale deploys the range of the mediating third, which is the symbol of the *soul*. To identify in the chromatic scale the twelvefold cycle of the third harmonic is to discover that what the scale "plays out" is the deployment of the archetypal mediating unit—the third, the son of the Two—into a twelvefold musical whole (octave). It is to discover the *12fold character of harmonic mediation*. It is to catch the

---

15 Huseman, *The Harmony of the Human Body*, pp. 115–116 (italics added).
16 Rudhyar, *The Magic of Tone and the Art of Music*, pp. 63–64.

II. Overture: The Mandorla of Mediation

The third harmonic held in the arms of the octave—"spiritual" arm of the fundamental or first harmonic, and "material" arm of the octave or second harmonic—is the acoustic and mathematical "child" of the fundamental Two. The harmonious, radiant "sol" (G) evokes the universal mediator: the soul. Pythagoras regarded the twelve-tone unfoldment of this third harmonic within the grand seven-octave cycle of the "fundamental" as a model of world creation, intimating a vision of the universe as the chamber of resonance of a primordial sound. In the symbolic language of harmony, the 12fold-ness of the Third within the Octave speaks indeed to the principle implicit (lit. folded) in the resounding "self-resonance" of the primordial sound of Being: the soul principle of reflective consciousness, the ground of the world's inwardness.

Geometry offers a similar insight into twelvefold-ness as the natural deployment of the primordial triad into a next "oneness." Indeed, 12 is the greatest number of evenly distributed points on a single spherical surface that, when connected by lines, result in equilateral triangles.

Wise sources across the ages have expressed the notion that God created the universe to know himself. "The universe is God regarding himself."[20] The fundamental inwardness of "regard," resonance, and reflection is the primordial milieu of the world soul. An octave-like field of self-resonance opens the eye-like consciousness ◯ of the world being. "God," the world-being, "appears" into a cosmos—and appears to itself as cosmos—by resounding and deploying its vibrant being into a self-resonant, self-reflective whole of vibratory space–time.

The vesica like Egyptian hieroglyph for the supreme creator, ◯, evokes the shape of a vibrating string, suggesting the vibratory nature of the mandorla archetype. The twelvefold held in the arms of the fundamental Octave conjures up an imagination of the world matrix as a vibrant mandorla ovum, whose harmonic sub waves generate a spiraling "serpent."

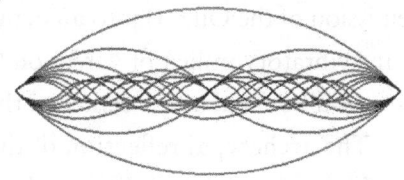

Fig. 12. Diagram of a standing wave

Remarkably, the laws of harmony give us direct and vivid intimations, based on acoustic perception and symbolic intelligence, of the 12fold structure of space and soul referred to in esoteric teachings.

✪ In this graphic representation of the waveforms underlying the harmonic series of overtones associated with a fundamental tone (fig. 12), we see at once a vibrating string between fundamental and octave, a standing wave, a vesica piscis, a mandorla, all alluding symbolically to the inwardness of soul space opened by the harmonious interplay between the primordial polarities of unmanifested being and manifesting cosmos. ✪

---

20 Lawlor, *Sacred Geometry*, p. 63.

THE GENESIS OF THE TWELVEFOLD
AND ITS TWIN, THE SEVENFOLD

When we penetrate the reality of the 12fold scale and its formation, we discern an operation that will prove of great significance. We see that what determines the twelvefold character of the scale is the interplay between the octave series of the fundamental and the mediating series of the fifth—the third harmonic, the Son of the Two. The octave replicates the fundamental at a lower or higher level of pitch and the octave series replicates this octavian unit of musical space. The fifth is the fruit of the harmonious interplay between the two ends of the octave, and the series of fifths replicates this mediation–fruition. So, the 12fold scale results from the intersection of the 2/1 series of octavian *self-replication* of a fundamental with the 3/2 series of *mediation* between fundamental and octave. The coming to coincidence of the two series defines a "metaoctave" that blends into one metaunit a 7fold space of fundamental *wholeness* (octave) and a 12fold rapport of harmonious *mediation* (fifth)—in other words, a dual whole of space and soul.

In stunning parallel with the iconic representations of world-self figures such as Mithras or Christ set in zodiacal mandorlas, the 12fold cycle scale of harmonic mediation reveals itself as both cosmos-like and self-like—*zodiac-like and soul-like*. It is cosmos like in that it embraces the polar ends of the octave, the two ends of self-replication—spirit and matter. It is self-like in that it is filled with the mediating third.

The contraction of the two series within a single octave yields a mandorla-like space evocative of the fundamental vibration of the world being, ◯, the primordial emission of the One (1) into an octave replication of its self (2), which opens the space-time vibratory milieu of "creation" (3)—the self-resounding, mediating milieu of the "son god" or soul at the center of the mandorla.

The archetypal reflection of the solar god mandorla in the meta-octave reveals in the iconography of the mandorla a blend of spatial self-duplication and soulful mediation: On one hand, the meta-octave of the scale illumines in the oval of the mandorla the fundamental space of an octave vibration filled with 12 notes (signs) etched around its periphery; on the other hand, it reveals how the "son" of Being (the 12fold soul) is wrapped in the serpent like standing wave of a 7fold chakric energy body—a 7fold octave. Like its vesica archetype, the mandorla holds, intertwined into one, the two of soul and cosmos, time and space, 12fold and 7fold.

Cosmologically, this resonance between visual mandorla and acoustic meta-octave suggests that the primordial, world-creative, self-resounding of Being brings forth the dual reality of world-space (objective) and world-self (subjective) as a 7fold/12fold harmonic field. This field is zodiac like in its partitioning of the

II. Overture: The Mandorla of Mediation

cosmic egg of space, and soul-like in the mediating process it represents (3rd harmonic). And it is cosmos like in the unit of wholeness that the octave represents. Cosmos is the space of interplay between Being and matter. Soul is their harmonizing milieu.

✪ The formative underpinnings of the harmonic 7fold/12fold *unit* captured in the musical scale open a whole new dimension of insight into the cosmological mysteries displayed symbolically in the mandorlas of the sons of God and articulated in esoteric statements about zodiac and soul. The intersection of the two series highlights simultaneously the two numbers that dominate cosmology and music: 7 and 12. Their joint emergence from the intersection (by a comma) of the cycle of the 2/1 ratio with that of the 3/2 ratio—the ratio of duplication and the ratio of mediation—points to a significant fact about this mysterious couple: *their "twin birth" out of the same womb process of harmonic unfolding.* The intersection between the number 3-based series of fifths (mediation harmony) and the number 2-based octave series (associated with duplication–proliferation) will prove to be a cosmological cornerstone. ✪

The twin character of the two numbers appears also discreetly in the fact that the seventh harmonic defines the 12th interval (the semi-tone), and the 12th harmonic is at the center of the 7 octave series.

By uncovering the acoustic laws of harmony while defining the soul as harmonizer of worlds, Pythagorean philosophy offers a bridge between metaphysics and physical facts. Harmony speaks at once to a principle ("fitting together"), a soul experience and an acoustic phenomenon. In the notion of harmony, the border between spiritual and physical realms becomes blurred—metaphysics blends with physics. Acoustic harmony reverberates metaphysical harmony and reveals it to the senses. It also points to the *soul* as an intermediary space of *awareness of harmony*—being the milieu where harmony is experienced as well as generated. We begin to sense how the musical, mathematical, and geometric grounds of harmony laid out by Pythagoras offer insights into the reality of the "soul" as a harmonizing milieu between matter and spirit, cosmos and life.

The Circle of the Fifth Principle

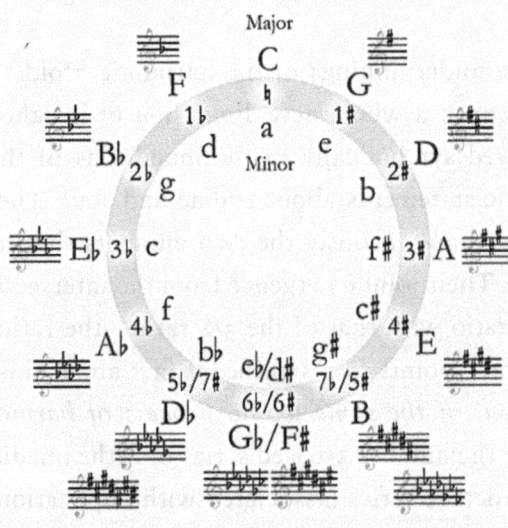

Fig. 13. Circle of fifths showing major and minor keys

The twelve fifths, which define the twelve notes of the chromatic scale, are the basis of the twelve keys of music composition. In music theory, they are arranged in a circle, referred to as the *Circle of Fifths,* a term that originated with Pythagoras. Starting clockwise from any note, the circle ascends from fifth to fifth through the twelve notes/keys, the twelfth being almost the same tone (but for a comma) as the first. As the diagram shows, the progression of the keys is accompanied by a regular increase in sharps on one side and flats on the other side of the tonic (C), indicating the modulations necessary to replicate the scale, starting from any of the twelve notes.

The twelvefold character of the series is striking enough for the parallel it offers with the zodiacal partition of time–space and the structure of the causal body. But it is its circular arrangement into the "circle of fifths" that makes the correspondence between the twelvefold scale and the circles of soul and zodiac most compelling. The similarity of structure (a 12fold circle) illumines the mediating, harmonizing function they share, and the circle of musical harmony lands into a mathematical fact. As the circle of fifths mediates between the tonic and the octave (first and second harmonic), the 12-petal lotus of the soul mediates between the extremes of spirit and matter and the 12-sign zodiacal wheel of time–space mediates between eternity–infinity and here and now.

II. Overture: The Mandorla of Mediation

If, instead, one arranges the circle according to the chromatic sequence of the scale, then links the notes by intervals of fifths, one obtains a 12-pointed star—a figure evocative of a 12-petal flower. In reverse, tracing the chromatic sequence within the circle of fifths yields a star dodecagon:

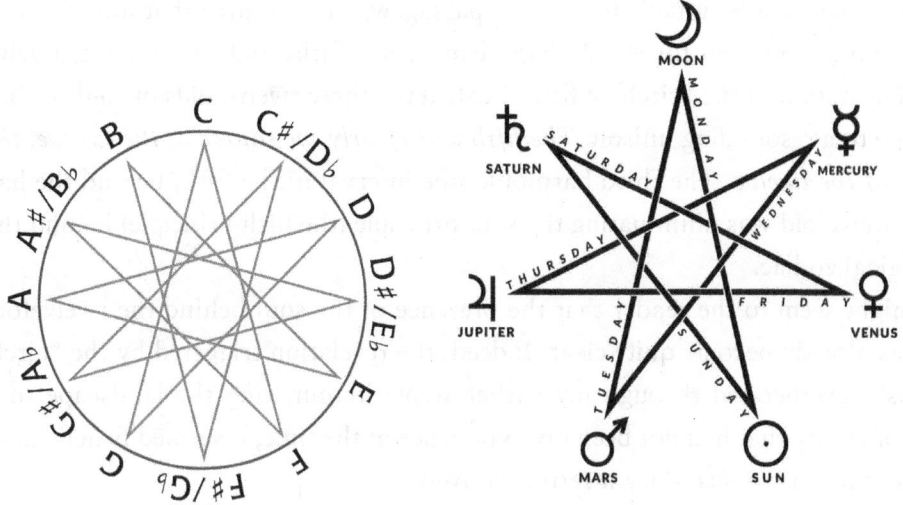

Fig. 14: (left) The circle of fifths forms a twelve-pointed star in the chromatic circle; (right) The sequence of days forms a seven-pointed star in the circle of planets

Similarly, if one arranges the planets around a circle in the order of their distance from the Earth (from Moon to Saturn, as shown in fig. 14), then links them according to the succession of the corresponding days of the week, one obtains a 7-pointed star.

In both cases, human order (chromatic or weekly) perfectly internalizes divine order. The human "stellates" the divine. This stellating process opens a deeply suggestive perspective on "selfhood" as a stellation of cosmos, a contraction of cosmos into a star like unit. Reminiscent of Steiner's insights into the human heart, the soul reads as an "introjection" in progress, a drawing-in of the cosmic, zodiacal whole, into a star-self—an individualization of the universe.

It is only after encountering the Pythagorean notion of the soul as mediation, and of the musical proportion as epitome of harmony, that I became aware of this classical arrangement of the series of twelve fifths in a "circle of fifths." The notion struck me with the magic of a missing key. It is no doubt a striking formula if one has in mind the inner reality associated with "five": *quint*-essence. The quintessence of the human being is the soul-self principle. To see this "twelvefold circle of *fifths*" drop in the midst of an exploration of the structural similarity of zodiac and soul was jolting. The jolt came from the sudden and "matter of fact" blend of 12fold circle and fifth—zodiac and soul. Written on the wall of music theory was a potential confirmation that the twelvefold circle is essentially the circle of *"the fifth" essence*—the circle of the soul.

Prior to this moment, I had certainly been intrigued by the parallel between the series of twelve fifths and the twelvefold framework of astrology. I had turned around it many times, unable to find the passageway to a source that might illuminate a common ground for the Pythagorean twelve fifths and the zodiacal twelve signs. The notion of the "circle of fifths" called the three twelvefolds of soul, zodiac and scale to a resounding unison. *The fifth was clearly the source of the twelve, the essence of the twelve.* The third harmonic (the interval of the fifth) behind the harmonic twelvefold was illuminating the soul principle (the fifth principle) behind the astrological zodiac.

It might seem to the reader that the presence of the soul behind the twelvefold scale has already become quite clear. Indeed, the revelation triggered by the "circle of fifths" reverberated through my earlier steps, illuminating the landscape to a degree of clarity that had not been my experience at the time; it seemed beneficial to the reader for me to introduce it retrospectively.

To understand the ways in which the notion of the "circle of fifths" might be revelatory, we must pause for a moment to uncover the fabric of cosmological connections between fifth principle, twelvefold circle, soul and mind. These connections are rooted in inner realities veiled by symbolic correspondences and signaled by cosmological facts—specifically, the symbolism and cosmological ground of number five, and its connection with the planet Venus.

Esoteric teachings refer to the soul principle—the core of what makes us human—as the "fifth principle." The context for this notion of a fifth principle of mind and self is the ancient cosmological framework, common to many wisdom traditions and world cultures, of the 7-leveled constitution of the universe and the human being—macrocosm and microcosm. What cultures across the globe have referred to as seven (+3) sephirot (Kaballa), seven climates (Islamic theosophy), seven spirits around the throne, seven heavens (Judeo–Christianity), corresponds to what nature displays in its physics. The 7 colors of the rainbow, the 7 major tones of the diatonic scale, the 7 planets of the solar system all point to these seven dimensions of being.

As we saw earlier, theosophy offers a psychospiritual delineation of these seven levels in terms of dimensions and stages of consciousness, ranging from the densest level of etheric–physical manifestation (7th) to the most subtle, first level of "Logoic" or divine identity. Between the two extremes, and identifiable to a degree in the spheres of human experience—which includes the inspiration offered by great human beings—the 6th level corresponds to astral–emotional consciousness, the 5th to self-awareness, mind and individuality, the 4th to enlightened love-wisdom (*Buddhi*), the 3rd to universal will-to-good (*Atma*), the 2nd to identification with divinity (*Monad*). Among these, the denser levels are readily recognizable in the corresponding "kingdoms" of nature: 7th level of physical–etheric as minerality and vegetality (the vegetative, physiological dimension in humans); 6th as animality (instinctual, emotional domain); 5th as mentality–humanity. In its principle, humanity is the fifth essence, the quintessence of the natural world. Clearly, however, everything in us humans is not "quintessential." Like the universe it replicates in a "microcosmic" form, humanity has "elemental" levels of being, and its quintessence is the 5th principle of "I AM-ness," or soul-self.

By definition, the quintessence is the essence of the four elements that make up the physical world (earth, water, air and fire). These four elements have their correspondence in the four vehicles of the human being: physical, etheric, astral and mental bodies. A quintessential feature of the human race is its unfolding, out of a fourfold nature (physical–etheric–emotional–mental), of a fifth dimension of soul wisdom, love, and will. The lotus flower of the soul unfolds out of the existential challenge to conjugate nature and spirit—"elemental" forces and mindful intent. The winds of the mental nature, the fires of animal vitality, the watery currents of vegetative sentiency, the solidifying forces of physicality are in constant process of being organized, refined, shaped and penetrated by the purposeful intelligence of the universe at work in the depths of the human self. In traditional cosmology, humanity is regarded as the universal mediator who binds together intelligible and sensible, eternal and temporal, mind and matter.[21]

The "quintessentially human," the fifth principle of mindfulness and soul, is associated cosmologically and symbolically with Venus. The planet of love and union, harmony and beauty, Venus is the great uniter and unifier, a function that resonates deeply with the keynote of the soul—mediation. The causal lotus of the soul, the quintessential flower of the human plant, is the radiance of wisdom, love and power garnered from the union and communion between personal elements and universal spirit. Venus engages polarities into relational partnerships, interplay and intercourse. It elicits the esthetic rapport we experience as beauty, and the moral rapport we perceive as fairness. Venus bridges, "binds and fits" opposites in friendships, unions, and weddings—all words commonly used by Plato as well as Kepler in reference to celestial relations. It is the planetary principle of harmony and proportion, rapport and ratio. In Greek mythology, Harmonia was the daughter of Aphrodite/Venus and Ares/Mars. The fifth musical interval, the archetypal

---

21 Fideler, *Jesus-Christ, Sun of God*, p. 213.

harmony from which the musical scale unfolds, induces a psychic response akin to the blissful unity and joy the soul experiences in the presence of beauty, love, peace, harmony. "The fifth is a beautiful sound, writes Stephen Ian McIntosh in *The Harmonic Lyre*, because it demonstrates how the universe works...The fifth is an archetypal expression of harmony that demonstrates the 'fitting together' of microcosm and macrocosm in an inseparable whole."

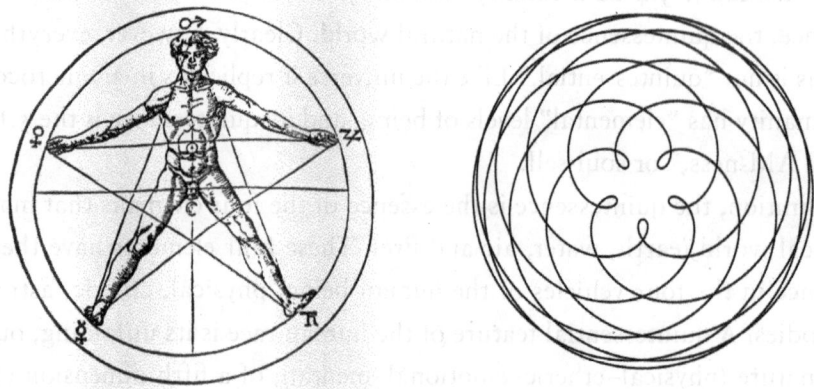

Fig. 15: (left) Agrippa Von Nettesheim's "pentagrammatic man"
Fig. 16: (right) Five-pointed star formed by the 8-year cycle of Venus-Sun conjunctions

Humanity is Venus-like in principle, and it is Venus-like in form. The "beautiful" human form, celebrated in mythology and art, is precisely the form of the *mediating* realm humankind represents between the kingdoms of nature that precede the emergence of individuality, and the bodiless realms of higher levels of consciousness. It is quite literally *the physical form of the soul*. While the animal does not stand or speak (or very rudimentarily so), the angel does not have a body. The human being stands between the two, self-reflective and self-determined. Venusian "beauty" alludes to the archetypal beauty of the human form as a microcosmos of harmonious and rhythmic proportions fit to harbor the nascent self-awareness of the fifth stage and kingdom of the world being. What distinguishes it among all other forms in nature is the informing presence in its midst of the quintessence of all forms: selfhood.

Venus is associated with the *five*-pointed star. The connection is more than symbolic. The planet's 8-year cycle of five inferior conjunctions with the Sun during its retrograde motion forms an actual pentagram in the sky (fig. 16). Viewed from the Earth, the orbiting pattern of Venus around the Earth resembles a *five*-petal rose. The rose, the equivalent of the Eastern lotus, is a major symbol of love and soul.

A traditional symbol of the human being dating back to Babylonian times, the five-pointed star or pentagram was popularized in the West by Renaissance

II. Overture: The Mandorla of Mediation

hermeticist Agrippa von Nettesheim (fig. 15) and artist Leonardo Da Vinci, whose drawing of the Vitruvian man was based on the correlations between ideal human proportions and geometry described by the ancient Roman architect Vitruvius (fig. 17). In the drawing, two superimposed positions of the human being suggest on one hand the four branches of a cross (also alluded to in the L-shaped "square" position of the feet) and, on the other hand, the five branches of a five-pointed star.

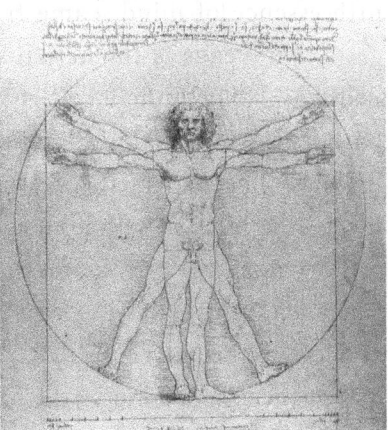

Fig. 17. Leonardo De Vinci's Vitruvian man

The human form is keyed to four and five—fourfold of human "nature" and fivefold of soul. The number 4 is associated with the quarter-based cycle of the Moon (the "lunar nature"), 5 with the cycle of Venus (the soul dimension). Superimposed, they allude to the universal function of humankind: to integrate circle and square—the two frames of the Vitruvian man—spirit and matter.

Five is also a fundamental number in the make up of the human body. The torso has 5 appendages: arms, legs and head. In turn, each of these has five appendages: fingers and toes and 5 openings on the face. We have 5 senses: sight, sound, touch, taste, and smell. The pentagram was the emblem of the Pythagorean brotherhood, an emblem most congruent with its philosophy of harmony centered on the interval of the fifth.

Like the *12fold circle of fifths*, the human form is both fivefold (pentagram) and twelvefold (Egoic lotus of the soul). *The 12fold lotus of the Ego is the actual "circle of the fifth,"* the human quintessence—selfhood, the soul.

Modern theosophy brings deeper grounds of ancient knowledge to the notion of a Venusian essence of humankind. Blavatsky and Bailey indicate that the principle of selfhood was "implanted," millions of years ago, into what was then an animal like humanity, by advanced solar beings from Venus.[22] According to the same sources, Venus is the prototype of the Earth's higher self. Interestingly, Kepler

---

[22] See Bailey, *Esoteric Astrology*, pp. 179: "When it had reached a certain development, the instinctual nature came under the direct influence of the Hierarchy of the planet in a new way, and then, under the stimulation of energies from the planet Venus, a fusion took place that resulted in the emergence of individual self-conscious man"; also, p. 680: "The *juncture* of spirit-matter and mind, or manas, [the very principle of "man"], was effected during the third root-race, and the definitely human family became present upon earth from that date. This was brought about by the coming in bodily presence of certain great Entities, who came from the Venusian chain, achieved the necessary juncture, undertook the government of the planet, founded the occult Hierarchy. The remainder have returned to their originating source."

describes the relationship—musically, the "interval"—between Earth and Venus as a marital one. This relationship-interval varies, he says, between the "masculine sixth" of G# – E and the "feminine" one of Gb – E.[23] The musical interval of the "sixth" is obtained by dividing a string in the fraction 5/8, which ratio of 5 to 8 speaks precisely to the 8-year cycle of 5 conjunctions that is the pattern of Venus, seen from the Earth.[24] In saying that the principle of self-awareness, the fifth essence of mind, came from the "pentagrammatic" planet, theosophy suggests that Venus is the actual source, and not only the symbol, of humankind's "selfhood." The human self is Venus like in that it "binds" nature and spirit and harbors the eventual union of square and circle. It "fits" nature for divinity, with divinity. The real birth of Venus from the sea is the emergence of the human form—the birth of self-awareness and soul on the planet.

The same esoteric sources associate Venus with the "fifth ray"—one of seven fundamental energies or "rays" that inform every living system, every dimension and level of being. The fifth ray is referred to as the energy of "concrete knowledge and science." It is akin to the fifth principle of mind associated with the fifth (mental–causal) level of Being. The association of Venus with the fifth ray may surprise initially, considering the more familiar associations of the planet with the affective and esthetic domains. However, every planet has a range of meaning that covers the entire spectrum of human experience (physical, emotional, mental, soul or spirit). One can easily follow the spiraling up of the meanings of Venus from its most concrete association with money and assets ("values") to the notions of love and beauty, from the concepts of uniting and harmonizing, which evoke the activity of the soul, to the more abstract notions of "proportion and ratio," which speak to its mediating function. The mind is indeed what *binds* world and self, objective and subjective worlds into percepts and concepts. And the real mind is the soul. In Alice Bailey's words, "The soul is the true thinker." Soulfulness is mindfulness. The categories of mind and heart fuse in Venus as they fuse in the soul. The higher mind is a heart-full, unitive mind, while the love of the soul is mindful and wise.

The Venusian character of the ray of science alludes to the unitive spirit of science, at work in the striving to integrate observations, experimental data and mathematical equations into lucid paradigmatic wholes—in a quest for what Einstein, Hawkins and others have called a unified field or a theory of everything. Venus hints at the aspiration of science to extract a quintessential understanding, a single formula that would capture the intelligent code or "mindedness" of reality and see into what Stephen Hawking himself, despite his affirmed belief that there is no

---

23   Godwin, *Harmonies of Heaven and Earth*, p.147.
24   Kepler, *Harmonices mundi*, book 5.

"god," refers to as "the mind of God."²⁵ Although what he means by "god" has little to do with a religious figure, and everything to do with a meta-"law of nature," the formulation implies nonetheless the notion of a meta-intelligence at work in the law. Such a "quintessential" understanding would unify the many unresolved facts and theories of science and provide a mediating ground between the complex multiplicity of objective data and the intuition of oneness humankind carries at the heart of its mind and in the functioning of its cells.

A unifying understanding is essentially—beautiful, and beauty is ultimately the driving force behind theoretical science—the Venusian heart of science. "Beauty is the splendor of the truth," says the Latin proverb. "True Science," Simone Weil echoes, "is the study of the beauty of the world." Einstein himself was convinced that beauty is a guiding principle in the search for important results in theoretical physics.²⁶

In essence, *the circle of fifths reveals the thread that links the twelve*. This golden thread is the 3/2 ratio of the fifth interval—the principle of mediation and harmony—and the fifth principle of mind–soul. The core of "twelve-ness"—the third harmonic—is the quintessence of harmony. Inner reality of mediation and objective facts of harmony meet in the 12fold circle of fifths, intimating what lies at the core of the structural similarity of zodiac and selfhood. What stands behind the twelve-fold formations we have been contemplating is the archetypal circle of harmonizing, mediating intervals. Harmonious conjugation of extremes is factually, not only intuitively, the key to their structural identity. It is the formative principle they share.

The circle of fifths proves to be a remarkable symbol of the twelvefold vessel of the soul—at once revelatory and grounding. More than simply illuminating with symbolic associations the harmonizing milieu that the soul represents, it grounds the notion of mediating harmony and "soul" in experiential facts of harmony, and it offers a reflection of the world of meaning in the mathematics of means. "Meaning" speaks to the wisdom distilled in the "midst" of human experience, out of the interplay between our spiritual essence and the matters of existence. Jung regarded the creation of meaning as the soul of human existence. "As far as we can discern," he wrote, "the sole purpose of human existence is to kindle a light of meaning in the darkness of mere being."²⁷

---

25 In his bestselling book, *A Brief History of Time*, physicist Stephen Hawking states that when physicists find the "theory of everything" that he and his colleagues are looking for, then they will have seen into "the mind of God."

26 As stated by mathematician Hermann Bondi.

27 Jung, *Memories, Dreams and Reflections*, ch. 2.

⊙ The twelvefold partitioning of the time–space octave-like milieu in which we come to existence is not arbitrary. Twelve is indeed—and not only in esoteric lore or zodiacal "fantasy"—an organizing principle intrinsic to our reality. While there are approximately twelve lunations in a cycle of the Sun around the zodiac, there are, archetypally speaking, 12 mediating notes in the octave "cycle" of a fundamental. Like the astronomical twelvefold, the musical twelvefold exists in the hidden mathematics of the world, which the mind can extract from the study of nature's rhythms and vibratory laws. Born of the fifth interval of the third harmonic, the twelvefold pattern is rooted in three, the son–soul principle, and in five, the number principle associated with mind, "self," humankind. Together they point to the *mediating realities of the world being: space–time and mind.* ⊙

It is important to note, Rudhyar points out, that the twelve notes are not derived from the harmonic series, but from the cycle of one specific harmonic: the third harmonic (or fifth interval)—even though, he adds, "going to very large numbers, one finds a series of twelve fifths whose frequencies measure in terms of whole numbers and therefore can be considered part of a harmonic series."[28] What Rudhyar emphasizes is that the 12 notes are not laid out primarily as a continuum by the laws of nature. They are derived instead from a conceptual operation: the replication into a man made musical scale of the archetypal mediation represented by the fifth. The space of the soul is unveiled to the mind by the mind—as the medium of mind—the fifth principle.

As life lays down the grounds of our destinies and guides us from site to site through echoes and resonances, the revelatory sighting of the circle of fifths, and the connection it lit up between soul, zodiac, and the world of harmony, called to memory the charge of intriguing information about the harmonic construction of the universe I had encountered years ago in the Platonic dialogue of the *Timaeus*. I pitched a tent in the ancient grove and started reading. Days passed, of probing and pacing, sorting out and trimming—till the thick grove turned into a certain clearing.

---

28 Rudhyar, *The Magic of Tone and the Art of Music*, p. 99.

III. THE HARMONIC COMPOSITION OF THE UNIVERSE

THE PLATONIC COSMOGONY: CREATION OF THE WORLD SOUL BY MEANS OF PROPORTION AND HARMONY

> *The cosmos is a single Living Creature that contains all living creatures within it.... It has come into existence as a Living Creature endowed with soul and reason.*—PLATO[1]

So far, the kinship we have been pondering between circle of fifths, zodiac, and soul has been based on wisdom teachings about the 12fold structure of the causal body and the chakric nature of the zodiac, in association with the Greek geometry and mathematics of harmonic mediation. Without the insights offered by esoteric teachings into the inner reality of self and zodiac, the similarity of pattern between musical and zodiacal 12fold amounts to little more than an intriguing echo, faintly orchestrated by the fact that they are both intimately involved with time and space. The zodiac provides a framework of time and space, and harmony is based on the relation between frequencies of sound vibrations (itself a time–space factor) and spatial measurements.

This intriguing echo gathers "soundness," however, when brought in resonance with Plato's intricate description, in the *Timaeus*, of the "construction" of the world according to the laws of musical harmony. What esoteric statements and intuitive inquiring have led us to see as a compelling correspondence between zodiacal and Egoic structure on one hand and formulas of musical harmony on the other hand, Plato spells out as cause and effect: Harmony is the foundation of soul and cosmos, what guides the constructing, structuring hand of the demiurge. World soul and world body emerge as twelvefold/sevenfold structures, and these structures are the direct expression of the laws of harmony. The *Timaeus* does not present this fact straightforwardly, as its focus is on the mathematics of world beginning—not the question of twelve. However, the evidence can be gently pulled out, bringing up with it, live, the root of the twelvefold, which will turn out to be the very germ of the creational operation. Not only does the cosmology presented in the *Timaeus* confirm the kinship between zodiac and soul that their analogous structure led us to

---

1   Plato, *Timaeus*, 30b, 30d3–31a1.

suspect, but it also reveals its essence, which is the principle of harmonic mediation epitomized in the 12fold circle of fifths.

The great surprise, as I turned to Plato's presentation of world beginnings, was to find that his cosmogony begins precisely with the emergence of a twelvefold milieu. Instead of knocking as we did on the door of fundamental structures (zodiac, soul, heart, harmonic circle), listening to resonances and intuiting a common source, Plato starts from the source itself—the "divine" hub of world creation.

We will see that Plato's musicological cosmology grounds in natural laws the metaphysical understanding of ancient wisdoms while it also lights up the intelligence of harmony that shapes the physical world. However, from the point of view of the modern mind, Plato's physics is closer to an abstract ground of metaphysical mathematics than to what we regard today as physics. It exposes a view of the essence of the physics of the world, which for the contemporary mind reads as a rather rudimentary and naïve "physical science." Yet, the profound stakes of the *Timaeus*, and the resonance between its mathematics and our own uncoverings along the path of "twelve," make it a compelling endeavor to gaze at the gap between Platonic and modern physics, and see if, out of centuries of post Platonic physics, stepping stones might not be gathered to support Plato's cosmological claims all the way down to our modern scientific consciousness. In this case, Plato would prove to be the precious bridge he was meant to be—a great herald of the nascent intellect of humankind, bridging the ancient world wisdoms and the scientific consciousness of our times. On the smaller scale of the present exploration, he would be the precious link between the esoteric teachings we started from and the "solid" ground of modern science.

The elaborate confirmation we are about to find in the *Timaeus* of what our inquiry into twelve has led us to envision—namely the harmonic structure of soul and cosmos—should not come as too great a surprise, given Plato's spiritual lineage. If anything, it illumines it. Pythagoras left no writings, but a century later, Plato expounded a philosophy of harmony whose foundations point directly to the teachings of the Samos philosopher, and through them to the old Egyptian and Babylonian knowledge-wisdom he was initiated into during his extensive travels and involvement with mystery centers. Bertrand Russell, in his *History of Western Philosophy*, even contended that the influence of Pythagoras on Plato and others was so great that he should be considered the most influential of all Western philosophers.

Plato is regarded as a great philosopher—not an "esoteric" voice. The Stanford Encyclopedia of Philosophy describes him as "...one of the most dazzling writers in the Western literary tradition and one of the most penetrating, wide ranging, and influential authors in the history of philosophy." Yet his Pythagorean heritage and

III. The Harmonic Composition of the Universe

initiation in the mysteries point to the same line of perennial wisdom as modern theosophists and esotericists. The sources that nurtured his philosophy, and whose cosmological premises echo each other across world cultures and religious divides, go back to ancient Egypt, India, Sumer, and, beyond them, to the enlightened ancestors who implanted their knowledge of universal laws in world mythologies, music and architecture, astronomy and astrology. The theosophical understanding of man and universe reintroduced in modern times by Blavatsky and Alice Bailey, the spiritual science of Rudolf Steiner, or the intuitive insights of Edgar Cayce, all tap into the same universal ground as the knowledge Plato received from the Pythagoreans. What Plato did is bring it down to conceptual rationality, and elaborate it in ways that served the evolutionary purpose of his time: the development of intellectual faculties and philosophical thinking.

Written toward the end of Plato's life (c. 355 BC), the *Timaeus* relays a conversation among Socrates, Plato's teacher; Critias, Plato's great-grandfather; Hermocrates, a Sicilian statesman and soldier; and Timaeus, a Pythagorean philosopher–scientist. The dialogue introduces a cosmology based on the Pythagorean philosophy of the soul as mediator "between the essence of the universe and the universe itself"—between spirit and matter. Timaeus exposes the operation of world creation that supports this view, not only philosophically but scientifically—*metaphysically because physically*.

Plato's world soul is a mediating ground analogous to the Pythagorean representation of the geometric intersection of the circles of spirit and matter—the vesica piscis. He describes the world soul as a "third form of Being"—a compound that unites and harmonizes the "indivisible, unchanging Existence" he calls "the Same," and "the divisible, changing existence...the Being that becomes in the bodies—the other"—into an intermediate "synthesis that links and unites the Essence with the world."[2]

> Midway between the Being that is indivisible and remains always the same and the Being that is transient and divisible in bodies, he [the demiurge] blended a third form of Being compounded out of the twain, that is to say, out of the Same and the Other...And He took the three of them, and blended them all together into one form, by forcing the Other into union with the Same, in spite of its being naturally difficult to mix.[3]

The first operation in Plato's cosmogony is the creation of the world soul, which involves a blending of indivisible and divisible Being—a compounding of

---

2   Ibid., 35a–36d.
3   Ibid.

Same and Other. Curiously, Timaeus has already begun his account of creation in the preceding pages by describing the formation of the world body, which, as Plato points out, is clearly not the actual sequence of creation. Indeed, he says, "God would not have permitted the elder to be ruled by the younger.... He constructed Soul to be older than Body and prior in birth and excellence, since she was to be the mistress and ruler and it the ruled."[4] Plato seems to make light of this permutation in the order of creation, attributing it to the "part of randomness and chance that exists in ourselves, hence in our discourses." However, it may be wise to question "randomness" in a topic of such magnitude and follow one's impression that there is a good reason for the permutation. The fact is that in these preceding paragraphs, a principle has been introduced that will prove central to the whole operation of creation—the principle of proportion and its mean, the golden mean. The introduction of this key factor of the perfect mean of "mediation" required an as of yet inexistent geometric ground—making it necessary to speak of the world body even before the first act of creation, which is the creation of the soul—hence the ambiguous sequence. Geometry pertains by definition to the relatively earthly (geo) domain of forms and measurements (metry), not to the divine realm. So, the presence of the "corporeal world" was needed to understand the "mean" used by the demiurge to create the soul—in other words, "to put together" the indivisible "essence" of the world and its divisible elements.

> Two things cannot be rightly put together without a third; there must be some bond of union between them. And the fairest bond is that which makes the most complete fusion of itself and the things that it combines; and proportion (analogia) is best adapted to effect such a union. For whenever in any three numbers, whether cube or square, there is a mean, which is to the last term what the first term is to it; and again, when the mean is to the first term as the last term is to the mean, then the mean becoming first and last, and the first and last both becoming means, they will all of them of necessity come to be the same, and having become the same with one another will all be one.[5]

What binds two extreme elements in the fairest way is a rapport of proportionality through a mediating third. Proportionality represents the fairest bond of union between the two and among the three. Introduced in the context of the world body, this "mean" is an abstract, conceptual reality, highly suggestive, in its principle, of the intermediating operation, about to be described on the next page, of compounding the world soul. On the other hand, nothing could be more helpful for the mind to grasp the intermediate reality of the soul than have it

---

4   Ibid., 34c.
5   Ibid., 32a.

III. The Harmonic Composition of the Universe

described in conceptual terms—"midway" between divine and material realms. Clearly, compounding a "third form of being out of the Same and the Other" is the essence of the geometric operation that has just been introduced, the golden mean. The "third form of Being" is a bond of union—a bond of "friendship"—the ideal expression of which, geometrically, is the proportionality of the golden mean. "The body of the world was created, and it was harmonised by proportion, and therefore has the spirit of friendship."[6]

Like the soul itself, the golden mean is suspended "midway between indivisible and divisible worlds"—and it is the secret of the mediating operation. The unitive principle of the mediating third opens in the primordial oneness of the One a milieu of harmonious self-bondedness. The notion of unitive rapport is at once mathematical and psychological, geometric and metaphysical. The reality of the third as a threefold One is the *experience of internal bondedness and union* of which the golden mean is the signature.

In its simplest geometric expression, the golden mean is the one and only way to divide a line into two segments so that the length of the whole line is to the large part in the same ratio as the large part is to the small part. This unique division creates what is referred to as *a continued geometric proportion.* $a/b = (a+b)/a$. The beauty of the golden mean is that it generates a continuous rapport of identical proportion between the smaller part and the larger part, and the larger part and the whole: a is to b as a+b is to a. The whole is to the large part as the large part is to the small part.

The golden mean is essentially a *relation*—a uniquely fair, proportionate, and rhythmic relation between parts and whole—a relation of relative identity. "All of them of necessity come to be the same, and having become the same with one another will all be one."

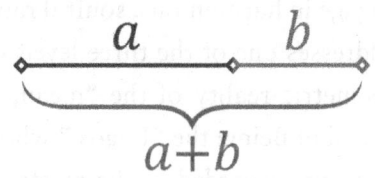

Fig. 18. *The golden section: a+b is to a as a is to b*

In the divine "third," the "son" of the Two, which the golden mean represents geometrically, the primordial oneness of Being gives way to a secondary oneness characterized by "golden" bondedness. The Third activates an experience of self-relatedness within the One. The new oneness is an *internal experience* of "sameness with one another" and identity in union. Proportional sameness reveals the mean for the One to be "Other" in a "similar" way, to be "Same" in a "different" way. Whereas oneness simply IS—a pure Act of Being—union is an experience, which opens within Being an *experience of itself*—the experience of "selfhood."

---

6   Ibid., 32c.

The fourteenth-century Persian Hadith of the *Hidden Treasure* voices the metaphysical event: "I was a Hidden Treasure and loved to be known. Therefore I created the Creation that I might be known." The 3fold one is the oneness of Being lit up with an internal experience of its wholeness, an experience of itself as a unitive whole—Being, aware of its self-bondedness. Self-bondedness is the primordial ground of self-awareness—the ground of the soul of the world.

Indeed, what the soul or self is, macrocosmically and microcosmically, is the *experience* of "being"—an awareness (3) of self (1) through reverberation in "otherness" (2). The soul is self-awareness arising from a rapport between the extremes of the world being—Indivisible Act of being on one hand and divisible forms of being on the other hand: "Same and Other."

The golden rapport of proportionate sameness takes duality to tri-unity. It introduces the dimension of rapport and union between polarities. Oneness reappears beyond duality in the form of a tri-unity, which defines the soul. The third awakens in the One the experience of itself as "self"—the reality of "soul."

◯ What makes the "golden" section the uniquely perfect section it is, is that the *division* it creates also generates *union*—oneness. This unique property expresses the central principle of Plato's cosmology, and makes the golden proportion the geometric–mathematical–metaphysical key to the essence of the soul, the principle of consciousness, and the creation of cosmos. ◯

The Greeks used several names for the three-term "mean" by which the extremes engage in harmonious, soulful rapport. They called it soul, mind, logos. Each notion addresses one of the three levels of reality engaged: the soul reality of "mind," the geometric reality of the "mean," and the metaphysical reality of the third being issued of Being: the "Logos," who personifies the soul of the world. His emblem, the Sun, was regarded as the meeting point between the spiritual and physical orders of existence. Humanity, the repository of the principle of "mind" and "soul," is a creative rendering-in-progress of the Logos.

The primary threefold compound—the soul—is then, Plato continues, distributed by the demiurge into portions, which are a blend of indivisible Being, divisible Being and the Mean between the two:

> And when with the aid of Being He [The Demiurge] had mixed them, and had made of them one out of three, straightway He began to distribute the whole thereof into portions; and each portion was a mixture of the Same, the Other, and Being... And He began making the division thus: First He took one portion from the whole; then He took a portion double of this; then a third portion, half as much again as the second portion, that is, three times as much as the first; the

III. The Harmonic Composition of the Universe

fourth portion He took was twice as much as the second; the fifth three times as much as the third; the sixth eight times as much as the first; and the seventh twenty-seven times as much as the first.[7]

What we are about to see, as we follow Plato into this forbiddingly elaborate (or so it appears initially) fractioning of the world soul, is that this apportioning of the soul "compound," followed by a "mediating" operation presented in the next paragraph, produces the Pythagorean ratios of the musical proportion. The demiurge "apportions" or differentiates the soul according to what amounts to musical intervals. In fact, this first operation will be found to expose the mathematical process from which these ratios originate. The Platonic Demiurge's mysterious (but quite simple, in the end) apportioning holds the key to the mathematical logic of the major musical intervals, and, within it, to the metaphysical logic of world creation. It will soon become quite clear how this is so, and how the secret of creation smiles in a divinely simple operation. Behind the abstract spelling out of numbers is the logical process of deploying "1"—the primordial Oneness—into the multiplicity of manifestation.

First we need to gain an understanding of the "apportioning" operation that Plato just described, keeping the mathematical facts to the minimum indispensable to appreciate the precise steps of "divine logic" that support Plato's cosmology, and fully grasp the principles engaged in the operation of Creation.

The demiurge's fractioning of the whole follows a mathematical pattern displayed in the figure below (fig. 19). He makes a total of seven portions ("First He took one portion from the whole; then...a portion double of this..."), which are defined by two geometrical progressions generated by the simple powers of 1, 2, and 3, raised to their own powers ($1, 2, 2^2, 2^3$ and $1, 3, 3^2, 3^3$): 1, 2, 4, 8 and 1, 3, 9, 27. The figure, known as Plato's Lambda (because of its resemblance with the Greek letter lambda, Λ), is a triangular array of numbers, which displays the unfolding, out of 1 (out of "the One"), of a geometrical progression by 2 to the left and a progression by 3 to the right.

✪ We note how world creation proceeds by reproducing at a next level of mathematical complexity (the *powers* of 1, 2, 3) the metaphysical opening of the One into Same, Other, and Being (1, 2, 3—same, other, mean), and how both processes are reverberated in the phenomenon of harmonic unfurling of a musical tone into fundamental (1), octave (2) and third harmonic (3). In fact, to listen to these first three harmonics inwardly is the royal way to access the inner, experiential reality behind the abstraction of numbers.

---

7   Ibid., 35a–b–c, 36d, 37a.

The powers of 1, 2, 3—the only numbers, raised to each other's powers, present in these seven numbers of the Lambda—constitute the mathematical and philosophical hub of the cosmogonic picture, as the great principles of world creation.

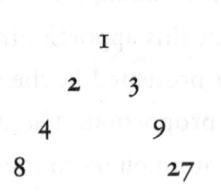

Fig. 19. Plato's lambda

They are the archetypal "powers" behind the ontological trinities of gods central to many wisdom traditions: Trimurti of Shiva, Vishnu, Brahma in India; Father, Son, Holy Spirit in Christianity; Tao, Yin, Yang in Taoism. Even the world Ash, the world tree of Norse mythology, is said to have three roots going in three directions. Language expresses this threefold root of the world tree in the core trinities of persons (I, you, it) and syntactic code (subject, verb, object) that form its basic scaffolding. ✪

Lao Tsu captures in a simple formula this archetypal unfolding of the universe contained in the first three numbers: *The Tao produces unity; unity produces duality; duality produces trinity; the triad produces all things.*

The apportioning of the world soul into seven parts based on the numbers 1, 2, 3 and their mutual "empowering" to one another's power (2 to the power of 3, 3 to the power of 2) is the first operation in the cosmogonic deployment of the world soul (the threefold one) into world creation. However abstract these numbers and Lao Tsu's formula may appear, a simple descent into the reality of each number will show that we know what these primary "powers" are. We know their inner, psycho-energetic realities intimately, for they are fundamental aspects of our own being, which mirror the three aspects of divinity Plato refers to as Same, Other, and Being:

The power of One is the pure power of "being"—the power to Be "one self," to express, act, and create—what translates in human experience as *will*, the indivisible, free, spirit core of "self." The Tao is "That which exists through itself...the One Being, the One primal spirit."[8]

The power of Two is the power of (self-) division and polarization, the power to separate, distinguish, distance (this from that, "Other" from Self)—the discerning, discriminating, reflective *intelligence* aspect of our being. The divisible realm of matter and body.

The power of Three is the power of reunion and harmonization, the power to inter-mediate, triangulate and unite, to bind and bond polarities, to unify diversity—the *love* aspect of being, and the essence of the threefold oneness of the world soul (Plato's "Being"). Soul.

---

8 The beginning of the treatise, *The Secret of the Golden Flower:* "The Master Lu Tzu said, 'That which exists through itself is the Tao. The Tao has neither name nor form. It is the One Being, the One primal spirit.'"

III. The Harmonic Composition of the Universe

Will, intelligence, love are the three fundamental aspects of being, to which the tripartition of our physical constitution bears witness: head (intelligence), limbs (will, action), torso with heart and lungs (exchange, relation, love).

The threefold third is the principle of self (1) con (2) sciousness (3) (lat. cum, with; and scire, to know), or reflective (2) self (1) awareness (3)—the reality of self-aware selfhood.

We now see from a new angle the path from Three to Seven (and next to Twelve). The seven portions correspond to the simplest *self-multiplication* of the threefold one, its multiplication by its self and its three powers: the seven combinations of the powers of 1, 2, 3.

The next gesture-operation of the demiurge, which Plato describes in the paragraph that follows, involves filling up the intervals between the multiples of the soul powers, which the first seven portions represent, with portions-intervals that *mediate* between each pair of the first portions. Every pair becomes a secondary 1–2, which elicits mediating "means."

This entire step simply takes to a next level the fundamental process of golden bonding of the two that resulted in the "third being." In other words, the process of mediation epitomized in the (threefold) third—the golden mean of soul—replicates itself in every mediating ratio between the 7 powers of 1, 2, and 3.

This whole apportioning operation shows what an exact mathematical operation of mediation, hence a perfectly mediation-defined (i.e., *soul-filled*) metaphysical process, presides to the cosmogonic unfoldment of oneness into multiplicity—of essence into existence, of Being into cosmos—from the opening of Being into a self-reverberating and self-bonding "milieu" (the world soul), to the proportionate fractioning of this mediating reality.

> After that He went on to fill up the intervals in the series of the powers of 2 and the intervals in the series of powers of 3 in the following manner: He cut off yet further portions from the original mixture, and set them in between the portions above rehearsed, so as to place two means in each interval [one harmonic, one arithmetic].... And whereas the insertion of these links formed fresh intervals in the former intervals, that is to say, intervals of 3:2 and 4:3 and 9:8, He went on to fill up the 4:3 intervals with 9:8 intervals. This still left over in each case a fraction, which is represented by the terms of the numerical ratio 256:243. And thus the mixture, from which He had been cutting these portions off, was now all spent.[9]

---

9 Plato, *Timaeus*, 36a-b.

Once reduced to their simplest denominator, the means inserted between the terms of the double geometric progression (by 2 and 3) give three types of intervals: the ratios of 3:2, 4:3, and 9:8. We recognize in these ratios the fundamental intervals of musical harmony: the fifth, the fourth, and the tone—arithmetic mean, harmonic mean, and their relation. The numbers 1, 2, and 3, on which the two progressions are based, are reflected in the musical fundamental, octave, and fifth, which are the first three harmonics—the fundamental trinity of harmony—and the notes obtained by dividing the mono-chord (1) by 2, then by 3. The other two ratios-intervals (4:3, 9:8) are derived from the fifth: the fourth as its complement within the octave, the tone as the interval between ascending and descending fifth.

Figure 20, showing the Table of Lambda by Nicomachus of Gerasa, illumines the whole operation, and shows the thoroughly harmonic composition of numbers-intervals obtained.[10]

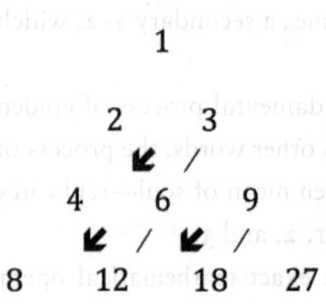

Fig. 20. Table of Lambda

Consecutive horizontal numbers are to each other in the ratio 3/2, defining the perfect fifth, whilst consecutive, diagonal numbers linked by the diagonal arrows (8 to 6, 12 to 9) are in the ratio 4/3, the musical fourth, and numbers linked by diagonal lines are in the ratio 2/1, the musical octave.

In other words, all the relations between these numbers, which represent *archetypal extremes and their means*, are musical. The diagram presents the mathematical matrix of the world to be—the process of harmonic differentiation of the soul of the world.

In the light of this harmonic cosmogony, one begins to fathom the scope of Steiner's statement, that "what the ear experiences as sound is the wisdom of the world. In the perception of sound, one hears the wisdom of the world." Lawlor points out that in ancient Egypt, the audial sense, which permits the immediate perception of the proportional laws of sound and form, was regarded as the epistemological basis of philosophy and science.[11] The Advaita Vedanta School of philosophy discreetly affirms the same when it calls "Shruti" the body of most authoritative, ancient religious texts comprising the central canon of Hinduism. In Sanskrit, Shruti means, "that which is heard." It refers to the gift and capacity of intuitive knowing—and, most eloquently, it designates the smallest musical interval.

In summary, the second step in Plato's cosmogenesis consists in fractioning the world soul compound, the third of the 1-2-3 trinity of same-other-Being, according to its own constitutive principles or "powers" (1, 2^2, 2^3, etc.), then filling the

---

10   Lawlor, *Sacred Geometry*, p. 83.
11   Ibid., p. 89.

III. The Harmonic Composition of the Universe

intervals between the resulting portions with their respective (arithmetic and harmonic) means. The entire process is one of proportionate "secting," then harmonious bonding of the sections with their "means."

What the last operation reveals is that the creation of the world—in other words, the externalization of Being into cosmos—happens by a *fractioning of Oneness according to the ratios of the musical proportion*. The mathematical apportioning of the world soul by Plato's demiurge parallels the Pythagorean production of harmonious sounds ("harmony filled" tones) by dividing a monochord according to the simple ratios of 1 (fundamental), 1/2 (octave, the other), 1/3 or 2/3 (the interval of the fifth).

The Platonic world opening is a marvelously unitive self-fractioning process, harmonic through and through. Born of the "golden" relation of the One to itself, the world soul differentiates itself through a similar self-fractioning and mediating operation (between fractional extremes). By self-dividing, it multiplies itself into a primordial "compound," in which every part is in complete harmony with every other part and with its source. What gives this compound its remarkable coherence is the principle of mediation. The mathematical/geometric operation of secting-uniting, which underlies the cosmogonical unfolding of the one into the many, parallels the acoustic-geometric unfolding of musical intervals from a fundamental. It offers a different perspective on the same harmonic fractioning of oneness.

✪ The significance of this primordial process of harmonic fractioning-bonding of the world Being is that it carries into every part the *internalized wholeness* of the One. Indeed, says Plato, "each portion was a mixture of the Same, the Other, and Being." Not only is the entire matrix "sound" with the primordial golden bonding that makes up the soul compound, but the bonding principle that defines the soul also governs its distribution into harmonious portions. In the end, every part is a sound, harmonious fraction of the fundamental threefold "identity" at the core of creation: the world soul. ✪

Each portion being a harmonic fraction of the world soul, *the soul or self-hood of the whole is in each portion*. The harmonic fractioning of the soul compound allows the "soul-fulness" of Being, the milieu of self-resonance opened by its "fundamental" sounding out, to propagate and be present as the soul of the whole in each of its parts.

Summarizing these first steps of Plato's cosmogony will help to appreciate the mathematics of unitive differentiation that ushers the One Being into manifestation. If soul is essentially mediation, this process is literally soul-ful mathematics. First, the world soul compounded by the demiurge as a third form of being "midway between indivisible and divisible Being" is a unit of relation and a "logos" of union—a *tri-unity* of relational harmony. Second, the critical step of "fractioning" this world soul unit proceeds with the demiurge making seven portions out of this triune soul compound. Every portion is defined exclusively by the "Powers" of the tri-unity (1, 2, 3). The means or secondary portions subsequently placed between them replicate the mediating operation of tri-union (soul). In the end, every single portion is pervaded with "tri-unity," and is thus a particular replica of the soul, a "part" that carries and expresses the internal wholeness of Being, the unity of the One. Oneness takes a new form indeed—a triune form—in which pure Being experiences itself in the form of *internal unity*, in the form of an internal bond of union. Spirit experiences itself as the soul of the world it is creating.

At work in this whole unitive fractioning is a way, literally a "mean," to allow the propagation into every part of the universe-to-be of the secondary one-ness that the third represents (*he made of them one out of three*), the tri-unity that is the "soul" of the world, so the unfolding world may be a soul-filled world—a world fully ensouled by a soul made world.

The beauty of Plato's mathematical model of creation is that it lays out a rational process of wholistic fractioning of oneness. According to the *Timaeus,* the world about to manifest is "whole" in every one of its parts by virtue of the harmony of its soul matrix. Should Plato's theory be supported by scientific findings, it could be regarded as a radical source of insights into what makes the world in "deed" and in mathematical fact "one" in all its parts—resonant as one soul and alive with one Life.

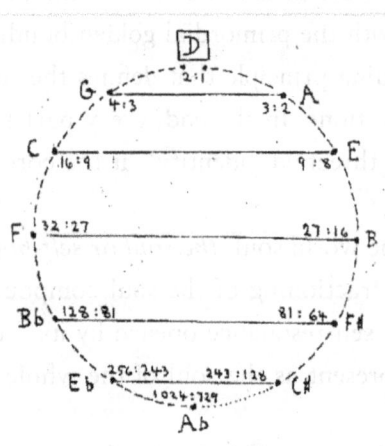

Fig. 21. *The circle of fifths viewed as a circle of ratios combining a progression by 2 with a progression by 3*

The ratios for the twelve chromatic notes of the circle of fifths are approximate. The division of the octave into fifths and fourths, for instance, is off-centered, and sharps do not exactly coincide with flats. The notes emerge as approximate ratios—"intervals." One way to assign exact value to these ratios is as a series that combines the two geometrical progressions of 2 and 3. Plotting them around a circle gives a direct recognition of the fact that the twelve-tone (untempered) scale is built on two geometrical progressions (fig. 21): a progression by 2: 1 : 2 : 4 : 8 : 16 : 32 : 64 : 128 : (512) : 1024; and a progression by 3 : 1 : 3 : 9 : 27 : 81 : 243 : 729.

III. The Harmonic Composition of the Universe

In the engraving reproduced in figure 22, a sixteenth-century allegory of arithmetic, the lambda of the archetypal progressions of 2 and 3 branching out of 1 is superimposed on the legs of a woman. Lawlor comments how "the geometric progressions act as a sort of copulation."[12] The figuration illustrates indeed quite remarkably the notion of the primary operation of creation as a vibrational *conjugation* of archetypal extremes. It suggests the metaphysical "conception" at work in the mathematical copulation of the two primal numbers-symbols—1 and 2, male and female—which generates the threefold third, the son-sol of Being, the soul-sun of the world. The one soul (1) of the world then differentiates itself through the "intercourse" of the powers of 2 and 3.

The feminine figure represents the "mathe-matrix" of the world being. The world is the child born of the harmonious interplay of archetypal polarities—be they called Same and Other, spirit and matter, yang and yin, two and one, male and female, fundamental and octave, then octave and fifth. Offspring of the harmonic matrix, the world is filled with the primordial vibrancy of One life.

Regarding the notion of male and female numbers, McClain offers this luminous comment, "The number 2 is female in the sense that it creates the matrix, the octave, in which all other tones are born. By itself, however, it can only create 'cycles of barrenness,' in Socrates'

Fig. 22. Arithmetica; from Margarita philosophica *(Pearl of Wisdom) by Gregor Reisch (1504)*

metaphor, for multiplication and division by 2 can never introduce new tones into our tone-mandala. In musical arithmetic, the powers of 2 generate cyclic identities."[13]

The Allegory of Arithmetic shows on the right side Pythagoras (c.580–c.490 BC) using a counting board, and on the left side the Roman philosopher Ancius Boethius (AD c.480–c.524) using Arabic numerals and mathematical symbols. Boethius, the author of *The Principles of Music*, synthesized ancient wisdom and carried classical

---

12   Ibid., pp. 7, 83.
13   McClain, *The Myth of Invariance* pp. 19–20.

theory into the Middle Ages and the Renaissance. The mysterious image brings together on each side of "Arithmetic" (the science of numbers), the Ancient Greece of the sixth century BC, and the ending Roman age of the sixth century AD. Pythagoras is on the left side of the woman, the side of the progression by 2 (octave) inscribed on her left leg. Boethius is on the side of the progression by 3 (fifth-like). The two philosophers epitomize two sides of the science of numbers, one expressed by a replication of the "same" (an addition of units, of "ones") and the other by the use of diverse ciphers (the Arabic numerals).

The allegory suggests how the *sameness* of "units of one" that adds up to numbers becomes hidden in the *diversity* of Arabic numbers. The reality of numbers as replications-additions of "one" disappears, abstracted behind the various ciphers. The co-existence in numbers of ontological sameness (multiples of the same "one") and external, practical diversity (distinct entities) illustrates the existential duality of the world itself. Similarly, the musical scale may be viewed as a series of *distinct* tones as well as a group of intervals implying a *common* unit (fourth, fifth, third)—a distinction that will gather meaning as we go.

The juxtaposition of the two philosophers, one millennium apart, reveals the evolution of the collective consciousness from a predominant awareness of ontological unity to an increasing absorption in the world multiplicity. An icon of the world creatrix, this personification of the Platonic mathematico–musical formula of the descent of the one Being into the diversity of forms belongs to the treasure of world cosmogonies.

SOUL AND SELFHOOD—A PRIMORDIAL TWELVEFOLD UNIT

The previously quoted paragraph from Plato's *Timaeus* ends with a significant detail:

> He went on to fill up the 4:3 intervals with 9:8 intervals. This still left over in each case a fraction, which is represented by the terms of the numerical ratio 256:243. And thus the mixture, from which He had been cutting these portions off, was now all spent.

The ratio of 256:243 is the interval of the semitone—one twelfth of an octave. That this interval is the ultimate wholistic portion of soul mixture indicates that the fractioning of the world soul on its own terms (the first three numbers-powers and their means) reaches its limit with the division by 12. 12foldness is the ultimate harmonic apportioning of the 3fold oneness (the "soul mixture") by the demiurge and the 12th interval represents the smallest harmonic portion. There are no more fractions "left over." There is no further *divine* fractioning of One.

III. The Harmonic Composition of the Universe

The semi-tone is precisely the smallest interval defined by the 12 fifths of the circle of fifths when they are transposed within an octave to yield the chromatic scale. The semi-tone is also the first harmonic interval to multiply itself quasi-perfectly into a single octave, as the octave divides itself quasi-perfectly into 12 semi-tones. This is not the case for the fifth, the fourth, or the tone (the interval of 9:8). There is not an exact number of fifths, fourths or tones within the octave, but there is an exact number of semi-tones.

The ratio of 256:243, which defines the chromatic 12foldness of the octave, is the first ratio to fraction the octave harmoniously *and completely* ("the mixture was now all spent")—holistically. Smaller intervals/larger fractions could accomplish the same, but the semi-tone, and the resulting 12fold, is the simplest, fullest, and thoroughly homogeneous harmonic fractioning of the primordial whole that the octave represents.

The archetypal deployment of the tri-unity of the world soul is a 12fold "unit." Twelve emerges as the most perfect wholistic fractioning of the triune oneness, which is itself the wholistic fractioning of One. The fact that 12 contiguous spheres, and 12 only, fit perfectly around a central sphere, and that 12 is the greatest number of equilateral triangles to fit on a spherical surface are geometric expressions of this harmonic, pleromatic fractioning.

Plato's mathematical path and Pythagoras' harmonic path converge to reveal in twelvefoldness the perfectly harmonious differentiation of the triune oneness that defines the world soul. Plato provides the mathematical underpinnings of the Pythagorean principles of harmony and discloses their cosmological ground when he describes world creation as a "sound" (holistic) opening of the One into a tri-unity, followed by the fractioning of the 3fold "one" into an archetypal 12fold matrix. The same 12fold matrix emerged from the Pythagorean conjugation of the musical extremes of the octave interval (2/1) into a third interval (the "fifth," 3/2), whose unfolding within the octave cycle led to the 12 semitones of the chromatic scale. What Platonic cosmology does is reveal the inner reality of these twelve folds or tones: they represent twelve types of mediation between same and other—twelve steps of integration (3) of difference (2) and identity (1).

Plato's mathematical metaphysics illumines in twelvefoldness the ideal of differentiated wholeness and integrated multiplicity. Only a perfectly sound fractioning of the whole can allow its vibrancy to pass and Be—alive—in all its parts. A wholistic fractioning of one, twelvefoldness is uniquely fit to vehiculate the wholeness of life—the indivisibility of being. We saw earlier that the reason "musical" sound involves exclusively sound waves with half-wavelengths in whole number ratios of each other is the constraint of constructive interference. We now see the mathematical and metaphysical logic at work in this physical fact. Whole number ratios are *unit*-based fractions of one—partial, yet whole sub-units. What whole numbers, or integers, share is the integrality of "unit." What qualifies the integer as unit is that it fractions (thus multiplies) the whole into equal wholes. Whole number ratios are *whole* parts, keyed to the wholeness of "1," hence apt to host the indivisibility of life. Every fraction that is a division of units by units is related to "one." The basis of harmonics and harmonic intervals is the wholeness or "unity" of a fundamental ("unit"), of which they represent whole fractions/ multiplications. In the words of Theon of Smyrna, a second-century Platonic teacher:

> Unity is the principle of all things.... It is indivisible and it is everything in power. It is immutable and it never departs from its own nature through multiplication ($1 \times 1 = 1$). All that is intelligible and cannot be engendered exists in it: the nature of ideas, God himself, the soul, the beautiful and the good, and every intelligible essence, such as beauty itself, for we conceive of each of these things as being one and *as existing in itself*.[14]

Wholeness is the key for the oneness of being—Life—to "exist" in forms. And *harmony* is the mean to foster and preserve wholeness in multiplicity. Twelvefoldness is the archetype of a form alive with the wholeness of *being*—the archetype of a living creation and creature. The word Plato uses to refer to this "whole" live creation is *dzoon* (ζῷον). The word, whose root *zoe* means life, is generally translated as "animal."[15] The same etymology has led to the interpretation of the word *zodiac* as the circle of "animals." Yet the word used by Plato is the same as the word used in the New Testament to refer to the living god: "He said to them: 'Who do you say that I am?' Simon Peter replied, 'you are the Christ, the son of the *living god*'"—*o uios tou theou tou dzontos* (Matt. 13:15–16). The Pythagorean–Platonic world is a "living" being. A few paragraphs down, Plato confirms: "He created the world a blessed god."[16]

The first phase of world creation ends with a complete picture of the differentiated soul compound as a form in motion—a form "revolving back upon itself." The world soul compound of Same, Other, and Being "is then proportionately divided and bound together and revolves back upon herself."[17] A new feature of the world soul emerges: the image of a wheel—in Sanskrit, a cakra. The Platonic picture of the world soul is becoming increasingly evocative of the esoteric depiction of the human

---

14   Theon of Smyrna, *Mathematics Useful for Understanding Plato*.

15   Plato, *Timaeus,* 33b; compare Jowett's translation, "to the *animal* which was to comprehend all *animals*," with Luc Brisson's translation of the same: "To the *living* one who must encompass all *living* ones."

16   Plato, *Timaeus,* 34b.

17   Ibid., 37a.

III. The Harmonic Composition of the Universe

soul as a twelve-petal lotus and of the zodiac as a 12fold chakra. The fundamental information it brings to the esoteric picture is the notion of its harmonic formation. What differentiates and binds the primordial oneness into a 12fold revolving wheel–chakra is the harmonious interplay of its waves and subwaves of internal resonance. The revolving back upon itself of the world soul suggests the fundamental dynamics of *self*-hood—the turning upon oneself that allows internalization of being.

The Platonic description of world creation illumines the condition indispensable to the presence of "being"—oneness of life, identity of self—in any form: harmony. It also illumines in selfhood an archetypal milieu of differentiation and harmonization—and a unique mean to participate in the universal creative process of blending the sameness of the whole with the otherness of the part.

※

The wholistic fractioning of the triune *"world soul"* opens a 12fold archetypal milieu of vibrant, differentiated bondedness, which precedes and transcends the separation of *soul and world*—the duality of "self" and "cosmos" characteristic of the human experience. In the primordial overture of Being, Sameness "resounds." It projects itself *out* into "Otherness," and this duality generates a triune oneness of divine "selfhood." In other words, externalization into world and internalization into self happen together—like out-sounding and resounding. Subjectivity and objectivity—soul consciousness and "compound revolving upon itself"—emerge as one space–time–mind reality, in the same way than a child's sense of self emerges hand in hand with looming intimations of "others" and "world."

The ontological third—reflected in the mathematical, geometric and harmonic thirds—epitomizes the advent of a self-aware world—the *"epiphanein"* of Phanes. Out of the primordial trinity, a 12fold milieu of self-consonance unfolds, which is the archetypal form of selfhood—at once subjective *self* and objective *form*. The creation of the soul, the self-consonant medium of Being, implies the creation of a self-revolving "unit"—quite literally, a "uni-verse"—which *turns* oneness into differentiated units.[18] This "world soul" becomes the archetypal ground of the next step of world creation, the "corporealization" of Being.

The notion of a 12fold archetypal milieu preceding the separation between soul and cosmos, while "toning" and intoning their common ground, hints at a potential foundation for a cosmos-based science of the soul. The connection between the creation of soul-and-cosmos and the astrological science of their joint reality will emerge more vividly in the background of the next step of Plato's cosmogenesis, the formation of the cosmic spheres.

---

18 *Universe*, from Latin, *uni-*, "one," and *versus*, "turned" (past participle of *vertere*, to turn).

Corporealization of the World Soul into a Soulful World

Having completed the harmonic differentiation of the world soul compound, the demiurge proceeds to divide it lengthwise in two strips, which he lays out across each other like a great X and bends in two circles, intersecting at two opposite points. He calls the outer circle "the circle of the Same"; the inner circle "the circle of the Other." He leaves the outer circle undivided (in accordance with the nature of the same). He divides the inner circle six times to form seven circles, which he places at proportionate distances from each other according to the harmonic intervals intrinsic to the soul matrix: the intervals of the three powers of the soul raised to each other's powers (1, 2, 3, 4, 8, 9, 27). He then places on these seven orbits the seven celestial bodies—Moon, Mercury, Venus, the Sun, Mars, Jupiter, and Saturn—a sevenfold wheel of revolving bodies with the Sun at the midpoint. 1 represents the distance of the Moon, 2 the Sun, 3 Venus, 4 Mercury, 8 Mars, 9 Jupiter, and 27 Saturn. He makes the two circles revolve uniformly on the same axis but in opposite directions. The creation of these two circles replicates the primordial generation and combination of Same and Other.

The two intersecting circles have been generally interpreted to refer to the circle of the ecliptic (the perceived path of the Sun and planets along the zodiac) and the circle of the equator (the celestial projection of the belt of the Earth), crossing at the equinoctial points. The sevenfold division of the inner circle points to the *revolving* circle of the planets. The "sameness" of the outer circle speaks to the axial *rotation* of the entire world sphere experienced by the rotating on itself of the Earth (equator belt). The outer circle is a circle of fixed stars that, "brought to revolve uniformly on the same axis as the inner circle, but in opposite direction," coincides with the zodiacal wheel. Through the circle of her celestial equator (rotation), the Earth participates of the sameness of the world soul "revolving back upon itself." Through the circle of her planetary orbiting around the Sun, she participates of the variances and "wanderings" of the world body.

The crossing—the Chi or "X"—of the two circles of 12fold zodiac and 7fold planetary system evokes the crossing, in the lambda (/\ ), of the progressions of powers of 2 and powers of 3—cycle of the second harmonic (the 7fold cycle of octaves) and cycle of the third harmonic (the 12fold cycle of fifths). Their "intersecting" speaks to the "intercourse" of the womb-like space of the 2nd harmonic, the octave, with the mediating consciousness of the 3rd (the fifth)—which gives birth to the world. The central significance of this point will continue to emerge as we move deeper into the inner realities involved. McClain's reflections on this mysterious X in *"The Pythagorean Plato"* further substantiate this perspective. McClain points

III. The Harmonic Composition of the Universe

out that Plato describes a chi (χ) rather than a lambda formation, and he finds in the earliest treatise devoted to a Platonic dialogue, a work by Crantor entitled *on Timaeus*, an approach that restores the table to its original form by taking reciprocals of the lambda for the upper half of the chi. To understand the passage from lambda to X, one needs only to note that, based on their tonal, harmonic underpinnings, Plato's numbers all have reciprocal values "as unit-fractions": *Tones engage numbers interchangeably in fractions and multiplications of 1.* The reciprocal relation between wavelengths and frequencies illustrates this fact, as does the simple notion, most evident in cell division, that self-fractioning is self-multiplication. The lambda becomes one half—the lower half—of Plato's chi (X), with the "reciprocals" of the lower half constituting the upper half (fig. 23).[19]

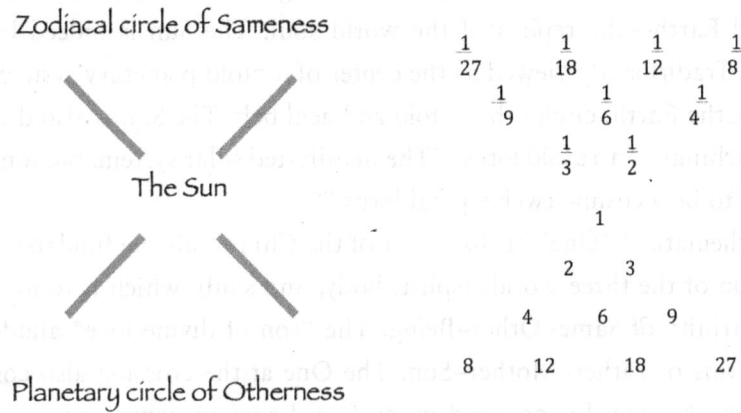

Fig. 23: Plato's X of world corporealization

This mathematical–harmonic figure of cosmogenesis illumines a profound statement from Carl Jung's *Seven Sermons of the Dead*—which in turn illumines the meaningfulness of the figure: "All things that are brought forth from the Pleroma by differentiation are pairs of opposites."[20]

McClain's insight confirms that this first corporealizing operation of the Demiurge represents an actualization of Plato's harmonic lambda into a celestial

---

19 McClain, *The Pythagorean Plato*, p. 65. McClain also explains, "In the number theory expounded by Plato, Aristotle and Nicomachus, integers have reciprocal meanings as Egyptian 'unit-fractions.' Nicomachus reminds us repeatedly that in ratio theory, numbers can be interpreted as multiples and submultiples of a basic unit. 1. As submultiples, the integers actually represent divisors, or 'Egyptian unit-fractions,' the reciprocals 1/2, 1/3, etc. The Crantor lambda then can be interpreted as a symbol for either upper or lower 'arms' in Plato's cross (*Chi*). The upper and lower arms of the cross symbolize the opposite, or dialectical functions of the natural numbers, fractions not being employed in formal ratio theory until Archimedes introduces them in the century after Plato." McClain, *A New Look at Plato's Timaeus*, 351.

20 Hoeller, *The Gnostic Jung and the Seven Sermons of the Dead*, p. 49.

dynamics of two intersecting circles, which embody the lambda's archetypal blend of the Octavian series of 2 and the fifth series of 3.

The division of the soul compound into the two circles replicates the primordial opening of Being into Same and Other. The two circles step down the metaphysical polarity of Same and Other, Indivisible and divisible, Intelligible and sensible, into the cosmic polarity of a universal circle of fixed stars revolving uniformly in one direction and a 7fold circle of planetary orbits revolving in the opposite direction with "different degrees of swiftness...infinite and various wanderings...and reversals of motion."

To the extent that the Sun pertains to the world of stars as well as to the planetary system, the circles symbolically "cross" in the Sun. The planet-star (our Sun) epitomizes the "1" in the X/Chi figure, the Logos who bonds Same and Other, Heaven and Earth—the replica of the world soul. The Sun is indeed both 7fold and 12fold. Traditionally viewed as the center of a 7fold planetary system, its path, viewed from the Earth, circles the 12fold zodiacal belt. The Sun is also described in spiritual teachings as a 12fold lotus. "The manifested solar system, the son of divine love, is said to be a cosmic twelve-petal lotus."[21]

The mathematical "One" at the center of the Chi reveals the fundamental, ontological set up of the three worlds (spirit, body, and soul), which is to manifest the primordial trinity of Same–Other–Being. The "son of divine love" alludes to this trinity in terms of Father–Mother–Son. The One at the crossing also conjures up the prefigure of a Sun-Logos—and many Sun–Logoi to come—arising from the depths of history (the "Otherness" of Being) to accomplish the reverse and complementary arc of creation—the "crossing" (X) and interpenetration, in his own being, of human and divine; Earth and Sun; planetary and stellar life. Not unlike the crucified Christ, whose body, outstretched between Heaven and Earth, opens a passageway between the unlit sub-solar, planetary world and the radiant world of the stars, Odin, the chief god of the Norse, hangs in self-sacrifice upon the world tree. The most illuminating caption for the Chi is undoubtedly Plato's own statement that "the soul of the world is crucified on the body of the world." This picture suggests the spiritual "science" of creation and evolution embedded in world religions.

The sevenfold division of the inner circle carries into the constitution of the cosmos the same apportioning of the whole according to the seven powers of 1, 2, 3 as served the creation of the world soul "so [that] the harmony thereof pervaded all her substance." The same harmonic fractioning of the whole "by itself" that created the soul engenders and pervades the solar system. Plato divines in the surrounding

---

21 Bailey, *Treatise on Cosmic Fire*, p. 1204.

celestial order the display—and the playing out—of the ontological mediating and harmonizing of Same and Other, which birthed the Triadal core of the universe.

The cosmological division into 7 circles and one 12fold circle parallels the musical division of a fundamental into 7 octaves and a 12fold cycle of fifths. We see how all three instances of 7-folding (7 soul portions, 7 notes, 7 octaves, 7 planets) go hand in hand with the production of a 12fold structure (12 semi-tones, 12 fifths, 12fold zodiac). Clearly, what Plato introduces in the *Timaeus* is another approach to the differentiation of the "Milieu" of Being, which Pythagoras captured musically in the octave interval of duplication of the "One" (the 1/2 ratio) and the fifth interval of mediation (3/2 ratio). What Plato presents is the mathematical-metaphysical counterpart of Pythagoras' harmonic-acoustic 7fold duplication and 12fold mediation of Oneness. The divine construction of cosmos is pervaded with the dynamics of harmonic mediation.

This Model being an eternal Living Creature, Plato continues, "The celestial bodies are generated as living creatures, having their bodies bound with *living bonds*."[22] The living bond par excellence is the bond Plato described early on as the fairest bond of union between two extremes, the golden mean. The golden mean is indeed well known as a key ratio in the structure of *living* organisms. We will later look at numerous facts of nature, which show that the notion of "living bonds" is more than a lovely idealistic vision typical of our metaphysical ancestors. It is a mathematical and physical reality.

❂ What makes the golden mean a living bond is that it is *the mean*, or *the way*, for the oneness of the whole to pass into the multiple systems of nature and cosmos through the "sameness" of *proportionality*. "When the mean is to the first term as the last term is to the mean, then the mean becoming first and last, all of them come to be the same, and *having become the same with one another will all be one*." In unifying or at-oning the rapport among the parts, and between parts and whole, the golden constant carries the oneness of Being as a discreet identity of proportion in all forms of nature. The unique capacity of the golden mean to engender constancy of *proportion* makes it the principle and agent of oneness in cosmos—the vehicle of the world soul. ❂

The golden mean points to the potential hub of a wholistic physics that would involve the "meta-physicality" of life and center its premises on the universal bond that conducts life in all parts and bodies of cosmos. The grounds of such a unitive science of a *live* cosmos filled with bodies that are "living creatures" would

---

22 Plato, *Timaeus*, 38e.

transcend the split between subjective-metaphysical and objective-empirical reality, between idealistic and scientific points of view, which dominates modern consciousness. Such is the ground that Plato proposes and delineates.

The Platonic stepping down and corporealization of the world soul into a soulful world proceeds according to a continuous process of harmonic bonding of oneness and diversity, of "life and bodies," of Being and forms of being. Proportionality is the discreet stairwell that the indivisible oneness of Being and life takes to descend into multiple separate forms, indwell world systems, and hold together tiny physical organisms. Proportion is a form of wholeness: it is wholeness in forms.

Indeed, releasing and binding are the prerogative of deity and the defining deeds of creation. "Canst thou *bind* the sweet influences (harmonies) of Pleiades, or *loose* the bands of Orion?" God asks Job, thereby pointing to the fundamental dual gesture of creation (Job 38:31). The key to the livingness of the Platonic universe is the unitive secting-bonding that pervades all its formations.

PHYSICALIZATION OF THE WORLD SOUL INTO FIVE PLATONIC ELEMENTS-SOLIDS

The Platonic descent of the world soul into a world body proceeds as a step-by-step "fitting together" of polarities into proportionate forms—polarities involving increasingly denser levels of reality. After the compounding of the world soul out of the primordial dualization-and-bonding of Same and Other, followed by the harmonic fractioning of the soul compound into a 7fold-12fold matrix, then a second dualization of "all this that He had put together into two parts lengthwise"—to form a circle of the same (zodiac) and a circle of the other (planetary system), the world manifesting process reaches a fourth step of physicalization. "When the construction of the Soul had all been completed to the satisfaction of its Constructor, then He fabricated within it all the Corporeal."[23] The fabrication of the corporeal proceeds according to the same harmonic secting-binding of polarities—the "mean," we saw, to create and ensure a pathway of wholistic unity and harmonic constancy for the oneness of life and Being to "spread out" in manifold manifestation—for the many forms to conduct and transmit life—Life.[24]

---

23 Plato, *Timaeus*, 36b–d.

24 The term *life* is intentionally capitalized and will be again when necessary to highlight the ontological indivisibility of *being* implicitly present in the casual use of the term *life*. The capital letter emphasizes, beyond the usual reference of the word to biological life, a reference to the ontological Life of self-aware *beings* at work in human bodies and cosmic systems. It remains that the term *life* includes and implies Life, which generally made it the most inclusive, immediately meaningful choice throughout the context of this book.

III. The Harmonic Composition of the Universe

What takes place in this next phase is a bonding of the world soul with the "mother-matter" receptacle, the replica of the primordial "other," to whom Plato refers, evocatively, as "the nurse of all Becoming."[25] "It is proper," he writes, "to liken the Recipient to the Mother, the Source to the Father, and what is engendered between these two to the Offspring... [or] the Becoming, that 'Wherein' it becomes, and the source 'Wherefrom' the Becoming is copied and produced."[26] This secondary binding produces denser forms of mediation, which are geometric shapes of an exceptionally harmonious character: the elements.

The diffraction of the world soul matrix into quasi-material, elemental forms is at the same time a process of concentration-organization of matter into geometric shapes. These geometric forms are informed, says Plato, by harmonic ratios. They are the so-called "Platonic solids," and Plato correlates their geometries with the elements (fig. 24).

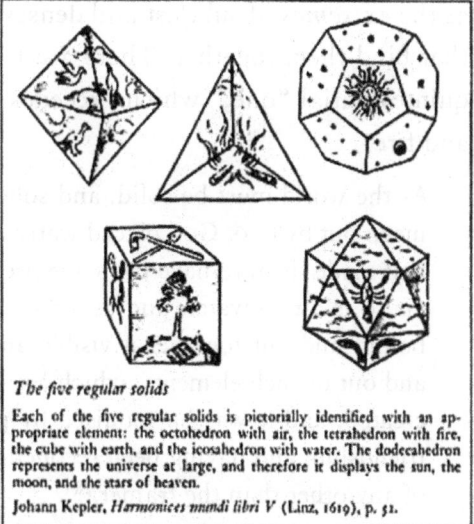

The five regular solids

Each of the five regular solids is pictorially identified with an appropriate element: the octahedron with air, the tetrahedron with fire, the cube with earth, and the icosahedron with water. The dodecahedron represents the universe at large, and therefore it displays the sun, the moon, and the stars of heaven.

Johann Kepler, *Harmonices mundi libri V* (Linz, 1619), p. 51.

Fig. 24. The five Platonic Solids (Heninger, Touches of Sweet Harmony)

What are the Platonic solids? They are the only five perfectly regular polyhedra with equal sides, equal regular faces, and equal solid angles, inscribable in a sphere. To qualify as a Platonic solid, each form must have the same shape on every side, every one of its lines must be of exactly the same length, and every internal angle must be the same as every other. Each shape must fit perfectly inside a sphere, with all its points touching the edges of the sphere. Only five shapes have all these characteristics: the tetrahedron, the cube, the icosahedron, the octahedron, and the dodecahedron. They represent the only five possibilities of fractioning Euclidian space equally. Plato calls them "the four fairest bodies, the four bodies that excel in beauty, with a fifth one encompassing them all"—the dodecahedron.[27] Although they are referred to as "Platonic" forms, archaeological findings show that they were already known in Neolithic times. A group of five stone objects excavated in Scotland, for instance, shows these very forms, deliberately shaped by human hands over five millennia ago. For Plato, these perfectly harmonious bodies represent the supreme expression in form of the harmonic principle underlying the construction

---

25   Ibid., 52e.
26   Ibid., 50d.
27   Ibid., 53b–e.

of the world. He regards these five solids as the building blocks of the physical world, and the models of the elements.

Early on in his *Timaeus*, Plato introduces the four elements (fire, air, water, and earth) as the fundamental "means" to harmonize matter and soul—substantiate the world soul and en-soul matter—and the building blocks of a soulful physical cosmos. Furthermore, he views the elements as replicating among themselves the fundamental dynamics of dualization-mediation at the foundation of the world. Based on logical necessity and reasoning, he points to fire and earth as the *extremes* of subtlest and densest matter, while air and water are the *means* that bind them together. There are five solids, so there is also a fifth element, a quintessential "one," which precedes and transcends the polarization into earth and fire.

> As the world must be solid, and solid bodies are always compacted not by one mean but by two, God placed water and air in the mean between fire and earth, and made them to have the same proportion so far as was possible (as fire is to air so is air to water, and as air is to water so is water to earth); and thus he bound and put together a visible and tangible heaven. And for these reasons, and out of such elements, which are in number four, the body of the world was created, and it was harmonised by proportion, and therefore has the spirit of friendship; and having been reconciled to itself, it was indissoluble by the hand of any other than the framer.[28]

Further on in the text, Plato comes back to the elements in reference to the construction of the physical cosmos and describes their formation as the last step in the descent of the world soul into world solidity. As the soul descends into physical formations, matter "ascends" or evolves by organizing itself into elements, "shaped into their distinct nature," he says, "by means of forms and numbers." The harmonic shaping of matter into elements represents at once a *corporealization* of the world soul and an *en-souling* of matter. Harmonic geometries carry soulfulness into the building blocks of cosmos. The two processes go hand in hand. Physicalization of the soul matrix ("order of the universe") implies organization of matter into harmonic forms.

"Before the birth of heaven," Plato elaborates, "the elements possessed only some traces of their own natures. The work of setting the Universe in order [turned them from] a state devoid of reason or measure" into distinct elements that, shaped unto the four "fairest bodies," could constitute the harmonious ground of the universe.[29]

---

28 Ibid., 32b-c.
29 Ibid., 53a-b.

III. The Harmonic Composition of the Universe

Also on the basis of logical necessity and reasonable likelihood, Plato establishes correspondences between the mediating ladder of the four elements and the four perfect shapes of "solidity," the Platonic solids, which he regards as their formative principles: From subtlest to densest, giving logical reasons for his associations, he connects fire with the tetrahedron, air with the octahedron, water with the icosahedron, and earth with the cube. The fifth solid, the dodecahedron, he ascribes to the structure of the whole: "There was yet a fifth combination, which God used in the delineation of the universe."[30] Plato's student Aristotle later added a fifth element, ether, and postulated that this was the element of which the heavens are made. Although he did not match it with Plato's fifth solid, we shall see that other sources support this connection.

Plato's presentation of the elements may seem far removed from what a contemporary reader would regard as rational, even sensible considerations about the universe. But is it that far? Not really, thought Werner Heisenberg, one of the key pioneers of quantum mechanics. In a series of public lectures in 1955, he spoke about "the similarity between Plato's ideas and modern atomic physics" and compared modern particle physics with the theory of elementary material components laid out in the *Timaeus*. He would recall how as a teenager at school, he found Plato's *Timaeus* to be a memorable poetic and beautiful view of atoms.

> I think that modern physics has definitely decided in favor of Plato. In fact the smallest units of matter are not physical objects in the ordinary sense; they are forms, ideas that can be expressed unambiguously only in mathematical language.[31] The mathematical forms by which we nowadays represent the elementary particles...are more complex than the geometric forms postulated by the Greeks. But essentially, in both cases these forms originate in certain simple mathematical basic requisites....The most fundamental kernel of all that is material is for us, as well as for Plato, a [mathematical] form and not some material content.[32]

The *Timaeus* advances, indeed, that the elements are geometric forms shaped according to mathematical ratios, and that everything in the universe, all sensible particulars, are made of elementary components that are mathematical entities. However, Plato adds a major dimension of insight to Heisenberg's general recognition: these elemental formations are *specific* mathematical forms; they are based on harmonic ratios. As such, and only as such, do they lay stepping-stones for the descent of the One into world multiplicity—of Being into physicality—as well as the evolution of matter into forms of being.

---

30   Ibid., 55c.
31   Quoted in *The New York Times Book Review* (Mar. 8, 1992).
32   Heisenberg, as cited by Brisson and Meyerstein in *Inventing the Universe*.

First Groundings of Plato's Harmonic Universe: Kepler's Celestial Physics and Bode's Law

*Geometry is the archetype of
the beauty of the world*—KEPLER

Before proceeding further with Plato's account of world creation, we will leap forward in time and take a brief look at Kepler's works, for it will give us the beginnings of an objective "sighting" of Plato's symphonic world construction in the actual physics of cosmos. This first corroborative mirroring of Plato's cosmology in Kepler's "celestial physics" will energize our going back to Plato to explore the "scientific" tenability of his description. The dialogue between ancient and modern insights thus initiated will reveal whether Plato's account of the harmonic overture of Creation can find echoes in what the measurements of physics and calculations of astronomy tell us about the construction of the world.

We can only hope, as travelers do, that our anticipatory glimpses prove accurate, and that the middle path that Kepler landscaped, we are about to see, between idealistic theory and practical observations, will represent a first step for the mediating cosmology of the Greeks to gather "grounds" of descent to the center stage of contemporary inquiries about the relationship between the "metaphysics" of consciousness and the physics of matter. On such a re-centered stage, the polarities of spirit and matter, the great world parents donned in the alien formulas of Platonic metaphysics and modern physics, could engage in a new creative intercourse, and in such intercourse reveal their mysterious octave like sameness. For now, there lies between them the chasm of a quest and the steps of an adventure.

So far, the notion of world creation according to the laws of harmony, however mathematically sound, is not much more, to the modern mind, than a marvelous theory. Whereas Plato claims that harmonic ratios inform the planetary spheres and the elements, he does not demonstrate it. Nonetheless, his harmonic cosmogenesis has a compelling coherence, which presses us onward to investigate whether it might not prove, *in fact,* to underlie the actual physics of the universe.

Centuries after Pythagoras and Plato, inspired by the harmonic order still toning the metaphysical ground of European culture, and stimulated by Copernicus' revolutionary insights, the German mathematician, astronomer, and astrologer Johannes Kepler had a remarkable epiphany about celestial geometry.

During his school years, Kepler had been exposed to Copernicus' work and intuitively felt the accuracy of his revolutionary view of the universe. He studied the two systems of planetary motion—Ptolemaic and Copernican—and

III. The Harmonic Composition of the Universe

felt inclined toward heliocentrism both from a theoretical and a theological perspective, affirming, in a student disputation, that the Sun was the source of power in the universe. "He based his defense of heliocentrism," science historian A. Koyré explains, "on a conception of the universe that, by an inspiration both Pythagorean and Christian, saw therein an expression of the divine Trinity, and caused the concept of the Cosmos, hereafter condemned, to shine with supreme brilliance..."[33]

In 1595, the 23-year old Kepler experienced a revelation, which was to be the source of his first major astronomical work, *Mysterium cosmographicum* ("The Cosmographic Mystery")—also the first published defense of the Copernican system. Guided by his conviction that there is a rational plan to the cosmos, Kepler had been pondering why there should only be six planets, "instead of twenty or a hundred." He spent many months trying to make sense of their distances and velocities. In a moment of illumination that released a flood of tears, it dawned on him that the spacing among the six Copernican planets could be understood in terms of the five Platonic solids nested in each other and contained within a sphere that represented the orbit of Saturn. He found that if he inscribed the five solids in a spherical orb and nested them within one another in the following order—Octahedron, icosahedron, dodecahedron, tetrahedron, cube—it produced six layers, corresponding to the orbits of the known planets: Mercury, Venus, Earth, Mars, Jupiter, and Saturn (fig. 25). Not only did this account for the number of planets, but it also reflected closely the proportional distances advanced by Copernicus. Kepler wrote in his introduction to *Mysterium cosmographicum*:

> The Earth is [the circle that is] the measure for all other orbits. Construct a dodecahedron round it. The sphere surrounding it will be Mars. Round Mars construct a tetrahedron, the sphere surrounding it will be Jupiter. Round Jupiter construct a cube. The sphere surrounding it will be Saturn. Now construct an icosahedron inside the Earth. The sphere inscribed within that will be Venus. Inside Venus inscribe an octahedron. The sphere inscribed within that will be Mercury. There you have the explanation of the number of planets.

Discovering precise harmonies (lit. "fitting together") in the geometries of the solar system overwhelmed Kepler with tears of bliss—the bliss of seeing revealed, in actual cosmic measurements, the harmony his soul intuited, and the bliss of being the conduit for the marriage of Plato's ideas with celestial physics. He wrote at once to his teacher: "I wanted to become a theologian; for a long time I was

---

33   Koyré, *The Astronomical Revolution*. In *The Music of the Spheres*, p. 126, Jamie James points out that heliocentricity had already made its debut in Western thought 2,000 years before Copernicus, in the philosophy of Pythagoras himself. "For Pythagoras," he says, "the concept was at once good science and good theology, for the two were the same thing."

restless. Now, however, behold how through my effort God is being celebrated in astronomy."

Although the elliptical character of the planetary orbits, which he later demonstrated into law, would appear to cancel his beautiful scheme, Kepler never relinquished the Platonist polyhedral-spherical cosmology of *Mysterium cosmographicum*. Even though it proved only to approximate astronomical data, it was too significant to be accidental. Interestingly, the discovery by Russian scientists that the Earth's crust shows a dodecahedral pattern echoes Kepler's association of our planet with the dodecahedron.[34]

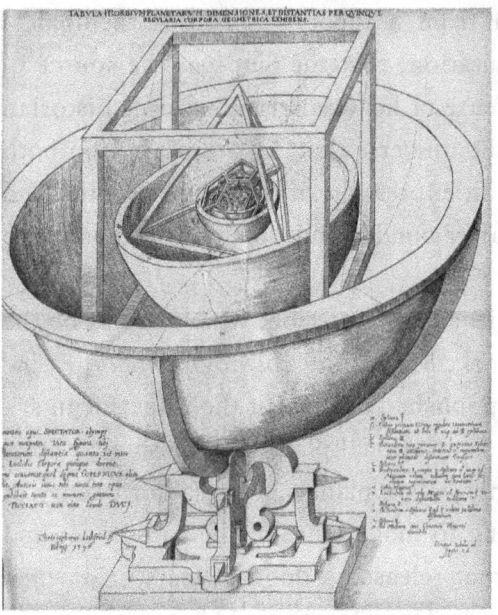

Fig. 25. Kepler's geometrical model of the solar system. The five Platonic solids are nested inside one another and separated by a series of inscribed and circumscribing spheres

Kepler is best known for his three laws of planetary motion, which continue to be part of the foundation of modern astronomy and physics. The first of the three states that the orbit of a planet is an ellipse with the Sun at one of the two foci. The third law, published in 1619, is the one of most interest in our context, for it carries in its very formulation Kepler's intuition of the harmonic geometry of the universe. It states that the square of the orbital period of a planet is proportional to the cube of its orbit radius (or average distance from the Sun). In demonstrating a proportional constancy between orbital periods and orbital distances, Kepler's third law offered the first inkling that Plato's mathematico–metaphysical model of cosmic harmony might have scientific relevance. Kepler enunciated his law in musical terms, and it became known as the *harmonic law*.

> It is absolutely certain and exact that the ratio that exists between the periodic times of any two planets is precisely the sesquialternate proportion (i.e., the ratio of 3:2) of their mean distances...[35]

Kepler's aspiration to demonstrate the "music of the spheres" in precise laws had been met. *Harmonices Mundi (Harmonies of the World)* brought together more than

---

34 See "Is the Earth a Large Crystal?" in *Khimiya i Zhizn* (chemistry and life), Dec. 1973.

35 Kepler, *Harmonices Mundi*, book 5, ch. 3.

two decades of investigations into the archetypal principles of the world construct: geometrical, musical, metaphysical, astrological, and astronomical. Arthur Koestler called it "a mathematician's song of songs to the chief harmonist of creation":

> What Kepler attempted [was] to bare the ultimate secret of the universe in an all-embracing synthesis of geometry, music, astrology, astronomy and epistemology. It was the last attempt of this kind since Plato, and it is the last to our day. After Kepler, fragmentation of experience sets in again, science is divorced from religion, religion from art, substance from form, matter from mind.[36]

Up to Kepler's times, there was no real distinction between astronomy and astrology, and a complete separation existed between astronomy, regarded as a branch of mathematics, and physics—a branch of natural philosophy. The celestial was not really physical. Kepler's new astronomy, which he referred to as "celestial physics," transformed this perspective by treating astronomy as part of universal physics. "The deep inspiration of his work," Koyré reflected, "was to substitute a 'philosophy' or 'celestial physics' for the theology or metaphysics of Aristotle, and to prove that 'celestial physics' and 'terrestrial physics' do not constitute two 'physics,' but one only.... Kepler was the first, and the only one up to the time of Descartes, to ask for a physical explanation of celestial dynamics."[37]

To wed astronomy and physics, the celestial and the physical, represented for Kepler a jubilatory revelation of the Logos in cosmos. However, as is often the case with cultural giants, the motive and mindset that guided his discoveries, and the mystical joy that resulted, did not survive what evolution was to make of the facts that he brought to light. The soul life that yields the discovery is often flushed out by the grand wave of the collective's evolutionary design. We will see how history dealt a similar twist on Fechner's findings. Every human work belongs ultimately to a collective arc of unfoldment that transcends the sphere of individual intentions. Newton wanted to prove God's existence, and his findings fed the stream that would lead to a "death of God."

In Kepler's case, what he experienced as the grand nuptials of Heaven and Earth was to become part of the Copernican revolution and set the stage for a gradual separation of the earthly from the heavenly, as well as a divorce between astro-*logy* and astro-*nomy*—the metaphysical language (*Logos*) and the physical laws (*nomos*) of the stars (*astro*).

---

36   Koestler, *The Sleepwalkers*.
37   Koyré, *The Astronomical Revolution*, p. 120.

What entered the stream of evolution as a glorious depiction of the harmony of Heaven and Earth, a wedding of Platonic metaphysics and physical measurements, was to contribute to a gradual and irresistible alteration in the perception of the celestial vault of the world—not in the direction Kepler experienced and intended. With the planets, including the Earth, now revolving around a physical Sun, and the system of orbiting powers shown to be an impressive nesting of geometric solids, the heavens, until then populated with angelic hierarchies, slowly hardened into a purely physical expanse of starry bodies—a majestic but material dome. Our modern mindset is so used to perceiving the celestial bodies as physical objects in space that to grasp the perceptual–conceptual shift that took place between the sixteenth and eighteenth century requires a genuine act of imagination.

As the parallel with the Platonic solids suggests, Kepler's marriage of Heaven and Earth had nothing to do with a marriage of planet Earth with the physical heavens. It had everything to do with a revelation of the marvelous bond, harmonic and geometric, between God and creation—between the trans-physical source of the universe (the trinity) and physical cosmic order. What Plato's solids lay out, fundamentally, is a stairwell of mediating means between the indivisible essence of Life and its highly divisible manifestation, between the metaphysical oneness and the physical multiplicity of the world we live in—and the beings we are. In this onto-cosmological context, the solar system is not a middle ground between planet Earth and the physical starry heavens, but a *mediating reality* between the earthly world of physical forms and the *soul matrix* of our world being, which the zodiacal sphere points to.

To better take hold of the distinction between middle ground and mediating function brought forth, and also blurred, by the Copernican revolution, and discern the inconspicuous collapsing of dimensions that Kepler's works unwittingly contributed to precipitate in the collective psyche, it may help to ponder an analogous distinction, commonly blurred and equally significant to restore. This distinction concerns the heart and the "middle ground" it represents. On one hand, the heart is a *middle ground, physically,* between head and limbs. On the other hand, it ensures a *vital mediation,* as the "vital organ" par excellence, between body and self. It harbors and anchors the *soulful function of mediating* between our physical and psychospiritual being. This distinction highlights two different perspectives on the heart: one in terms of its "middle position," the other in terms of its mediating function. The "collapsed" view of the heart as the "middle" chakra in a linear sequence of seven, while logical from a one-dimensional topological perspective, limits the understanding of this center of centers. Instead of a dynamic hub between center and periphery of consciousness, within and without of experience, the heart chakra tends to be viewed as simply the fourth one in a linear ladder of seven. Yet, its 12-petal structure hints at the fundamental mediating role it is invested with, which is central to the essential trinity we are (spirit, soul, and body), beyond a 7fold etheric-physical structure—a trinity in which the heart is precisely the great mediating-uniting agent.

The reductive conception of the heart parallels and illustrates what happened to our collective view of the universe as we moved away from a world conception that included the nonphysical dimensions of spirit and soul. The multi-dimensionality of the beings we are collapsed into one dimension: physicality. The last fifty years have seen a collective awakening to the subtle dimensions of the human constitution: energetic, emotional, mental, soul, spirit. A corresponding recognition of the subtle planes of the world being is slower to emerge.

Similar to the heart amidst the seven chakras, the Copernican Sun surreptitiously lined up with the seven planets between the Earth and the *physical* skies. Still viewed as the star-center of the system, the Sun was so as a *physical* star, one stepping "stone" to more distant physical stars. By contrast, in the eyes of the ancients, and in those of Kepler, the Sun was less an astronomical middle ground between Earth and stars than an intermediary and mediating celestial power—between physical and metaphysical realms. The Sun was an *epiphanic* being. *Phanes* was one of his names as solar "god" and Word-Logos of world manifestation—at once heavenly and endowed with an archetypal human form, at once immortal and engaged in all of creation as its Life-source. Kepler's favorite biblical passage was the line from John (1:14): "And the Word became flesh and lived among us." He regarded the spatial configuration of corporeal entities as the direct expression of the Logos-Word, whose triune formative powers he saw reflected in the sphere as the best geometric symbol of the threefold mystery of God: Father (center), Christ the Son (circumference), and Holy Spirit (intervening space).

Ironically, Kepler's geometric binding of Heaven and Earth, his paean to the harmonious bonds of the creation with its Creator was to help shift humankind's gaze from the spiritual heavens to the physical skies, and close the portal of the Sun to the spiritual spheres. And yet, one can also imagine welling up within Kepler's tears of bliss, the longing of the universe to perceive its fundamental harmony in the "objective" chamber of the scientific mind. In such paradoxes lies the complex beauty of transitional moments of history and thresholds of evolution. The culmination of a previous cycle and intention offers itself as seed for the next—and in this seed, dies.

※

The awareness of cosmic order and harmony, which pervades the wisdoms of old, went hand in hand with the sentient communion the ancients experienced with the surrounding cosmos. Pythagoras and Plato epitomized the emergence of an intellectual consciousness, which their conceptual, rational cosmology based on the mathematics of harmony contributed to foster. What Kepler did, two thousands years later, was to open the way for a factual, and not only rational, demonstration of the Pythagorean-Platonic cosmos, in keeping with the emerging *scientific* consciousness. While Kepler's blend of astronomy and astrology, empirical observation and metaphysical intuition, marked the end of an era, it also contained the vital seed of a bridge between science and metaphysics. His cosmograph and his third law stand as a unitive beacon above the current gaping abyss between materialistic science and the "meta-physicality" of mind and consciousness. The unity of astronomy and astrology, geometry and music, within which Kepler conducted the thinking that led him to his harmonic law, represents a moment of unified science worth contemplation by the modern mind in search of a unified "theory of everything." Kepler was both a famous astronomer and astrologer, and the last to blend the two. Carl Sagan described him as "the first astrophysicist and the last scientific astrologer."

Kepler's intuition of the geometric majesty of the solar system is iconic of the transitional period in which he lived and the brilliant blends of metaphysics and physics it yielded, as humankind's consciousness was responding to new yearnings to bring metaphysical reverence for the cosmos and nature under the eager curiosity of its sharpening eye, brain and senses. New urges to discover how nature works and discern God's order in the arrangements and operations of cosmos were arising from within the theocentric landscape of European culture, precipitating what would amount to the emergence of modern sciences.

Alexandre Koyré introduced the term *Scientific Revolution* to refer to this pivotal historical period of the sixteenth and seventeenth centuries, which he regarded as central in the history of science. He saw in Kepler one of its foremost representatives—for his theorization, he added, rather than his empirical work. By all accounts, this revolution began with the publication in 1543 of Copernicus' *De revolutionibus orbium coelestium* (*On the Revolutions of the Heavenly Spheres*) and came to completion with Isaac Newton, whose *Mathematical Principles of Natural Philosophy*, first published in 1687, laid the foundations for what we know as classical physics, often referred to as Newtonian mechanics. During this period, major developments in mathematics, physics, astronomy, biology, anatomy, and chemistry transformed the collective theoretical viewpoint, leading to fundamental changes in humankind's relation to cosmos and nature. The result was a major paradigm shift from a theological–metaphysical worldview to a scientifically minded outlook, which would lead to a complete emancipation of science from both philosophy and religion. Galileo, Kepler's contemporary, was one of the first modern thinkers to clearly state that the laws of nature are mathematical, and that philosophy is written

III. The Harmonic Composition of the Universe

in the grand book of the universe in geometric characters such as triangles, circles, and other geometric figures.

The philosophical itinerary of these major figures of the period is suggestive of the general turn, in the collective reverence for a primary source of knowledge, from ancient metaphysical grounds to physical experimentation and calculations. Both Kepler and Galileo grew up in a cultural context of religious philosophy, which defined their early sense of life purpose. Galileo seriously considered the priesthood as a young man, and Kepler planned to become a theologian, until, in his own words, Divine Providence guided him to the study of the stars. By contrast, Newton abandoned orthodox beliefs and asked to be exempted from the ordination as Anglican priest usually required from fellows at the Trinity College in Cambridge where he taught. However, very early on and up to the publication of the *Principia* when he was 45, he delved in the Hermetic and alchemical tradition, pondering his scientific questions in the light of these ancient sources of wisdom. While Newton is most famous for his formulation of the laws of motion and universal gravitation, which were to dominate science's view of the physical universe for the next three centuries, he also assiduously studied the Bible and sought to decode its symbolic and mathematical language in an attempt to rediscover the occult wisdom of the ancients. He was interested in the sacred geometry of Solomon's Temple, from spirals to golden sections and conic sections. He felt that architecture, just as the Bible and the writings of ancient philosophers, contains hidden wisdom that, when deciphered, could yield scientific information and reveal knowledge of how nature works. Rosicrucianism greatly influenced his thought. Had his deep studies of the esoteric traditions not been forced into secrecy by the threat of scrutiny and punishment, and had those who discovered its evidence after his death not hidden it away, his spiritual scientific path could have paved a royal way beyond the split about to emerge between perennial wisdom and science. Since the rediscovery, in the middle of the twentieth century, of Newton's alchemical papers, confiscated after his death as "not fit to be printed" by the Royal Society, most scholars concede that this founding father of the mechanistic science of modern times was first and foremost an alchemist. Keynes, the British economist who purchased Newton's papers after they were released in 1936, remarked,

> Newton spent the first phase of his life (up to the *Principia*), the period of life in Trinity when he did all his real work, with one foot in the Middle Ages and one foot treading a path for modern science...In the eighteenth century and since, Newton came to be thought of as the first and greatest of the modern age of scientists, a rationalist, one who taught us to think on the lines of cold and untinctured reason. I do not see him in this light. I do not think that any one who has pored over the contents of that box that he packed up when he finally left Cambridge in 1696...can see him like that. Newton was not the first of the

age of reason. He was the last of the magicians, the last of the Babylonians and Sumerians, the last great mind that looked out on the visible and intellectual world with the same eyes as those who began to build our intellectual inheritance rather less than 10,000 years ago.[38]

Indeed, if Newton formulated a number of the physical laws underlying natural phenomena, opening the way to a new mastery of the physical world, he also asked,

> Whence arises all that Order and Beauty that we see in the World?...Does it not appear from phaenomena that there is a Being incorporeal, living, intelligent, omnipresent, who in infinite space, as it were in his Sensory, sees the things themselves intimately, and thoroughly perceives them, and comprehends them wholly by their immediate presence to himself?[39]

In 1766, almost two centuries after Kepler, Johann Daniel Titius made the significant observation that the orbits of the planets in the solar system follow quite closely a simple arithmetic rule. His insight was published in 1772 by Johann Elert Bode, initially without attribution, hence the name Bode's law. Bode later credited Titius for his earlier remarks on "the astonishing relation which the six known planets observe in their distances from the Sun." Titius wrote,

> Let the distance from the Sun to Saturn be taken as 100 [parts], then Mercury is separated by 4 such parts from the Sun. Venus is $4 + 3 = 7$. The Earth, $4 + 6 = 10$. Mars, $4 + 12 = 16$. Now comes a gap in this so orderly progression. After Mars there follows a space of $4 + 24 = 28$ parts, in which no planet has yet been seen. Can one believe that the Founder of the universe had left this space empty? Certainly not. From here we come to the distance of Jupiter by $4 + 48 = 52$ parts, and finally to that of Saturn by $4 + 96 = 100$ parts.

In other words, by taking 4 as Mercury's mean distance from the Sun and adding the series $3 \times 1, 3 \times 2, 3 \times 4, 3 \times 8, 3 \times 16, 3 \times 32, 3 \times 64$ (which involves again a blend of 1, 3 and powers of 2), one obtains figures very close to the actual mean distances of the planets from the Sun: Mercury, 0.387 (4) astronomical units (AU); Venus, 0.723 (7); Earth, 1.000 (10); Mars, 1,524 (16); Jupiter, 5.203 (52), Saturn, 9.539 (96), Uranus, 19.18 (191), Neptune, 30.06 (300), Pluto 39.53 (395), etc. In the end, the distances minus the constant 4 are in the ratio of $3 : 6 : 12 : 24 : 48 : 96$ (fig. 26)[40]

Titius' mathematical insight revealed an empty space between Mars and Jupiter, which correctly "predicted" the orbit of the asteroid Ceres, in "the space where no

---

38   From the lecture, "Newton, the Man," in Keynes, 1936.
39   Newton, *Optics*, 1704.
40   Koyré, *The Astronomical Revolution*, footnote, p. 386.

planet had yet been seen." It also predicted the orbit of Uranus (discovered in 1781) as well as that of Pluto, which was discovered in 1930 at a distance that matched the position mistakenly assigned to Neptune. The law only failed to account for the position of Neptune's orbit.

Planet	Titius-Bode Prediction	Actual
Mercury	0.4	0.39 AU
Venus	0.7	0.72 AU
Earth	1.0	1.00 AU
Mars	1.6	1.52 AU
X (Ceres)	2.8	2.80 AU
Jupiter	5.2	5.20 AU
Saturn	10.0	9.55 AU
Uranus	19.6	19.2 AU
Neptune	38.8	30.1 AU

Fig. 26. Titius–Bode Law

In *Harmonies of Heaven and Earth*, Joscelyn Godwin notes, "Translated into musical terms (with Mercury here assigned to the top note of the piano), the Bode numbers represent the progressive approximation of a perfect octave," thus bringing firm empirical ground under the Pythagorean idea that "the planetary spheres are spaced at intervals comparable to those stoppings of a string that produce a scale." In this context, Godwin introduces the work of an eminent twentieth-century crystallographer, Victor Goldschmidt, who found "a principle which related the stage-by-stage growth of crystal to a series of nodal points comparable to those which sound the harmonics of a string or pipe.... By a simple set of arithmetic operations, [he] succeeded in turning the planetary distances into a series of harmonious octaves and fifths."[41] Other systems, cited by Godwin, reveal significant harmonic relations embedded in the time-periods of the planets' rotations (Thomas Michael Schmidt), their mean distances from the Sun (W. Kaiser), and the golden section of planetary distances (Alexandre Denereaz). Although the correspondences are never completely exact between ideal ratios and actual phenomena, Godwin comments, "just as the perfectly average human being is never to be found, whenever the astronomical data are investigated with a Pythagorean attitude, cosmic harmony never fails to manifest."[42] What these modern investi-

---

41 Godwin, *Harmonies of Heaven and Earth*, p. 116.
42 Ibid., p. 117.

gations suggest is that, far from being only an inspiring "esoteric" idea, cosmic harmony may well be a scientific fact.

MODERN INSIGHTS INTO THE HARMONIC STRUCTURE OF PLATONIC SOLIDS AND ATOMIC ELEMENTS

Kepler's cosmograph of nested spheres and polyhedra brought to light the structural analogy between the last two steps of Plato's cosmogony: harmonic unfurling of planetary spheres and structuring of the elements according to the five regular polyhedra. The cosmograph transposed Plato's fivefold system of elements (in correspondence with the Platonic solids) to the structuring of the solar system. Kepler's accomplishment was to transfer the fivefold cosmogonic key introduced by Plato from the organization of matter to the structuring of cosmic space. Each solid became associated, not only with an element, but with a planetary sphere, supporting the harmonic setting of the solar system affirmed yet undemonstrated by Plato.

Kepler's transposition can embolden a similar cross-dimensional leap along structural correspondences. This leap, which will prove to be a simple fact of psycho-cosmology, is from the structuring of the macrocosm to the organization of the microcosm—from the harmonic ordering of the constituents of cosmos to the ordering of the human constitution. A structural parallel emerges readily between Plato's dodecahedric fifth element and the quint-essential constituent of the human being, the 12fold Egoic body of the soul. The four Platonic solids are to the fifth what the four vehicles are to the fifth (causal) body of the human being. The causal body—the quintessential human "solid"—encompasses four vehicles, each made of one of the four mediating substances-elements: physical body/earth; etheric body/water; astral body/fire; mental body/air. The higher three planes of the mental plane harbor the dodeca-petal flower of the soul, the fifth principle.

Theosophy validates this extrapolation of Plato's understanding of the elements when it refers to the human vehicles as "elementals." Vehicles, or bodies, are forms of elementary energy-consciousness, which carry out physiological functions, instinctual urges, emotional drives and defenses, mental dispositions and patterns. In the same way that the macrocosmic elements of nature corporealize and physicalize the world soul, microcosmic (human) "elementals" represent the agencies that the individuality needs to exist and function, body and soul, in the world. A combination of sublimated matter and rudimental mind, these "servants" or "daimons" of the soul are personalized along vibrational patterns carried over as character tendencies from past lifetimes. Elemental patterns attract their vibrational match in circumstances and people, which provide the opportunity to attune or retune, refine or remold undifferentiated or inharmonious formations. Through events and experiences, destiny (the universal Self within) alerts us to the pull of one elemental or another, and the evolutionary necessity to heal or resist its self-absorbed, autonomous, and discordant grip to bring about a new balance and "proportional" rapport with the soul of life, and so access a next gradient of wholeness—in other words, a new gradient of integrality, indivisibility and individuality.

Plato's cosmology gives precious intimations about the human constitution and the harmonizing apparatus it represents. It suggests the fitting and binding of extremes at stake in our elemental vehicles (or "solids") and the increasing presence of the soul's quintessence that the fourfold harmonization process makes possible. While Being and material "dust" are wedded externally into planetary spheres (macrocosmically), they are bound internally through vehicles of consciousness (microcosmically). Like the cosmos, the human constitution is a precise composition of structures that mediate and "wed" (a favorite notion of Kepler's) Indivisible and Divisible, same and other. The human being is a wedding-in-the-making of spirit and nature.

Having introduced the correspondence between the four elements and the four perfect solids as their formative principles, Plato endeavors to explain how these bodies that excel in beauty are proportionately constructed, and how they can engender each other—for if we succeed herein, he says, "we shall grasp the truth concerning the generation of earth and fire and of the proportionate and intermediate elements between them."[43] To demonstrate that the solids are composed harmonically and that they proceed from each other, as "we see the elements passing on to one another in an unbroken circle the gift of birth,"[44] would establish that the generation of cosmos is a continuous process of mediation and creative harmonization between spirit and matter, cascading down ever-denser octave-levels of manifestation. It would show that all components of cosmos are, in one way or another, harmonic (soul-full) mediators between essence and manifestation, life and forms, and that the entire cosmos is indeed a harmonic composition, pervaded through and through with the sound oneness of Being—the oneness of a living being.

---

43 Plato, *Timaeus*, 53e.
44 Ibid., 49d.

The ideal "fairness" of the Platonic solids certainly suggests harmony. However, a mathematical demonstration of what Plato affirms, on the basis of such "fairness," to be their harmonic generation is essential if the modern mind is to give serious consideration to the idea of a "soul-full" structuring of cosmos.

While Plato clearly spells out the mathematics of harmony involved in the compounding and differentiation of the world soul, and while he correlates elements and solids in a way that integrates them in his cosmology of proportional bonding, he does not demonstrate "the truth" concerning the harmonic generation and composition of the Platonic solids. So, the descent of Being into corporeality by way of harmonic steps presents a beautiful theory of world genesis, yet one lacking the mathematical factuality necessary to ring a serious bell of truth in the modern ear. One is left wanting a convincing demonstration.

This question takes us back to Kepler. Kepler demonstrated the harmonious nesting of Platonic solids and planetary spheres, and established a law of harmonic proportionality between the orbiting periods of the planets and their distance from the Sun. However, he did not provide the conclusive demonstration of the "kinship between the Harmonic Ratios and the Five Regular Solids," which he set out to reach in the pages he wrote under this title. He did highlight a number of signs of kinship in simple ratios derived from the outward form, side constructions and spheres circumscribing the solids, and he rightly pointed out, for instance, that "the ratio which is called divine (i.e., the golden ratio) rules in various ways throughout the dodecahedral wedding."[45] These insights did not result, however, in the specific demonstration necessary for Plato's description of the harmonic "involvement" of spirit in forms of matter to land on its feet.

An elegant mathematical and geometric unveiling of the harmonic ratios humming indeed in the Platonic solids is proposed by Jean Le Mee, a Professor of Engineering at the Cooper Union in New York. In his *Adquadratum Construction and Study of the Regular Polyhedra*, Le Mee presents an ingenious way to derive the Platonic solids from a single drawing, and demonstrates how each of the five structures is based on a particular harmonic ratio.[46]

While the emphasis in the construction of the Platonic solids, he explains, is generally on their regular surfaces, vertices, and angles, he thought of another way

---

45 Kepler, *Harmonies of the World*, p. 273. "This kinship [*cognatio*] is various and manifold; but there are four degrees of kinship. For either the sign of kinship is taken from the outward form alone which the figures have, or else ratios which are the same as the harmonic arise in the construction of the side, or result from the figures already constructed, taken simply or together; or, lastly, they are either equal to or approximate the ratios of the spheres of the figure."

46 Le Mee, *Ad Quadratum Construction and Study of the Regular Polyhedra*.

III. The Harmonic Composition of the Universe

to consider their regularity. From this new vantage point, the musical ratio-nality of the solids revealed itself effortlessly. Comparing his move to the bold step of Copernicus "who sat himself on the Sun and saw the Ptolemaic universe dissolve into a well-ordered ballet of heliocentric orbits," Le Mee sat himself at the center of the polyhedral universe and, from this internal viewpoint, considered "the symmetries generated for each Platonic form by the radii issuing from the center of the circumsphere to two adjacent vertices respectively [and defining their so called Maraldi angles]. The lines bursting from the center create, by impact on the circumsphere, the regular polyhedra. They delimit within the sphere regular pyramids with a triangular (tetrahedron, icosahedron), square (cube) or pentagonal (dodecahedron) basis as the case may be, all with a common apex at the center of the circumsphere" (fig. 27). The tetrahedron is constituted of four such pyramids, with a triangular basis; the cube of six pyramids with a square basis; the octahedron of eight with a triangular basis; the dodecahedron of twelve, with a pentagonal basis; and finally, the icosahedron of twenty, with a triangular basis. The picture, reproduced from *Adquadratum*, gives an experiential sense of the unique way in which each Platonic solid perfectly "fractions" space.

Through a meticulous mathematical and geometric demonstration, Le Mee shows that "the entire geometry of the Platonic forms results, like the Pythagorean musical scale, from the simple ratio of the first three whole numbers, which are found to govern their internal or 'Maraldi' angle" (the angle created by any two consecutive diagonals in each figure). He demonstrated on his models that the internal angle ratios of 1/1, 1/2, 1/3, and 2/3 give rise in turn to the octahedron [1:1], the icosahedron [2:1], the cube [1:3] and tetrahedron, and finally the dodecahedron [2:3]. "These numbers, and none others," he says, "are needed to account for the fundamental shapes from which all others are derived." These ratios, which correspond to the musical intervals of unison, octave, and fifth, are embedded in the Platonic forms and define them. "Starting with the geometric point as origin," Le Mee concludes, "the Platonic solids constitute five steps toward the complete sphere that, becoming the seventh step, links them as in a musical scale in an octave progression." The little-known fact, he marvels, that "the entire geometry of the

Fig. 27. The Platonic solids, viewed from the perspective of their Maraldi angles
(*Jean Le Mee*, Ad Quadratum Construction and Study of the Regular Polyhedra)

Platonic solids results, like the natural musical scale, from the simple ratios of the first three whole numbers—one, two and three—is of such elegance that it is a perpetual source of wonder and deserves to be better appreciated." Le Mee is indeed placing his finger on the central cosmological marvel.

Another researcher, Eric Rankin, offers a similar demonstration by playing as frequencies the respective numbers of degrees of the solids' Maraldi angles. The outcome is a striking line up of perfectly consonant tones related by simple harmonic ratios—1/2, 3/2, 5/4.[47]

Solid	Maraldi angle	Tone
Tetrahedron	4x180° = 720	F#
Octahedron	6x240° = 1440	F# (octave, 1/2)
Cube	8x270° = 2160	C# (fifth from F#, 3/2)
Icosahedron	12x300° = 3600	A# (third from F#, 5/4)
Dodecahedron	12x540° = 6480	G# (fifth from C#, 3/2)

To translate degrees into frequencies in order to appreciate interval-relations between the solids may seem like an odd proposition. However, the categories of angle and frequency ratios, and their actual values, are more connected than one might initially think. "I give the degrees of the objects seen by the eye as the musician does the notes heard by the ear," wrote Leonardo Da Vinci.[48]

McClain points out that the division of the circle in 360 degrees is clearly tied to astronomical cycles, "even though we do not know when 360 days actually became 360 degrees." The approximate number of 360 days marking the completion of a Sun cycle (as viewed from the Earth) defines a frequency of 360 cycles per year, hence a certain cosmological equivalent between degrees and frequency. The cosmological frequency of 1:360 is repeated in the precessional cycle, in which the annual lag of 50 seconds of arc amounts to 1 degree in 72 years and 360 degrees in 25,920 years, the great Platonic year of the precession cycle. McClain also points out that the division of the calendar is based on astronomical cycles, and that it is tied to the musical scale. He cites the Rig Veda (10.85.5): "The Moon is that which shapes the years." He continues, "The month of 30 days and the 12-month schematic year of 360 days are tied to the scale, as shown by the diatonic scale in smallest integers in the 30:60 octave and by its derivative chromatic scale in the 360:720 octave."[49]

---

47 See http://www.sonicgeometry.com.
48 Richter, *The Notebooks of Leonardo da Vinci*, vol. 1, p. 102.
49 McClain, *The Myth of Invariance*, p. 95.

III. The Harmonic Composition of the Universe

The last sentence sounds rather obscure to the modern mind, who might be tempted to dismiss the seemingly antiquated notion it contains, of attributing value to the smallest range of whole numbers (or integers) to express a set system such as a scale. Yet, it is a notion that informs consistently the number world of ancient traditions, and recurring numbers in sacred Scriptures have been found to coincide with such frequency numbers. This systematic identification of the smallest array of *whole* numbers to express a system like the scale lights up with significance when one realizes that what such an array extracts and emphasizes is again the now familiar principle of a most primordial blend of *wholeness and particularity, indivisibility and divisibility.*

Given the principle at stake, it seems important to lay out the numbers that McClain presents. He first notes that the smallest integers to fit the rising Hindu-Greek diatonic scale or its falling reciprocal scale lie within the octave double 1:2 = 30:60, and he points out how 30 and 60 are fundamental numbers in the measurement of time and space.[50]

Rising notes	D		G		A		D
Falling notes	D		A		G		D
Ratios	6	:	8	: :	9	:	12
Smallest Integers	30	:	40	: :	45	:	60

He also notes that the lowest frequency on the basis of which the ratios of the eleven tones of the chromatic scale (missing only the tritone) can be deployed as a scale of integers is 360. "The smallest integers which can define the eleven tones in chromatic order lie within the octave double 720:360."[51] In other words, the lowest octave frequencies for which the eleven tones acquire integer (whole number) values in chromatic order are 360 and 720.

Rising scale:	D	Eflat	Fsharp	E	F	G	A	Bflat	B	C	Csharp	D
Falling scale:	D	Csharp	C	B	Bflat	A	G	Fsharp	F	F	Eflat	D
Ratios:	1	15/16	8/9	5/6	4/5	3/4	2/3	5/8	3/5	9/16	8/15	2
Smallest integers:	360	384	400	432	450	480	540	576	600	648	675	720

360 thus finds itself associated with the number of degrees of a circle, as well as the fundamental frequency (number of cycles per second) of a chromatic scale of whole-number frequencies. In the end, degrees and musical-frequency ratios are simply two ways, astronomical–geometric and harmonic, to express *the wholistic, cosmological fractioning of wholeness.*

---

50   Ibid., pp. 12, 26.

51   Ibid., p. 33.

The "primary" chromatic scale defined by the octave interval of 360:720 suggests a relative equivalence between a semitone and 30 degrees, a tone and 60 degrees. The division of wholeness by 60 or 12 thus makes full cosmological sense, based on the fundamental harmonic differentiation of one into 12. Hence also, the division of the day in 12 hours of night and day, of the hour in 60 minutes, of the minute in 60 seconds—of the degree into 60 minutes (of arc), and the minute into 60 seconds (of arc). We may also note that 12, 36 (12 x 3) and 72 (12 x 6) are common denominators of the Maraldi angles of the Platonic solids.

What Plato presented as metaphysical fact: the 1-2-3 triune matrix of the world soul and the subsequent formation, out of its deployment into the 12fold/7fold soul-cosmos, of the five building block–elements according to the major harmonic ratios, Le Mee demonstrates to the modern scientific mind: "In the adquadratum method for generating the Platonic forms, only 1, 2, and 3 are used."[52] A simple geometric demonstration attests to the harmonic structuring of the "beautiful" solids, in convincing parallel with the simple ratios associated with the "pleasant" acoustic experience of musical intervals. The attractive proposition that the world (and everything in it) is born of a "fitting together of polarities" via harmonic means is acquiring the features of a realistic hypothesis.

Furthermore, Jean Le Mee shows how the golden ratio—the archetypal Mean—is central in the generation of the polyhedra. His demonstration involves another construction of the Platonic solids, based on three golden rectangles.

However, Robert Lawlor offers even simpler and more luminous evidence that the golden mean is the supreme key to the unfolding of the Platonic solids (fig. 28).[53] Taking its start from the golden ratio itself, Lawlor shows how the five polyhedra can be simultaneously obtained from the creation within an initial circle (radius=1) of a smaller concentric circle related to the first by the phi proportion (radius=1/Phi). "*Phi*" (or "φ") is the mathematical equivalent (1.618) of the Golden Mean ratio and the term most frequently adopted to designate the golden section, although it is important to note that the essence of the golden mean is a *rapport*, not a number. The term *Phi* refers to the architect Phidias, who designed the Parthenon—well known for its abundance of golden proportions. *Phi* is the Latin transliteration of the Greek letter "φ."

Once the smaller circle related to the first by Phi is created, the diameters joining the points of a hexagon inscribed within the first circle create six points

---

52 Le Mee, *Ad Quadratum Construction and Study of the Regular Polyhedra*, p. 144.
53 Lawlor, *Sacred Geometry* pp. 98–101; illustration reproduced from p. 102, fig. 9.5.

III. The Harmonic Composition of the Universe

of intersection on the inner circle (fig. 28). Joining all the points (three dimensionally) reveals the volume of an icosahedron. Linking all the vertices of the icosahedron will then generate intersecting points that define the form of a dodecahedron suspended within the icosahedron. In turn, the dodecahedron leads to the cube, easily defined within it by 8 of its vertices. Next, the diagonals of the faces of the cube form a star tetrahedron. Last, the octahedron appears in the volume enclosed by the star tetrahedron. This process shows most organically how

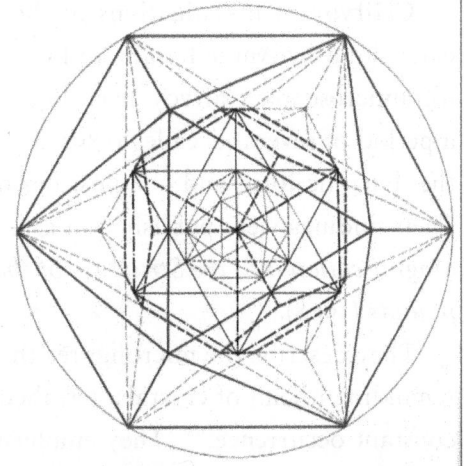

Fig. 28: The Platonic solids, derived from two concentric circles in a Phi ratio

the five solids "grow" from the one "seed," indeed, of the Phi ratio. The Phi proportion is the key to the generation of the five solids from the one primordial sphere. Lawlor calls it "the divine seed."

Lawlor points out how the construction illustrates the Buddhist idea of the codependent origination of the archetypal principles of creation—a notion that has presented itself to us in the interdependent emergence of 12fold and 7fold, and, even more essentially, in the jointly arising principles of separation (2) and union (3) at work in the simultaneous secting-and-bonding operation of the Golden Proportion.

In each of the two demonstrations above, the choice of the starting point—the center—is key to revealing both the structural harmony of the Platonic solids and their harmonic generation.

Along with Kepler's works, such demonstrations of the harmonic geometry of the five solids lend significant ground to Plato's description of the harmonic structuring of the solar system. Since the Platonic solids are harmonically built and interrelated, so that they do, indeed, "bind" the spherical oneness of the whole (universe) with the fivefold polyhedrality of its parts, the planetary spheres may be regarded as well, based on Kepler's cosmograph, to "bind" Heaven and Earth harmonically.

The next question is, can any corroboration be found for Plato's claim that the structure of the elements replicate the Platonic forms? Should the correlation of the elements with the five solids be confirmed, the previous demonstrations would imply that the elements are also organized on the basis of the major musical intervals.

Clairvoyant investigations of the minute (atomic) components of the elements carried out by Annie Besant and Charles Leadbeater, two theosophists gifted with ultramicroscopic clairvoyance, do support what Plato advances in this regard. For a period of several decades (1895 to 1933), they conducted a careful exploration of the dynamic form and composition of sixty-five of the known chemical elements. Their findings were published in *Occult Chemistry: Investigations by Clairvoyant Magnification into the Structure of the Atoms of the Periodic Table and Some Compounds* (1908).

They describe the microfigures they perceived as "three-dimensional, and often reminding [them] of crystals; tetrahedral, octagonal, and other like forms being of constant occurrence."[54] They emphasize that "lines do not exist as stable walls or enclosing films, but rather mark limits, not lines, of *vibrations*."[55] In the end, they found that the chemical elements fell into groups that reflected the Platonic shapes and paralleled the classes arrived at by renowned physicist and chemist Sir William Crookes on the basis of valency.

Reproduced on page 99 are two periodic tables: Crookes' 1886 table (fig. 29), and a 1933 table entitled "after Crookes" and published in *Occult Chemistry* (fig. 30). The second table shows Platonic shapes (tetrahedron, cube, octahedron) substituted for Crookes' diatomic, triatomic, and tetratomic categories, which categories refer to the principle of valency, the capacity of chemical atoms to combine with other atoms. Valency tells how many "arms" each element has, with which it can bond with others. Hydrogen, for instance, has a valency of one, oxygen of two, and carbon, four. Correspondences were revealed between the Platonic forms sighted clairvoyantly and the categories defined by the chemical principle of valency. "In diatomic elements," Leadbeater and Besant explain, "four funnels open on the faces of tetrahedra; in triatomic, six funnels on the faces of cubes; in tetratomic, eight funnels on the faces of octahedral. *Thus we have a regular sequence of the Platonic solids*." Except for the dodecahedron and isosahedron, the Platonic shapes were thus showing their profile behind the symmetry groups defined by mainstream chemistry. "The question suggests itself," Leadbeater points out: "Will further evolution develop elements shaped to the dodecahedron and the icosahedron?"[56] We will return to this significant observation and question.

Physics cannot verify information of this nonphysical nature and has hardly any context to even consider them apart from intuitive theoretical insights from its best scientists, such as Heisenberg's musing quoted earlier, which resonates with

---

54   Besant and Leadbetter, *Occult Chemistry*, p. 29

55   Ibid., p. 23

56   Ibid., p. 22.

III. The Harmonic Composition of the Universe

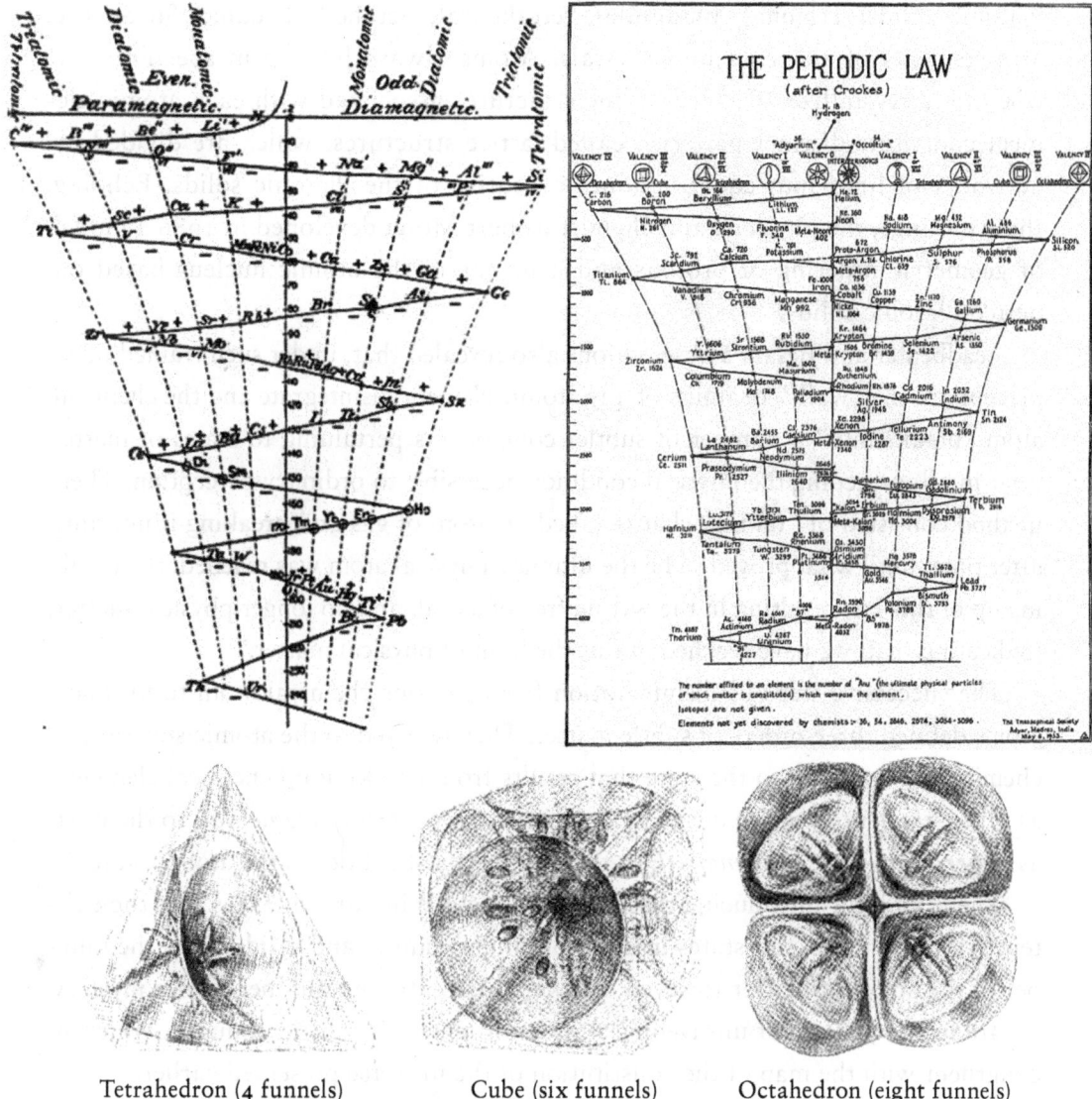

Fig. 29. (above left) Crookes' periodic table (1886);
Fig. 30. (above right) Leadbeater and Besant's periodic table (1933);
Below: Platonic solid correspondences with Crookes' valency categories

the character of this "occult" research: "The most fundamental kernel of all that is material is for us, as well as for Plato, a [mathematical] form and not some material content." Nonetheless, corroborations such as those just mentioned between occult and mainstream findings are powerful enough to warrant the attention of science.

Whereas modern science has not sought to define the shapes of the atomic elements and offers no corroboration in this regard, the structures of crystals and molecules, which represent a next stage in physicalization, do suggest a similar harmonic structuring. Charts of molecular structures show figures akin to the

Platonic solids: Trigonal, pyramidal, tetrahedral, octahedral, cubic. So do the arrangements of elements into crystals. Atoms always line up in specific ways when they crystallize. The crystalline structures associated with each atomic element constitute distinct patterns, called lattice structures, which are divided by networks of lines into equal volumes evocative of the Platonic solids. Echoing these findings, the physicist and chemist Robert Moon developed in 1986 a model of geometric ordering of protons and neutrons in the atomic nucleus based on nested Platonic solids.

Leadbeater and Besant's observations also revealed that, under such trained ultra-perception, the "walls" or limits of any atomic element disintegrate and the chemical atom "breaks into" a number of subtler components pertaining to states of matter (four in all) preceding the physical condition accessible to ordinary perception. Their method consisted of "taking what is called an atom of gas and breaking it up, time after time, until what proved to be the ultimate physical atom was reached, the breaking up of this last resulting in the setting free of astral, and no longer physical matter, [indicating that] we have reached in this the limit of physical matter."[57]

The successive steps of disintegration from gaseous chemical atom to ultimate atoms defined three orders of subtle matter. They referred to the atomic state of the chemist as *elemental*; to the state that results from breaking up chemical elements, as *proto-elemental*; to the next higher state, as *meta–proto–elemental*; to the next, as *hyper-meta–proto–elemental*; to the last and most subtle, as the atomic state.[58]

The reason for introducing these "occult" details in our context is that the existence of four etheric sub-states in addition to gas, liquid, and solid brings the number of the states of matter to seven, the key differentiating number whose mystery we are in the process of unraveling. A view of matter is thus introduced, which is congruent with the map of the constitution of the universe presented earlier.

Remarkably again, this "occult" method of investigation led to results that proved to coincide with the mainstream scientific view inherited from Mendeleev. In summary, setting the elements free from their gaseous condition and raising them to subtler states revealed anterior, less integrated structuring stages, up to an unexpectedly large number of ultimate atoms (from 18 in hydrogen to 418 for Sodium, 1,945 for silver, 3,546 for gold), yielding weight-numbers that closely approximated those assigned to chemical elements by Mendeleev. In ordinary chemistry, hydrogen, the lightest of the known elements, is taken as 1, and all atomic weights are multiples of this base-weight. For the authors of *Occult Chemistry*, hydrogen was still the basis, as it contained the smallest number of ultimate atoms they found in

---

57 Ibid., p. 8.
58 Ibid., p. 18.

III. The Harmonic Composition of the Universe

a chemical element: 18. With hydrogen as the unit base of 18 ultimate atoms, they obtained "number-weights" for each element by dividing the total number of atoms they would count in each by 18. After counting the number of "ultimate atoms" entering the composition of 65 elements, they noted a consistent correspondence between the number of ultimate atoms they found to be contained in each element and what is referred to by mainstream science as its atomic weight. In other words, each successive element revealed a number of "ultimate atoms" equal to its atomic weight multiplied by 18.

The demonstration that the Platonic solids are structured harmonically and the intimations that they inform the most elementary arrangements of matter certainly advance our "grasp of the truth" concerning the harmonic construction of cosmos. To uncover the harmonic structure of solar system and elements is to begin to demonstrate the soul-full character of cosmos. It is to support Plato's proposition that the body of the universe is the manifestation of its harmonious soul matrix; that center and periphery, soul and world, are respectively the seed and the deployed form of a principle of harmonic rapport at the root of the universe. It is to suggest that the entire universe, from soul to cosmic systems to elements, from subtle matrix to envelope, from within to without, is one harmonic continuum—which is what Plato affirms in his depiction of the world soul:

> And in the *center* he put the soul, which he diffused throughout the body, making it also to be the *exterior environment* of it.... The soul, interfused everywhere from the center to the circumference of heaven, of which also she is the external envelopment, herself turning in herself, began a divine beginning of never ceasing and rational life enduring throughout all time.[59]

Golden mean and musical ratios—the geometric key of proportional harmony and the mathematical signatures of acoustic harmony—are thus the essential instruments in the structure of the elements and the solar system, supporting Plato's words that "the body of the Cosmos was harmonized by proportion, brought into existence...united in identity with itself [and] indissoluble by any agent other than Him who had bound it together."[60]

Scientific and philosophical minds alike have intuited this harmonic ground of the world. "What we perceive as various qualities of matter," Bertrand Russell writes, "are actually differences in periodicity." Amstutz, a professor at the Mineralogical Institute in Heidelberg, echoes: "Matter's latticed waves are spaced

---

[59] Plato, *Timaeus*, 34b, 36e (Italics added).
[60] Ibid., 32c.

at intervals corresponding to the frets on a harp or guitar. The science of musical harmony is practically identical with the science of symmetry in crystals."[61]

As the kinship between Pythagoras' laws of harmony and Plato's mathematical cosmology leads one to suspect, the golden mean or Phi ratio (1.618) is very close to the musical fifth (1.667). They represent respectively the geometric and musical expression of the principle of tri-unity that defines the world soul—the "third," child of the Two, whose archetypal unfolding yields the 12fold circle of fifths musically, and the dodecahedron geometrically.

The intriguing fact of a quasi perfect, yet inexact co-incidence between major articulations of the time–space organism of the world being, such as the coincidence, but for a comma, between the cycle of 12 fifths and 7 octaves, or the coincidence, but for 11 ¼ extra days, between 12 lunation cycles and one yearly cycle of the Earth around the Sun, this quasi-coincidence comes up again here in the gap of a few decimals between fifth and golden mean. Such proximities do not cease to attract the mind with the uncanny convergences they represent in the sophisticated body of nature, while the associated gaps repel the modern intellect with what it reads as "unacceptable" margins of difference. Hidden in these margins is the invitation to allow a third path to emerge between the sharp discriminative order of the intellect and the unitive order of metaphysical intuition—a path of creative inquiry into the *relative* precision of the universe's workings. The "divine" gaps of nature invite the mind to cultivate a mediating course between the repelling effect of the hiatuses and the compelling pull of the vibrant proximities. If the analytical sight dares suspend its blinding one-sidedness and accommodate its sharpness to approximative circles and slightly gaping cycles, seeking the unitive picture beyond the disconcerting gaps, it may discover an order that does not negate empirical measurements but softens them enough to allow divine coincidences to sweep the gaps into an integrative world picture.

A number of scientists whose insights heralded historical thresholds in our understanding of the world have acknowledged that intuitions pressed them to find the empirical details, not the reverse. "On principle," Einstein states unequivocally, "it is quite wrong to try founding a theory on observable magnitudes alone. In reality the very opposite happens. It is the theory which decides what one can observe."[62] The notion of "placing oneself at the center" (hence, the question of where the center is) emerges again as a master key in the quest for truth. After studying Newton's manuscripts, John Maynard Keynes, whose intellect Bertrand Russell described as "the sharpest and clearest that I have ever known," made the following remark: "It

---

61 Lawlor, *Sacred Geometry*, p. 4.
62 Cited in Heisenberg, *Physics and Beyond*.

was his intuition that was preeminently extraordinary. The proofs, for what they are worth, were dressed up afterward—they were not the instruments of discovery.... His experiments were always, I suspect, a means not of discovery but always of verifying what he knew already."[63]

The mathematics of nature is not exact, and this inexactitude blurs access to its wholistic order for the purely analytical intellect. Undeniable gaps obscure the gates to a coherent grasp of cosmos but for those depth scientists who follow their intuition to take hold of both ends of the world reality—physical and nonphysical, parts and whole—and allow them to meet, tentatively and approximately, in the only "milieu" where they can cognize each other as one: the human mind. Trying to assemble with analytical data exclusively what no human has assembled and to figure out from their isolated parts the truth of worlds no human has given form to, is the unwitting hubris of the prevalent scientific attitude of our time. Taking one's start instead from universal patterns and intuitions often proves revelatory of hidden regularities behind apparent anomalies. The case of an empty space (or "irrational distance") between Mars and Jupiter is an example of the discovery of particulars induced by pondering universal structures.

Blending intuition of the whole and analysis of the parts opens the way to ponder the approximations of nature as places of adaptation of indivisible to divisible, transrational to rational, infinite to finite, eternal to temporary. If that which is without beginning and end is to create a finite, physical world, it must devise ways to enter time and space and land in forms that have beginnings and ends. Like Pi (=3.14159265358979...), the golden number is a transrational number (ϕ = 1.618033988749895...). In order to land the trans-physical harmony of Being in physical forms conditioned by vibrational geometries and geo-finitudes, the divine proportion would have to fit itself into whole fractions (such as 3/2).

A fuller understanding of the gap between golden mean and musical fifth will emerge later and confirm the essential identity of harmonic and geometric mediation. However, to insert at this time a vista that was to present itself only much later in this journey would be to step out of its experiential trail and compromise its particular unraveling of the cosmological landscape. Suffice to say that in more than one instance, what appeared initially as an unreconcilable gap or discrepancy later proved to be a reasonable and rational similarity. In the end, trusting a noble logic fostered gains in noble trust.

---

63   See http://www-history.mcs.st-and.ac.uk/Extras/Keynes_Newton.html.

THE GOLDEN FIFTH, THE LIVING BOND OF CREATION

The five Platonic solids emerge from the "divine seed." However, only two of them, the dodecahedron and the icosahedron, feature the golden ratio as a *structural* characteristic.

What distinguishes these two solids is that they are the only ones with pentagrammatic and dodecahedral (fivefold and twelvefold) structures. The icosahedron has 12 vertices, 20 faces, 30 edges; the dodecahedron 12 faces, 20 vertices, and 30 edges. They are duals or reciprocals of each other, which means that each can be derived from the other by translating the faces of one into the vertices of the other. Connecting the center of every face of the dodecahedron, as shown on the diagram on the left, produces the outline of an icosahedron.

The dodecahedron, which Plato associates with the fifth element, is made of 12 pentagons, and thus intimately related to the 5-pointed star or pentagram, the Pythagorean symbol of the fifth principle of soul and higher mind. The basic shape of the pentagram shows a side-to-diagonal ratio of Phi. All its lines, as they bisect each other, define segments that are related to each other by the golden ratio: a to c; c to b; b to d; a to 1; 1 to b. Also, all the angles of the five-pointed star are in the proportion of 72° : 108° = 2:3. The pentagram is simply "made of" golden ratios; it is the ideal two-dimensional expression of the golden mean.

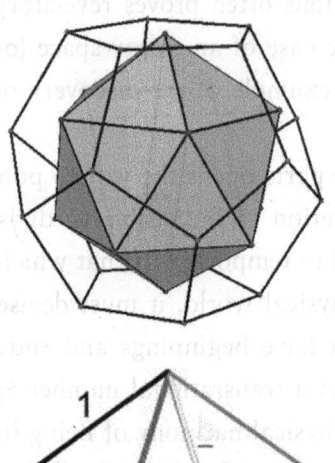

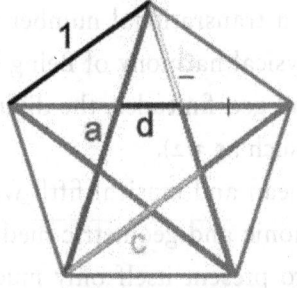

Fig. 31. (top) Icosahedron within a dodecahedron; (bottom) Pentagram with "golden" bonds highlighted

The dodecahedron is a three-dimensional 12fold multiplication of the pentagram. This unfoldment of the pentagram to a quasi-sphere is the geometric analog of the circle of "fifths." The dodecahedron is the equivalent in *form,* and the manifestation in space, of the 12fold circle of fifths. The unfolding of the *golden mean* into a quasi-sphere is the geometric expression of the unfolding of the ideal *harmonic ratio* (the third harmonic/fifth interval) into a chromatic octave-cycle. While the circle of fifths is the archetypal circle-scale of musical harmony, the dodecahedron is the ideal three-dimensional geometry of harmony, the golden volume.

Unsurprisingly, yet marvelously, the numbers 5 and 12, associated, we saw earlier, with mind, soul and Venus, come back in association with the golden mean—the principle of mediation between whole and part, the bond between life and forms. Together, they speak to the (12) harmonic intervals at the soul-full core of cosmos and the (5) harmonic forms (Platonic solids) at its periphery.

Phi is in fact closely related to 5. Its mathematical value, 1.6180339, is computed from the square root of 5 multiplied by .5 plus .5. Both Phi and five are central in the cycle of Venus—the planet of harmony and love, associated with the fifth principle of soul and mind. In addition to tracing out a five pointed star in the sky, thus displaying in its cyclical relationship with the Earth the cosmic principle of "golden proportion"—the supreme definition, indeed, of beauty, the Venusian gift—Venus takes 1.618 years to complete each side of the star pentagram. So, Phi and five penetrate not only the geometries of the cycle of Venus but its timings, which magnifies the gesture of perfect harmony the planet enacts in our skies. Far from miraculous coincidences, such harmonic relationships are the manifestation of a universe produced by harmony. "Wherever there is number, there is beauty," says Proclus.

These ideal geometries "corporealize"—to use Plato's word—the vibrational laws of harmony. We saw how the golden thread that links the twelve notes of the circle of fifths is the 3/2 ratio of the fifth interval—the principle of mediation and harmony—expressed in the fifth principle of mind–soul. Similarly, what links the twelve faces of the dodecahedron, the geometry associated with the fifth element, is Phi, the golden ratio, and its 5fold, pentagrammatic, two-dimensional expression.

The fact that the golden mean is present in the structure of the fifth solid and its dual (the icosahedron), but not directly in the other three polyhedra, lights up with meaning when it is exposed to the geometries of nature. Observations have shown that pentagrammatic and dodecahedral structures are predominant in *living* organisms.

While Kepler contributed to ground Plato's harmonic cosmology in the celestial physics of the solar system, the research of sacred geometrists, inspired naturalists, occult and non-occult scientists will prove critical to further solidify a "staircase" for his cosmogonical theory to descend into physical facts.

In his *Art and Geometry of Life*, the Romanian mathematician and philosopher Matila Ghyka observes how a certain preference for pentagonal symmetry, a symmetry connected with the golden section and unknown in inanimate systems, seems to exist in the animal realm as well as in botany.[64] The pentagon and the dodecahedron, he points out, while never appearing in inorganic crystalline

---

64   Ghyka, *The Geometry of Art and Life*, p. 91.

systems, play a predominant role in the shapes of living organisms, and in the diagrams of living growth.[65]

"Crystals also," notes Hugo Kükelhaus, "know nothing of the number five and remain in the forms of the triangles, squares, and hexagons that resound through their domain... The formal equilibrium, the static symmetry of crystals, changes in plants and animals to a 'dynamic symmetry,' whose archetype is the pentagon."[66]

These remarkable facts illumine in the golden ratio, its two-dimensional pentagrammatic and three-dimensional dodecahedral deployment, the vital bond that carries *Life*, the one principle of "being," into the multiplicity of forms. It is the key to the "en-souled" character of cosmic systems and organic forms of nature. "Divine proportionality" is the inconspicuous "conductor" of life—in the electrical and the musical sense—and the carrier of the soul–mind of the universe (the fifth principle) throughout its body. It is the secret formula of its oneness.

That "life," the principle of life, implies indivisibility may be the one realization needed to embrace the wholesome simplicity of this unitive perspective. Life is: be-ing, and the first attribute of "being" is "life." In the Scriptures, God is referred to as "the living one." Life is beyond the purview of the human. It transcends our perception, escapes our science, baffles our mind and confounds our reason. We can analyze forms; we cannot "ex-plain" life. The wisdom at work in our bodies can reproduce it, but our ingeniousness cannot create it. The wisdom of our bodies, like the wisdom of nature, is indeed "divine." We cannot create a living system, no more than we can "bind the Pleiades." The only statement we can make about life, because it is all we know, is that it is the oneness of which we participate, the being-ness we are as a cosmos. What Plato invites us to consider is that if life, the transcendent beingness of the whole, can pass into forms, it is by "means" of perfectly unitive, harmonic fractions of one. If the indivisible energy of life can pass into multiple forms, it is by shaping them according to proportionate geometries that allow its *indivisible* thrill to seed itself into wholesome (harmonic) portions of "oneness"—"living" forms.

It is interesting to note how the line that Ghyka's observations draw between organic and inorganic structures concurs with Leadbeater and Besant's finding that the shapes of the chemical elements fall into classes that correspond to the Platonic forms of cube, tetrahedron, and octahedron, but do not show dodecahedric and icosahedric structures.

---

65   Ibid., p. 9.
66   Kükelhaus, *Urzahl und Gebärde*.

III. The Harmonic Composition of the Universe

The discovery of a new form of carbon molecule structured like a truncated icosahedron (12 pentagonal and 20 hexagonal faces) would appear to modulate both Ghyka's and Leadbeater's conclusion. In 1985, Harold Kroto, Robert Curl, and Richard Smalley discovered a new class of molecules, the C60 molecules—also referred to as buckminsterfullerenes, after Buckminster Fuller—composed entirely of carbon, in the form of a hollow sphere, ellipsoid or tube. The smallest fullerene is the dodecahedral C20. The discovery, for which the three scientists were awarded the 1996 Nobel Prize in Chemistry, came about during an investigation of the constituents of astronomical dust, carbon rich grains expelled by old stars. It was later found that minute quantities of the fullerenes, in the form of C60, C70, C76, and C84 molecules, are produced in nature, hidden in soot and formed by lightning discharges in the atmosphere. They have since been found in a family of minerals known as Shungites in Karelia, Russia. In 2010, fullerenes (C60) were discovered in a cloud of cosmic dust surrounding a distant star 6,500 light years away. Sir Kroto commented: "This most exciting breakthrough provides convincing evidence that the buckyball has, as I long suspected, existed since time immemorial in the dark recesses of our galaxy."[67]

Thus, pentagonal symmetries, though rare, are not foreign to the mineral kingdom. Given the *star*dust provenance of fullerenes, one might venture the hypothesis that they represent forms of matter more *alive* and solarized than the matter of our planet—than planets in general. Stars emit their own light, whereas planets only reflect the light of the star around which they revolve. Stars contain their own self-sustaining principle, while planets orbit around a Sun-star that is their center and "principle." Similarly, the diamond body of the human "I" (and its 12fold structure) is of a radiant solar-like "substance," whereas personality vehicles (from physical to lower mental) are referred to as "lunar" to indicate partial dimensions of selfhood, elemental rather than soulful, re-active rather than emittive and life giving. They speak to aspects of our nature, rather than to the essence of our being, in the same way that the planetary bodies represent partial vehicles of the solar Logos. Such an understanding could explain the unusual, for the mineral kingdom, 12fold constitution of the fullerenes.

Astronomer Letizia Stanghellini echoes this hypothesis when she advances that buckyballs from outer space might have provided seeds for *life* on Earth. This association of the 12fold molecules with life is consistent with Ghyka's point about the predominance of 5fold/12fold structures in living organic forms. It resonates as well with the solarizing and enlivening of the Earth, which took place, Rudolf Steiner observed, when the crucified Christ, the 12fold envoy of the solar–star

---

67 See https://www.nasa.gov/mission_pages/spitzer/news/spitzer20100722.html.

Logos, who said of the planet Earth, "this is my body," resurrected–solarized his physical body

This first glimpse into the harmonic structure of inorganic and organic forms opens a landscape of mathematical and geometric facts in which the Platonic cosmogenesis begins to catch its reflection and stir in our dulled perception of the forms of nature inklings of a subtle wave-ground, frozen over by the spell of the materialistic eye.

The secret physics of Life: bonding elements into 12fold mediating molecules—Chlorophyll and DNA

The understanding we have gained is that the golden ratio is the fundamental signature of organic forms, because it is the unique mean to fraction oneness into wholeness-carrying, life-conducting parts. The divine seed, indeed, of all forms, the golden mean proportion "represents philosophically the seat or basis of the created worlds."[68] It guides the compositions of nature along harmonic proportions apt to transmit the world generative thrill of Being. Intrinsic to the world soul, Plato's "fairest bond" of concord ushers life into separate, yet *whole, self-bonded* "organisms"—from solar systems to human, animal and plant systems.

The involutionary descent of life into separate forms is, by the same token, an evolutionary ascent of matter into forms of life. From this complementary perspective, the golden mean models the integration of elements into *units* of life (plant, animal, human). This integration is what we call evolution.

The golden "descending and rising" at work in creation evokes Goethe's contemplation of the "sign of the macrocosm" at the beginning of Faust, highlighting the spiritual scientific insights that laced his artistic and philosophical inspiration. Indeed, it will become increasingly clear that golden mean and golden fifth are the sign and signature of cosmos.

> How all things live and work, and ever blending,
> Weave one vast whole from Being's ample range!
> How powers celestial, rising and descending,
> Their golden urns ceaselessly interchange!
> Their flight on rapture-breathing pinions winging,
> From Heaven to Earth their genial influence bringing,
> Through the wide whole their chimes melodious ringing![69]

---

68   Lawlor, *Sacred Geometry*, p. 55.
69   Goethe, *Faust*, lines, 447–453.

III. The Harmonic Composition of the Universe

Teilhard de Chardin had a similar intuition of these descending and ascending celestial powers when he spoke of matter and spirit as polarized and complementary directions of life—powers of materialization and powers of spiritualization exchanging urns of golden harmonies.

※

Like organic systems, chemical elements are harmonically structured forms. However, they do not show the penta–dodecahedric structures characteristic of living, organic forms, the structure associated with the wholeness of the fifth principle. They represent an early stage of organization of matter and corporealization of life—as of yet unseeded with self-reproducing capacity—which is to say, not endowed with a self-reproducing "instrumentation" or equipment. The cell is the smallest unit of life that can replicate itself independently. While elements and atoms represent organized units of matter, they are not *living* organisms or units of life. Atoms aggregate and crystals replicate their structure, but they do not reproduce themselves.

The mechanism of reproduction for all cells and living organisms is based on the DNA molecule, the universal structure that carries the genetic code, or set of instructions, for cellular development and maintenance. A remarkable fact is that the DNA is structured like a ratcheted dodecahedron. *Ratcheted,* in geometry, means spun at regular degrees, in this case in the pattern of a concertina with 12 turns of 30 degrees for each stage—evocative of a spiraling horoscope. So the presence of an essential 12fold structure distinguishes living, organic forms from nonorganic forms. This distinguishing factor is associated with fundamental differences between mineral elements and cellular organisms (plant, animal, human): not only do cells and living systems self-reproduce, but they have their *own ("individual") cycle* of growth and decay, self-unfolding and self-reproduction. The apparition of life *cycles* comes with the apparition of *units* of life. Cycles are the correspondence in time to separate forms of life in space.

The cell and DNA stage of evolution marks the internalization of the "oneness" of life in an autonomous form and cycle—the housing of life in one physical unit of form. "Own cycle" and "self"-reproductive independence point *in fact* to the association of twelvefoldness with the principle of "self." The contrast between the quasi-nonexistence of penta–dodecahedric structures in the mineral world and the structure of the DNA, corroborates the notion that twelvefoldness is the archetypal structure of the "self" principle.

What we contemplate in the apparition of twelvefold structures is the emerging capacity of matter to harbor the mediating principle of "self" and soul—by "con-forming" its organization to the harmonic geometry of the fifth principle. Organic forms are units of mediating interplay between life and matter, Sun and Earth. Crystals are forms of matter that have not yet internalized a *mediating* milieu of autonomous life such as living entities, be they vegetal, animal or human, express in their cycles of growth and decay, birth and death. What makes a physical formation an "organism," what distinguishes a plant from a heap of mud or a mountain, a cell from an atomic element, is the presence of a "self-organizing principle."

One may wonder if this principle of self-organization of life associated with number 12 is not already peeking through one atomic element: carbon. Carbon, the element with atomic number 6, is the basis of all organic molecules. It is *the* element essential to life on Earth, the element without which life as we know it would not exist. Could there be more than chance to the fact that the most common natural isotope of this "element of life" is carbon-12, and that it is the basis for the scale of atomic mass units? Since 1961, the standard unit of atomic mass has been one-twelfth the mass of an atom of the isotope carbon-12. The association of the element of life with number 12 and a 12fold scale of units is intriguing enough, in its cosmological resonance, to warrant the thought that there may be more to it than a fortuitous coincidence. "Organic"—living—compounds are compounds that contain carbon atoms.

In addition to the DNA, another twelvefold structure at the heart of the plant kingdom illumines the same advent of the fifth or quintessential principle of mediation as the fundamental characteristic of living systems: the chlorophyll molecule. Chlorophyll is the biomolecule critical in photosynthesis,[70] the process by which plants absorb energy from light and convert it into chemical energy, which will fuel the synthesis of carbohydrates out of carbon dioxide and water. The chlorophyll molecule is atomic matter organized in such a way that it allows light energy to get "involved" in chemical bondings and reactions and produce substances charged with life: food nutrients.

"If the eye were not Sun-like," Goethe wrote in the preface to his *Scientific Studies,* "the Sun's light it would not see." In the same way that the eye is made by the sunlight for the sunlight, biomolecules and organisms are shaped and "put together" by the energy of life, for life to dwell in a (living) world.

The unique arrangement that makes possible this involvement of light in matter, and evolving of matter into light (by means of color), is a bonding of elements that

---

[70] A molecule is defined as a group of atoms bonded together. It represents the smallest fundamental unit of a chemical compound that can take part in a chemical reaction. The term *photosynthesis* comes from the Greek φῶς, phōs, "light," and σύνθεσις, synthesis, "putting together."

replicates the archetypal 12fold differentiation of "one"—in this case, the oneness of light energy. Lawlor remarks, "Plant can carry out the process of photosynthesis only because the carbon, hydrogen, nitrogen and magnesium of the chlorophyll molecule are arranged in a complex *twelvefold* symmetrical pattern, rather like that of a daisy." He concludes, "It seems that the same constituents in any other arrangement cannot transform the radiant energy of light into life substance."[71] In other words, the 12fold pattern is the only way, the incontrovertible "mean" to funnel the wholeness of light energy into the bio-molecules necessary to supply all the organic compounds and most of the energy necessary for life on Earth.

Julius Robert Mayer, the nineteenth-century German physician and physicist who contributed the last piece to the story of the chlorophyll by recognizing that plants convert solar energy into chemical energy, captured in one brief remark the archetypal significance of the chlorophyll. "Nature has put itself the problem of how to catch in flight light streaming to the Earth and to store the most elusive of all powers in rigid form. The plants take in one form of power, light; and produce another power, chemical difference."[72] Contained in these words is the cosmological dimension of the plant kingdom's opus of mediation: turning the elusive energy of sunlight into rigid form; turning the sameness of light into different chemicals, and chemical diversity into one green-ness; materializing sunlight and synthesizing elements of matter into a photoreceptive organ.

The 12fold molecule that corporealizes sunlight by organizing chemical matter into a receptor of light replicates in the plant kingdom the great matrix of the world soul, the "light of the world"—at once externalizer of Being and centrating organizer-"incorporator" of chaos into a world milieu. Lawlor's daisy like drawing (fig. 32) illumines the reflection, in the chemical depths of physicality, of the ontological ground of the world. The molecule of chlorophyll emerges as a first epiphany of the mediating process that will eventually lead to the formation of the 12fold etheric heart and egoic lotus of the human being. Is it not our intuitive perception that there spreads out, in the greenness of the chlorophyll, the peaceful, vibrant soul of the plant world?

Twelvefoldness points to the unique arrangement of elements that makes it possible for matter to internalize the energy of life,

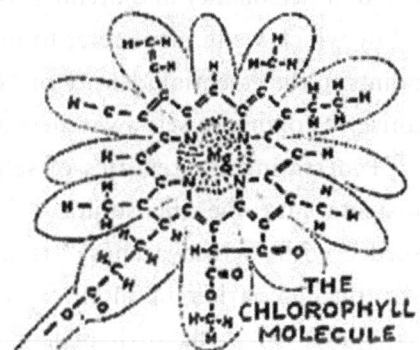

Fig. 32: *The Cholophyll molecule*
(*Lawlor,* Sacred Geometry)

---

71 Lawlor, *Sacred Geometry*, p.5.

72 Mayer, *Die organische Bewegung in ihrem Zusammenhange mit dem Stoffwechsel.*

mediated by sunlight, in a chemical form apt to synthesize the nutritional compounds that sustain "all life on Earth." Twelvefoldness displays the archetypal housing of being (life) in time and space, the core pattern of mediation to which living organisms conform as they gather themselves out of atomic dust and elemental chaos at the peripheries of the world being. The archetypal structure of selfhood (individual and cosmic), "dodeca-hood" points to chlorophyll as the "selfhood" of the plant kingdom.

The dodecahedron is the quintessential form of *self*-organized matter—the prototype of living systems as "principled forms." Indeed, what characterizes such systems and distinguishes them from inorganic compounds is that they are seeded with the principle of their formation, growth, and self-reproduction. This line of distinction between inorganic "compounds" and organic "systems" sheds light on an analogous difference between animal and human kingdom. Like plants, animals harbor the self-reproductive principle characteristic of living entities on a biological level. In addition, they also have an instinctual component, an ensouling, "astral" principle, which "moves" them with desire and fear and harbors their character traits, gifts, and functions *as a species*. This ensouling principle, however, is a collective one. The tiger does not harbor his tigerness as an individual principle within. His identity is a collective one. Animals are not endowed, as humans are, with a 12fold causal body hosting a "self" principle.

The instinctual, emotional, and mental elemental forces of human nature are to the soul as nature and animals are to human beings—or as atomic elements and minerals to the seed—unprincipled. What the harmonic backdrop suggests is that the daily challenge to recenter one's consciousness from "elemental" drives to Egoic mindfulness is the challenge to shift identification from partial harmonics to the full "sol" of the soul, from partial intervals to the *life-giving* fifth, the heartful blend of personality and divinity. However little we may stop to contemplate it, the "I" by which we humans refer to our unique individuality and quintessential being, points to the essential identity of "selfhood" we share—which is the identity of the universe coming to self-awareness and expression in the human "I."

Penetrating the realities of self-reproduction and "self-principle" offers new insights into the theosophical indication that animal-stage humankind was seeded with "self-awareness" 18 million years ago by "solar angels" from Venus. We touched earlier upon the spiritual–scientific understanding that the planet associated with the fifth principle was also the originating place of the great "sowers of selfhood," referred to as solar angels. These beings underwent human evolution in preceding eons. The full flowering of selfhood they achieved would have implied a capacity to "self-reproduce" and implant their Egoic "seed" in a new "self evolution." They brought their solar seeds of selfhood into the Earth evolution, allowing the animal-like creatures of humankind-to-be to take their next step into self-aware beings. One could say that the solar beings "reproduced" themselves in humankind.

Similar twelve-toned matrices in Indian, Chinese, and Celtic cultures

To view the dodecahedron as the prototype of "principled" forms is to regard it as the ideal form of a living universe—which is what Plato suggests when he says that the demiurge used the fifth perfect solid for the construction of the world. Closest to a sphere, the ultimate harmonious solid is celestial perfection in form—wholeness made solid, soul made form.

Plato's statement about the dodecahedron echoes ancient sources of wisdom. In one of the primary Upanishads, the *Prasnopanishad,* one reads, "Some say that He is the Father with the twelvefold form with a fivefold support having His (vital) currents in the upper half of the electric sphere." L. G. Hallett comments, "The twelvefold form corresponds to the twelve signs of the zodiac, which indicate twelve creative powers in nature. The Universe is here viewed as a Dodecahedron. The pentagonal nature of each side is the fivefold support or the five vital airs."[73]

The "twelvefold form" of the Vedas is also flanked by a sevenfold: "The Father with five feet and twelve forms, they say, dwells in the higher half of heaven full of waters. Others say that he is the clear-seeing one who dwells below in a sevenfold wheel that has six spokes."[74]

Sir John Woodroffe, (alias Arthur Avalon), a British orientalist who, while a judge in India in the early part of the twentieth century, studied Sanskrit and became interested in Tantric scriptures and Yogic practices, describes the nature of the five currents:

> In the Tantric darsan (a vision of God), nature (*prakriti*) is differentiated into five forms of motion. Ether (*Akasa*) fills space with the "Hairs of Shiva," nonobstructive motion radiating lines of force in all directions, sustaining the space in which the other forces operate. Air (*Vayu*) is a transverse motion and the source of locomotion in space (from the sanskrit root *Va,* "to move"). Fire (*Tejas*) is an upward motion, giving rise to expansion. Water (*Apas*) is a downward motion, giving rise to contraction. Earth (*Prithivi*) is a motion that produces cohesion and obstruction, the opposite of the nonobstructive Ether.[75]

---

73 Quoted in Hallett, *Studies in Ray Correspondences.* Adyar pamphlet no. 142.

74 Asya Vamasya Hymn I.164.

75 Woodroffe, *The Serpent Power,* pp. 77-78.

This Hindu view of the elements mirrors Plato's. Ether, the quintessential element associated with the dodecahedron, is the *element of life*, the "element" through which the wholeness of Being, the world soul, "radiates" and "vibrates" as omnidirectional currents (the hair of Shiva) underlying and sustaining the physical world. The description of "ether" as "radiating lines of force in all directions and sustaining the space in which the other forces operate" points to its matricial character. The etheric plane is the plane of vital energy, as the etheric body is the body of *life*—human and macrocosmic.

The ancient Chinese understanding of the harmonious relation between Heaven, Earth, and Human existence also involves twelvefoldness, and their conception of world creation is centered on the unfolding series of the primary numbers 1, 2, 3—in close parallel with the Platonic cosmogony. In their system, 3, the third, is also a 12fold.

Like most ancient cultures, the Chinese regarded cosmic sound as the primordial event of world creation. They called the first vibration Wu Chi—the equivalent of the Sanskrit OM. Wu Chi differentiates into Tai Ch'i, or Two Tones (yin/yang polarity), which then manifests as 12 Tones, the Lu (1+2=12=3). The 12 Lu are the 12 fundamental pitches within the octave, related to each other by specific ratios.[76] Of the twelve Lu, only the seven "major" tones, and even more regularly five of the seven (they rarely made use of the two semi-tones), were part of their musical scale, referred to for this reason as "pentatonic." So, almost five thousand years ago, long before the Greeks, China saw in the mathematics of harmonious sound the principles that govern Creation.

In striking convergence with the mathematical philosophy of the Greeks, one of the five classics of the Confucian canon, the *Record of Rites*, states, "Since 3 is the symbolic numeral of heaven and 2 that of Earth, sounds in the ratio of 3:2 will harmonize Heaven and Earth." Clearly, the Chinese 12 tones were born of the same womb of harmonic mediation between Heaven and Earth.

Ancient Indian and Chinese sources orchestrate the harmonic metaphysics of Pythagorean-Platonic thought in regarding the 12fold circle-sphere as the archetypal unfolding of the "Third," the world soul, into the space–time matrix of the universe. This notion of a trinity at the root of the 12 is also found in the theological and metaphysical foundations of Judeo-Christianity and Islam: Judaism features the three great patriarchs and the twelve tribes of Jacob-Israel. Christianity features the trinity of Father, Son, and Holy Spirit, with 12 apostles surrounding the Son. Islam features the prophetic trio of Mohammed, Ali, and Fatima, whose revelation

---

76 Leviton, *Ley lines and the meaning of Adam*.

is to be carried into the future by the 12 Imams. The voice of the living "One"—the sound of Being—is dodecaphonic.

In *The Secret Power of Music*, David Tame explains how the Chinese associated "the twelve Cosmic Tones, the emanations and aspects of the Primal Sound, with the twelve zodiacal regions of the heavens. Astrology," he says, "was originally conceived on the basis of the twelve tones and the influences that their vibratory frequencies exerted over the Earth. In all lands, astrology began in ancient times as the study of Cosmic Tone.... It was the science of celestial sound.... The twelve Tones of the zodiac, radiating their super-physical vibrations unto the Earth, were believed to affect psychological states, the phenomena of Nature and so forth.... The monthly changes from one sign of the zodiac to another indicated cosmic modulations in the pattern of celestial harmonics.... The harmonic relationship between the twelve Tones was also believed to change with the changing of the phases of the Moon and the hour of the day.... With each new stellar configuration, new Tones inundated the earth, bringing with them new tendencies in thought, new moods, different behavior patterns and different activities in the Nature kingdoms."[77] Chinese musicians would adjust to these cosmic tonal shifts and change the key in which they played. Astrology was the science of tone and vibration, and music was a sacred science of invocation of cosmic harmony into the affairs of the Earth. Its purpose was to align civilization with celestial vibrations and harmonies.

In the course of his fascinating depiction of the relation of Chinese culture to sound and music, David Tame asks the question: "How did they maintain the same basic culture and society largely unaffected by all the events that threatened them, over thousands of years?" He ventures the hypothesis that the exceptional longevity and stability of the Chinese civilization, which went from the third millennium BC to 1912, was tied to their intelligence of the power of vibration and their practice of a scientific art of tonal alignment and entrainment of society to celestial harmonies. Knowledge of cosmic mathematics and harmonic proportions infused not only their music and the design of their musical instruments, but their earthly affairs, which they sought to attune to the changing modes and tonal configurations of heaven.

China is called *Zhōngguó*. The first character *zhōng* means "central" or "middle," while *guó* means "kingdom" or "nation"—literally "Middle Kingdom" or "Central Kingdom." The fundamental concern of the Middle Kingdom for a vibrant alignment between below and above that would foster a golden civilization was expressed in the central attention it gave to the huang chung, the foundation tone. The proportions of the pipe used to produce this perfect tone were the object of utmost care, and they

---

77 Tame, *The Secret Power of Music*, pp. 37 and 51–52.

provided the standards for the Chinese system of weights and measurements.[78] This fundamental pitch from which everything moral as well as physical in China received its number, measure and weight, was not immune to slight changes. After all, Tame points out, "if the kung of the previous dynasty had been in perfect harmony with the eternal principles of the universe, how could the dynasty have ever ended?"[79] With the wisdom of balance that famously guided them, the Chinese also saw the danger of excessive rigidity and the importance of balancing their "celestial" regulations with a willingness to alter and adjust the fundamental on which they rested. Interestingly and eloquently, the "divine" monarch regarded as the "inventor" of music, emperor Fu-Hsi, who reigned around 2850 BC and was the first of five rulers credited with the emergence of Chinese civilization, is also the most likely author of the *I Ching*, the wise book of changes. The majestic wisdom of celestial fundamentals is based on understanding and adapting to change as the only permanent feature of reality, which applies as well to the perfect pitch of the kung. The foremost concern on the mind of the first emperor of a new dynasty accessing to the throne was to correct the kung.

The deterioration of the ancient Chinese civilization and the end of the imperial house that governed China for almost five thousand years happened concomitantly with the decline of music during the final Ch'ing dynasty (1644–1912).[80] In a suggestive parallel, Plato, who claimed that "the forms of music precede and determine forms of government," attributed the corruption of Athens to the decadence of music.[81] The replacement of the Chinese imperial house by a republic coincided with a worldwide toppling of autocratic governments precipitated by war chaos and met by a rising wave of totalitarianisms, whose tyrannical enforcing of unity–uniformity carried seeds of a new world order rooted in democratic values and global thinking. The time China lost her dynastic tonal alignment was also the time when a number of nations were pressed out of an elitist, nationalistic and imperialistic style of governance, and ushered into a new global paradigm of democracy and exchanges.

The thought arises, in revisiting this history, of how much inspiration could be drawn from the Chinese practice of tonal governance by a world trying to bring its many factions to work together and create viable forms (of fair trade, international justice, energy policies, etc.) to host its emerging global self. While the latent tone of humankind is largely trapped, for now, in coarse economic forces, and absorbed more than channeled by electronic technologies, it is also, however feebly, gathering

---

78 Ibid., p. 56.
79 Ibid., p. 58.
80 Ibid., p. 68.
81 D'Olivet, *Music explained as Science and Art* (p. 24. in French edition).

itself through overarching entities such as the UN, as well as numerous disseminated "harmonics," attuned to global betterment and the necessity to restore harmony with humankind's chamber of resonance: nature and cosmos.

The musical cosmology of the Chinese lives on in the principles of their traditional medicine, most clearly so in acupuncture. The charting, by Chinese acupuncturists, of the body's 12 primary energy meridians emanates from the same science that informed its musicians and wise rulers—the knowledge of a primordial pattern of wholistic differentiation of oneness at the heart of all living forms, macrocosmic and microcosmic. Each meridian has a yin/yang polarity. Each is associated with a particular organ and element—heart with fire, liver with wood, kidneys with water—as well as a yearly-daily time of heightened activity, propitious for treatment. The heart, for instance, is associated with summer and the noon-to-2-p.m. period of the day. The 24-hour day is viewed as a full cycle of the two-hour "high tide" of the 12 meridians. The body is a microcosm constituted of the 5 elements (wood, fire, metal, water, soil) and keyed to the 12fold macrocosmic cycle of hours. Health lies in sustaining and restoring harmony between microcosm and macrocosm. Healing is based on a science of the 12fold vibrational reality that is their common ground.

Richard Leviton's spiritual scientific research into the Earth's electromagnetic grid indicates that the central 12fold design is present not only within the human being (heart chakra and causal body) and around him in the great expanse of cosmos (the zodiac), but also around the energetic body of the Earth. In *Leylines and the Meaning of Adam*, he describes how "the Earth is enveloped in a lace doily of electromagnetic fields and pulsing points of light that move the incoming intelligent cosmic/stellar energy streams through their grid." Specifically, 12 major Oroboros Lines encircle the globe, making 12 great circles, each carrying a different solar energy characterized by one of the 12 zodiacal attributes (e.g., a Taurus Oroboros Line). Three minor lines carry what is ordinarily designated as the qualities of male, female, and neutral. The Oroboros Lines correlate with the 12 Chinese Lu and 12 organ meridians, while the 3 minor lines echo the balancing of Chi along "governing" (yang) and "conception" (yin) vessels. These lines find their reflection, Leviton suggests, in the iconic 12 knights of King Arthur's Round Table, each knight representing one station of this spinning zodiac. "The Oroboros Grid," he says, "represents the 12 Knights of Gaia's Round Table, the means by which Gaia can experience the differentiated energies of the Solar Logos, of which She is an integral part."

Every Oroboros Line was at some point anchored in the Earth by special vibrating stones positioned at key sites around the planet to form a global hermetic megalithic calendar aimed at enhancing energy flow for the benefit of all living beings. The 12 stones resonated with each other and every stone vibrated with a special intensity at times determined by the overlapping zodiacal schedules, from the 12-hour cycle of day and night to the 12 precessional ages of 2,160 years of the Platonic Year. Similar to the Chinese foundation tone, the essential function of ancient sites and stone circles in general was to attune the Earth to its cosmic evolutionary tones.

Like the acupuncturist rebalancing the human microcosm, ancient and modern geomancers (stone circle geomancers, shamans, Feng Shui practitioners) seek to facilitate the flow of life energy (Chi) and cosmic rhythms to reharmonize global and local environments and enhance their spiritual vitality. While the Chinese rulers kept their civilization abreast with the harmony of the spheres and old Fengshui masters created architectural spaces alive with cosmic energy, the Druid geomancers, says Leviton, "kept Gaia's biosphere humming with a steady stream of life-supporting, life-enriching, life-evolving energy transmissions from above." Other researchers into geomantic mysteries, such as John Michell and Freddy Silva, have shown how ancient sites, temples, megaliths, as well as medieval churches, are placed on power spots along electromagnetic lines—reminiscent of acupuncture needles placed at key points along the body's meridians—which sheds spiritual scientific light on the reverence for the healing power associated with such sites. "A great scientific instrument lies sprawled over the entire surface of the globe," notes John Michell. "The vast scale of prehistoric engineering is not yet generally recognized."

The Dodecahedron—The Quintessential Form of the Universe

A convergent picture of the universe as a dodecahedric form with a 12fold harmonic matrix emerges from these ancient cultural streams. The pentagonal character of each facet is similarly associated with the five elemental realities (vital airs, elements or elemental motions) variously acknowledged in each tradition as constitutive of the advent of "solidity."

Blavatsky adds the voice of theosophy to that of Plato, India, and China. "Our universe," she says, "was constructed on the geometrical figure of the dodecahedron." The notion of a dodecahedral universe has actually appeared in the scientific world relatively recently. "*Is the Universe a Dodecahedron?*" (an article in the science journal *Nature,* Oct. 8, 2003), offers the following: "The standard model of cosmology predicts that the universe is infinite and flat. However, cosmologists in France and the US are now suggesting that, based on measurements of cosmic waves left over from the 'Big Bang,' space could be finite and shaped like a dodecahedron instead."[82] The science journal cover illustrates the findings with a

---

82   See http://physicsweb.org/article/news/7/10/5#polygon.

III. The Harmonic Composition of the Universe

baseball: 12 curved pentagons tiled together into a sphere. Interestingly, this most concrete comparison is echoed in Plato's view of the Earth. The "true Earth," he writes in the *Phaedo*, "if looked at from above, is many-colored like the balls that are made of twelve pieces of leather." Less graphically, Lawlor remarks, "The point of view of modern force-field theory and wave mechanics corresponds to the ancient geometric–harmonic vision of universal order as being an interwoven configuration of wave patterns."[83]

The most sphere-like polyhedron, imbued through and through with the harmonic fractioning–bonding ratio characteristic of the mediating principle of "soul," the dodecahedron is the ideal embodiment of the world soul, the most majestic form—"sound" with the full resonance of Being.

In his short statement about God's use of the fifth solid—the dodecahedron—as the model of the universe, Plato inserts in fact a little more information. He uses the word *diazographon*, which intimates the association of the dodecahedron with the zodiac. Translations vary from: "There was yet a fifth combination which God used in the delineation of the universe," to, "The fifth figure, the god used for arranging the constellations on the whole heaven" or "God used this solid for the whole universe, embroidering figures on it."[84]

The word *diazographon* includes three notions: across (*dia*); life (*zoe*); tracing, design or sign (*graph*). To say that God delineated the universe with the dodecahedric (zodiacal) criss-crossing design, or signature, of life could be regarded as an exact translation. The very word *zodia-kos* appears, reversed (*diazo-*), in the term used by Plato. The reversal in the ordering of *zo-dia* has the effect of animating in *"zo-diac"* a primordial designing *gesture*—a *dia-zo-graph*-ing of space; a cross-delineating-wiring of the ground of the universe by the vibrational waves of life. Dia-zo-graphy animates within the static notion of "zodiac" a dynamic matrix of "life" currents underlying our space–time universe and all its forms. The similarity between the twelvefold design en-souling the universe and the twelve "animal signs" of the zodiac is significant. It proposes a widened interpretation of the animal circle as an "animating," en-souling circle. Dodecahedron and zodiac merge to represent the quintessential form (indeed, the fifth solid) of the universe and its twelve "signatures."

Manly Palmer Hall offers an intriguing distinction between dodecahedric universe and twelvefold zodiac, which he says touches upon "one of the most arcane of ancient secrets." He refers to two zodiacs—one fixed, one movable—and their relationship. "The fixed zodiac," he says, "is described as an immense dodecahedron,

---

83 Lawlor, *Sacred Geometry*, p. 4.
84 Plato, *Timaeus*, 55c.

its 12 surfaces representing the outermost walls of abstract space. From each surface of this dodecahedron a great spiritual power, radiating inward, becomes embodied as one of the hierarchies of the movable zodiac, which is a band of circumambulating so-called fixed stars. Within this movable zodiac are posited the various planetary and elemental bodies."[85] In a similar vein, or perhaps sighting this arcane secret, Robert Powell points to *The National Geographic* map of the Milky Way Galaxy, and the greater 12-sided structure of the galaxy it makes visible, as the expression of the archetypal dodecahedric structure of the universe.[86]

The dodecahedron is the externalization of Plato's world soul into a relatively "solid," etheric form. A supreme expression of differentiated-and-bonded wholeness, this "sphere of fives" is at once the archetypal form of the soul and the "quintessential" etheric body of the world; the soul core and all-encompassing form of our universe; its pleromatic heart and its zodiacal periphery.

Clearly, Plato connects the 12fold harmonic "apportioning" of the primordial Oneness with the soul matrix of our cosmos, and the dodecahedric form with its corporealization in a zodiac contained universe. He speaks to world soul and zodiac, soul and cosmos, as two sides of one reality, the latter being the etheric "embodiment" of the former. The world soul is at once within and without, soul and universe, core and circumsphere:

> And *in the midst* thereof He set Soul, which he diffused throughout the body, making it also to be the *exterior* envelope of it; and he made the universe a circle moving in a circle, one and solitary, yet by reason of its excellence able *to converse with itself* [self-resonate], and needing no other friendship or acquaintance. And because of all this He generated it to be a blessed God.[87]

Twelve-pointed star and twelve-petal lotus are alternative versions of this core structure of all living forms—from zodiac to solar system to Earth grid macrocosmically, from causal body to etheric heart within the human being, from cellular DNA to chlorophyll molecule in live organisms—all replicas of the fundamental "selfhood" of the universal Being we are.

We now see quite distinctly how the congeniality of zodiac and soul intimated by their 12fold structure points to their common ground—the harmonic fractioning of the "fundamental One" into a 12fold matrix of space-time-consciousness—which

---

85 Hall, *The Secret Teachings of All Ages*, pp. 174–175.

86 See Powell, "The Year 2019 and the Opening to the Angelic Realm: The New Star Wisdom" in *Cosmology Reborn: Star Wisdom*, vol. 1 (SteinerBooks, 2018).

87 Plato, *Timaeus*, 34b (Italics added).

is the nondual ground of the existential dualities of consciousness and cosmos, self and world.

The phenomenon of creation is the archetype of all phaeno-mena, the "appearing" of Being in form, the "Phanein" that gives its name to the God. What we call universe is the externalization of Being. What we call self is the appearing of being to itself in time and space. It happens by way of an intermediary dyad of selfhood-and-cosmos: *vibration*. Phanes is indeed referred to as the duad, anterior to the triadal Logos. The primordial vibration of creation, the soundless self-"resounding" of Being, is at once the latent milieu of *consciousness* (the world soul) and the time-space harmonic matrix of *cosmos*.

The self-resounding of the One, the resonance of "one" with its "double" (2:1), which gives birth to the triune third (3:2), which then deploys into a 12fold, is what Plato contemplates philosophically when he speaks of the universe as one in itself and *able to converse with itself—a self-blessed god*. Displayed in zodiac and Egoic lotus is complete harmonic self-bondedness. This vibrant milieu of soulful bonding of the one with its "other" is at once the self-resonant temple of Being and the self-bonded template of selfhood.

THE ZODIAC, ETHERIC MATRIX OF THE SOLAR SYSTEM AND TIME FRAME OF THE SOUL

New realizations about the zodiac arise from these insights into the foundations of the universe. *Dia-zo* suggests the spreading out (*dia*) into space and time of the vibrancy of life (*zoe*); *Dia-zo-graphein*—its "designing" effect. The differentiation of the triune soul of life defines twelve secondary forms of harmony. Twelve tonalities of soul ring out, which will en-soul or animate the forms of the world—twelve "animating" signatures, twelve "animal signs." Analogous to musical intervals, the zodiacal signs symbolize different qualities of harmony. Each sign portrays a "relatively" whole signature of the world soul. Each expresses a particular "soundness" of Being, colored by a specific rapport-ratio to the fundamental vibration.

The origins of the zodiac have not been established. The *Rig Veda*, the oldest Vedic text, contains clear references to a chakra, or wheel, of 360 spokes in the sky. The number 360 and its related numbers, such as 12, 24, 36, 48, 60, 72, 108, 432, and 720 (all multiples of 12), occur commonly in Vedic symbolism. The hymns of the great Rishi Dirghatamas are particularly rich in cosmological references. For Dirghatamas, and later Vedic astronomy, the main God of the zodiac is the Sun God Vishnu, who rules over the highest heaven and is sometimes identified with the pole star. Vishnu had 360 names or forms, one for each degree of the zodiac.

"With four times ninety names, he (Vishnu) sets in motion moving forces like a turning wheel (cakra)."[88]

Aristotle called the zodiac "the circle of life," supporting the idea that the *"dzoon"* of *"zoo-diac"* refers to "life" and Living Ones rather than "animals." His formulation explodes with fresh significance when the zodiac is viewed as the quintessential dodecahedric sphere of the universe's life—an etheric *BIO*-sphere, structured harmonically by the imperative wholeness of life, thus apt to shape matter into life-conductive forms. A quasi-corporeal manifestation of the soul of the world, the zodiac is both a spatial and a temporal matrix: a space in which further "corporealization" of the world being into planetary spheres, elements and bodies, can take place, and a time matrix for the evolution of matter into forms of "self," and of selfhood into universality of being—from element to cell and cell to seed, from self to soul, and from soul to zodiacal consciousness. It is our archetypal space–time matrix.

Alice Bailey's statement that the zodiac is an etheric wheel or chakra—the heart-in-the-head chakra of the cosmic being in whom our solar system has its being—echoes this view of the zodiac. So does the ancient Hindu association of the Dodecahedron with the Universal Mother, Prakriti. The same context associates the icosahedron, its dual, with Purusha, the Universal Father. Together, they form the parental pair.

This entire perspective illumines in the zodiac the etheric "halo" of our universe. In icons of saints, the halo suggests the heart-in-the-head center, the auric, etheric radiance of the monadically realized human being—what Steiner refers to as *Geistesmensch*, or spirit body.[89] This zodiacal aura of the universe is charged with the primordial life giving, world-intoning 12fold sound of Being. The "circle of life" is the matrix of the involutionary descent of Being into forms of matter and evolutionary integration of matter into organic replicas of the world soul—from biological structures to causal bodies. The dual movement of involution–evolution reveals itself in this light as the corollary, *in time*, of the structures of mediation nature displays in *space*. The conjugation of life and matter is a twofold opus: a differentiating of the one life into multiple forms of matter and a unifying-integrating of matter into forms of life—a secting and a bonding.

The bio-logical process of photo-synthesis illustrates beautifully this archetypal pattern of the descent of life into forms and integrative ascent of the elements of matter into unit-"beings." As we saw earlier, photosynthesis represents literally an integration (a *syn-thesis*) of matter and light (*phos*), which produces the lightening

---

88   *Dirghatamas Rig Veda*, I.155.6.

89   C. Bamford's translation of *Geistesmensch* as "spirit body" rather than "spirit man" seeks to convey with greatest clarity what the term refers to, which is the physical body transformed by the "I."

III. The Harmonic Composition of the Universe

of matter and substantiation of light we marvel at in the yearly leafing and greening of the Earth. "In mythological thought, Lawlor remarks, twelve most often occurs as the number of the universal mother of life and so this twelvefold symbol is precise even to the molecular level."[90] At this point, the notion of "symbol" pales in the light of the factual "pattern" that the chlorophyll molecule replicates. If twelve is a symbol of the great mother of life, it is because it is the actual pattern of the universe's *matrix*. We find 12fold symmetry as the life-giver, or womb, that transforms light into the basic spectrum of organic substance. Chlorophyll is indeed the secondary mother of life on Earth, the great nutrient-maker and nurturer of the animal and human kingdoms.

Bio-logy means, literally, the "logos of life." In this deeper sense, biology reveals the soul logic of life at work in the harmonic composition of the molecule central to the vegetative organization of matter—in other words, the organization by the solar "Logos" of the vegetative "life" of his being (his bio-logy). The 12fold womb that turns sunlight into organic substance and "involves" life energy into chemical nutrients is analogous to the primordial 12fold soul matrix that turns Being into world—the harmonic hub that allows the vibrancy of being to pass, a-live and *as life,* into separate lives, by organizing matter into forms of *being*—living organisms. The secret of this "organism-making" is the conjugation of life-transmitting waves of solar light with their reflection from the periphery of the solar "system"—into the evolving geometries of elemental and organic forms. The centrifugal radiance of Being and the centripetal assembling of chemical forms "exchange their golden urns" and partake of each other in the stemming and "treeing up" of the Earth under the Sun.

Lawlor suggests that the perception of the sweetness of the rose is less a response to chemical components than to internal geometries of proportion and harmony. Intuitively attuning senses and mind to this suggestion opens the way to perceive the sweetness of the rose as perfect harmonic vibrancy. The rose, a key symbol of the soul, is the Western counterpart of the lotus, the traditional Eastern representation of a chakra. The fragrance of the rose is what emanates from ideal self-bondedness, a pure intimation of the "blessed god He generated the world to be." This psycho-sensory perception of the harmonic order exuding from nature fragrances is not unlike the intuition of world foundations that the ancients accessed acoustically—when, says Lawlor, the audial sense was regarded as the epistemological basis for philosophy and science. "In the perception of sound," Steiner echoes, "one hears the wisdom of the world." We will see that facts of human perception do provide the beginnings of a scientific orchestration of Lawlor's musical olfaction.

---

90 Lawlor, *Sacred Geometry*, p. 5.

The dimensional leap from zodiacal matrix to Egoic lotus and rose highlights an archetypal lineage, which the association of rose and lotus with the full radiance of heart, soul, and cosmos alludes to in world mythologies, literature, and medieval architecture.

Dante's journey culminates in a vision of the universal rose of the courts of Heaven, deploying the rays of the creator to "its extremest leaves," their effulgence mirrored and multiplied by the thousands of ranks of enlightened beings surrounding it, its odor a fragrance of "praise to the ever-rising Sun."

> There is a light above, which visible
> Makes the Creator unto every creature,
> Who only in beholding Him has peace,
> And it expands itself in circular form
> To such extent, that its circumference
> Would be too large a girdle for the Sun....
> Ranged aloft all round about the light,
> Mirrored I saw in more ranks than a thousand
> All who above there have from us returned.
> And if the lowest row collect within it
> So great a light, how vast the amplitude
> Is of this Rose in its extremest leaves!...
> the Rose Eternal
> That spreads, and multiplies, and breathes an odor
> Of praise unto the ever-vernal Sun (CANTO XXX)

Opening of the world and culmination of matter, Dante's rose evokes the circular rose stained-glass windows, which solarize the dark naves of Gothic cathedrals and immaterialize their walls. Scriptures and songs have mused on the beginnings and fulfillments of matter, which make the Prakriti-matter of worlds, the "mother-of-God," akin to the sophisticated simplicity of the rose. Associated with the mysteries of Isis, the rose intimates the completion of the alchemical work of transmutation of matter. In its later association with the Virgin Mary, the most recent, Western incarnation of the "mother of the world," the rose inspires a contemplation of human nature so harmonized and unified that it may host the light-of-the-world. The alchemical rose is 5fold—the pentagrammatic seal of Venus. The master of the alchemical opus is the soul, the fifth principle, and its eventual achievement is the radiant stone of "selfhood," fragrant with conquered harmonies.

The vibrant "synthesis" of spirit life and matter associated with the twelvefoldness of the soul opens a view of the zodiac as the harmonic womb of the universe, a "toned" milieu and *toning incubator,* in which inorganic elements of matter gather to form self-sustaining vegetative units (plant organisms), organic systems assemble and leap into flesh to embody animal souls, and bodies stand up to host the rose-lotus blossoms of human individualities. Reminiscent of a Tibetan bowl turning frictional noise into "cosmic" sound, the great zodiacal chakric rose turns elemental matter into organic forms and soul-ful bodies. All by means of harmony and harmonization, the gestational cycles of time turn (e-volve) inert forms of matter into forms of being by involving (in-turning) Being ever more deeply into increasingly complex forms.

Zodiac and Soul, "Means" of Harmony Making, and Milieus of Evolution

Zodiac and soul emerge as the two facets, cosmic and psychic, of the milieu of creative interplay between essence and existence that founds the universe. The zodiac is the quintessential body of the world soul—its externalization into the time space matrix of the universe. Together, soul and zodiac constitute the great *mean(s)* and "milieus" of the dual process of evolution: descent-involution of Being into forms of matter in space, and evolution of consciousness into forms of being in time.

This sighting of the zodiacal wheel as the circular loom on which the soul weaves, in time, the fabric of its unfolding petals lights up what may be the true ground of astrology as a science of soul-making. Configuring milieu of cosmos and harmonizing milieu of the psyche, the zodiac points to the ontological ground of a unitive apprehension of the two sides of human experience—subjective consciousness and objective reality. For the Heaven-and-Earth science, which Kepler last practiced in union with astronomy, world and soul are not radically distinct but mirror each other as two poles of one reality.

The "harmonic" structure of soul and zodiac indicates that the essence of the involutionary adventure of Being and evolutionary labor of the soul is the harmonization of matter and spirit into a mid-world of creative self-awareness—a process of humanization for the universe and universalization for the self. In traditional cosmology, humankind is the great mediator who binds together intelligible and sensible, eternal and temporal. The zodiac is the gestational matrix in which souls incarnate in time and space to engage in the harmonizing of elemental forces and divine spirit that produces the stupendously creative field of individual and collective consciousness. Plato's psycho-cosmology illumines the process of evolution as a process of *harmonization*. From photosynthesis in plants to soul "synthesis" in humans, the inner reality of becoming is harmonization.

Plato's mathematical metaphysics of harmonic mediation resonates with the intuitive apprehension we have of the soul as a mediating reality between our essential being and existential matters—and a creator of "mean-ing." If we lose meaning, we lose the real living ground of our becoming. What Plato's harmonic universe also suggests, however, is that, in any abyss of meaning lost, a harmonizing imperative awaits, eager to hand its golden key to a next step in becoming.

Harmony making is the heart of the "living creature" we are, as a universe. The divinely toned, harmony-making milieu of world unfolding, which is also the harmony-inducing milieu of soul becoming is the quintessence of cosmos.

Blavatsky marveled at the way Plato grasped and described the reality of the soul, and how later philosophers intuited it:

> How precise and true is Plato's expression, how profound and philosophical his remark on the (human) soul or Ego, when he defined it as "a compound of the same and the other... for the Ego (the "Higher Self") when merged with and in the Divine Monad is Man, and yet the same as the "other," the Angel in him incarnated, as the same with the universal Mahat. The great philosophers must have felt this truth, when saying that "there must be something within us which produces our thoughts. Something very subtle; it is a breath; it is fire; it is ether; it is quintessence; it is a slender likeness; it is an intellection; it is a number; it is harmony." (Voltaire)[91]

The "solar Angel" who implanted the fifth principle in the human being-to-be, remains his/her overlighting "other self" during the long process of soul development—until the human being becomes "same" as his great other—a "solar" being.

How interesting to see the words, charged at once with metaphysical and mathematical insight, subjective and objective references ("self, ether"), that Voltaire uses to convey his insight into the essence of the human mind, and how strikingly close to Plato's intimations: quintessence, ether, number, harmony, intellection. The impression emerges of a quintessential, ethereal "fifth" principle (a "solar angel") of lucid fiery harmony, a breath whose dual current, down into the body and up into spirit, harmonizes our components and tunes our beings into potentially fragrant and vibrant creatures, exuding rose or lotus like divine proportionality.

The gradual silencing of the harmonic borderland between Heaven and Earth

The harmony that shapes and "geo-metrizes" living organisms is immaterial. Its formulation, mathematical and geometric, pertains to the pattern forming realm of the transitional milieu by means of which the Logos "becomes" visible or "appears" as cosmos. That harmony is the likeness of the Logos in the forms of cosmos is the

---

91. Blavatsky, *Secret Doctrine*, vol. 2, pp. 88-89.

golden secret that Plato whispers in the ear of the future. Physical senses alone cannot perceive harmony. Only the mind, using the senses, can see and hear its effects in the physical world of sounds and forms. Only the mind can conceive it, the soul experience it, and science verify it. The mind–soul is itself, precisely, the likeness of the "Logos" in humans. The harmonizing chamber between outside world and inner subjectivity, it is the replica of the world soul. Rudolf Steiner's reflections on thinking suggest this golden mean character of the mind:

> The activity exercised by man as a thinking being is not merely subjective. Rather is it something neither subjective nor objective, which transcends both these concepts.... Thinking is an element which leads me out beyond myself and *connects me* with the objects. But at the same time it *separates me* from them, inasmuch as it sets me, as subject, over against them.[92]

Harmonic bonding, on the other hand, is the "mind" of natural forms. Harmonic geometry is the signature of the universal "mindedness" at work in their organization.

Randolph Stone, the founder of Polarity therapy, spoke in similar terms of this organizing intelligence and in-forming mind at work in the forms of nature and cosmos:

> Mind energy is the first essence of matter... It is the pattern energy of geometric proportions in the atomic fields of matter as the shape of things to be... The whole body is but a duplication of these patterns in a more dense form and a lower vibration key of action.... Diffused mind energy rules every cell of the body.[93]

John Woodroffe points to the same reality, called *prana*, in Hindu philosophy:

> Another form of the mental principle is Prana or Life. Though not specifically called mind, it is nevertheless the aspect of mind that is wholly immersed in matter as the directing consciousness of the material energies of the body. Prana is Paramatma, that is the Supreme Being, beyond and in bodies as their controller and director.[94]

The immaterial, gossamer-like harmonic patterns underlying physical forms reverberate the immaterial character of the zodiac—its *etheric* character. When we identify in the zodiac an etheric "structure," we understand why constellations, though composed of visible stars, are not visible *as such* to the physical eye, yet why our ancestors knew about the zodiac. They perceived it. They perceived it with the participatory consciousness that characterized humankind before the emergence of thinking and intellect during the last millennium before Christ. Prior to the Greek era, human beings experienced themselves as part of a vast unitive

---

[92] Steiner, *The Philosophy of Spiritual Activity*, ch. 4 (italics added).
[93] Randolph Stone, quoted in Burger, *Esoteric Anatomy*, p. 246.
[94] Woodroffe, *The World as Power*, pp. 103–105.

reality. They lived in symbiosis with a nature and cosmos filled with entities and gods. The organization of the thinking faculty, which began to unfold in ancient Greece, marked a significant step in the collective evolution of consciousness. The dawn of the discriminating and personalizing faculty of thinking induced a more objective perception of self and world, and a growing separation between the two. This development was accompanied by a gradual shrinking of what used to be an expansive etheric "aura" around the physical body, and a withering of the experience of being part of a universal field of life. Personalization and physicalization led to a sharpening of perception, which became increasingly reduced to physical realities. Humankind lost contact with the subtle living grounds of nature and cosmos. We no longer perceive, or only faintly, the etheric life of the world we live in. We do not perceive the zodiac. Nor do we hear the harmonies of the spheres.

Two historical figures tower above this transitional era from sentient clairvoyance to intellect, and from the wisdom traditions of the East to the philosophical reasoning of the West: Moses and Akhenaton. Before Plato, Kepler, and Newton, they pertain to a lineage of transitional individualities whose unique aptitude to take on a particularly deep breath of evolution catapulted them upon the crests of significant waves of history. Their destiny contributed to deliver a whole new phase of collective development and articulate a major leap in consciousness. This is how Steiner characterizes Moses' forerunner role, "What humankind of later epochs owes to Moses is the power to unfold reason and intellect, to be able to think intellectually about the universe out of self-awareness in a fully awake state of mind."[95]

The emergence of the thinking faculty, which would not reach full force until the eighth century BC, started as early as 1800 BC, around the time when the Vernal point entered Aries, and it was to last for the approximate duration of the zodiacal Age of Aries—from 747 BC to AD 1413.[96] Michelangelo's sculpture of Moses features his forehead crowned with the two horns of a ram (Aries), while Botticelli's painting

---

[95] Steiner, lecture in Antworten, Mar. 1911, quoted in Emil Bock, *Moses*, p. 93. Following is the full quotation: "Moses stood with his soul being before the totality of the seven soul forces, but had the task above all to imbue human evolution with a single one of them.... This was the soul force that gathers the other soul forces, previously pictured separately, together into uniform inner soul life, into the life of the "I"... "I"-consciousness, intellectuality, rationalism; reason and intelligence that are directed toward the outer sense world were to be established in humanity instead of the old clairvoyance."

[96] Vreede, *Le ciel des dieux*, p. 132. "The vernal point was in Aries from 1800 BC until the Mystery of Golgotha, but the Age of Aries lasted from 747 BC to AD 1413. Each time, there were transition periods where the Sun found itself already in the new constellation at the beginning of the spring, but could not yet extract its forces fully" (tr. by the author).

III. The Harmonic Composition of the Universe

shows two streams of light radiating from his brow, hinting at the fading organs of clairvoyance. The story of the burning bush is the story of Moses' awakening to the "I am"-ness of his being, a central theme of the symbolic path of Aries. The exodus of the Israelites under Moses' leadership heralded a taking leave of the magical practices and mystery knowledge of Egypt, which was emblematic of a larger collective step toward a next cultural age of conceptual self-awareness. This evolutionary step included the emergence of monotheism, also associated with Moses. The dawn of a single omnipotent and omniscient God in the collective psyche mirrored the arising of "I Am-ness" and individual thinking—out of the earlier sentient communion with a soulful world inhabited by deities and spirits of nature. The many divinities worshipped in the forces of nature and cosmos began to recede behind a One God figure. Commandments and laws of conduct spoke to an analogous impulse to bring instinctual drives under a "divine eye" of self-awareness.

Around the same period of the fourteenth to thirteenth centuries BC, the Egyptian pharaoh Akhenaton, whom historian Henry Breasted interestingly refers to as "the first individual in history," also promoted the cult of a supreme solar deity, Aten. Through the prototypical individualities of Moses and Akhenaton, the Logoic presence that pervades the solar system was beginning to "appear" to humankind as "One god" or "I AM"-ness—reflected in the human being's strengthening sense of "self." The undifferentiated, archaic oneness with cosmos characteristic of an earlier stage of consciousness was gradually concentrating into one transcendent source, as a concomitant centering of self was making it possible for the human psyche to "conceive" it.

Monotheism introduced a first cleavage between divinity and nature. The one god in the heights superseded the immanent divinities of the pantheistic consciousness. God was no longer in the all-ness (pan-) of nature, no longer ubiquitous in the many forms of life. This cleavage marked a major first step in the gradual separation between self and world that the development of the intellect was to bring about over the next millennium. The looming of intellectual reasoning, conceptual abstraction, and logic epitomized in the philosophy of Plato and Aristotle actualized the evolutionary impulse carried out by Moses. Aristotle, Plato's most famous pupil, took the evolutionary stream to its next step, as he departed from the metaphysical emphasis of his teacher to focus on objective knowledge and lay the first building blocks of an empirical science based on the perception of the world as separate from self, of matter as separate from consciousness. The first germ had been planted for what Max Weber would refer to, many centuries later, as the gradual disenchantment of the world. "The fate of our times is characterized by rationalization and intellectualization and, above all, by the disenchantment of the world" ("Science as Vocation," Munich University, 1918).

The Greeks held a beautiful balance between self and world, metaphysical and physical. The profoundly mediating note of their Logos—at once rational and mystical, metaphysical and geometric-mathematical, spiritual and objective—bridged the nascent gap with the graceful ethereal poise that pervades their architecture and sculpture. The perfect balance of the discerning power of light and the bonding power of tone associated with the Greek Logos initially vivified and illumined the ground of what was to be the next installment of monotheism: Christianity. Like the Greek Logos, the Christian Logos would have the meaningful distinction of being trinitarian, thus carrying into history the majestic cosmological descent of the One (Old Testament God) into a Father-Son-Spirit trinity centered on a "son of god" and his twelve disciples. Although these numbers, which are the numbers of the cosmogonical ladder, were already quietly "figured" between the lines and lineages of the Old Testament, they were not yet part of a doctrine of divinity. The trinity of Christianity ushered into theology the triune logos of the Greek. For the early Fathers of the church, the Christ was the Logos, the incarnation of the "golden mean." This living mean between the spirit and matter of God would soon find itself surreptitiously abstracted from the matter of nature and the flesh of human beings and co-opted by the mental faculty to fit the ideal imperative of a "divine" Son of God. What emerged was a Christianity bereft of an essential dimension of the Logoic "proportionality" that the God-Man meant to introduce in the evolution of humankind and the very midst of the Earth. Although the central theme of the Christ was the "incarnation" of divinity, and his central word "this is my body," the gradual separation of self and world was soon to extend to the connection between soul-spirit and body, severing what the Greek held beautifully bonded. The gradual loss of a genuine connection with cosmos, nature and body as the *dwelling of God*, was to be compounded by the obliteration of any "rational mean" to connect god and world, Heaven and Earth. Faith would be the only mediating agency acceptable. Intelligence of the divine would go underground and be referred to as "esoteric."

All these separations clearly obeyed an evolutionary necessity for human dimensions (of body, soul, spirit) and faculties to gain differentiation through phased alliances—first between mind and spirit (Christianity), then between mind and world (the last few centuries of science), to be followed, only recently, by a distinct movement toward a reunion of mind and body, and mind and nature—all stages of the archetypal path of human growth consigned to rich pictorial narratives by Renaissance alchemists. To these we will return later on.

III. The Harmonic Composition of the Universe

Set in this context, Platonic philosophy stands as an archetypal borderland—between the fading intuitive humanistic wisdom of old brought to Greece by such ambassadors as Pythagoras, and the nascent intellectual reasoning and sense-focused perception about to lay the ground for the physicalist sciences of later centuries. The task and gift of the Greeks was to transmit the cosmogonic mysteries to the emerging intellect of humankind by translating them into a conceptual–mathematical language. "The universe is present in the Temple in the form of proportion," says the ancient Hindu architectural sutra. In Greece, the intuitive perception of the harmonic ground of the world was still present, only beginning to recede in the background of a new metaphysical reasoning, as the collective consciousness grappled with the realm of "thought" and the development of reasoning and logic. On one hand, Plato frequently rests his arguments on the notion of the "reasonable"—in other words, the ability of human reasoning to discern and define. On the other hand, as Steiner points out, Greek sculptors "had a feeling—speaking as concretely as we do nowadays, this is how we have to put it—a feeling for how the etheric body, with its vital energy and mobility, provides the basis for the forms and movements of the physical body, how the etheric is reflected and revealed in the forms of the physical body, how the forces that move in the etheric are expressed in the movements of the physical body."[97] Steiner notes how Goethe became convinced that, in the creation of their works of art, the Greeks proceeded according to the same laws by which Nature herself proceeds. And it was art that would carry into the future, nested in its classical tradition of *proportionality*, the secret pulse of the world. Medieval architecture would do so most spectacularly, hosting the harmonic heart of creation in the golden geometries of its cathedrals, while painting, up until the seventeenth century, would involve a prior structuring of the canvas in proportional units. In the seventeenth and eighteenth centuries, however, this systematic method was removed from the arts. Teaching proportion in art academies was even prohibited in France. Like science, art was emancipating itself from its cosmic matrix and declaring independence.

What Plato grasped and formulated into a reasonable understanding of the world are the laws and facts of a formative etheric milieu between physical and transphysical reality. The mediating world of "ratios" and reason at the center of his cosmology hosted in a metaphysical and geometric framework what the Greek were still experiencing as the vibrant, soulful, harmonic milieu of human existence between the spiritual ground of being and the physical world.

---

97   Steiner, *Art History as a Reflection of Inner Spiritual Impulses*, p. 210.

Such a transphysical borderland is a significant reality to ponder at a time when, in a slow turning of its vast tide, the collective consciousness is challenged by science itself to release its exclusive reliance on quantitative measurements and let go of what has been its foundation for centuries: the separation, so clear to our physical senses yet fuzzy and unreal to quantum physics as well as psychological experience, between object and subject—hence cosmos and soul. Advances in the minute realm of subatomic particles have confronted our perceptual assumptions with the confounding immensity of the "small," the illusory compactness and trans-physical wavy reality of solid "matter," while the relativity of time and space, the elusive nature and wave like behavior of elementary particles have defied the classical ground of physics and challenged the most basic intellectual categories of modern science. The dissolution of the border that kept the world and its science cleanly separate from "self" has led scientists themselves to inquire into ancient paradigms and consider new theories that could accommodate this emergent trans-dual ground of reality and offer a unified understanding, not only of macro and micro physics (astrophysics and particle physics), but of matter-consciousness.

By hindsight, one can see how the wave of scientific discoveries that challenged the perceptual divide between subjective and objective world was part of a vaster wave of unifying consciousness sweeping through the entire world culture of the twentieth century. On the world stage, two devastating wars resulted in new steps in global awareness, attitudes, and concerns, leading to the creation of world organizations and the emergence of global thinking. Advances in technology opened a new sphere of instantaneous global exchange of information via *one* ubiquitous screen. Everyone on Earth can be "out" on this worldwide screen, and this screen can be "within" everyone's sight. On the inner stage of consciousness, the pre-war awakening to the existence of the "unconscious" precipitated new intimations about the human self, which challenged the psychological divide between "internal" and "external" reality. Depth psychologies began to bring to consciousness the "unknown/unconscious" forces that the psyche tends to project onto others and encounter through outside events and agencies. The rise of unifying spiritualities in the later part of the twentieth century carried the same collective prompting to step beyond outgrown divisive borders and take in the unitive breath of a whole, global being. These powerful waves of scientific, psychological, cultural and socioeconomic awareness continue to fill with the emerging facets of a more wholistic paradigm the gaps between man and nature (ecological awareness), inside and outside (depth psychology), nationalistic and global interests (political and economic globalization), person and planet—as well as the gap between divisible particulate world and indivisible field-source of reality. David Bohm remarked how he frequently had the impression that the electron sea was "alive."[98] The gates of cosmos have reopened within us, while the cosmos we perceive shimmers again with systemic intelligence (Gaia). A growing interest in ancient wisdoms, sacred geometries of nature or electromagnetic lines of the planet (to name but a few) is yielding new cohesive spiritual scientific insights to light up the vast order looming beyond the great divides.

While Plato and the Greeks were taking hold of discriminative reasoning, the unitive life of cosmos and soul trailing, still aglow, behind their blend of metaphysics and mathematics—a blend the modern mind would no doubt regard as "nebulous"—we of the twenty-first century are brought to release our hold on the exclusively physicalist thinking and particulating grids into which we have so finely laced the world reason and silenced the world being. The findings of science are pressing for new reasoning grounds beyond the limited "rationality" of the Cartesian division of reality. The subtler Reason that could account for our world-and-consciousness, matter-and-self reality would have to be grounded in understanding their "bond"—for which the Greeks offer their magnificent "ratios" and "proportion" based insights. Relation, rapport, ratio, reason are variations on the same root meaning, which points to the subtle unitive ground of our existential reality: mind, logos, soul.

The puzzling undissociability of particle detection from detecting observer, of object from perceiving subject brought center stage by quantum physics, is only one aspect of the undissociability of the multiple systems of the world from its one beingness, alive in the depths of what we call "self." Oneness is simply a profound intimation, at the heart of the mind, of the indivisibility of Being intrinsic to every living organism—and this oneness is tirelessly catching up with us, breaking down barriers to assemble new concordances, unleashing destructive extremisms to elicit new gradients of union, and opening doors for materialistic science to uncover the immaterial mysteries of solid matter and the elusive indivisibility of atomic depths.

Despite the nearsighted lenses placed by science on the eyes of humankind, the zodiac continues to shimmer in the archaic layers of our retina as a mysterious borderland behind the path of the Sun. On a purely physical basis, we can only regard the zodiac as an imaginary border between our solar system and a vaster astronomical beyond, a calendar-like background to the path of the Sun, a circle

---

98 Talbot, *The Holographic Universe*, p. 3.

of constellations whose particular groupings have no other reality than the imaginative fantasy of ancient cultures projecting their mythologies onto the heavens. Awakening to its etheric reality and the cosmic life it conveys, however, has the consequence of restoring a borderland between the physical and transphysical dimensions of the universe. It has the effect of resurrecting, from a new post-physicalist ground of awareness, the pre-Keplerian heavens and their cosmic lives cast out of the astronomical skies by the scientific revolution. The key factor, if we are to make sense of the zodiac, is to realize that it is a nonphysical organism–matrix, that it organizes physical stars into (nonphysical) groups, similar to how the etheric body and its seven nonphysical centers organize cells into organs, organs into systems that embody different functions—skeletal, nervous, digestive, circulatory, respiratory, etc. Not unlike the zodiac, such systems are not perceptible to the senses per se. They can only be discerned by that which transcends our senses and organizes sensations into meaningful wholes: the human mind, its reason and intuition.

Lit and toned anew by the Platonic cosmology, the zodiacal figures slowly passing behind the horizon of the physical world hint at the etheric, *formative* source of every visible system in cosmos. This space–time womb of harmonies is at once our "fish bowl" and a grand Vesica Pisces between the physical universe and the nonphysical dimensions of the cosmic Being whose manifestation it is. Acknowledged yet not perceptible as such, spatial yet immaterial, the zodiac nudges the mind beyond the dual world of space–time physicality ushered into existence through its harmonic whorl. Radiating through this shimmering auric sphere is the vibrant presence of the transphysical essence of the universe—the universe's being.

The 12fold/7fold harmonic composition of the human body

The reification of this harmonic auric field into the astrophysical vault of modern times is but the other side of the noble potential of a science acutely alert to physical reality—the potential to discern and verify if the uniquely coherent psycho-cosmology presented by Pythagoras and Plato is reflected in physical facts. Can we actually see in the physicality of world and body evidences of the fundamental ratios and intervals, which, according to Plato, play out the process of creation and constitute its very ground?

Physical measurements have indeed consistently shown that, as the 12fold Logos pervades the organic structures of nature with its golden mean signature, the 12 petal lotus at the heart of the human "plays out" its harmonies in the proportions of our bodies. A net of dodeca–pentagrammatic proportions in our physical and etheric organization shows that the soul of the world informs and in-shapes our constitution.

III. The Harmonic Composition of the Universe

A number of researchers have focused on the "sacred geometry" embedded in the forms and bodies of nature and concluded that the ideal human proportions lie in the canon defined by Phi, the golden section. "Each normal skeleton," says Matila Ghyka, "reveals a perfect symphonic design or harmonic theme, a symphonic commodulation of proportions either in the golden section theme or in that of the consonant √5 proportion." He also emphasizes the important part the dodecahedron plays in the symmetry of the human body.[99] The construction of the Dodecahedron is based, we saw, on the internal angle ratio of 2/3 (or 3/2), which ratio, very close to the musical fifth (0.667 versus 0.618), will prove to be identical in essence. The great harmony of Heaven and Earth is embedded in our physical form.

The divine proportion is famously present, for instance, in the rapport of arm length to fingers to knuckles. In the average human hand, if the length of the smallest segment of any finger is given the value "1," the length of the second segment will measure approximately 1.618 times longer than the first; the third segment approximately 1.618 times longer than the second; and the fourth segment in the back of the hand 1.618 times longer than the last. The proportion between forearm and upper arm or lower leg and upper leg is also Phi. Phi ratios abound in the ideal human face: length to width of face, lips and eyebrows to length of nose, length of face to tip of jaw and eyebrows, length of mouth to width of nose, distance between pupils to distance between eyebrows. Leonardo Da Vinci found that the total height of the human body, from toes to top of head, and the height from the toes to the navel are in golden ratio. So is the height from toes to navel, to the height from the navel to the top of the head. A study involving 207 students in Munster, Germany, confirmed these proportions: the value of 1.618 was obtained for the former ratio average and 1.619 for the latter. This value held for both girls and boys of similar ages.[100]

German composer Peter Hamel points out how, while the navel divides the body according to the 3:2 ratio, the nipples divide it roughly according to the 4:3 ratio, which, he says, "corresponds not only to the musical fourth but also to the relationship between the synodic periods of Mars and Venus. For every proportion of the human body that has a musical counterpart, a corresponding relationship between two or three planetary cycles can be demonstrated."[101]

Logically and marvelously, the human body is the physical expression of the quintessentially human: the soul. The traditional representation of the body stretched around the zodiac suggests an ancient awareness of the subtle 12fold form

---

99 Ghyka, *The Geometry of Art and Life*, pp. 97, 50.

100 Davis and Altevogt, "Golden Mean of the Human Body."

101 Hamel, *Through Music to the Self*, p. 122.

of the human being, an awareness that microcosm and macrocosm are one identity, and that soul and zodiac are two facets—the core and the circumference—of this identity. How eloquent that the first central formation to appear in the median region of the embryo is the group of twelve ribs arching around the heart.

According to the eighteenth-century Master Hakuin Ekaku, one of the most influential figures in Japanese Zen Buddhism, a twelvefold pulsation animates the human body. "The pulses of the body are twelve-branched," he taught, "corresponding to the twelve months of the year and the twelve periods of the day."[102] On the other hand, graphs of the heart rhythm show that the ventricles reset themselves at the golden ratio point of its cycle.[103] Given the essential link between golden ratio and twelvefoldness, these apparently unrelated observations point to one and the same reality, which is the "mean" and milieu of the soul principle, the human "identity," the human "I."

The golden ratio is not only the fundamental proportion of the human skeleton, but of the skeleton of all living creatures. From insects to the largest creatures upon the planet, all are in-formed by the epitome of harmonic proportion—3:2—Phi. The works of the Hungarian architect Gyorgy Doczi demonstrate this fundamental harmonic structuring most stunningly. We will come back later to his detailed measurements of plant forms, insects, fishes, animal and human bodies, which show how every living form is "bound together" by harmonic proportions.

The proportionate geometries of the physical body—in other words, the Earthly (*geo-*) measures (*-metry*) of beings—display the repercussion in the organization of matter of the etheric currents that transmit the vibrancy of life and modulate the relation between psychospiritual and physical realms. In all living beings, the connection between the "animating" realm of the soul and the four elements of physicality (solidity, liquidity, air, and warmth) is the etheric "quintessence." In animals, the psychospiritual dimension is not individualized, which accounts for the different—though equally harmonic—organization of their systems. In humans, the etheric body is at once the quintessence of physicality and the quasi-corporeality of the "self." It "binds" body and psyche through the seven major chakric centers. From the Platonic perspective, which views the universe as a corporealization of the world soul, it would be logical to find that our etheric organization is structured in accordance with the major harmonic ratios. The research studies that follow will confirm that the seven chakras form a 7fold unit-structure, analogous to the 7fold

---

102   See http://www.scribd.com/doc/60202616/Yasenkanna-Compare.
103   See http://www.british-israel.us/34.html.

III. The Harmonic Composition of the Universe

diatonic scale, harmonically binding our Heaven and Earth—the crown and base of our being—into a human octave.

Before exploring the harmonic character of the chakric structure, we must clarify, in anticipation of the reader's experience of a perhaps confusing shift from the expected revelation of a 12fold structuring of the etheric to that of a 7fold harmony, that the numbers 12 and 7 are equally essential, as *a pair*, in the harmonic binding of polarities that results musically in the diatonic and chromatic scales. The combination of 12 fifths and 7 octaves is what generates both scales. At the hidden center of this 7fold-12fold pair is the "divine proportion" of the third harmonic. A later phase in our exploration will shed light on the respective specificity of seven and twelve in terms of the realms they organize, by taking us, through another trail, to the root of the perplexing blend of the two that confronts any inquiry into the structures of time and space, world and consciousness.

The most direct access to the harmonic character of the 7fold ladder of chakras is through the analogical evidence of correspondences. While the analytical approach offers the kind of particulate evidences suited for the divisible worlds, the way of correspondences and analogy is the suitable approach for the indivisibility of life's compositions—the way to grasp and correlate the parts from the perspective of their homologous function in the "whole" they are part of. "Traditionally," Fideler writes, "the Book of Nature is read through the faculty of analogia or proportional insight, the formal expression of which is the Hermetic science of correspondences."[104] The notion of proportion (logos) at the core of the word *ana-logia* illumines in "analogy" the royal method of approach to the inner workings of a world governed by proportion. Analogy is literally the logic that traces the world up (*-ana*) to its Logos.

> Analogy is the most beautiful bond, muses Proclus...it imparts a power that causes all things to have sameness and union. For through analogy the universe is completely rendered one, this having the power of making things that are divided to be one, of congregating things that are multiplied, and connecting things that are dissipated.[105]

The correspondence between the 7 major centers and the 7 notes of the scale has become a familiar one in meditative practices involving energy clearing and chakra balancing. These correspondences extend to the color spectrum, the visual counterpart of the harmonic spectrum (7 major colors of the rainbow and 5 minor "tones"), and are generally viewed as follows:

---

104 Fideler, *Jesus-Christ, Sun of God*, p. 20.
105 Proclus, *Commentary on the Timaeus of Plato*, 2.16, 2.27.

C-do	root	red
D-re	belly	orange
E-mi	solar plexus	yellow
F#-fa	heart	green
G-sol	throat	turquoise
A-la	brow	indigo
B-si	crown	violet

As in all systems of correspondences, variations exist in this attribution. Esoteric teachings frequently present different schemes of correspondences appropriate for different levels of consciousness, for instance. Manly Palmer Hall remarks that two facts support the accuracy of the above arrangement: 1) The three fundamental notes of the musical scale—the first, the third, and the fifth (do, mi, sol)—are in correspondence with the primary colors—red, yellow, and blue; 2) the seventh, and least perfect, note of the musical scale corresponds to purple, the least perfect tone of the color scale.[106]

Doczi's investigation of the same archetypal correspondence offers another set of results, yet a similarity of pattern. Sound vibrations of one octave of the keyboard, from G to F, tabulated together with the seven spectral colors show the correspondence in pattern. Color wavelengths are measured in angstrom units, sound vibrations in cycles per second (fig. 33).

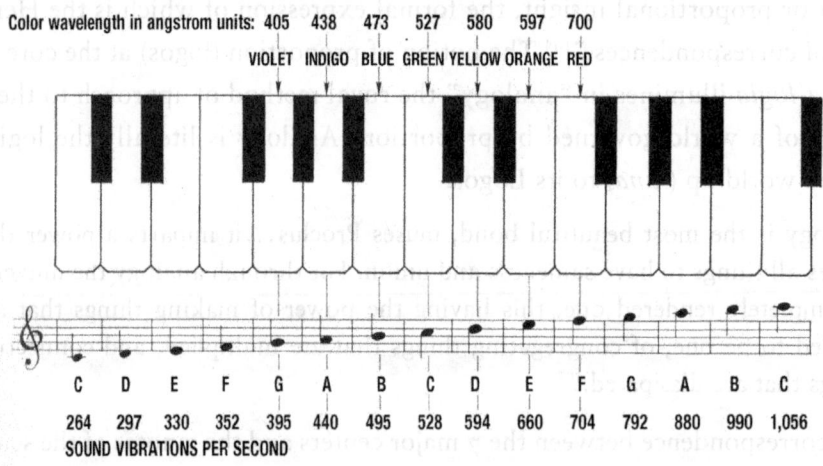

Fig. 33: *Correspondences between musical tones and colors*
(adapted from Doczi, The Power of Limits, p. 50)

A simple way to perceive the harmonic setting of the chakras is to experience the bodily vibrations associated with sounding the vowels. The resonance between vowels and regions of the body on one hand, and the correspondences established

---

106 Hall, *The Secret Teachings of All Ages*, p. 259.

III. The Harmonic Composition of the Universe

between vowels and the harmonic series on the other hand, provide tangible evidence of the harmonic character of our chakric "spheres."

The connection between vowels and harmonics can be easily experienced in the process of vowel formation associated with the production of overtones. Anyone trying to produce overtones while singing a bass-note will find that the emission of higher or lower pitched overtones depends on the vowel sung, or rather on the mouth position used to produce a certain vowel. The formation of vowel sounds engages small spaces in the mouth, which act as resonators, differently sized for different harmonics. Traditional or meditative forms of chanting involving overtone production, such as the throat singing from Tuva, Mongolia, and Tibet, or David Hykes' Harmonic Choir, also expose this connection between harmonics and vowels. In researching Mongolian and Tibetan chant, Peter Hamel found "that the gradual transition from the lower to the higher overtones corresponds to the process of vowel-formation leading from U (OO), via O, A, and E to I (EE). There is thus a correspondence between the overtones, the mathematical ratios 2:1, 3:1, 4:1 and so on, and the vowel-sequence U, O, A, E, I." Hamel's measurement of the average frequencies associated with each vowel confirmed their lining up along a harmonic spectrum.[107]

Vowels and Average resonance frequency (Hertz):

Vowel	Hz
U (Oo)	300 Hz
A (Ah)	500
Ae (Eh)	1,000
Oe (Er)	1,350
O (Aw)	1,550
Ue (French u)	1,800
E (Ay)	2,100
I (Ee)	3,000

Joscelyn Godwin explains that when we hear vowels, what the ear actually registers is harmonics. "Somewhere in our perceptive mechanism the harmonics of vocal tones are analyzed and presented to us as vowels.... It is after all our most direct perception of the numbers which underlie not just the acoustical world, but the whole physical universe."[108] On the same basis of acoustical physics, the tone quality of an instrument is tied to the particular harmonics it emphasizes: fundamental for the flute, second harmonic for the French horn, odd-numbered harmonics for the clarinet.[109]

---

107 Hamel, *Through Music to the Self*, pp. 102, 124.
108 Godwin, *Mystery of the Seven Vowels*, pp. 16–17.
109 Ibid., p. 12.

Meditative practices have revived the ancient knowledge that each vowel opens a certain region of the body. The production of overtones through the mouth-positions associated with specific vowels, while concentrating on the resonating place in the body, facilitates an awakening of the corresponding psychic centers. "When imagined during the in breath and spoken inwardly, it allows the breath to penetrate more deeply in the region concerned.... The U fills the lower regions of the body, the I the uppermost regions, etc."[110] Experiments carried out in schools in Germany have shown the following correspondences:

I:	Head cavity
E:	Throat and upper chest (but also the sides)
A:	Chest cavity (but also the body as a whole)
O:	Abdomen (as far as the navel)
U:	Pelvis and lower body

Peter Hamel concludes, "since the vowels which are themselves associated with given parts of the body correspond to the harmonic series, it seems reasonable to assume that *the bodily proportions between their associated inner centers also reflect the harmonic series.*"

What emerges from Hamel's experiments is that the chakras represent a fundamental harmonic differentiation of the etheric body into 7 major energetic centers, analogous to the prismatic refraction of the light into 7 colors and the harmonic differentiation of the solar system into a planetary sevenfold, thus grounding in detectable facts Plato's account of the 7fold polarity of the process of corporealization of the world soul.

Experiments such as Hamel's shed light on the ancient reality of the 7 vowels, each of which was traditionally associated with a planet. These seven vowel sounds formed the basis of the sacred sounding by which the ancient Egyptians initiated their rituals. "They confirmed their sacred songs to the seven primary sounds," explains Manly Hall, "forbidding any others to be uttered in their temples." One of their hymns contained the following invocation, "The seven sounding tones praise Thee, the Great God, the ceaseless working Father of the whole universe."[111] The names of God were conceived from combinations of the seven planetary harmonies. Similarly, it was with the seven vowels, the early Roman historian and theologian Eusebius points out, that "the Jews sought to express the name of God that cannot be spoken, but they reduced these to four for the

---

110 Hamel, *Through Music to the Self*, pp. 115–126.
111 Hall, *The Secret Teachings of All Ages*, pp. 255–256.

use of the multitude." The same Eusebius quotes the following verse from a Greek ode, "The seven vowels celebrate me, the Great imperishable God, indefatigable Father of all. I am the indestructible lyre, having tuned the songs of the celestial vortex."[112] We will later see how strikingly accurate the term *vortex* is. Many early instruments did have seven strings, which had their correspondences in the human body and the planets.

The 7fold "vowel-ing" of the human voice and the 7fold chakric system reproduce on a small scale the structure of the solar system—the 7fold planetary body of the solar Logos, the body of "God." By intoning the vowels to invoke the planetary gods, the Egyptians sought to tune their etheric and astral bodies, as well as the whole etheric-astral field of their culture, to the solar systemic harmony, in a way reminiscent of the ancient Chinese. Practices similarly aimed at harmonizing Heaven and Earth were also central in the Chaldean culture. In parallel, the architectural conception of the Egyptian temples demonstrated the priests' superior knowledge of the principles underlying vibration. Manly Palmer Hall points out how special sound chambers were constructed for the invocations central to their mystery rituals. "A word whispered in those apartments was so intensified that the reverberations made the entire building sway and be filled with a deafening roar."[113]

Essentially, the seven major musical intervals underlie what ancient traditions worldwide refer to as seven fundamental levels of being and consciousness—seven levels of the fundamental "one." This archetypal sevenfold set is reflected in the human constitution: physical/base center, vital/sacral center, emotional/solar plexus, mental self-awareness/throat chakra, soul intelligence/ajna center, universal love/heart chakra, spirit will/crown chakra. Plato's lens relates the seven to the four elements (the first four), their quintessence (soul/heart) and the two core dimensions whose polar interplay begets the quintessential trinity of being. Here again, it must be emphasized that there are no absolute correspondences, only fluid and relative parities between major differentiations in forms and systems whose common ground is that they constitute indivisible wholes: a plant, a human body, an octave, light. Rather than fixed equivalents, correspondences are archetypal resonances across different realms of reality whose seven dimensions give expression to an ontological order that transcends its many variations. The essential correspondence lies in the pattern of differentiation rather than the differentiations per se.

---

112 Eusebius, quoted in Godwin, *The Mystery of the Seven Vowels*, pp. 21-22.
113 Hall, *The Secret Teachings of All Ages*, pp. 255-256.

A remarkable similarity emerges also between the petal structure of each successive chakra and the cyclical pattern created by a particular planet's alignments to the Sun—from the Earth's perspective—as it cycles around the zodiac.

> The 2fold/4fold pattern of conjunction-opposition of the Moon with the Sun is evocative of the 4-petal pattern of the base chakra.
> The 3fold/6fold pattern of superior and inferior conjunctions of Mercury with the Sun, as the planet completes a cycle around the zodiac, is reflected in the 6-petal sacral center.
> The 5fold/10fold pattern of superior and inferior conjunctions of Venus with the Sun over its eight year cycle around the zodiac mirrors the 10-petal structure of the solar plexus.
> The 12fold cycling of the Sun around the zodiac corresponds with the 12-petal character of the heart center.
> The 8fold/16fold pattern of Mars' alignments with the Sun reflects the 16-petal nature of the throat center.

As for the last two planets, Jupiter and Saturn, the number of petals of the corresponding chakras—2-petal Ajna center and 1,000-petal crown center—seem to reflect primarily the fundamental Two-ness and Oneness that precede the corporealization of the world soul. The 12fold pattern of Jupiter is also reflected in the 96 (48 x 2 or 12 x 8) petals of the Ajna center.

Chakras have been described as rotating color wheels or circular concentric interference wave patterns. The paths of the planets speak to chakric dynamics of our solar system. Like musical notes, chakras–vowels–planetary orbits are vibrant whorls of harmonious wave conjugation. The simplest ratios of wave interference yield the seven diatonic notes, which represent the major intervals of integrated differentiation of the world soul. Lesser intervals do not resonate in us with the same power. The unique quality of each note expresses a special "rapport" to the whole—a facet of "wholeness," a fundamental modality and function of being. As the Platonic solids illustrate, each interval organizes matter in a singular geometric fashion. The glands and organs associated with each chakra, and the purposes associated with the corresponding planets, offer intimations about these functions.

> *When the Buddha was born, he walked seven steps forward and with each of his steps, a lotus flower appeared on the ground.*

As a whole, the etheric vehicle bonds body and spirit—the human octave—through seven psycho-physiological spheres. As a unit, each chakric whorl attunes a certain domain of body-psyche to its macrocosmic prototype, shaping organs, organizing physiological systems and energizing their functions so as to inform and entrain an individual expression of the macrocosmic sevenfold.

The question is: Is there any scientific data to support the idea that harmonic waves of vibration might shape physical organs? A number of experiments have indeed brought into the realm of noble logic the notion of a vibratory energetic body shaping and vitalizing the physical organism. The power of sound to organize substance was first demonstrated by the German physicist and musician Ernst Chladni. His most famous experiment, now a classic in high school physics, consisted in drawing a bow over a piece of sand-covered metal until it reached resonance, at which point the vibration caused the sand to concentrate along "nodal," still lines, in patterns referred to as *Chladni figures*. Buckminster Fuller, and later, the Swiss doctor and scientist Hans Jenny demonstrated the universality of Chladni's discovery by extending his experiments to gas and liquids, and showing the power of sound vibration to organize matter into harmonious designs reminiscent of natural patterns—from mandala structures evocative of snowflakes to the early stages of cell division. Jenny was especially impressed by the observation that vocalizing the ancient Sanskrit syllable OM (regarded by Hindus and Buddhists as the *sound of creation*) would cause the lycopodium powder he experimented with to form a circle with a central point—the traditional graphic representation of the Sun: ☉. Photographs of standing wave patterns created by vibrating a medium have shown visual analogues of the distribution of spots on a leopard, the geometric designs of plants and flowers, even the intricate patterns on the shell of a tortoise.

These experiments led to the creation of a field of research called Cymatics (after the Greek word *kyma* meaning "wave," and *ta kymatica* meaning "matters pertaining to waves"). By exploring the effect of sound waves on matter, cymatics has been orchestrating the principle that vibration is the source of all forms of nature.

The works of the Swedish engineer and water researcher Theodor Schwenck also speak eloquently to the formative power of waves, showing the organic-like forms generated when sound waves are sent into a body of water or air: from mushroom to spine to larynx looking formations. We will look at photographs of his experiments in the fourth part of the book.

Masaru Emoto, the author of *The Hidden Messages in Water*, went further, engaging more subtle levels of vibration to demonstrate the formative influence of waves. He discovered that the crystal structure formed when water droplets freeze could be significantly impacted by exposing the water to vibrational energy in the form of music and emotions. Exposing crystallizing water to harmonious music dramatically affected the shape of the crystals: delicate crystals emerged with Mozart, distinct geometries with Beethoven, and elegant details with Chopin.

Similarly, exposure to positive emotions and words led to symmetrically shaped crystals, while negative thoughts produced malformed, disjointed snowflakes. His experiments demonstrate *physically* what many recognize intuitively: that emotions are vibrations.

THE THREEFOLD GROUND OF THE ARCHETYPAL SEVENFOLD

Musical scale and color rainbow extend a parallel archetypal ladder between the multiplicity of forms and their one source. Like harmony and intervals, light and colors provide not only a field of observation for the archetypal differentiation of "unity" into cosmic multiplicity, but an avenue to trace this differentiation back to its earliest model: the very first step of the world opening—from Oneness to tri-"unity"—from one to two to three. Goethe was the great diviner of this primordial phenomenon of tri-union at the root of the emergence of colors.

When Goethe said, "Colors are the deeds and sufferings of the light," he gave expression to the dual gesture of cosmogenesis he experienced intuitively in the transition of the light into colors: externalized cosmos ("deed") and internalized soul ("suffering"). His well-balanced, scientific, and poetic mind experienced colors as an objective diffracting of the light and a subjective differentiating of the soul.

The great distinction between Goethe and Newton's approach to color is that Goethe sees colors arising from the encounter and interplay of darkness and light. Color is a certain *rapport* and "ratio" of light to dark. "It didn't take long," he notes, "before I knew that a border was required for color to be brought forth." By contrast, Newton sees colors as intrinsic components of light disclosed by the prism. For Goethe, light is the most homogenous, undivided reality. For Newton, light is composed of colors. Goethe's experience of colors as different proportions of dark and light brings subjective impressions to the mathematics of harmony: "Yellow," he says, "is a light which has been dampened by darkness; Blue is a darkness weakened by light."

Revealing itself in these impressions is the cosmological process of spirit dampened by matter and matter lightened by soul. Goethe was a phenomenologist in the "divine" meaning of the word. In his approach to the mystery of color or plant, he drew near the very "phainein" of phaenomena: the divine mystery of "appearing," the passing into visibility of the invisible, the wrapping of a God in the layered garments of creation. Goethe's scientific method of identifying with the realities of nature as "appearing" advents allowed Phanes to reveal his pathways into visibility, tangibility, physicality. The royal way into manifestation of the God who is himself

III. The Harmonic Composition of the Universe

a "third," an offspring of the world parents, is indeed a blending of polarities—light and darkness. "God separated heaven from Earth. And there was light."

The threefold unity of the "light of the world" issued from the primordial ground of "Heaven and Earth" deploys the colors of cosmos by combining itself with its complementing other: darkness. As the third emerges from the interplay of same and other, so do the 7 colors of creation and 12 colors of the soul arise from the interplay of 3fold light and 4fold matter.

Again musical harmony powerfully illustrates the notion that the elemental 4fold world is the "other" of the trinity, by showing how the fourth harmonic, which is also the fourth interval, is the "other side" of the third harmonic (the fifth interval)—the descending interval implied by the ascending fifth within the octavian whole. As notes are born from the interplay of fundamental and octave, colors are born out of the interplay of light and dark, and forms of consciousness from the interplay of spirit and matter. The 1–2–3 metaphysical cosmogony of the Pythagoreans plays itself out in the domain of light as it does in the realm of harmony. It is the great "phaenomenon" behind all phenomena.

Not only did Goethe see polarities at play in natural phenomena, he also felt their inner reality as the deeds and sufferings of the soul of the world. In the stepping out of Being into different *forms, tones, and colors*, he experienced the concomitant internalization of different *states* of consciousness.

The soulful radiance of the world alive in Goethe's mind and alight in his intuition of the threefold reality of light-darkness-color is also the source of his insights into the archetypal entity of the plant. What allowed him to experience in the colors the enactments and moods of the light is also what led him to intuit in the plant seven stages of unfolding of one archetypal plant-being (*Urpflanz*): stem–leaf–calyx–petal–stamens/pistils–fruit–seed. In each case, he grasped the underlying dynamics of an essential interplay of polarities (light and dark, Earth and Sun) and "divined" how key proportions of such interplay yield a 7fold whole of colors or plant stages.

Echoing Goethe's findings from a different perspective, cosmologist Arthur Young has pointed out how the angiosperms (flowering plants) show in cross section 7 layers of tissue—from outside to inside: Sepal, petal, stamen, pistil, ovary, ovule, gametophyte, egg nucleus. "Like the seventh-row atoms," he remarks, "the flowering plant has 7 layers, or 'shells.' The layers have an additional significance in that they also represent stages in the life cycle of plants. Thus the plant starts as a seed, it grows, it blossoms, it is fertilized, it fruits, and finally the fruit decays, leaving the seed which still contains the germ."[114]

---

114 Young, *The Reflexive Universe*, p. 107.

Rudolf Steiner, who was the science editor for the Kurschner edition of Goethe's works, highlighted the core feature of Goethe's insights: magnetic polarity. "Goethe pictures to himself that light and darkness relate to each other like the north and south pole of a magnet. The darkness can weaken the light in its working power. Conversely, the light can limit the energy of the darkness. In both cases color arises."[115]

Goethe's archetypal eye saw in the shades of colors and the stages of plants the spectrum of creative rapport and harmonic bindings between the great extremes of light and dark (colors), Sun and Earth (plant phases)—a spectrum of seven deeds and sufferings in the relation of spirit and matter.

The stages of the plant also revealed to his passion for the "appearing god" a remarkable interplay of qualities associated with solar and earthly forces: expansion and contraction. He recognized how "the cotyledons begin in a *retracted* state, then how the main leaves, and their spacing on the stem, mark a *first expansion*. The bunching of leaves to form the sepals at the base of the flower mark a *second contraction*, and the elaboration of petals a second *expansion*. The drawing in of leaf size to form pistils and stamens marks a third *contraction*, and the formation of fruit the last and most exuberant *expansion*. The contracted seed within the fruit then starts the cycle again in the next generation." He saw in these different expressions of the plant one organism unfolding varying functions and forms. "The same organ that expanded on the stem as a leaf and assumed a highly diverse form will contract in the calyx, expand again in the petal, contract in the reproductive organs, and expand for the last time as fruit.... The organs of the vegetating and flowering plant, though seemingly dissimilar, all originate from a single organ, namely, the leaf."[116]

Between the one and its sevenfold deployment, Goethe grasped the primary threefold dynamics of the phenomena of color and form. In color and form, he saw a "Third" born of Two, and he discerned the intercourse of the Two from which the third is born: color from the interplay of light and darkness; form from the interplay between the contracting forces of the Earth and the expansive forces of sunlight.

The physical eye alone could not have perceived this threefoldness. Only the inner eye "senses" deeds and sufferings, only the soul's sight sharpened by the existential interplay of actions and feelings can recognize its own passive and active modalities in the color happenings of the world. Only the soul's eye sees the between of things, where the divine "milieu," the world soul, appears. Similarly,

---

115 Steiner, *Goethean Worldview*, part 3.
116 Goethe, *The Metamorphosis of Plants*, lines 115–119.

III. The Harmonic Composition of the Universe

the ear may hear the notes, but it is the soul that resonates with the intervals, and interval is where music lives.

Only a unitive sight, at once objective and subjective, could behold the deeds and sufferings of the world soul in the dark and light blends of the world body—and unveil their common cosmological Logic. Only the threefold light of the soul could offer a suitable milieu for the triune light of nature to be "seen," and for the world soul of which it is the radiant shroud to be felt in the colors of the world.

The contrast between the Goethean and Newtonian perspectives illustrates the transition from an organic rapport with the world as the psycho-cosmological manifestation of a being, to an abstract representation of the world as an assortment of objects, cut off from inner dynamics of formation and becoming, and spread out for dissection on the table of physics. Time, the realm of harmonies, imperceptibly collapses into space as the Newtonian laws open a new chapter in the history of the world mind. With Newton, the fluidity of becoming that characterized the ancient understanding of the world as a descent of Being into spatial forms by way of harmonic means is about to freeze into spatial simultaneity under the cold bright scrutiny of sense perception and quantitative measurements that will define modern physical sciences.

The ladder of chakras is the expression in the human constitution of the archetypal sevenfold of colors, leaves, notes, and vowels. The old stanzas of Dzyan re-introduced by H. Blavatsky state it most unequivocally, referring to the human being as "saptaparna," the *seven-leaved plant*. The symbol of the human septenary (sevenfold) is a triangle (or *triad*) above a square (or *quaternary*)—reflecting the four elements and their 3fold quintessence, "the heart of the man-plant."

> When the one becomes two, the "threefold" appears. The three are (linked into) one; and it is our thread, O Lanoo, the heart of the man-plant, called *Saptaparna*.[117]

Vegetal and human plant play out successive phases in the unfolding of the archetypal scale that binds the great polarities of Heaven and Earth, indivisible being and visible forms, into bio-psychospiritual sevenfold units of life. In turn, each phase-kingdom actualizes on its own terms the sevenfold ladder. Every entity of every kingdom initiates this 7fold differentiation by recapitulating the cosmological opening of the One into a threefold, which, via a twelvefold differentiation, becomes the quintessence of four elements. With the inner three, the four add up to seven.

In reference to the triad, Steiner introduces new terms that speak to the inner edification of the human sevenfold: "The square," he says, "is the symbol of the fourfold nature of man: physical body, ether body, astral body, and ego. The triangle

---

117 Blavatsky, *Secret Doctrine*, vol. 1, stanza 7, p. 231.

is the symbol of the higher human: spirit self, life spirit, and spirit man [or spirit body]." The higher three are the transformation of the lower three by the self at the center: spirit self points to the universalized mental self, or "I"; life spirit to the transmuted, lighted astral body; spirit body to the spiritualized etheric–physical.

The essential threefoldness of being exalts in every sevenfold three major centers: two extremes or poles and a mediating center. In the musical scale, the fifth has the quintessential status of the "dominant." In the chakra system, the heart is the "third" of the major three: head center, base center and heart center replicate the triad of Same, Other, and Being—Father, Mother, Son. Indeed, Alice Bailey states, "The thread of life is anchored in the heart"—the *triune* thread of life is anchored in the heart. Similarly, humankind is destined to be the heart of the seven kingdoms of Earth, the (soulful) "dominant," seeded with the golden Logoity, whose inner reality, we are about to see, is love. Love—the soul thread of the universe—is anchored in humankind.

The golden mean—
The geometric vessel of love, the soul of life

> *Love is the most universal, the most formidable and the most mysterious of cosmic energies... one pretends to ignore it, but it is surreptitiously everywhere. Considered in its full biological reality, love—that is to say, the affinity of being with being—is not peculiar to man. It is a general property of all life and as such it embraces in its varieties and degrees, all the forms successively adopted by organized matter.* —Teilhard de Chardin[118]
>
> *Behind all outer seeming, the motivating power of the universe is love.* —Alice Bailey

The simplest and most stunning evidence, to ancient and modern minds alike, that life uses the harmonic blending of polarities to pass, whole, into a *"form* of life," is the conception of a human being through the intercourse of masculine and feminine energies embodied in complementary forms of physicality.

The transmission of life through the vibrant secting-bonding ("sexual-love") of human intercourse illumines the inner realities of the golden mean core of the world: life and love. Life is the spirit of the golden mean, love its soul. Golden sected bonds vessel the love that is the Life of our universe. The triunity of Creative "Father,"

---

118 De Chardin, *Phenomenon of Man*, p. 264.

III. The Harmonic Composition of the Universe

Receptive "Mother" substance, Bonding "Son" consciousness is reflected in the spirit, body, and soul of its geo-metric expression, the golden mean.

Wilhelm Reich observed, musingly, "we all owe our existence to the enigma of love." The mystery Reich intuits in this strikingly simple and universal fact is indeed nothing less than the aperture into the grand symphonic secret of the cosmos.

Carl Jung plumbed the same cosmological depths in the mystery of love when he remarked that, while he had never been able to explain what it is, "we are in the deepest sense the victims and the instruments of cosmogonic love." He saw in Eros the grand cosmic force that generates beings and creates consciousness. "Eros is a kosmogonos," he writes, "a creator and father–mother of all higher consciousness."[119]

Alice Bailey precipitates these intimations in the arresting statement that at the heart of every atom is love. "The principle of buddhi, of cosmic love," she says, "is in a mysterious way the principle found at the heart of every atom."[120] "For the positive central life of every form is but the downpouring of a love which has its source in the Heart of the Solar Logos, itself an emanating principle from the One above our Logos, He of whom naught may be said."[121] At all levels, heart centers represent mediating hubs of creative intercourse between the fundamental polarities engaged in any "particle" or system of cosmos.

The mathematics of the Pythagorean means shed interesting light on the human realm of mediation and love. As mentioned earlier, the "means" capture three mathematical-musical possibilities of mediation of the octave. The harmonic and arithmetic means give the fifth and the fourth, and the geometric mean produces the tritone interval. We also saw that, in order to express the major intervals and means in whole numbers (integers), the Pythagoreans represented the octave by 12:6 (half a string length).[122] In this setting, the harmonic mean 8 produces the fifth (12:8) and the fourth (8:6) and the arithmetic mean 9 produces the fourth (12:9) and the fifth (9:6). The third or geometric mean divides the octave exactly in half (six semi-tones or three tones). It is helpful to experience these means by simply playing or singing the corresponding notes, which, in a C major scale, would be G or sol (fifth), F or Fa (fourth) and F# (tritone).

---

119 Jung, *Memories, Dreams, and Reflections*, p. 353.
120 Bailey, *The Rays and the Initiations*, p. 415.
121 Bailey, *Treatise on Cosmic Fire*, pp. 1225–1226.
122 See earlier section, "Modern insights into the harmonic structure of Platonic solids and atomic elements."

It is not indispensable for the reader to grasp in detail what may feel like abstruse arithmetic. The mathematical facts are only meant to show that, from the Pythagorean harmonic perspective, there are three kinds of mediation. Each of these means, when pondered, proves to carry a different "meaning." Deployed in the fifth, the fourth and the tritone are three archetypal ways to negotiate the polarities of matter and spirit: the ideal, generative mediation of the golden fifth, the grounding mediation of the fourth, and the quantitative division of the tritone. We will gain clarity about the meaning of the fourth at a later turn of our journey. For now, we will look at the tritone and the fifth.

The tritone harbors a precious insight into mediation. Our modern mind would readily think of it as the most normal mean of mediation: a division in half. Yet, the note obtained is very dissonant. It was referred to as "diabolus in musica"—the devil in music—and was avoided, even prohibited from musical compositions for centuries. This geometric, half-and-half division—as a mean to "mediate"—evokes the provocatively "diabolic" mediation proposed by Solomon when two women came to him for judgment, the Bible recounts, each claiming to be the mother of a baby they brought to the king. Each had given birth to a son a few days earlier. One child had died during the night. One woman alleged that the other substituted her dead child for her own healthy infant. The second woman insisted the living child was hers. Solomon, who, the Bible says, was granted by God "a mind able to discern between good and evil," and "whose wisdom surpassed all the wisdom of Egypt," asked for a sword and ordered that the child be cut in two. One half should be given to one mother, one half to the other.

> The woman whose son was alive was deeply moved out of love for her son and said to the king, "Please, my lord, give her the living baby! Don't kill him!" But the other said, "Neither I nor you shall have him. Cut him in two!" Then the king gave his ruling: "Give the living baby to the first woman. Do not kill him; she is his mother." When all Israel heard the verdict the king had given, they held the king in awe, because they saw that he had wisdom from God to administer justice. (1 Kings 3:16–28)

The "geometric" mean to resolve the argument, clearly the most "unfitting" mean to mediate the indivisibility of life and the relatedness of love, did however what nothing else could have done so powerfully: it revealed the truth, the heart of the truth—which always touches on life and love. Not only did the spontaneous self-renouncing response of the mother identify her as the true mother, but it illumined the truth that love is the only viable "milieu" and mean of life. True love, willing to cut off its separate claim in order to foster the wholeness of life, exposed the selfish, "evil" eagerness of the second mother to deprive the first of the child

III. The Harmonic Composition of the Universe

she no longer had. Good and evil appeared, a simple line drawn in the sand of human nature by Solomon's sword of wisdom, between that which fosters wholeness, and the egotistic force that compulsively grabs its inevitably "dead end."

While the purely dividing, geometric mean speaks to a lethal absence of bond, it emerges in the end as a revelator of the truth. Lucifer continues to be the bringer of light. Humankind becomes enlightened, Jung wisely pointed out, not by seeing beings of light but by bringing light into the dark recesses of the psyche. Any appeal to divisiveness will reveal the worst—and conjure up the best. History has shown, horror after horror, what the Biblical story suggests, that evil eventually elicits and fertilizes the core truth of humanity: brotherhood. Brought to the dead end of separate urges, humankind reawakens to itself. The better wakes up out of its fall into the worst. The "devil in music" is indeed a mean to harmony, and the potential for evil an aspect of the quintessentially human, a corollary of freedom. The freedom to side with self-centered drives and impulses is a necessary part of the evolutionary adventure of "selfhood."

Separation per se, however, cannot host the wholistic essence of life. If an exclusive regard for physical data guides science, science is at risk of emulating the heartless mother, ignorantly cutting-killing the life of nature in order to take hold of it. The discrete indivisibility that runs, in the realm of harmony, from fundamental to octave (as its sameness) to fifth (the "division that unites"), hints at the hub of the mystery of the divisible worlds tackled by sciences. Embracing the wholeness of phenomena is the only mean to elicit their unveiling. As love forbade the division of the body, the only viable way to approach the realities of nature is to hold their indivisible core in the heart of the mind—which heart cannot find its beat in theories that exclude the mystery of life and consciousness.

In essence, what the golden mean binds and bonds, thus preserving the indivisible at the core of the divisible, are the principle of division and the principle of union. Love is the psychospiritual "inside" of the differentiating-and-unifying intelligence at work in the geometric and acoustic harmonies of nature. To balance secting and bonding, to hold together the imperative of separation and the imperative of union is the great apprenticeship of love, whose unfolding in human matters is suggestively encapsulated in its geometric expression. Human love emulates the divine capacity, at the source of the world's creation, to build harmonious relationships and vibrant wholes (family, communities, societies, nations) out of the two fundamental dynamics of separation and union.

Born of separation, the bond of love forever includes separation, as the Buddha gently reminded his monks: "All of you Bhikshus do not be grieved or distressed. If I were to live in the world for a kalpa, my association with you would still come to an end. A meeting without a separation can never be."[123]

When separateness refuses unity, divisions, devils, and evils arise. When the unitive denies separation its "fair" due, more radical divisions arise in reaction—or totalitarian forms of evil. The potential for love and evil are simply inscribed in the cosmogonical formula of the universe.

How instructive to ponder love as a separation-that-relates, a secting-that-bonds, a bond that differentiates. Not only does this formula shed light on the central challenge of relationships, but it also offers a fount of insight into the creative effects of loving bonds. Teilhard de Chardin pondered this theme to great depths.

> By observing around us the creative effects of love, we are led to accept this paradoxical proposition, which contains *the final secret of life*: true union does not fuse the elements it brings together, by mutual fertilization and adaptation, it gives them a renewal of vitality.... Union differentiates. The fundamental law of being, as the physical sciences and history teach us, [is] that "union differentiates."[124]

A balance of union and separation, love is the catalyst of the differentiating process of individuation, which leads to a deeper capacity to connect with others. Under the generous and generative gaze of love—be it mother, teacher, partner, friend, therapist—uniqueness of self is vitalized and the heart is quickened to hearten the life of the world.

"Far from threatening our ego," Teilhard continues, "the gift we make of our being must have the effect of completing it." This paradox of life suggests, by analogy, the "completion" of its "self" that the work of creation and evolution represents for the world being.

Goethe's wholistic eye and artistic being grasped this universal intelligence of love at work in the separating, "interfusing" golden dynamics of Nature. In one sweeping insight, he marvels at what he calls the "crown" of nature.

> Nature's crown is Love. Only through Love can we come near her. She puts gulfs between all things, and all things strive to be interfused. *She isolates everything, that she may draw everything together.* With a few draughts from the cup of Love, she repays for a life full of trouble.[125]

---

123 Sutra on the Buddha's bequeathed teaching.
124 De Chardin, *Human Energy*, pp. 63, 64 (italics added).
125 Goethe, *Maxims and Reflections* (Italics added).

III. The Harmonic Composition of the Universe

Islamic theosophists expand on how the cosmic spheres revolving around their stellar or galactic center are *moved by love*—for the being that grants them life and inspires their becoming. At the end of *Paradise* (canto 33), Dante echoes:

> But yet the will roll'd onward, like a wheel
> In even motion, by the Love impell'd,
> That moves the sun in heaven and all the stars.[126]

The love that moves the spheres and assembles nature's formations is the same force that draws the distracting impulses and tendencies of human nature to revolve around the soul, modulates "disproportionate" drives, attunes solipsistic parts to their central diapason and creates a consonant presence of being. While the golden mean is the life-channeling structuring principle of physical forms, love is the only "viable" organizing principle of the psyche, individual and collective—the one and only *life sustaining* principle.

As evolution proceeds from mineral to vegetal, from animal to human, the love-harmony of the world soul, which shapes elements and kingdoms of nature, reaches awareness of itself in the self-reflective capacity of the human soul. The human being is the first complete "resonating chamber" of the world soul—the first replica of the primordial milieu of "self-awareness" of Being. Through the five fingers of the human hand, the universe begins to take hold of itself and experience its creative intelligence.

Rudolf Steiner states most strikingly, "The whole human being from inside out is a fifth, he is constructed inwardly as a fifth. That is something which permeates the whole person."[127]

In the soul, harmony becomes aware of itself as the challenge and longing to negotiate world and self, and the imperative to generate mediating grounds. This imperative translates culturally in the fundamental of social harmony, the "golden" rule: "Do unto others as you would have them do unto you."

In humankind, love becomes aware of its power to "move the spheres" and organize life. The daily experience of attractive bonds and distancing forces points to the deep forces of love-harmony that shape destiny and character, as they shape cosmos and nature. Dante's love for Beatrice leads him to the vision-experience of the organizing principle of cosmos in what he describes as the rose of the cosmic courts of heaven—how rows upon rows of orbiting circles of beings connect the all to its divine source, the life-giving Sun, through fragrant bonds of attraction and distance, radiant tides of "praise" to the "ever vernal Sun."

---

126 Dante, *The Vision*, p. 570.
127 Steiner, lecture, April 30, 1924.

Steiner recounts how Dante's teacher, Brunetto Latini, returning from Spain to his native Florence, "came upon a hill in the midst of a desolate forest and on this hill saw the Goddess Natura weaving at her loom."[128] This initiatory experience was transmitted to Dante, Steiner says, and became the source of his *Divina Commedia*.[129] Within the circles of the rose of heaven, the culminating vision of the comedy, lies indeed the dynamic secret of the world weave: the golden bond that does not fuse, the unfolding rows that do not separate.

Under the Goetheanum, the majestic architectural building that Steiner designed in Dornach, Switzerland, as a world center for anthroposophy (from anthropos, human, and sophia, wisdom) he placed a foundation stone in the shape of a dodecahedron. He spoke of this stone as "the dodecahedral foundation stone *of love* which is shaped in accordance with the universe and has been laid into the human realm." Unequivocally, Steiner associates the dodecahedron with love and pronounces it the core of the human. The "foundation stone laid into the human realm" evokes the subtle foundation of the human self, the causal body, the twelve-petal lotus–rose of soulful Egoity laid in human beings by solar Angels—and its etheric replica, the heart chakra. These foundational twelvefold structures, which mirror the cosmic matrix, point to what the human realm is "building": universal selfhood within, dodecahedron-like human "solidarity" without. The bronze basin of the temple of Solomon, the king famous for his wisdom, rested on a foundation of twelve bulls, in symbolic resonance with the temple "not made by hands, eternal in the Heavens,"[130] the twelve-gated heavenly Jerusalem experienced by John at the end of the Book of Revelation. In conformity with its archetypal lotus-model in the subtle realms, the brim of Solomon's temple basin was, the Bible says, "like the calyx of a lily blossom."

The dodecahedron is pervaded through and through with the archetypal harmony of the golden mean. So is the universe shaped by a principle of harmony and love, which entrains the inert particles of matter to assemble in compositions organized according to the unique instrumentation of life: golden mean and harmonic intervals. The singular ratio apt to vehiculate the vibrant soul milieu opened by the self-resounding of Being, the golden mean is the ubiquitous cradle in which the mysterious power of love comes to existence, as a son born of the Two.

The third harmonic (the ratio of 3/2) highlights in its geometric quasi-equivalent—the golden mean—a principle of perfect consonance (fundamental-octave-fifth) and harmonious union, "fit" for the triune principle of love. Third harmonic

---

128 Steiner, *True and False Paths in Spiritual Investigation*, p. 85.
129 Steiner, *Karmic Relationships* vol. 3, lect. 6.
130 Cited in Bailey, *The Rays and the Initiations*, p. 279.

III. The Harmonic Composition of the Universe

and divine proportion represent, musically and geometrically, the archetypal "unit" of wholeness, of which the unit-universe will be the expression and the experiment—a universe that is, *in principle*, a unit of love. A dodecahedric unit of love.

Harmony ushers the world soul in the body of the universe. The soul of harmony, love is the inner reality of the physics of constructive interference and harmonic conjugation by means of which the "Living One"-ness of the universe runs through its multiple forms and systems. The central feature of harmony, which presides to the appearing of the One in the visible world of forms, is proportionality—among the parts and in reference to the whole.

The effect of proportionality on the soul is the experience of beauty. Beauty lies in harmony and beauty is the great mediator of love. Plotinus defines beauty as "the translucence, through the material phenomenon, of the eternal splendor of the 'one.'"[131] This "one" is the triune one, whose countenance is harmony and whose face is proportion.

This threefold principle of consonance and proportion is the Logoic code embedded in all living systems as their soul core: the unity of their diversity. A "divine" geometry of *proportionate unity* beats the triune pulse of the world soul in all forms of nature. It provides the cosmic rider, the Logos, with the ubiquitous carriage necessary to exist in cosmos—as cosmos. The golden mean is quite literally the formula of creation and the triune heartbeat of the universe.

In the end, harmony is the subtle garment of the world being, the garment by means of which the love essence of the universe we are rustles within the shapes and rhythms of all creatures with the elusive, steady grace of a great breath undulating through the multiplicity of forms. Rendered translucent by the perception of its harmonious countenance, the most inconspicuous form of nature becomes a "diaphany" of Phanes.

The golden mean is first and foremost a ratio—a rapport. The first known calculation of the golden ratio as a decimal appears in a letter written in 1597 to Kepler by his former teacher Michael Maestlin. While the ratio 3/2 approximates the value of Phi, Phi itself cannot be written as a fraction of two whole numbers. An irrational number, its decimal digits (0.6180339887498948482...) go on forever and do not repeat.

In his *Divina proportione*, the Italian mathematician Luca Pacioli writes,

---

131 Cited in Wilber, *Quantum Questions*, p. 69.

It seems to me that the proper title for this treatise must be Divine Proportion. This is because there are very many similar attributes which I find in our proportion—all befitting God himself—which is the subject of our very useful discourse...just like God cannot be properly defined, nor can be understood through words, likewise this proportion of ours cannot ever be designated through intelligible numbers, nor can it be expressed through any rational quantity, but always remains occult and secret, and is called irrational by the mathematicians.

The term used by Plato when he introduces the golden proportion at the beginning of the *Timaeus* is *analogos*. In Greek, *logos* also means proportion, as well as reason, word, speech. *Logos* comes from *lego*, which means to gather, to put together, and later, to speak or say. Greek philosophers used the term *logos* not only in reference to the spoken word but also the unspoken word, the word still in the mind—reason—and the synthesizing ("putting together") that presides to reason and logic.

Heraclitus first used the term around 600 BC to designate the divine reason that coordinates and structures the entire universe, as he discerned in the cosmic process a reason analogous to the reasoning power in man. In Stoic philosophy, the Logos was the operative principle of the universe, the dynamic reason pervading and animating it—the *anima mundi*—and the "seed" generating it, the *"logos spermatikos."* Philo of Alexandria, a Hellenized Jewish philosopher, referred to the Logos as an intermediary divine being, a bridge between God and the material world.

A mediating being, the Logos of the Greeks is both transcendent and immanent, metaphysical and physical—divine reason and geometric proportion. As a being, he personifies the soul of the world. "Reason (logos) or speech pertains to souls, and to the order of souls."[132] What will become *theo*-logical belief is at first *cosmo*-logical knowing. Philo, who refers to the Logos as "the first-born of God," also writes, "the Logos of the living God is the bond of everything, holding all things together and binding all the parts, and preventing them from being dissolved and separated." A vital unity of meaning runs through these various notions of "mediation, bond, seed, ratio, reason, mind, first-born son of God," bringing all dimensions of experience—conceptual, mathematical, ontological, experiential, physical—to converge around the hub of the cosmogonical operation.

Monotheistic Jews used the term *Logos* to refer to God as the supra rational mind behind creation. For John, and Christianity after him, the Logos was the creative Word of the beginning, who incarnated in Jesus–Christ when Jesus became the vehicle for the divine Logos at his baptism by John—in other words, when he became the golden bond between "the Father" and his creation, divine

---

132 Proclus, *Commentaries on the Timaeus of Plato*.

III. The Harmonic Composition of the Universe

and human.[133] "In the beginning was the Word, and the Word was with God, and the Word was God. Through him all things were made; without him nothing was made that has been made. In him was life, and that life was the light of men" (John 1:1–5). "For in him," Paul continues, "all things were created, and in him all things hold together" (Col. 1:16–17). The Word–Logos of the beginning emits a world held together by "living bonds."

A similar notion of the Word of the beginning was present in Hellenistic Egypt, where the figure of Hermes was venerated as "the Word who has expressed and fashioned the things that have been, that are and that will be." His triune nature as *"trismegistus"* (thrice-great) parallels the trinitarian Logos of Platonic cosmology and Christianity. According to the *Pymander*, he was called Trismegistus "because he was the greatest philosopher, the greatest priest, and the greatest king"—in which one recognizes the three aspects of the world soul: Intelligence, Love, and Will. The following excerpt from the Emerald tablet attributed to Hermes and translated into English by the father of modern physics, Isaac Newton, conveys in a stunningly concise way the "word" of world foundations, spelled out in the simple terms one can fathom at the core of Plato's harmonic cosmology.

> As all things have been and arose from one by the mediation of one [*logos*]: so all things have their birth from this one thing by adaptation. ["by adaptation"— i.e., through harmonization].

As Word of the beginning, the Logos is the vibration of the world soul, the love essence of the harmony spelled out in the proportionate bonds of cosmos. Harmony is the "principle" of the world body, and proportion (*ana-logia*, logos) is its geometric vehicle—the "earthly measure" of harmony. The divine proportion points to the Logos as the unitive mind our journey has showed us at play, literally *at love*, in the bonding of natural forms. At the beginning of every"thing" in nature is harmonic binding. Harmony, the essence of the world, is the nature of the Logos, whose living principle is love. Harmony is the way love externalizes itself.

To allow the vibrancy of the Word-of-the-beginning to imbue the word *logos*, and the geometry of reality it commands, is to open human reason to the supremely "rational" bonding intelligence and love alive in the proportions we find everywhere in nature. In the same way that knowing the potential hidden in a Tibetan bowl invites us to ring out its space harmonizing tone, realizing the Logos' presence in the proportions of nature conjures up a subtle perception of its rhythmic vibrancy, opens the inner ear to its inherent harmonies, and ordinary reason to the grace and love emanating from the innumerable "bodies" of divinity.

---

133 Fideler, *Jesus-Christ, Sun of God*, p. 46.

With the rise of modern science and mathematics, the words reason and logic found themselves gradually emptied of their vital operative principle—of "gathering together" through proportion-harmony. The divine *phaino-menon*, the revelation of Being in the Logoic rhythms and harmonies of nature slowly collapsed and crystallized into rigorous logical processes (reason) tackling material realities. No longer did the "Logos" radiate within the reasoning mind, as it no longer shone in the marvelous intelligence of nature. And yet, what else but Logos operates in human reason, when, following its deepest intimations of an underlying unity, science seeks, as nature does, to bond what is separate and to meet a sameness beyond its heterogeneous data and partial "figuring out" of reality.

The key to ushering the divine oneness into science, and life into the light of thought, Emerson suggests, is to "kindle science with the fire of the holiest affections. When a faithful thinker, resolute to detach every object from personal relations, and see it in the light of thought, shall, at the same time, kindle science with the fire of the holiest affections, then will God go forth anew into the creation."[134]

If harmonic binding is at the beginning of everything in nature, the power to bind worlds harmoniously is at the core of every human being—as the spark of Logoity called "I." Matter harmonizes but does not know that it does. It executes the harmonizing of the gods. Humans, on the other hand, are endowed with the quintessential privilege to sect and bond—consciously. To be human is to enter a long training in harmonious, creative bonding—within oneself, with others, in the world—a training in harmonization. In essence, it means to learn love—the intelligence, wisdom, and courage of love.

This fundamental cosmological operation of harmonic binding sheds a new light on the words of the Logos-Christ to Peter: "I will give you the keys of the kingdom of heaven; whatever you bind on Earth will be bound in Heaven, and whatever you loose on Earth will be loosed in heaven" (Matt. 16:19). The keys entrusted to Peter are nothing less than the means to wield in "divine" proportion the principles of creation: separating and binding—the keys of harmony. The supreme key—the golden mean of love—is the ultimate secret of co-creative harmony with the "heavens." The key to the kingdom of soul is the key to the soul of the kingdom, which is precisely the principle introduced in humankind's history by the Christ—Love.[135] Inde-

---

134 Emerson, *Nature*, concluding section.

135 Bailey, *The Rays and the Initiations*, p. 415: "*until Christ came and revealed the love of God to humanity*"; also Bailey, *Serving Humanity*, p. 477: "...two great Sons of God, the Buddha and the Christ. As you know, One of them brought illumination to the world and embodied the principle of Wisdom—the Other brought love to the world and embodied in

III. The Harmonic Composition of the Universe

pendently of any church doctrine, love is the "key" to seeing, knowing and relating to "the other as self"—to the Other as Same. Peter's Church widens its walls to the community of humans of good will, oriented and attuned to the principle of right relationships.

The Old Testament God who had the sole prerogative, as he made clear to Job, to bind the Pleiades and loosen Orion, is handing his divine key to humankind through his son—as his son, the Logos–Christ. The keystone of the harmonizing opus of creation, the mediating principle of love-harmony is now kindled as a potential in the human soul. The Word that sects and binds has "made his dwelling" in humankind; the fifth principle has been "delivered" in its hands. For better and for worse, humanity, a divine musician in the making, is entrusted with the baton of its orchestra, and empowered to entrain its collective heart to one golden beat. The human being is "a universe in miniature," says Clement of Alexandria, an analogos of "the Celestial Logos, the all-harmonious, melodious, holy instrument of God...[Who] composed the universe into melodious order and tuned the discord of the elements to harmonious arrangement."[136]

Many centuries later, Blavatsky will fully articulate John's intimation. "The logos," she says, "is *Christos*, that principle of *our inner nature* which develops in us into the Spiritual Ego."[137] The Christ-among-the-twelve epitomizes the universalized selfhood, the "tabernacling" of the Logos in a self that is experienced as the divine itself. "There is nothing in my cloak but God," said the great Persan Sufi saint Mansur Al Hallaj, before being led to his torture and execution in Baghdad in 922 for heresy. His ecstatic experience of oneness with God and his public affirmation, "I am the truth," reflected the hosting of the Logos, which is the plenitude of the unfolded soul.

> I saw my Lord with the eye of the heart
> I asked, "Who are You?"
> He replied, "You."

Al Hallaj stands as a grand figure of the universalized "self," which he himself referred to as the self "reduced to oneness." "What is important for the ecstatic," he would say, "is for the One to reduce him to oneness." Attar's narrative of his death recounts his rapturous opening to the successive and horrific "reductions" of his body. "When his legs were cut off, he smiled and said, 'I used to walk the Earth with these legs; now there's only one step to heaven; cut that if you can.' And when his hands

---

himself the principle of Love."
136 Quoted in Fideler, *Jesus-Christ, Sun of God*, pp. 62, 101.
137 Blavatsky, *Secret Doctrine*, vol. 2, p. 230.

were cut off, he painted his face with his own blood; when asked why, he said, 'I have lost a lot of blood, and I know my face has turned yellow; I don't want to look pale-faced' [as of fear]...then they put out his eyes...cut his tongue, and he smiled..."[138]

Before they took him to court, a Sufi asked him, "What is love?" He replied, "You will see it today, tomorrow, and the day after tomorrow." They killed him that day, burned him the next, and threw his ashes to the wind the day after that. "This is love," says Attar.

Centuries before Christ, more than a millennium before Al Hallaj, Plato knew the initiatory course. "The just Man," he wrote in the *Republic*, "will be scourged, racked, bound, will have his eyes put out, and will at last be crucified." He spoke of the soul being nailed to the body by the passions. We saw earlier how the grand Platonic opening of creation involved the *splitting* of the world soul into the two axes of an X *bonded* by the Sun-Logos raying forth its mediation between planetary and zodiacal circles—divisible and indivisible worlds. This cosmogonical "crossing" of Logos into cosmos finds its evolutionary counterpart in the crucifixion of the human being into universality. Plato understood this crossing, which would be epitomized, centuries later, in the grand narrative of Jesus, Al Hallaj, and other advanced figures of humankind. The "passing across" the human into the divine is a momentous leap beyond the polarities of difference and sameness, divisible and indivisible, body and spirit, world and self. "I am God" is the purest expression of this crucial at-onement with the One on the cosmic X-cross: the Sun-Logos—"god." Both crossings, the first one of creation and the last one of evolution, are achieved *by mean of* harmony–love.

While the ecstasies of mystics point to temporary states of at-onement, the words of Jesus Christ ("I and the father are one") or Al Hallaj ("I am the truth") speak to the evolutionary stage of knowing one's "self" as the universal self, and being one with "god." To become identified with the universal self is to be one with the world. The duality of individuality and universe ceases to exist. The "miracle" of Jesus Christ calming the waters or changing water into wine is a manifestation of his "Logoic" intimacy with the elements of the earth—His body ("this is my body"). Francis of Assisi lives in brotherhood and sisterhood with Sun and Moon, wind and water, fire and earth, illness and death. Al Hallaj foretold his servant that when they threw his ashes in the river, the waters would swell and threaten to overflow the city of Baghdad. "Lay my coat on the bank of the river," he instructed him, "and the waters will calm down." So the servant did, and the turbulence subsided. That "there was nothing in his cloak but God" proved to be fact when his cloak wrapped the water body of God back into its earthly vessel.

---

138 Attar, *Muslim Saints and Mystics* (pp. 306–307 in French edition).

III. The Harmonic Composition of the Universe

If cosmology illumines spiritual history, spiritual history actualizes cosmology. The passage a-cross and beyond dual consciousness, which marks the apotheosis of the human journey, represents the evolutionary consummation of the separation of soul and cosmos "in the beginning." It heralds the achievement of a *conscious* reunion of soul and cosmos. The great "mean" of this union is the same golden "mean" of love that released cosmos into existence and bound existence with essence.

The event of the cross dawns on the initiate-to-be when the causal body reaches its plenitude of solar radiance. Only when hosted in such soul plenitude can the central seed of Logoic, solar life—the monad, "whose home is in the Sun"—begin to "fire" up the soul and entrain the denser worlds with logoic life (X). His twelve labors completed, Hercules is ready to "put on the divine" and walk to the pyre. Ignited with Logoic fire, the beautiful robe of the causal body turns into ashes in the solar cloak of the Logos. Logoic life has obliterated the individual "self" and "reduced" his body to universality. Individual mindedness has ceased to exist. Logoic consciousness is.

John's cryptic account of "what he saw and heard from the Christ at the time of the crucifixion" touches on the same cosmological depths, "hidden away" in "apocryphal" writings.

> The Lord showed me a cross of light...and said to me, "This cross of light is sometimes called logos by me for your sake, sometimes mind, sometimes Christ, sometimes a door, sometimes a way...but what it truly is, is this: it is the distinction of all things, and the strong uplifting of what is firmly fixed out of what is unstable, and the harmony of wisdom, being wisdom in harmony.... The cross is that which has united all things by the word...and which has compacted all things into one."[139]

This cross of light, the symbol of the great duality and its harmonizing mean—mind, love, logos—hovers above the evolutionary cross of human experience—the experience of self and world "crossing" in the soul. Initially lit by the flickering faculty to distinguish self from world, unstable in its rapport with the world and the forces of its nature, the cross of experience gradually strengthens and opens to the radiating warmth of love and the wisdom of harmony, until, increasingly filled with universalized awareness and will, it enters the consuming grounds of the sacrificial fire—in which "all things are compacted" and "reduced into one." The "Universal Man" who has passed through the cross becomes an outpost of the world-harmonizing Logos, a mediating link between Logos and cosmos.

---

139 *Apocryphal Acts of John. The Other Bible*, p. 419.

The Tibetan master Djwal Kuhl alludes beautifully and enigmatically to the harmonizing work of the head of the Spiritual Hierarchy of the planet. Referred to as Sanat Kumara ("Eternal Youth") in Eastern traditions, Lord of the World, or Ancient of Days in Judeo-Christian traditions, Al Khdir (Green Man) in Sufism, the Lord of the Earth is the *"analogos"* on Earth of "the Celestial Logos." As described by Djwal Kuhl, the daily ritual of the Lord of the World is a rigorous and vibrant "implementation" of love and beauty through tones and colors, a calibrated outpouring of cohering, unifying energy, which breathes harmony into the three worlds.

> Does it mean anything to you when I say that the ceremonial ritual of the daily life of Sanat Kumara, implemented by music and sound and carried on the waves of color which break upon the shores of the three worlds of human evolution, reveal—in the clearest notes and tones and shades—the deepest secret behind His purpose? It scarcely makes sense to you and is dismissed as a piece of symbolic writing, used by me in order to convey the unconveyable. Yet I am not here writing in symbols, but am making an exact statement of fact. As beauty in any of its greater forms breaks upon the human consciousness, a dim sense is thereby conveyed of the ritual of Sanat Kumara's daily living. More I cannot say.[140]

The daily toning of the planet by the Lord of the Earth echoes the ancient Chinese practice of attunement to cosmic tones, by which "the middle kingdom" kept its daily matters, social forms, and practical instruments in harmony with the whole. The ritual invocation of the planetary gods in Egypt reflects the same commitment to the fundamental world ordering and beautifying power of harmony.

Rather than a long gone theory of creation, harmony is an "exact statement of fact"—a fact of creation and a vital practice-ritual by which the guiding powers of humankind sustain and entrain the harmony-in-the-making that defines the evolutionary process. The work of the Lord of the Earth, as well as the gracious bonds that pervade nature, model for humanity its "logoic" destiny to breathe harmony into existential "formations"—to symphonize the many voices, chaotic movements, and cramped knots of the personal and social psyche. Clement of Alexandria writes:

> The union of many in one, issuing in the production of divine harmony out of a medley of sounds and division, becomes one symphony, following one coryphaeus and teacher, the Logos.... Let us who are many haste that we may be brought together into one love, according to the union of the essential unity.[141]

---

140 Bailey, *Rays and the Initiations*, pp. 246–247.
141 Cited in Fideler, op. cit. p. 43.

IV. THE SPIRALING DYNAMICS OF THE UNIVERSE

> *First of all, the twinkling stars vibrated but remained motionless in space. Then all celestial globes united into one series of movements.... Firmaments and planets both disappeared, but the mighty breath that gives life to all things and in which all is bound up remained.*—VAN GOGH, (on his painting *Starry Night*)

FROM HARMONIC STRUCTURES IN SPACE TO SPIRALING MOTIONS IN TIME

A new perspective opens up in our journey with the realization that the archetypal 12fold/7fold pattern defines not only the structuring of cosmos in space, but the dynamics of soul becoming in time. The fractioning of oneness into a consonant whole is a process of involvement of spirit into forms of matter, which engages a complementary process of evolution of matter into forms of being—from atomic elements and molecules to cells, plants, animal, and human bodies. In "corporealizing" itself into cosmos, the world soul *involves* itself *into spatial* formations, which represent *evolutionary* steps of matter *in time*. Cosmological order deploys itself into spiritual history. The descent of spirit into spatial forms organizes the ascent of matter, in time, into the complex form of a human being. The 12 intervals-petals of the soul inform 12 evolutionary labors of consciousness. The splitting of the world soul into body and psyche—planets and zodiac—lays out the cross of human experience on which consciousness eventually transcends duality and identifies with the universe. The *involution* of spirit into multiple forms of matter, and the *evolution* of matter into forms of being are two complementary polarities of one process, two intertwined currents of one dynamic.

Is it not logical then that the 12fold/7fold archetypal pattern that informs living structures in space would also be the organizing pattern of time: 12 lunation cycles to a year and four 7-day quarters to a Moon cycle (29 days); 12fold cycles of hours of day and night, months of the year and ages of the great Platonic year, and 7fold week cycles. The 12fold/7fold pattern of the matrix of the world being is also the framework of its unfolding in time.

The two facets of this pattern—time and space, consciousness and structure—astrology addresses as one. Astrology regards the subjective evolutionary experiences of a psyche and the objective cosmic configuration of its natal sky as two sides of its one 7/12fold framework. The recognition that the unfolding of consciousness is but the reverse side of cosmic structures, and that the inner reality of the cosmic wheels is the coming to self-awareness of the soul in time, lays a ground of reasonability under astrology's "implausible" equation of cosmos and soul, and restores tone and vitality to its maps. Looking back, the two-pillared gate into this journey—the 12fold wheel of the zodiac and the 12-petal body of the soul—has become a single two-sided portal into the mystery of cosmos and consciousness.

The correspondence between the structure of forms and the patterns of time illumines the fact that time is essentially the subjective, experiential side of the evolution of forms. Ultimately, all structures of nature and cosmos are temporary phenomena—momentary "appearings" of Phanes, the Logos of life, and "transversal cuts" in the unfurling motion of life. These appearances and disappearances, formations and transformations of life operate according to the principle of harmony—harmonic differentiation in space, harmonic integration in time.

Fleeting glimpses into "involvement" and "evolution" came in and out of view in the first phase of the journey, but our attention was predominantly on the structuring of cosmos, its harmonic composition. We focused with Plato on the harmonic "apportioning" that leads to the spatial geometries of creation. Bringing forms of cosmos and nature under the measurements and calculations of scientists and researchers, we found indeed that harmonic intervals, proportionate bonds, and constant ratios shape organic forms and bind cosmic systems.

This penetration into the harmonic structures of nature is now calling into view a less measurable feature of organic forms: the formative and transformative gestures of time inscribed in their shapes. To the attentive eye, nature's living forms reveal coherent gestures embedded in their temporary fixity. Looking at a plant, standing under a tree, contemplating a seashell, pondering the course of the embryo, our perception opens to the motions that design shapes and the dynamics that steer growth. A slow twist, an upward curve, a rippled surface, a smooth stairwell of expanding layers, the carved hollow of a spin, all expose the motions of life—Life—involving itself in a particular formation and shaping matter into yet another organism. Stilled in definite, albeit transient forms, these gestures suggest how life wraps itself in matter to land on the shore of the visible, tangible world.

The Platonic description of the formation of cosmos led us to the golden ratio as the unique geometry that makes it possible for the indivisible wholeness of life to pass into separate forms—to *ex-ist* in forms—by way of a "sound" bonding of

IV. The Spiraling Dynamics of the Universe

existential polarities such as spirit–matter, Sun–Earth, male–female. What we are going to discover in the next phase of this adventure is that the "mean" to differentiate oneness into wholistic structures *in space* is also the core dynamics, *in time*, of the involvement of life and evolution of consciousness through forms of being. We will see that the golden ratio informs not only the geometry of living forms but also the dynamic pulse of their formation and growth.

As the root of the words *evolution, volute, and involvement* intimates, and as nature abundantly illustrates, this involution and evolution of life follows a spiraling motion. The Sanskrit term for involution is *pravritti*, from *pra*, forth; and the verb root vrit, to turn. Involution means to spiral forth, or centrifugally. Evolution is *nivritti*, from *ni*, back; it means, to spiral back, or centripetally.[1] Spirals in nature move simultaneously from within to without (involution) and from without to within (evolution). In Latin, *volvere* means to turn, to spin, to enroll and roll out. The term *evolution* intimates an onward spiraling (*e-volvere*).

Jean le Mee says of the spiral that it is the "movement of generativity and iteration characteristic of life." H. P. Blavatsky speaks of "that immutable law of nature which is *Eternal Motion, cyclic and spiral.*"[2] The concomitant *in-volvement* and *e-volving* of the world being in the forms of cosmos is the intertwined "volving" visible in the swirling, vortical motion characteristic of cosmic systems and forms of nature: from the orbiting of planets to the onward spiraling of solar systems and the swirling of galaxies, from the growth pattern of plants to the flight of birds, from the shell of a nautilus to the fine structures of a human body. The process of plant growth from root to flower represents a joint spiraling down of Sun energy and spiraling up of earthly matter. As the plant *evolves*, turning matter into stem, leaf, and flower, the Sun's light *involves* itself in matter, shaping-lighting it to its resemblance—the flower representing the ultimate spinning of earthly and solar forces into a minute epiphany of the Sun. The involvement of sunlight into earthly matter evolves earthly matter into solar forms.

Blavatsky describes how the solar system spirals forward around a central point located in the Pleiades, in the neck of the constellation of Taurus. Intriguingly, the Pleiades are precisely the sevenfold set of stars mentioned in the Bible as the epitome of what God alone can "bind."

> The central group of the system of sidereal astronomy, the Pleiades are considered (Alcyone, in particular) as the central point around which our universe of fixed stars revolves.[3]

---

1   Burger, *Esoteric Anatomy*, p. 134.
2   Blavatsky, *Secret Doctrine*, vol. 2, p. 80.
3   Ibid., p. 582.

It is toward this eternal Esse that everything, as every being, is gravitating, gradually, almost imperceptibly, but as surely as the Universe of stars and worlds moves toward a mysterious point known to, yet still unnamed by, astronomy and called by the Occultist - the Central Spiritual Sun.[4]

Around the same period, Edwin Babbitt also wrote,

> Our Sun is moving around some greater Sun. This greater Sun is also moving onward, probably around some still greater center, and carrying our Sun with it. Under this double motion, then, our Sun must describe a vast spiral through the heavens. Again, our Earth moves around the Sun, and at the same time is carried by the Sun around its center, making a smaller spiral... Then, finally, the Moon makes its baby spiral around our Earth. Thus we have first the great solar spiral, then the telluric spiral around the solar, then the lunar spiral around the telluric, three distinct gradations on nature's favorite trinal plan.[5]

This description echoes Copernicus' third postulate, which modern astronomy, though based on his first two postulates, has not taken into account, Steiner points out. "Copernicus said that the Sun also moves. It advances in a spiral so that the Earth, following the Sun, moves in a complicated curve. The same is true for the Moon that revolves around the Earth. You see here how the spiral has significance for celestial bodies, and these describe a form with which men will one day *identify themselves*."[6]

Randolph Stone emphasized how this universal motion is at work in all macrocosmic and microcosmic systems:

> All our created objects are really *whirls* in space... curvilinear motion or vibration operated in space and created sphere after sphere until all was manifested, and the same energy maintains this creation everywhere. Motion is always in the shape of a curve, from the path of the planets in the sky to the very atoms of matter, and spinning electrons.[7]

The curving-spiraling shapes of nature reveal the whirling pattern of the descent of life in matter and ascent of matter into fuller vehiculations of Being—the process so aptly referred to as involution and evolution. When the self-manifesting impulse of Being emits the first breath of "Creation," a primordial swirl of time–space–consciousness is generated, which is the matrix–template of all whirls to come. Unfurling of solar systems into planetary orbs, twirling of leaves and flowers up the plants, spinning up of chakras along human spines, unfolding of kingdoms of

---

4   Blavatsky, *Collected Writings*, vol. 12, p. 46.
5   Babbitt, *The Principles of Light and Color*, p. 96.
6   Steiner, *Occult Signs and Symbols*, lect. 4 (italics added).
7   Randolph Stone; quoted in Burger, *Esoteric Anatomy*, p. 17.

nature along the axis of time, every organic system reflects, replicates, perpetuates the central whirling pattern of Being externalizing itself.

If it was serendipitous to encounter Plato while probing the harmonic mysteries of twelvefold structures, it is the works of the paleontologist, Jesuit priest and philosopher Teilhard de Chardin that loom behind the vista of universal evolving now spread out in front of us.

Whereas Plato was coming down the mountain of spiritual history, imbued with an ancient knowledge he would translate into the nascent mathematical–geometric intellectuality of his time, Teilhard de Chardin ascends from the opposite, physicalist extreme of history, up the now well-paved road of intellectual lucidity and scientific factuality, to prompt cultural history on its way to a more wholistic paradigm. With hammer in hand and intuitive vision in mind, extracting rocks and gathering fossils with one, inviting universal insights through the other, Teilhard arrives, in what he will experience as a central moment of his destiny, at the borderland where Plato stood, embracing a similar convergence of metaphysical and physical reality—seized by a unified understanding of external world and inner life.

The borderland Teilhard reaches is a unitive view of the universe as an evolving *psycho-cosmic* whole, analogous in essence to Plato's picture of world beginnings. Straddling the external clues of paleontology and the inner beckonings of spiritual insight, Teilhard comes to a realization of the universe as a live, intelligent, evolving organism, unfolding through increasingly complex and integrated stages of psychophysical actualization—from a *cosmosphere* of inanimate minerality to a biosphere of vegetality and animality to a noosphere of reflective humanity to a *Christophere* of Logoic unity. His orientation, clearly, is different from Plato's—opposite in fact—and complementary. Plato's perspective, rooted in the metaphysical knowledge of a "divine" ground of world beginnings, describes the externalization of Being in forms. Teilhard's perspective, which stems from the human pole of scientific observation and inner experience, leads him to a vision of evolution as a spiraling and concentrating ascent of the universe toward a point that, he will eventually realize, is—also—a *Being*. His spiritual scientific intuition of the spiral cyclic process of evolution uniquely complements Plato's description of the harmonic structuring of creation.

The realization of the ultimate oneness of physical evolution and psychospiritual becoming, which was to catapult Teilhard to the unitive viewpoint Plato started from, was the result of years of scientific research conducted in tandem with an equally intense spiritual life. His scientific works first led him to view evolution as a forward movement of convergent complexification of matter toward and around a central point he named the "Omega point." At the same time, and all along, the profound currents of his inner life were sweeping him to the heart of the Christ as to the spiritual center of the universe. The great event of his life, he would later write, was the moment when his scientific intuition of the concentration of matter and convergence of cosmos toward the Omega point came to coincide with his mystical experience of the Christ:

> A God who makes himself cosmic and an evolution which makes itself person... a Christ who was distinctly seen as "evolver" and a cosmic Center which was positively attributed to Evolution. Thus I reached the Heart of the *universalized Christ* coinciding with the heart of *amorized Matter*.[8]

Teilhard experienced this realization that Omega and cosmic Christ were one as "the astonishing phenomenon of a general global conflagration." The Omega point was none other than the organizing, attracting, and integrative power of the love center of the universe—the cosmic Christ. The evolutionary hub of the universe was the Logos. The Logos who makes himself cosmos and the universe that makes itself one—a "person"—were two facets of one reality. As the Two became one in the blazing Sun of Al Hallaj's soul, we see, in the filigree of Teilhard's experience, the two branches of Plato's cross fuse into one. The distinction between cosmos and soul vanishes in the universalizing self.

A bronze medal cast in 1951 captures Teilhard's vision of the spiral cyclic motion of life. It was engraved with his words, 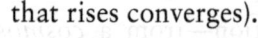 *"Tout ce qui monte converge"* (everything that rises converges).

Teilhard's intuition fills with experiential depth the outlines of the spiral cyclic motion of life we sighted around the last turn of our trail. If the Greeks led us to recognize, through the "progressive lens" of their word logos, that proportion, which governs the structuring of nature, is the *geo-metric* or "Earth-tailored" garment of the Logos and the silent articulation of the descending "Word," Teilhard not only unveils the spiraling motion of life at work in evolving matter, but also animates and enlivens it with the vibrant pulse of the heart of cosmos. His vision of evolution as a "cosmic vortex" drawing the universe to its

---

8   De Chardin, *The Heart of Matter*, p. 49 (italics added).

IV. The Spiraling Dynamics of the Universe

hub through a sequence of increasingly complex and inclusive spheres intimates the wave like, pulsating energy of the heart of a "Lord of the world." It reopens our perception of the cosmos to the soulful vibrancy that was the sentient experience of an earlier humankind, and that the guiding mind that Plato represented for humanity's emerging intellect brought into the objective light of rationality.

Plato's inside-out depiction of the harmonic deployment of the cosmic construct and Teilhard's outside-in view of its spiral-centrating evolution are grounded in the same unitive divine milieu defined by harmony (for Plato) and love (for Teilhard). Plato's view issued from it, Teilhard gradually discerns it as the ultimate destination and core of the evolving universe. The key to Plato's cosmology of world structuring is harmony, the moving power of Teilhard's world evolution is love—and these are precisely the two facets, we saw, of the one source and ground of reality—its essence and its vehicle.

The formative core of Plato's harmonic–geometric universe is the golden mean, and the attractive core of Teilhard's spiral–cycling universe is the cosmic Christ—the Logoic power of love. Plato's wholistic fractioning of the world soul by way of the golden mean meets Teilhard's view of the convergence of world dimensions under the attractive power of the heart of the universe. Teilhard's inspired pages on love as "the agent of universal synthesis...the affinity that links and draws together the elements of the world" echo, from the other versant and complementary current of our cultural history, Plato's "rational" description of the golden differentiation of the world soul into the perfect geometries of the elements.

As befitted the particular phase of history to which each individuality was born to contribute, Plato focused on the cosmogonical tide of involutionary differentiation and physicalization, and Teilhard on the evolutionary "tide of cosmic convergence"—of matter–consciousness concentrating and rising in "psychic temperature" from mineral to organic forms, to mindful beings, to unitive collective awareness. Plato's description of the enfolding of the world soul into cosmos finds its counterpart in Teilhard's intuition of the unfolding, onward spiraling of cosmos—from *cosmosphere* to biosphere to noosphere, toward an ultimate hosting of the world soul in humankind—a *Christosphere*.

A unique insight of Teilhard into the nature of matter and spirit further illuminates these cosmic tides. He regarded matter and spirit as two *directions* of evolution—the two cosmic gestures we have been contemplating. He saw in matter the movement of spirit toward manifestation; in spirit, the integrative motion of matter toward essentialization.

> Matter and spirit are not opposed as two separate things, as two natures, but as two directions of evolution within the world... two aspects of one and the same cosmic Stuff.... There is no such concrete reality as matter and spirit: there is only matter becoming Spirit.[9]

"There is only matter becoming Spirit" and, one might add, with Plato in mind, spirit becoming matter. This dynamic grasp of the two great polarities of human experience opens a new view of matter as an involutionary gesture of differentiation, separation, expansion and densification, and spirit as an evolutionary dynamics of integrative attraction to a center. That "there is no such concrete reality as matter" echoes the findings of quantum physics, which were indeed part of Teilhard's awareness:

> What is finally the most revolutionary and fruitful aspect of our present age is the relationship it has brought to light between matter and spirit: spirit being no longer independent of matter, or in opposition to it, but laboriously emerging from under it under the attraction of God by way of synthesis and centration.[10]

The notion that matter is the crystallized form, and spirit the subtlest essence, of one and the same reality, is familiar to world theosophies. "Esoteric philosophy," Annie Besant writes, "posits a primary substance, out of which kosmos is evolved, which at its rarest is Spirit, Energy, Force, and at its densest the most solid Matter."[11] Even though such understanding implies that matter and spirit are polarized states of one reality, the intellect tends to follow its propensity to reify each pole as a relatively separate reality and "static" state. Teilhard's understanding of concomitant and complementary processes takes us one dramatic step beyond this view of matter and spirit as alien realities. It releases one's thinking from the object focus of the analytical mind to the pattern and process-oriented, intuitive mind, whose blend of objective and subjective contents is not without evoking the ambiguity of quantum phenomena. We will go more deeply into these significant connections in chapter 5, where the course of our exploration will have opened the possibility to bridge the gap between the views of ancient cosmology and the findings of contemporary science.

---

9   De Chardin, *Science and Christ*, p. 51.
10 De Chardin, *Let Me Explain*, p. 103.
11 Besant, *Reincarnation*, p. 22.

IV. The Spiraling Dynamics of the Universe

Teilhard's majestic picture of evolution as a convergent swirling in of the universe into integrative spheres of world becoming offers a magnificent introduction to the second phase of our journey. The spiral cyclic evolutionary gesture it deploys arches over the upcoming landscape with the same emblematic power as Plato's twelvefold circle of the world soul hovered above the first part.

As we step out of our cosmological descent in the harmonic structure of cosmos, what presents itself to us as we pause and turn around, is the spiraling mountain surrounding the vertical axis of the mathematics of harmony we just left behind. Unfurling in front of us are the rounded slopes of the increasingly complex kingdoms of the world being, the cosmic mountain of evolution.

While Plato grounded the metaphysics of the world soul in harmonic solidity, Teilhard uplifts our "physics" with a dynamic vision of the universe's successive stages of integrative becoming—minerality, life, self-awareness. As the world soul and its harmonious "solidifications" (the Platonic solids) inform and pervade everything in nature—from chemical elements to cosmic structures—so does the cosmic spiral, we will find, swirl through the most minute and macrocosmic "particles" of the universe.

The Platonic cosmology introduced us to the core pattern of involvement of spirit in matter: the golden fractioning of oneness into a 12fold consonant whole. What remains to be fathomed is how this mathematical fractioning actually *in-volves* or "rolls in" the world soul into a cosmos, how it swirls stairwells of "descent" into forms and bodies, which are as many spiraling up of matter into vehicles of being. Unless the process of the descent of the god into his creation is grasped, the "turning of the Logos into cosmos" remains a mystery, however clearly we may discern his golden, musical footsteps in everything that "was made."

Plato stood at the historical threshold of a major step in intellectual development and conceptual objectivity, sustained by a sharpening of ego distinctness. This threshold is reflected in reverse in the current process of re-ensouling, "re-enchantment," and subjectification of the world that our times find themselves engaged in, as scientific findings are dissolving the border between matter and consciousness. At the same time, collective concerns, from ecology to globalization, are also inducing a decentralization of self. As the present juncture of our journey takes us from physical structures in space to formative processes in time—from the harmonic structuring of cosmos to its rhythmic becoming—we are stepping on a trail that is necessarily charged with the spirit of *this* time and its call to re-ensoul nature and cosmos and planetize our selves.

Having gained insight into the harmonic composition of the universe, our next step is to find out how cosmos "conducts" the shaping and evolving at work all around us in a way that sustains and perpetuates harmony. We seek to understand how the universe moves as one multifaceted whole through the complex motions of its many parts; how, from the minute cycles of plants to the vast cycles of stars and galaxies, the grand tidal gesture of spirit-to-matter and matter-to-spirit unfolds and refolds. We want to penetrate into the secrets of the spiral, to see if and how this fundamental motion relates to harmony.

If it does, the coherent picture of a harmonically composed, harmoniously moving cosmos will continue to gain solid ground—mapped out with what the factual lucidity of the modern mind can experience as surefooted trails. If it does, the spatial diffraction of the world soul into the many forms of the universe will prove to be a multidimensional stairwell of becoming, experienced inwardly by all beings as their most intimate processes of growth.

Logarithmicity: The "rhythm of the Logos"

The notion that the golden section may be at the core of the evolutionary motion of life emerges from the convergence of two factors: on one hand, a conceptual understanding that the golden "section" provides the mean for the indivisible world soul to indwell the divided world of forms—proportionality; on the other hand, the factual prevalence of the golden mean in the structure of living organisms. What is at this point but a possibility will be confirmed by what we find when we look at nature and cosmos with an eye for the motions hidden in their formations and a mind equipped with basic facts of physics and mathematics.

From spiral galaxies to seashells to the cochlea of the ear, from the growth of plants to the regrouping pattern of swarms of insects, schools of fish and flocks of birds after a scattering disturbance, from the motion of a fly approaching an object to a falcon circling its prey, nature patterns its motions on the spiral—specifically, a certain type of spiral, the logarithmic spiral. The logarithmic spiral is a spiral curve whose radius increases exponentially with every successive turn. In other words, the distance between the turns of a logarithmic spiral increases in a geometric progression. In an Archimedean spiral, by contrast, these distances remain constant.

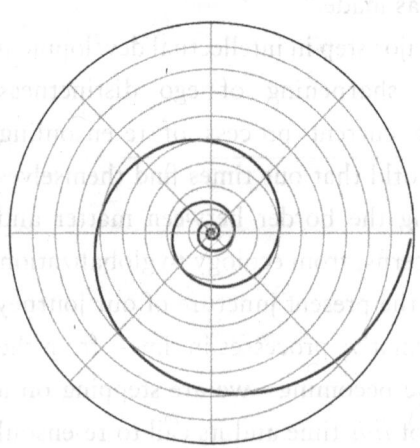

Fig. 34. Logarithmic spiral

IV. The Spiraling Dynamics of the Universe

First described by Descartes, the logarithmic spiral was studied extensively in the late seventeenth century by Jacob Bernoulli. He called it "spira mirabilis," the marvelous spiral. What fascinated Bernouli was its unique mathematical property of self-similar reproduction: while the size of the spiral increases with each successive curve, its shape remains the same (fig. 34).

The spiral is without a terminal point: It may grow outward or inward indefinitely, its shape remaining unchanged. It was Bernoulli's wish that his tombstone in Basel be decorated with such a spiral, accompanied by the words: *eadem mutata resurgo*: "Although changed, I arise the same."

This formula, "Although changed, I arise the same," echoes the Platonic model of cosmogenesis, with its generation of "different sameness-es" through a proportionate blend of Same and Other. Bernouli's formula articulates the unique marvel of a "golden" generation of new wholes, different yet similar *in proportion*. Self-similar replication is precisely the principle underlying the Platonic compounding of the world soul and the deployment of a soulful world. The living systems of nature display different yet similar forms of the fundamental "self-bonded whole" that is the vibrant soul-matrix of the universe. The Bible concurs: God created man "in his own image and *likeness*." World creation rests on the principle and process of "analogia." While Plato exposed the geometric structuring of self-similar forms, what the marvelous spiral displays is the dynamic *process* that generates "same yet different" forms.

Characteristic features of the logarithmic spiral offer different perspectives on its uniqueness. First, it is called "equi-angular," because at every point of the curve, as shown in the figure, the angle between tangent and radius is the same. Every point is in the same angle of rapport to the hub and periphery of the spiral (fig. 35). This angular constancy is a constancy of mediation. The onward motion of the spiral constantly mediates hub and periphery, inside and outside, contraction and expansion. Simply following the curve with an eye to its hub and an eye to the periphery gives an experience of this process of mediation in action. As the spiral expands, we feel how it also curbs its expansion to stay in balanced rapport to its "extremes." It constantly negotiates the pull to expand with the pull to concentrate.

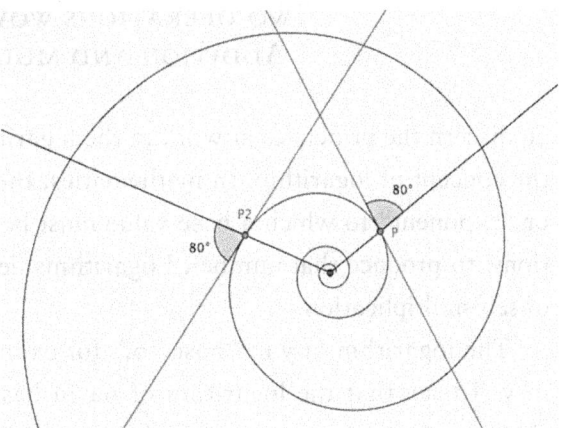

Fig. 35. Equiangular spiral with an 80° angle

173

In the case of the golden spiral, which is a special case of the logarithmic spiral, this constant angle is 72.968°—about one *fifth* of a circle. A golden spiral is a logarithmic spiral whose growth factor (or ratio of geometric progression) is Phi, the golden ratio. It gets wider (or further from its origin) by a constant factor of Phi for every quarter turn. The golden spiral will prove to be the archetypal logarithmic spiral, further highlighting the mediating character of the logarithmic spiral.

What of the mysterious term—*logarithmic*? Allegedly coined by the late-sixteenth-century Scottish mathematician John Napier, the word means, etymologically, "proportion number," from the Greek, *logos*, "proportion, ratio, word," and *arithmos*, or "number," literally "logos number." Interestingly, the word *arithmos* shares the same prehistoric Indo-European root (*ar-*) as the word *harmony*, "to fit together," which appears as the root of *order, ratio,* and *rhyme*.[12] "Number," says Iamblichus, "was called by the Greeks *arithmos*, as that which measures and orderly arranges all things and unites them in amicable league."[13] This definition of *number* comes very close to the world-organizing function of harmonic ratios introduced in the *Timaeus*. *Logarithm* could be translated literally as the world ordering "figure" of the Logos, the world-organizing formula of proportion. The "fitting together" of source and periphery displayed geometrically in the logarithmic spiral is indeed a dynamic expression of the Pythagorean notion of the soul and the harmonic order of the Logos. At this point, we can only regard the correspondence as intriguing.

Two Operations Woven Into One: Addition and Multiplication

To discern the principles at work in the logarithmic spiral, we need to briefly revisit the concept of logarithm. In mathematics, the logarithm of a number is the power or "exponent" to which a base value must be raised, via a series of self-multiplications, to produce that number. Logarithms designate orders of magnitude—orders of self-multiplication.

The logarithm of y in "base 10," for example, is the power of 10 that is equal to y. To say that the logarithm of 64 in base 2 is 6 is equivalent to saying that $2^6 = 64$ ($2 \times 2 \times 2 \times 2 \times 2 \times 2 = 64$; 2 multiplied by itself six times = 64, or 2 raised to its 6th power = 64). The exponent designates the number of multiplications of the base by itself. Logarithms are commonly used to simplify calculations. By substituting additions and subtractions of exponents to cumbersome multiplications and

---

12   Fideler, *Jesus-Christ, Sun of God*, p. 58.
13   Iamblichus, *The Life of Pythagoras*, p. 221.

divisions, they allow the reduction of large number series to smaller ones based on orders of magnitude or grades of intensity.

The defining feature of a logarithmic series is its blend of additive and multiplicative processes. It is at once an additive series of exponents and a multiplicative series of terms. It blends arithmetic and geometric progressions.

As we go a little deeper into mathematical details, the reader must rest assured that meaning will soon arise in association with such abstract notions as addition and multiplication, and re-infuse these operations with essential world dynamics as well as familiar modes of consciousness. These details are not exercises in mathematical precision, but necessary doors to access understanding.

An arithmetic progression is characterized by an equality of difference, but an inequality of ratio between terms. It contains the law for addition and its inverse, subtraction. The natural series of cardinal numbers (1, 2, 3, 4, 5, 6) is an arithmetic progression.

By contrast, a geometric progression is characterized by an equality of ratio but an inequality of difference. For example: 2, 4, 8 is a geometric progression, in that 4:2 = 8:4, but $4 - 2 \neq 8 - 4$. The sequence 2, 6, 18, 54, etc. is a geometric progression with common ratio 3. Similarly 10, 5, 2.5, 1.25, etc. is a geometric sequence with common ratio 1/2. The geometric progression contains the law for multiplication and its inversion, division.[14]

A logarithmic series is a blend of arithmetic progression of exponents (2, 3, 4) and geometric progression of terms (for instance: 2, 4, 8, 16). Instead of quantities (as in any typical addition), what the additive series is made of in the logarithmic series are the "times" that a "base" is multiplied by itself, in other words, raised exponentially to new magnitudes. It is an *addition of magnitudes*. This blend of addition and multiplication is what characterizes the logarithmic spiral mathematically. As shown in the diagram, the radius grows exponentially, or logarithmically, with every regular increase of the polar angle. In other words, the radius increases in geometrical progression as the polar angle increases in arithmetic progression.

In nature, shells display most tangibly this blend of additive and multiplicative processes in their addition of exponentially expanding and similarly shaped chambers. Figure 36 shows how the radius of every successive cycle of a nautilus shell increases in a geometric progression. Similarly, plants and trees move radially outward at a rate proportional to the stem's growth.

The omnipresence of the logarithmic spiral in the forms and motions of nature suggests the significance of "logarithmicity" as a fundamental principle in the

---

14   Lawlor, *Sacred Geometry*, p. 81.

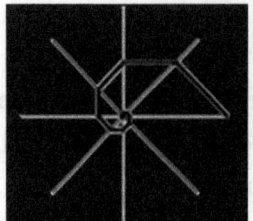  

Fig. 36. Logarithmic progression: diagram, seashell, snail (Gary Meisner, Golden ratio proportions in seashell; photo by Luke Stackpoole)

mathematics of life. This impression of significance intensifies when we discover, as we are about to do, that our sensory perception operates logarithmically as well.

The term *logarithmicity* does not exist in English dictionaries. However, it emerged as a necessary designation for the unique principle that began to appear as I turned around the intriguing husk of the word *logarithmic,* peeled off some of its dry mathematics, then heard fine explorers of nature expose its secret activity in the world we see and the world we are. The word, it became clear, was the exquisite envelope of a cosmological gem. That the term *logarithmicity* is not found in the dictionary might simply suggest that it has not been identified as of yet as the key principle it appears to be.

We will postpone exploring the ubiquitous presence of the spiral in the forms of nature, and turn first to the logarithmic principle itself. A fuller grasp of the way it operates will help us perceive what is involved in the formative motions of nature around us and the perceptive processes within us.

The Logarithmic Pulse of Forms of Nature and Processes of Perception

The discovery that our senses operate in a logarithmic fashion goes back to the nineteenth century and the works of Fechner and Weber. The German physician Ernst Heinrich Weber was a pioneer in the scientific measurement of human responses to sensory stimuli. Gustav Theodor Fechner, a physicist and early experimental psychologist, gave a theoretical interpretation of Weber's findings and demonstrated that there exists a nonlinear relationship between psychological sensation and the physical intensity of a stimulus. The Weber–Fechner law established that the relationship between stimulus and perception is logarithmic: "Equal relative increments of stimuli are proportional to equal increments of sensation."[15] In other words, sensation is a logarithmic function of the physical intensity of a stimulus.

---

[15] Fechner, *Elemente der Psychophysik.* Fechner's general formula for getting at the number of units in any sensation is $S = c \log R$, where S stands for the sensation, R for the stimulus numerically estimated, and c for a constant that must be separately determined by experiment in each particular order of sensibility.

IV. The Spiraling Dynamics of the Universe

To say that sensory perception operates logarithmically is to say that it blends additive and multiplicative processes. Fechner's law establishes indeed that for the intensity of a sensation to increase in arithmetic progression, the stimulus must increase in geometrical progression. We perceive as additions in weight, light intensity, volume of sound (arithmetic progressions) what are in fact multiplications (geometric progressions) in the magnifying of an external stimulus. In other words, a progressive multiplication of the magnitude of the stimulus is perceived as an additive series of intensification—as one, two, three times brighter, higher, louder, heavier. In fact, the *progressive augmentation of the stimulus has to reach a whole number multiplication of the base* (it must get to 2, 3 or 4 times brighter, higher, etc.) *to be perceived*, and the difference is registered as an incremental addition (visual, auditive etc.). Weber found for instance that the smallest difference in weight that a person can perceive is proportional to the base weight the experiment started with. The weight increase would have to reach "twice" the initial or base value to be "perceived."

Similarly, the eyes perceive brightness logarithmically. Stellar magnitude, for instance, is measured on a logarithmic scale, a scale based on exponential increases in perceived magnitude, not on actual quantitative measurements.

Hearing operates logarithmically as well. What we hear as equal differences in pitch are equal ratios of frequency. We perceive as equally spaced notes frequencies that are related by a multiplicative factor. Sound intensity is measured on the decibel scale, a logarithmic scale based on our perception of multiplicative processes as arithmetic progressions in sound intensity. A logarithmic scale uses a unit base of measurement for a particular quantity instead of the quantity itself. This unit base is referred to as the "logarithmic unit." *Decibel*, for example, is a logarithmic unit of acoustic intensity. By measuring differences in pitch or intensity in terms of logarithms of a specific base, logarithmic scales mirror the way our perception operates, which is in terms of quantum additions/subtractions.

The logarithmic blend of multiplicative and additive processes is an essential dimension of musical harmony—which is a perceptual phenomenon. For example, the octave of a fundamental is obtained by adding intervals in terms of string lengths (a fifth plus a fourth equals an octave) or by multiplying the corresponding frequencies (a fourth multiplied by a fifth equals octave: $4/3 \times 3/2 = 2$). In general, to add intervals, we multiply their ratios expressed as fractions. The fourth interval, for instance, which is an addition of tone + tone + semitone, equals $9/8 \times 9/8 \times 256/243 = 4/3$. We can also say that the note emitted by a plucked string rises by an interval of a fifth if the string is reduced to 2/3 of its length, and the interval of a fourth is obtained by reducing the string to 3/4 of its length. The *difference* between a fourth and a fifth is a tone, an interval whose ratio (8/9) is obtained by *dividing* 2/3 by 3/4. Mathematician John Webb comments, "That is probably the earliest manifestation of a logarithmic law."[16]

What we perceive as an additive succession of 12 notes is in fact a series of multiplications/ divisions of a base-unit of string length (1, 16/15, 9/8, 6/5, 5/4, 4/3, 45/32, 3/2, 8/5, 5/3, 9/5, 15/8, 2/1). The chromatic scale is obtained by transposing into an arithmetic progression, in the musical space of a single octave, the first 12 terms of the multiplication of 3/2, the fifth or golden ratio, by itself. What we perceive as an additive sequence of notes is the product of a multiplying-fractioning process.

A logarithmic scale of cents is commonly used to measure musical intervals in a way that mirrors our perception. The cent is the unit base of a scale of twelve equal divisions of the octave into 100 cents increments, each equivalent to a semitone in equal temperament. An octave equals 1,200 cents. Such a scale expresses the sequence of musical intervals or ratios in terms of equal differences in pitch, as a progressive multiplication of the base 100 by the first twelve numbers. The ratio of 3/2 for instance, measured in cents, is the chromatic interval of 700 cents (7 semitones).

Although we are accustomed to think of logarithms as a modern invention, McClain points out, any ancient culture that mathematized the scale actually possessed in the tone-cycle diagram a kind of circular logarithmic scale of base 2.[17]

The logarithmic character of the musical scale brings logarithmicity to bear upon the Platonic process of creation—the process of a Logos manifesting in cosmos—stirring up the meaning asleep in the etymological roots of the word *logarithm*.

Could logarithmicity be a core principle in the descent of the world soul into physical forms, hence in the pattern of our space–time–mind reality? The ubiquitous presence of the logarithmic spiral in nature would then be the logical expression of the Logoic en-souling of nature—the expression of an all pervading logo-arithmic model underlying creation and becoming.

The Dynamics of Logarithmicity:
A Threefold Bundling Process

The significance of the preceding question warrants a deeper look at the logarithmic blend of additive and multiplicative processes, in order to grasp what might be at stake in such a blend. Addition involves a linear sequence of terms distinct from each other. By contrast, multiplication is a unit-core centered process, an operation of concentric replication and "multi(re)plication" of a unit-base number (5 x 3 = 15 is

---

16   See http://plus.maths.org/content/perfect-harmony.
17   McClain, *The Myth of Invariance*, p. 19.

a 5-times replication of 3 as base unit). Addition is characterized by *distinctness* of terms, multiplication by *similarity* of self-replication. Addition juxtaposes, multiplication un-folds (Latin root *pli* literally means "fold"). "Every multiplicity," Plotinus writes, "is a multiplicity of unities, and therefore is predicated in unity." In merging additive and multiplicative processes, logarithmic series integrate a process of quantitative sequencing with a process of wholistic (whole-unit based), qualitative replication.

Logarithmic sequences are not series of quantities as much as sequences of magnification-reduction of a unit-base of quantity. They represent sequences of self-multiplication or self-dividing of a base by an additive series of exponents. They are sequences of proportional intensifications-multiplications of a core or base. We could say that logarithmic sequences "wrap-up" a multiplicative process into a succession of *units,* which represent proportional replications (multiplication-division) of the base. They "bundle" into successive terms the periodic saturation of a multiplicative or dividing process, the saturation being due to the necessity to replicate a unit-base (exponentially). The "saturation" corresponds to the limits imposed on the multiplication process by the (proportional) replication of a core unit (2^3, 2^4). The multiplication process is held in check by the necessity of replication.

✪ To sense the process from within, one could say that logarithmicity bundles a multiplicative process into an additive series of self-replications; it involves the periodic curtailing of a multiplying/dividing process by a self-replicating process. From the complementary perspective, one could say that logarithmicity fills the terms of an additive series with a self-multiplying–dividing process; it involves the rhythmic saturation or fulfillment of each term of an additive series by a self-unfolding (multiplying–dividing) process. In other words, *it represents the mutually defining encounter of self-replication and self-unfolding.* ✪

These formulations highlight the three components of logarithmicity: unit base, multiplication, and addition. Every logarithmic series involves a base-unit, which is the core "one" of the series. This unit is the "self" at the core of "self"-replication. In a logarithmic universe, core self points to the Logos, the world self, the "living God" of Judeo–Christianity, the universal notion of "god." Theosophies use the terms planetary and solar *logoi* to refer to the "self" of cosmic systems. Like the ancients, they hold every system in cosmos to be a living entity. So did, interestingly, Gustav Fechner, whose philosophical and mystical intuitions about nature and cosmos match and illumine his "psychophysics."

Fechner's philosophical writings reveal that his discovery of the logarithmic character of perception grounded in scientific facts the profound intuition he had of the world as a living, conscious Being, and of psyche and matter, mind and world as a continuum. The following excerpt from his work *Über die Seelenfrage* reads as a paean to the Logos of the Earth.

> On a certain spring morning I went out to walk. The fields were green, the birds sang, the dew glistened, the smoke was rising, here and there a man appeared; a light as of transfiguration lay on all things. It was only a little bit of the earth; it was only one moment of her existence; and yet as my look embraced her more and more it seemed to me not only so beautiful an idea, but so true and clear a fact, that she is an angel, an angel so rich and fresh and flower-like, and yet going her round in the skies so firmly and so at one with herself, turning her whole living face to Heaven, and carrying me along with her into that Heaven, that I asked myself how the opinions of men could ever have so spun themselves away from life so far as to deem the Earth only a dry clod, and to seek for angels above it or about it in the emptiness of the sky—only to find them nowhere.... But such an experience as this passes for fantastic. The Earth is a globular body, and what more she may be, one can find in mineralogical cabinets.[18]

Behind Fechner's law is the vibrant psyche of a man who felt the thrill of life and the presence of consciousness in plants and stars, Earth and universe. Behind the sharp eye of the empiricist and founder of psychophysics is the loving, "embracing look" of a spiritual scientist who sought to discover the connection between soul and body—psyche and physics—and demonstrate it in physical measurements.

Fechner viewed the human being as standing midway between the souls of plants and the souls of stars, which he called angels. He envisioned God, the soul of the universe, as having an existence analogous to men, and saw in natural laws the modes of God's unfolding perfection. His unique insights into life after death describe how, when the soul leaves the body, all sensory experiences dissipate with the dissipation of bodily boundaries and the human being passes away into an experience of himself as cosmic *being*:

> When man dies, with the destruction of his body, that combination [with the senses] is loosened, and, released from its bondage to it, the soul will now return to nature with full freedom. He will no longer be conscious of the waves of light and sound only as they strike eye and ear, but, as the waves roll forth into the sea of ether and the sea of air, he will not merely feel the blowing of the wind and the wash of the waves against his body, but will himself murmur in the air and sea; no more wander outwardly through verdant woods and meadows, but himself consciously pervade both wood and meadow and those wandering there.[19]

Fechner's intuition of the Earth's subjective being translated into the remarkably unifying power of his scientific demonstration. In showing the logarithmic character

---

18 Fechner, *Über die Seelenfrage;* quoted in James, *A Pluralistic Universe,* p. 164.
19 Fechner, *The Little Book of Life after Death,* p. 69.

IV. The Spiraling Dynamics of the Universe

of perception, Fechner placed his finger on a continuum of modus operandi between external nature and internal experience, perceiving subject and perceived object. However, neither Fechner nor the culture that adopted his discovery focused on the logarithmic principle at its core, or on the bridge our context suggests it represents.

A cultural landmark for the scientific connection it established between body and mind, the Weber–Fechner law was to lay the foundation of experimental psychology. The revelation that the "subjective" experience of perception has an objective, quantifiable dimension, which can be formulated mathematically, opened the door to the notion that other psychological functions and operations might lend themselves to scientific measurements. A new potential was there to take hold of human psychology with empirical tools and data.

The collective orientation toward empirical science, which prompted Fechner's research in the first place, bent to its bias the connection he uncovered between mind and body, consciousness and matter. His law became the basis of a physics of the psyche—but not of a psyche of physicality. The quantification of psychology introduced by his works was not accompanied by an awakening to the psychic life of the "objective" world such as inspired his vision, his science, and his life.

The psychologist William James, however, who delighted in Fechner's writings, and to whom many owe their awareness of his works, deeply resonated with the essence of Fechner's insights: "If the heavens really are the home of angels," he remarked judiciously in his introduction to the passage just quoted, "the heavenly bodies must be those of very angels, for other creatures there are none."[20]

In the present context, as our exploration of the harmonic proportions and ubiquitous spiraling of nature runs into Fechner's law, the mirroring between the logarithmicity of cosmos and the logarithmicity of human perception is immediately striking. The logos-like proportionality of cosmos reflects itself in the cosmos like "proportionalizing" of perception. A likeness of modus operandi surges between psyche and nature. The logarithmic continuum between world processes and subjective processes of perception intimates that physical and psychospiritual facts—and so, in a larger sense, Logos and Cosmos—are not radically alien to each other, but share fundamental processes. A subtle higher ground—a common ground—appears beyond the split between cosmos and psyche. With logarithmicity turning the two into one Möbius-strip–like continuity, perception and perceived become two sides of one reality. Beyond the great divide, the same principle appears to govern nature formations outside and human perception inside.

Remarkably, the psyche revealed its cosmos-like features to a mind whose fresh, contemplative gaze perceived the angel-Logos of the world. Although largely

---

20 James, *A Pluralistic Universe*, p. 164.

unrealized by Fechner himself, it seems, his law had the potential to provide a scientific footing for the notion that there might indeed be a "Logos" of the Earth, a living entity indwelling the "dry clod," wrapped in its organic spirals and nested in its proportions, as there was a self taking in the world along the "proportioning" perception of the human system.

It seems that this revelatory potential of Fechner's law has remained dormant, yet it is the potential that truly speaks to the spirit of his research—which was to demonstrate that physical facts and subjective experiences, world and consciousness, are two sides of one reality.

Although Fechner himself did not suggest it, his works provide a first scientific hint that logarithmicity might be a common modus operandi of universe and mind. What he uncovered is a scientific fact and mathematical principle in support of his experiences of the vibrant "Logos" of life in the graceful countenance, elements and beings of nature.

His unitive visionary science anticipates the wholistic view that has been emerging in the collective psyche for the past five or six decades. The Gaia principle formulated by the chemist James Lovelock in the mid-1960s and co-developed by Lynn Margulis in the 1970s echoes and reintroduces Fechner's splendid intuition. The Gaia hypothesis proposes that all organisms and their inorganic environment on Earth constitute a closely integrated self-regulating, self-preserving system—a living being. "The quest for Gaia," Lovelock writes in the first paragraph of his book, "is an attempt to find the largest living creature on Earth."

The Threefold Operation of Perception

What we just discovered with Fechner is that human perception registers variations in weight, light, or sound in terms of logarithmic increase or decrease, not in linear quantitative terms. What we perceive are not quantities but proportional increases and decreases in quantity. We serialize quantitative changes into progressive proportionate increments. The way the mind processes modifications in sensory input involves an exponential "bundling" of such input.

The discovery that perception operates logarithmically opens new perspectives on the phenomenon of logarithmicity itself. Logarithmicity, we saw, blends two types of progression: an arithmetic progression characterized by equal differences between terms (1, 2, 3, 4), and a geometric progression (such as 2, 4, 8, 16) characterized by equal rapport between terms and multi-replication of a core unit.

Presenting the geometric progression of 2 as 2, 4, 8, 16, highlights sameness of rapport (4 to 2 equals 8 to 4 etc.), while presenting it as 2^2, 2^3, 2^4, 2^5, highlights its

IV. The Spiraling Dynamics of the Universe

base unit, 2, and the equal difference between exponents. In mathematics, this core unit is called the "base." In geometry, it is the constant angle or ratio of growth of a logarithmic spiral. In cosmology, it is the Logos.

Arithmetic progressions, based on sameness of difference between terms, represent additions of separate quantities, as appropriate for the realm of *matter* and physicality. Geometric progressions, based on a sameness of rapport defined by their base or "seed" unit of growth, speak to the self-similar unfolding process associated with *organic life*. In blending the two progressions, logarithmicity merges two processes: a differentiating process of extension (in space and matter) and a binding process of integration (in time).

Fechner's demonstration of the logarithmic character of sense perception leads to new insights in the phenomenon of perception as a blend of two basic modes of apprehension of the world: "difference making" and "rapport making," with a third, core perceptual mode of "identification" with the object. If I look at a tree, I can focus on distinguishing its components (branches, leaves, trunk, and so on), I can also be aware of the wholeness held in its shape, and within this wholeness, of internal symmetries and gestures of growth. I may also "identify" with the central life force rising from the Earth, defying gravity, and projecting itself into the sky. The "self" in me perceives the tree's autonomous life.

Logarithmicity brings to light the three-dimensional character of perception:

1) A capacity to distinguish and separate visual, auditive, olfactive, gustative stimuli as isolated, objective phenomena that can be intensified and decreased, added and subtracted from one's environment. This distinguishing function is currently humankind's dominant form of perception. It supports the analytical dissecting, measuring, and counting of data, which science regards as its quasi-exclusive ground of evidence. Quantum physics however is challenging this mode of perception.

2) A wholistic capacity to sense rapport and relations between objective factors: harmony and disharmony in color and sound, proportions in forms, rhythm in language. Kepler stated it unequivocally, "The faculty which perceives and recognizes the noble proportions in what is given to the senses must be ascribed to the soul."[21]

3) An intuitive capacity to perceive an archetypal core: "fundamental" in a melody, theme in a musical composition, archetypal pattern in natural forms, vibrational identity in a person. An example of this third aspect of perception is Goethe's notion of the archetypal plant. Goethe identified a singular reality-entity at work in the metamorphoses of the seed from stem to leaf to fruit. "Whether the plant vegetates, blossoms, or bears fruit," he writes, "it nevertheless is always the same organ with varying functions and with frequent changes in form, that fulfills the dictates

---

21 Kepler, *Harmonices mundi*.

of Nature. The same organ which expanded on the stem as a leaf and assumed a highly diverse form, will contract in the calyx, expand again in the petal, contract in the reproductive organs, and expand for the last time as fruit."[22]

⊙ These three dimensions of perception: difference making, rapport making, and intuitive identification, may be viewed as the body (physical quantity), soul (relational quality) and spirit (core dimension of identification) of perception. They mirror the primordial threefold event of the Platonic creation: differentiation (2) and self-bonding (3) of Being (1). In the Platonic cosmology, this primordial step, which is replicated in the deployment of the world soul into a 12fold–7fold form, is the basis of all the formative processes of nature and cosmos. The blend of self-differentiating and self-bonding is clearly logarithmic-like. What is emerging, and will soon be confirmed, is that this threefold operation is the seed-pattern of the spiraling involvement of the world soul into forms. ⊙

When we realize that the "base" of the unfolding universe is its Logoic core, whose 3fold harmony is the very "mean" of deployment of the One in the multiplicity of forms, hence the one-mindedness animating them all, we realize that logarithmicity may well be, as literally as the word amazingly states it, the "arithmetic of the logos," the spiral-cycling "figure" of life. How insight is visited upon language and how language comes to host insight is an intriguing question and a marvelous fact.

Logarithmicity articulates mathematically the threefold metaphysical process of creation as an integrative (3) fractioning (2) of oneness (1)—an integrative diffracting of the *Logoic* soul "unit base" of our universe. This process pursues itself in the spiraling involution of spirit and evolution of matter visible throughout nature and cosmos. At work in this spiraling is the stepping down of Logos into cosmos, and the simultaneous gathering and centering ascent of cosmos into Logoic forms.

In the following pages, we will look more closely at actual instances of spiraling in nature. The archetypal perspective that, at the moment, continues to draw us to the hidden center of things will make for a deeper grasp of this fundamental shaping motion of nature, by revealing in it the physical trail of the breath of the world being—the exhale and inhale of the world soul in time and space, *into* time–space forms.

THE LOGIC OF WORLD DIMENSIONS AND LEVELS OF BEING

From the cosmological perspective of a live, self-ordered cosmos, every particle of the universe vibrates with the spirit of the world being, and basic forms such as atomic elements bear witness to the early organization of matter by "life." The principle of self or soul, however, only begins to abide in forms with the emergence of

---

22   Goethe, *The Metamorphosis of Plants,* line 115.

IV. The Spiraling Dynamics of the Universe

the self-organized, self-reproducing organic units of the plant kingdom. The apparition of *auto*-nomous forms of life, centered on their own seed-"self" cycles, heralds the involution of the world soul in physical forms.

This essential difference between inorganic and organic forms is reflected in their geometric depths. We remember how dodecahedric structures (Phi-filled) appear only in organic forms, and are not generally present in chemical elements. Similar to the line drawn between inorganic and organic forms by the presence in the latter of Phi and 12foldness, the apparition of a logarithmic-like dynamics of growth distinguishes organic life from inorganic processes.

Atomic elements combine and aggregate (add up) into molecular compounds, but they do not multiply and reproduce. Cells, by contrast, self-divide (i.e., self-multiply) then bundle this multiplication into "organs" and systems, which are vital parts of the plant, animal, or human life, and manifest different potentials or exponents, one could say, of the "entity." Also in contrast with atomic elements, cells and living systems have cycles of formation and decomposition, growth and decay, which point to an autonomous core of unfolding and refolding: a seed or a self. The multiplication-and-bundling of cells into tissues, tissues into organs, organs into systems, systems into bodies, and the recursive nesting of the same basic pattern at different scales of organization and growth, imply the "base-unit" characteristic of logarithmicity: a self-core informs the "exponential" unfolding of a "basic" sameness or identity. The externalization of a seed-self into a plant, animal, or human organism replicates the self-unfolding of a Logos into a cosmic system—planetary, solar, galactic. Both are logarithmic-like processes.

We see how exponential progression provides a model for organic growth as it illuminates the archetypal process intrinsic to any unfolding of life into form: the *replication* of a base-unit into *different* stages or dimensions of its "self." Something of this ordered nesting dynamics appears to deteriorate in the phenomenon of cancerous growth. With cancer, the process of multiplication-division emancipates itself from the replicating dimension. Cell multiplication is no longer contained within the healthy and wholistic bounds of proportionate "self"-replication. What regulates cellular growth and "organ-izes" it into organs is the same principle that blends additive and multiplicative series into a logarithmic series: the principle of proportional *self-replication*.

We can say that logarithmicity articulates in mathematical terms of exponential self-replication, and in geometric terms of spiral-cycling motion, the dynamics of self-unfolding characteristic of vegetal (Goethe's *Urpflanz*) and biological processes (from cell to organ formation). It also articulates, as we have seen, the process of physiological perception and, potentially, that of psychological progression—which, we all know from experience, proceeds by quantum leaps through the initiatory thresholds of biographical landmarks, rather than as a continuum. Every domain shows a replication of the primary process of cosmo-genesis: the spiral cyclic "turning," quite literally, of Logos into cosmos.

An archetypal illustration of this fundamental process is the simple logarithmic progression of $2, 2^2, 2^3$. This progression expresses mathematically the passage from one physical dimension to the next—from line (2) to surface (2^2) to volume (2^3)—and illumines in logarithmicity a *progression in "dimensionality."* Logarithmicity is, indeed, a mathematical principle of dimensionality. It captures the threefold operation of cosmo-genesis: the self-multiplication/replication of the Logos into dimensions of cosmos–consciousness. It spells out quite literally the steps (1, 2, 3) and rhythmic (exponential) flow of the expansion–reduction of a core unit, or "logos," into a series of dimensions.

What happens geometrically from line to surface (2^2), then plane to volume (2^3), has its correspondence in the progression from seed to root, stem, flower, and fruit organically, from one magnitude to the next perceptually, from information to understanding to insight intellectually, and from cosmosphere to biosphere to noosphere cosmologically.

Physio-logy, bio-logy, psycho-logy, and *cosmo-logy* point to a discreet "logos" at their core. The translation of "logos" in these contexts as "study" or science (of nature, life, psyche, cosmos) blinds one to the fact that understanding these fields, which is the aim of such "study," requires tapping into the logic *of* the field itself. And the logic of a field is the logoic intelligence at work in it.

Both as mathematical principle (logarithmic series) and geometric dynamics (logarithmic spirals), logarithmicity appears to articulate the internal logic of organic growth, as well as the cosmological unfolding of world dimensions along the spiral of involution and evolution.

❍ The spiral-cycling motion characteristic of the logarithmic spiral combines an *addition* (of self-similar bundles or cycles) with a progressive *multiplication* of a *base*—the unfolding of its *self* of a living entity. Self, multiplication, addition are the three key components of logarithmicity. The additive sequence of cycles represents an *externalization of* the unit base into a succession of (self-similar) replications-dimensions. The proportional expansion represents a self-unfolding process *internal* to each cycle. The multiplicative process is lumped into successive bundles by the constraint of proportional self-replication, which rounds out each cycle into a next step of the spiral. ❍

IV. The Spiraling Dynamics of the Universe

Mathila Ghyka points out that the Greek root of the word *rhythm*—*rhuthmos*—is distinct from the word *arithmos*, yet the Greeks used indifferently *ruthmos* and *arithmos* to mean "number." Both are derived from *rhein* (to flow). For the Greeks, the notion of number includes that of flow, proportion and rhythm. However, he explains, *rhuthmos* corresponds more specifically to the "symmetry," or proportional sameness, of a sequence of numbers, *arithmos* to the actual "measure" of each term of the sequence. How eloquent in the end that the word *log-arithmic* itself condenses into one the two aspects of sequentiality and proportionality characteristic of the principle itself and of the logoic operation of manifestation. Logarithmicity appears indeed to be the way the Logos "flows" into and throughout cosmos.

The Logarithmic Chain of Being

> *The spark hangs from the flame by the finest thread of Fohat. It journeys through the seven Worlds of Maya. It stops in the first, and is a metal and a stone; it passes into the second and behold—a plant; the plant whirls through seven changes and becomes a sacred animal. From the combined attributes of these, Manu, the thinker, is formed.*—H. P. Blavatsky[23]

The principle of logarithmicity illumines from "below" (the rational mind's point of view) the logic of cosmological unfolding presented by Plato. In fact, it unveils the mathematical logic underlying the models of the universe deployed in a number of world traditions—from ancient India to Egyptian Hermeticism to the Jewish Kaballa. We will see, indeed, that this logic is the logic of harmony, and that what makes logarithmicity the logic of the universe is that it is the logic of harmony.

The notion of dimensional unfurling, illustrated geometrically in the passage from line to surface to volume (an additive series of dimensions) and associated with the exponential progression from 2 to 2^2 to 2^3, is reflected for instance in Ken Wilber's presentation of the great Hierarchy of Being, the central tenet of the perennial philosophy, as a spiraling descent of the One into the Many—a descent visible, to the attentive eye, in the constructs of nature.

In a footnote to an excerpt from the works of mathematician, physicist, and astronomer Sir James Jeans, Ken Wilber presents the Great Chain of Being as an arithmetic, *additive/subtractive* progression of dimensions.

But first, passages from Sir James Jeans' *Mysterious Universe: A Universe of Pure Thought* reveal how modern voices of theoretical science echo the views explored in the present work and the ancient wisdom in which they are grounded:

---

23 Blavatsky, *Secret Doctrine*, vol. 1, p. 34.

> Today there is a wide measure of agreement, which, on the physical side of science, approaches almost to unanimity, that the stream of knowledge is heading toward a nonmechanical reality; [God is a mathematician, and] the universe begins to look more like a great thought than a great machine. Mind no longer appears as an accidental intruder into the realm of matter; we are beginning to suspect that we ought rather to hail it as the creator and governor of the realm of matter—not of course our individual minds, but the mind in which the atoms out of which our individual minds have grown exist as thoughts...The old dualism of mind and matter...seems likely to disappear, not through matter becoming in any way more shadowy or insubstantial than heretofore, or through mind becoming resolved into a function of the working of matter, but through substantial matter resolving itself into a creation and manifestation of mind. We discover that the universe shows evidence of a designing or controlling power that has something in common with our own individual minds...with the tendency to think in the way that we describe as mathematical. So at least we are tempted to conjecture today, and yet who knows how many more times the stream of knowledge may turn on itself?[24]

Interestingly, Jeans' last thought rings with logarithmic inspiration.

Ken Wilber comments how Jeans' idea of the physical cosmos as a "materialization of thought" is widely supported by the ancients' worldview that matter is a crystallization or precipitation of mind. The mapping that Wilber offers of this precipitation of the Great Chain of Being is of special interest in our context, because it emphasizes a step-by-step subtraction/addition of dimensions.

The 7fold model presented below expands on Wilber's model to reflect the 7fold structure that the Great Traditions concur to regard as the archetypal "spine" of the world being, macrocosmic and microcosmic—from the 7 days of Genesis to the 7+3 sephirot of the Kabbalah's tree of life, from the 7 planetary spheres of the solar system to the 7 chakras of the human being, from the 7 climates of Islamic theosophy to the 7 planes of Being of Eastern and Western theosophy, of which Alice Bailey explains that they represent seven qualities of differentiation of matter. "The cosmic physical plane exists in matter differentiated into seven qualities, groups, grades or vibrations."[25]

The world of harmony and music reflects this 7fold principle of extension of the universe. "Practically," Rudhyar points out, "music spans seven octaves of vibration, approximately the extension of a piano keyboard."[26] The 7-octave span of musical space, which is also that of the human ear, mirrors the 7fold cosmological range of

---

24 Jeans, *The Mysterious Universe*, p. 151.

25 Bailey, *Treatise on Cosmic Fire*, p. 116.

26 Rudhyar, *The Magic of Tone and the Art of Music*, p. 58.

levels of being. It would seem no mere coincidence that the standard extension of musical space, the range of the human ear and the dimensional spectrum of the world being are mirrored in the range of seven octaves defined by the combined cycle of the two primordial series associated with number 1, 2, and 3 (octave and fifth series).

The distribution and nomenclature of the hierarchical ladder of Being varies from one source to the next. However, it tends to be organized around a fundamental sevenfoldness. The main purpose of the model presented below is to highlight the notion of addition/subtraction of dimensions, and the logarithmic character of the descent/ascent of Being in forms of matter. Addition-subtraction, geometric progression of levels and unfolding of an ontological "base," are clearly the fundamental components of the ladder. Each kingdom of nature is a *different* kingdom of the one and *same* Life. Each addition of dimension yields a more complex form, which hosts a simpler "fraction" of being:

Matter : Mineral kingdom. Physical dimension. Elemental consciousness.

Matter + life : Plant kingdom. Biological dimension. Physiological consciousness involved in life processes.

Matter + life + sentiency : Animal kingdom. Instinctual dimension; autonomous motion; self-centered impulses. Sentient, astral, emotional consciousness.

Matter + life + sentiency + reflective self-awareness: Human kingdom. Mental, intellectual dimension of consciousness.

Matter + life + sentiency + concrete mentality + higher mind–soul: Kingdom of soul, sometimes called "Kingdom of God." Egoic, soul dimension. Love–wisdom consciousness of the individuality aware of the other as "self," motivated by the good of the whole and engaged in mastering mental and astral dimensions. Leads to the emergence of what Steiner calls "spirit self."

Matter + life + sentiency + concrete mentality + higher mind–soul + monadic consciousness: Universalized consciousness; will identified with world purposes; mastery of etheric forces. Emergence of "life spirit" (Steiner).

Matter + life + sentiency + concrete mentality + higher mind–soul + monadic + logoic consciousness: "Divine" dimension of the world creative intelligence of love that pervades every atom. Identification with the generative and organizing spirit of the world being. Experience of the world as "self" and self as "god." "I and the Father are one." Mastery of matter. Emergence of "spirit body" (Steiner)

This presentation of the chain of being shows how each new level supersedes and includes lesser levels by adding a dimension. Simple observation reveals the transformative "addition" of consciousness associated with each ascending kingdom of nature.

Islamic theosophists speak of "the fleeting, transient, ascending faces of God": "The intense perfecting of the mineral, the plant, the animal, the human being, who bear in their center this existence of their existence that they epiphanize... heralds the gathering in God."[27]

Ontologically, the hierarchy of being is a downward-outward deployment of the Logoic world soul into physicality, and this involutionary motion involves successive *subtractions* of dimension, which are successive fractioning of the Logoic "unit-base." Wilber comments:

> This subtraction process is a progressive precipitation of the lower from the higher, a process called "involution"; each junior dimension is thus a *reduced subset* of its senior dimension. The reverse of this subtraction, precipitation, or involution process is simply evolution, or the unfolding of successively senior dimensions from their prior or involutionary unfoldment in the lower domains.... This is why evolution, vis a vis the lower, is an *addition or creative emergence* of successively higher domains from (or rather through) the junior dimensions.... Matter is a reduced, subtracted, or condensed version of whatever Idea is.[28]

Ken Wilber's "arithmetization" of the Chain of Being into an additive/subtractive progression through dimensions of greater/lesser integrality of being corroborates the notion of a logarithmic unfolding/refolding of worlds, and of logarithmicity as a fundamental principle of world creation. Logarithmically is how the world soul deploys itself in cosmos, "scaling itself down" into seven dimensions of self-similar replication to manifest in physicality, while matter responds by shaping and transmuting itself into ascending vehicles of consciousness.

This grand archetypal spiraling cycle replicates itself at many different scales of time and space, yet the motion is the same and so is the 7fold template of stages. The brief season cycle it takes for Earth and Sun to climb the stairwell of root, stem, leaf, and calyx before the Sun can reveal itself in flower and fruit, is analogous (of the "same logos" indeed) to the eonic cycle it takes for spirit-and-matter to descend-ascend through atom, cell, plant, animal, before the Logos may begin to individualize in human units of Logoic life (monads). The monad's relation with a body engages an analogous cycle of individuation—human instead of Logoic—so that, after a number of incarnations, the monad begins to shine in the center of the soul's full blossom—as the "jewel in the lotus"—charging the radiant individuality with creative potency of a universal order.

Blavatsky emphasizes that, while matter is informed all along by universal energy, it is not specifically connected to a monad until the emergence of the human being, whose form represents the first involvement of a "monadic" spark of the Logos in a form of matter—a *human* body. Matter constitutes itself for the first time into

---

27 Molla Sadra Shirazi, in Christian Jambet, *Se rendre immortel*, pp. 39–40 (tr. by the author).
28 Wilber, *Quantum Questions*, pp. 152–153, footnote.

IV. The Spiraling Dynamics of the Universe

a unit-form of spirit life. Humankind is to the solar Logos what the Ego is to the monad. The passage is worth quoting in its entirety, as it clarifies the mutual process of differentiation in which matter and spirit are engaged. While matter becomes organized in complex forms, spirit individualizes as unique beings.

> The Monadic, or rather Cosmic, Essence (if such a term be permitted) in the mineral, vegetable, and animal, though the same throughout the series of cycles from the lowest elemental up to the Deva Kingdom, yet differs in the scale of progression. It would be very misleading to imagine a Monad as a separate Entity trailing its slow way in a distinct path through the lower Kingdoms, and after an incalculable series of transformations flowering into a human being.... The atom, as represented in the ordinary scientific hypothesis, is not a particle of something, animated by a psychic something, destined after æons to blossom as a man. But it is a concrete manifestation of the Universal Energy which itself has not yet become individualized; a sequential manifestation of the one Universal Monas. The ocean (of matter) does not divide into its potential and constituent drops until the sweep of the life-impulse reaches the evolutionary stage of man-birth."[29]

This important point highlights how, from the perspective of the descending Logos as well as that of the evolving world, the central landmark—literally, the *heart*—of every grand and small cycle of involution-evolution is the advent of the quintessential dimension—the fifth. The fifth is the third harmonic and third principle of the "son"—the soul—the "likeness" of the Logos of the beginning. Evolution is essentially a harmonizing process, which culminates in the quintessential conjugation of the great polarities of spirit and matter—epitomized in the third harmonic musically, the dodecahedron geometrically, and the soul-self ontologically. On the universal path of involution/evolution, this quintessential hub is the advent of humankind. On the evolutionary path of the human being, it is the awakening of the soul dimension that the interplay of spirit and matter intones at the heart of the human self when a critical proportion of other-awareness has been integrated in the self.

Blavatsky's next remark further illuminates how the human self, the fifth principle of mind, is at once the incarnating foot of divinity and the crown of matter and its elements.

> The two higher principles [Atma–Buddhi or the Monad] can have no individuality on Earth, cannot be man, unless there is (a) the Mind, the Manas–Ego, to cognize itself, and (b) the terrestrial false personality, or the body of egotistical desires and personal Will, to cement the whole, as if round a pivot (which it is, truly), to the physical form of man...Incarnate the Spiritual Monad of a

---

29 Blavatsky, *Secret Doctrine*, vol. 1, p. 178.

Newton, grafted on that of the greatest saint on Earth, in a physical body the most perfect you can think of... if it lacks its middle and fifth principles, you will have created an idiot—at best a beautiful, soul-less, empty and unconscious appearance.[30]

The cosmic function of humankind—the quintessence of the Earth being—is to harmonize spirit universality and earthly matters within its Egoic selfhood, and thus individualize the Logoic consciousness of the world being. The wise forms, which come out of the hands and will, ideas and words, of human souls, bear the harmonic signature of the Logos and replicate "on a human scale" the generation of nature and cosmos by the Word of the beginning. By virtue of the fifth principle, the human being is the first likeness of the Logos on the ascending scale of being—the 12fold soul-self replica of the 12fold world soul. "Man is the measure of all things," said the Greek philosopher Protagoras. "The human being is an integral of the universe," echoes Molla Sadra Shirazi.[31]

✪ The cosmic ladder of world dimensions and planes of consciousness could be referred to as *the logarithmic scale of being*. While the decibel or cent is the unit-base of the acoustic and musical scale, the unit of this scale is a Logoic magnitude: a "kingdom of being." From the perspective of forms, it is an ascending scale. In terms of energy, it is a descending scale. From divine to mineral, every kingdom of being represents a reduced degree of Logoic magnitude, a denser "bundle" and smaller fraction of the Logoic world Unit—as decreasing octaves correspond to decreasing fractions of a fundamental frequency (1/2, 1/4, 1/8). ✪

The core "unit" is the Logos indwelling every level of being and kingdom of nature. Each kingdom represents a partial "bundle" of Logoic unfolding, and the universe as a whole is a logarithmic unfolding of self-similar replications of the Logos-"base"—a grand organism of nested hierarchies. The terms we use to refer to the study of different dimensions of reality—*mineralogy, biology–physiology, zoology, psychology, theology, cosmology*—discreetly acknowledge the ladder of Logoic manifestation to which these different fields point.

SENSE PERCEPTION SHEDS LIGHT ON THE INNER REALITY OF LOGARITHMICITY

While logarithmicity sheds light on perception, the senses, we are going to see, illumine, in turn, logarithmicity. They help uncover what is truly at play in logarithmicity, hence at the core of the world construct. The nature and function of the senses, as well

---

30 Ibid., vol. 2, pp. 241–242.
31 Shirazi, in Jambet, Op. cit. p. 97: *"L'homme est une integrale d'univers."*

as their shapes, not only confirm the Platonic process of dualization–triangulation at the source of manifestation, but bring new insights into it.

The five senses (sight, hearing, smell, taste, touch) are essentially passageways of encounter and exchange between perceived world and perceiving self, object and subject. They are channels through which the soul takes in the outside world (its "other"), as well as impresses itself upon it. They allow the *blending of outside and inside* we call sensory experience. Via the senses, the outside world passes into self and "turns into" subjective experiences, which both structure and "universalize" our egoity. Via the senses as well, the soul passes into the world, informs and shapes it. Matter turns into subjective forms of sensory perception, and self turns into objective forms of matter, as the immense variety of human creations attests to. What we see when we contemplate gardens and cultivated lands, cathedrals and cities, is externalized human selfhood. Flying over inhabited lands can be a special source of moving impressions about the deployment of the human soul on the surface of the Earth.

Like the five elements in the macrocosmic body, the five sense organs represent the outposts of the self in the human body. They are the most physical "means" of self-and-world interplay and communion. While the elements represent the five forms of harmonic conjugation of world soul and matter at the physical basis of cosmos, the senses are the five forms of conjugation of outside and inside at the physical root of self-awareness. They do for the soul what the elements do for the universal Logos: they bind the extremes of spirit and matter—self and world— into spaces of creative interplay. They conjugate world and self into percepts. They "organize" the interplay of consciousness and physicality, mind and body. In that, they reproduce the very interplay (of same-other, then soul and matter) that brings them to existence. The sense organs epitomize the fitting together of descending soul and self-gathering world, the conjugation of the incarnated self with evolving matter. They are matter "turned into" organs of perceptive flesh, matter evolved to an "elemental" selfhood.

We saw that perception blends a faculty to discern, add, and subtract quantities of stimuli, with a faculty to grasp such quantities in terms of proportional increase or decrease in magnitude of a base-unit of the stimulus. It blends a differentiating process of quantitative distinction with a qualitative registering of proportionate rapport to a core "unit" of stimulus. Perception combines an *objective–external* apprehending of reality with a *subjective-internal* grasp. In weaving together sensing of difference and sensing of proportional sameness, perception carries out at once a distinguishing and integrative apprehending of reality—it apprehends *outer bounds and inner bonds*.

The blend of outside and inside intrinsic to the senses, and the corresponding blend of objective and subjective components inherent to sense perception, bring to light the conjugation at work in logarithmic operations. What this analysis of senses and perception reveals is that behind the blends of addition–multiplication and arithmetic/geometric progression characteristic of logarithmicity is a blend of objective addition and subjective, qualitative augmentation.

Fechner's discovery of the logarithmic character of sense perception intimates that a common dynamics presides to the subjective operations of the psyche (self) and the objective operations of nature (world). The light that his law sheds on the operation of perception also reveals key aspects of the principle of logarithmicity: dimensionality and inside–outside interplay.

○ These insights lend new perspectives on the process of involution ("creation") and evolution. What they illumine in the deployment of the Logos and the evolution of cosmos is a dual operation: differentiation of dimensions and self-similar replication of a core-identity or "self." In blending separation-distinction and proportionate sameness (other-making and replicating-of-same), logarithmicity conjugates an *outward process of objective extension (key to world-making) and an inward process of inner expansion of "self" (key to soul-unfolding)*. It articulates the joint deployment of outside (world) and inside (consciousness, soul), which constitutes the universe. Indeed, the perfect middle path that the logarithmic spiral conducts between hub and periphery is a *constancy of rapport to inside and outside.* ○

What geometry formulates as the "equi-angularity" of the logarithmic spiral (the constancy of the angle of radius to tangent) is a dynamics of constant and perfect negotiation between outward, centrifugal pull and inward, centripetal pull—between the impulse to unfold and grow and the imperative to stay within the boundaries of proportionate unfolding.

Logarithmicity emerges as a formula for the dual process of world formation and self-unfolding—involutionary deployment of Being into world and evolutionary "turning" of world into self.

On one hand, the spiral reflects the *core* process of centrifugal, involutionary unfolding of the Logoic matrix of the universe: dualization of Oneness into "same" and "other," inside and outside, spirit and matter, triangulated by a dynamic bond of harmonic, proportionate, "love" interplay (soul, logos).

On the other hand, it speaks to the *peripheral* process of centripetal, evolutionary at-onement of world and self, outside and inside, "otherness and same"—exemplified in sense organs and sense perception.

IV. The Spiraling Dynamics of the Universe

The inner reality of a logarithmic universe is an all pervading dynamics of vibrant nuptials of Heaven and Earth, self and world, into whirling forms through which a Logos (whose essence is indeed love) proceeds into cosmos and cosmos proceeds to become conscious of itself as a soulful being—a "self."

Cosmos and Nature's common sense

> *Pure mathematics is, in its way, the poetry of logical ideas. One seeks the most general ideas of operation which will bring together in simple, logical and unified form the largest possible circle of formal relationships. In this effort toward logical beauty, spiritual formulas are discovered necessary for the deeper penetration into the laws of nature.* —Albert Einstein

No wonder that the very shapes and structures of our sense organs—and not only their modus operandi—carry the logarithmic signature of the soul of the world. What comes to light in the swirling structures of our sensory apparatus is the operation that inspires and in-spirals the entire world process: the wedding of cosmos and Logos.

The ear offers the most visible example of a logarithmic spiral formation. The vortical shape of the external ear, evocative of a seashell, displays the formative power of the sound wave of life. The inner ear, which contains the receptors for hearing and equilibrium, is composed of the spiral formation called the cochlea (a spiraled, hollow, conical chamber of bone), the spiral organ of corti, equipped with the auditory sensory "hair cells," and the three semicircular canals. The term *cochlea* comes from the Greek *kokhlias* ("snail, screw"), from *kokhlos* ("spiral shell"), which highlights its coiled shape. The term *labyrinth*, used to designate the whole inner ear (bony labyrinth and membranous labyrinth), suggests the convoluted path between periphery and core, outside and inside, traveled by the cosmos to "turn" into "self."

With regard to the eye, scientific studies have established that the corneal nerves of the sub-epithelial layer terminate near the superficial epithelial layer of the cornea in a logarithmic spiral pattern.[32] Other researches have shown that corneal epithelial cells in mice also display logarithmic spiral patterns. "These observations of spiral patterning in distinct layers of the cornea are suggestive, researchers conclude, of a global systemic force that regulates formation and maintenance... There is currently no consensus on a best explanation."[33]

---

[32] "Transgenic corneal neurofluorescence in mice: a new model for in vivo investigation of nerve structure and regeneration," *Invest Ophthalmol Vis Sci.* 2007 Apr. 48(4): 1535-42.

[33] See http://onlinelibrary.wiley.com/doi/10.1002/cplx.21562/full.

The main components of the internal structure of the nose, the organ of smell, are the turbinate bones or nasal "conchae." These are thin, scroll like bones shaped like an elongated seashell, which protrude into the breathing passage of the nose. The term *turbinate* means "shaped like a spinning top or inverted cone"—from Latin *turbo, turbin*, spinning top, whirl. In humans, turbinates divide the nasal airway into three groove-like air passages. Explanations of an adaptive, teleological order generally accompany the description of these forms, indicating how these air passages "force inhaled air to flow in a steady, regular pattern around the largest possible surface of cilia and climate controlling tissue." The logarithmic spiraling shape points to a deeper, ontological logic, which, far from contradicting scientific considerations on the marvelously adaptive character of such bridging formations, illumines in them not a secondary process but the very principle of their formation: mediation and conjugation.

Harmonization is the divine logic of adaptation, the ontological reason behind the "adaptive" forms of nature. The logic of the Logos is a logic of harmony. The "constant" harmonization of inside/outside embedded in the logarithmic design of the sense organs reproduces the overarching logic of love-harmony from which the world sprung. The garment of the Logos, harmony is the subtle fabric of cosmos and nature. Harmony is how the Logos resounds and "plays" itself out into forms, and harmonization is his dynamics of love in every formation of nature, of which the human body is an instance. This love dynamics of harmonization guides all formative and growth processes.

The nasal passages are perfectly adaptive formations shaped by outside air and breathing self. In the turbinate formations of the nose, outside air moves in along the "wound up" bony flesh of a being. This "winding up" of bodily matter in the nasal conchae is a perfect example of logarithmic blend of inner self and outer cosmos. Linguistically, to wind (past: wound) means, as an intransitive verb, "to take or to move in a twisting spiral course"; as a transitive, "to repeatedly twist or coil around itself or a core." As a noun, "wind" is defined as "the perceptible, natural movement of the air." Wind can be a substitute for "breath." So, archetypally, the motion of the air outside and of the breath within is conceived as *naturally* whirl-like, spiral-like. A tornado is a spectacular version of this natural winding down/up of air between Earth and sky. Like galaxies or nautilus shells, hurricanes approximate the shape of logarithmic spirals. At work in the "wound up" bones of the nose is the same archetypal motion of the air. The nasal cavity houses the sense of smell. From the nasal cavity, the air passes through a thick layer of mucous to the olfactory bulb where smells are identified. As of yet, no spiral formation seems to have been identified in the olfactory epithelium.

The receptors for gustation are found in the taste buds, which are described as *flask-like* in shape. They contain two kinds of cells: supporting cells and gustatory cells. The *gustatory (taste) cells*, which occupy the central portion of the bud, are *spindle-shaped*.

In conclusion, sensory organs show both spiral-like anatomies and logarithmic physiological response curves, which has led pathologist Frederick A. Hottes to advance that "the response curve of the central nervous system is probably also logarithmic."[34]

The whirlpool-like formations of sensory organs are the physical tracks of the monad, a Logoic spark, in the human body. The logarithmic dynamics of this involvement spin matter into organs apt to "turn" objects into percepts, world into impression, and subjective impulses into objective forms. Through the senses, cosmos becomes self (-aware) and self becomes world.

The spiral shapes of the sense organs reproduce the spiraling forms of nature. The human body is indeed a form of nature, as generically specific as an animal or a plant form. The archetypal motion of life epitomized in the stilled vortices of the senses illumines in logarithmicity the *common sense* of nature and cosmos—in the two meanings associated with "common sense."

Logarithmicity first points to the "basic" overarching intelligence of *becoming* that courses through nature in a dual operation of progressive embodiment and progressive "lightenment," self-deployment and self-integration, externalization and internalization. This "common sense," this (bio-) logic of nature speaks to the one-mindedness referred to earlier and spoken of in the Upanishads. The expression in nature of the Logoic core of the universe, it is the common intelligence that guides the making of organs and the unfolding of living systems in a perfect articulation of outer extension and inner expansion. From the spinning of matter into atoms, cells into organs, earth into trees, solar systems into galaxies, to the spinning of etheric currents into chakras, and mental energy into soul lotuses, the universe is a logo-rhythmic deployment of integrative unfoldings. The spiraling motion of life is the archetypal dynamics of the Logos—the fundamental dynamics of bio-logy, psychology and cosmo-logy.

Logarithmicity is also a common "sense" in that it is a sensing organ, one could say, common to all living forms, being the dynamics by means of which the world soul apprehends itself through forms of matter, and cosmos senses its "self."

This long detour through sense perception has brought us new insights into the "common sense" of the universal intelligence that pervades nature—the logarithmic intertwining of the *outgoing and incoming flow of the world being*. The spiraling

---

34 See https://prezi.com/aspjr9pv4nqe/fibonacci-in-the-human-body.

deployment of a plant between Sun and Earth and the psychospiritual growth of a human being through the integrative engagement of self and world are both expressions of the harmonizing intelligence of the Logos of life.

This common sense for evolving, growing, becoming—common to world and self—is what astrology takes hold of and frames into an organ of observation and a lens of insight. The spiral cycling middle ground of growthful interplay between Logos and cosmos is what it maps out to study human becoming. Astrology's single horoscopic eye grasps as one reality celestial configurations and psychological dispositions. It reads as one logic circumstantial cosmic factors and personal experiences. It deciphers in the changing harmonies between cosmic and natal configurations (cosmos and self) the archetypal script of the psyche. Celestial alignments and individual character, planetary transits and subjective "transitional" experiences converge in one common cosmic hub, which is an intuitional organ of perception of the soul's evolution, charged with a symbolic language based on seven and twelve.

Human beings share an innate sense of "evolving," a "common sense" about growth, a generic drive toward greater fulfillment and self-actualization. It "goes without saying," says our common sense, that willy-nilly we evolve, stumbling through mistakes, groping our way through better choices, learning to meet opportunities. Given this "evolutionary sense" we share, as we espouse our destinies, it is arresting to realize how little it is tended and explored. Many would affirm that the ups and downs of destiny have no script or score. Many, on the other hand, feel that "there is no randomness." Astrology proposes that there is indeed a spiral cycling script for this "evolving," which gives access to the archetypal tides of our destinies. The noble aura of the word *destiny* implies a creative course of relation between a person and the circumstances of their lives brought up by the "outside" of world, body, unconscious. It hints at the opportunity presented by the "fate" of an illness or the challenges of a relationship, to weave a next gradient of selfhood through a deliberate intercourse with this "other." While depth psychology seeks to bring to awareness and conscious interplay the "other" whose encounter with "self" is bending one's destiny, astrology offers an organ of insight into this very otherness of the self, and an archetypal common sense to understand the harmonizing at stake. What astrology puts forward is nothing less than a "sense organ" of insight into our evolving psyche-cosmos—individual and collective.

The unique value of the astrological mapping lies in the access it gives to a logic of psychospiritual unfolding—a psycho-logic—that transcends the dichotomy between subjective psyche and objective circumstances, conscious and unconscious.

The Fundamental Dynamics of Creation and its Reflection in the Human Mind

It lays out the cosmic-psychic score of changing harmonies and challenging disharmonies that elicits the development and "universalization" of individuality.

> *Eventually, we must all create our own systems of universal understanding, though our present life may not be the time for this. Metaphysically speaking, this will be the gift of knowledge that we return to the Mind who made us and who knows its own universe through the unique knowledge of every being.* —Joscelyn Godwin[35]

Folded in the transient spatiality of forms, unfolding itself in the course of time, is the marvelously graceful step of the Logos in cosmos, a cosmos that rounds itself and revolves upon itself to involve its Logos in orbiting spheres, spinning chakras and evolving kingdoms.

The logarithmic enfolding of the world soul into cosmos and the concomitant unfolding of chaos into Logoic forms is, in essence, a process of outward self-differentiation bundled up by inward self-integration. This process, which gives birth to the successive orders of being and multiple forms of cosmos, originates with the world soul itself. The world soul is the reality of this process, since it is, according to the *Timaeus,* the primordial interplay of love-harmony between the *objective* Other and the *subjective* Same, the creative engagement between the *without* of matter and the *within* of spirit launched by the initial "out-sounding" of Being.

The logarithmic deployment of Logos into cosmos (cosmology) and evolution of cosmos into self (biology and psychology) reproduce the primordial threefold dynamics of Creation: Oneness, differentiation, integration; Identity (1), externalization (2), internalization (3)—which the three dimensions of point, surface, and volume epitomize in the physical world. The three logarithmic parameters of unit-base, additive chain of dimensions and multiplication-deployment of self offer a remarkably simple mathematical model for the unfolding of the universe in space and time. In the end, the concrete world of natural forms reveals the same "Logoic" gestalt in the threefold spiraling gesture of any shell or growing plant: forward (unitive), cyclical (additive), expanding (multiplicative). The plant turns and so do we. In the recognition of this Logoic dynamics behind all the "turns," vortices, and wheels of cosmos lies the potential for a common "sense" of the world.

---

35 *Harmonies of Heaven and Earth*, p. 154.

The unraveling of the way sense perception operates has lifted a corner of the veil of Isis, the countenance of nature and cosmos, and revealed, suspended in its spiraling curves and exponential rounds, muted yet glorious, the sphinx like smile of the Logos spread on the face of the world. Vast and wide open in its differentiating outgoingness, enclosed to a mystery in the integrative binding of forms, this smile is the essence of the human smile. How magical a balance the human smile strikes indeed, as it pours a being out into the world and at the same time enshrines him or her in a glowing enclosure.

The logarithmic principle captures mathematically and geometrically the *differentiating-outgoing* and *integrative-binding* smile that pervades the peripheries of the universe with the soul life from which it sprang. The "golden" smile of the world soul, which runs through the universe as the expression of its unicity of being, is the quintessence of the universe. It is the quintessence of the universe *as we know it*, since it reflects nothing more than our human capacity to "know." Indeed, while the presence of the Logos pulsates, quasi-visible, almost palpable, within the proportions and rhythms of the world, it also forever withdraws, as the *One* forever does, of necessity, before the *dual* eye of the mind. However, the human capacity to know is also nothing less than the fifth principle of mind, soul, and selfhood—the objective *quinte* essence of the universe. The mind is "self-similar" in structure, and analogous in principle to the world soul. It mediates. Our perception mirrors the logarithmic dynamics of life. Logarithmicity, the "rhythmicity of the Logos," is the way the world soul unfolds into cosmos and it is the way the mind enfolds the world back to oneness by perceiving it and making sense of it.

In his incompleteness theorem, mathematician Kurt Godel demonstrated that we are necessarily limited in our capacity to know a system of which we are a part. Our discursive mind emerged from the same process as the universe and cannot step "outside" of it to view its own origin. Nonetheless, as the Logos articulates itself in the forms of nature, lending its *"arithmos"* to the shape of a shell and the growth of a plant, so does it express and discover its "self" through the human mind—and this never more definingly than when the mind apprehends and "figures out" the logic of the universe in mathematical or geometric formulas that capture a "likeness" of the whole. Einstein's equation ($E=MC^2$), to which we will return later on, is a stellar example of the continuously emerging scientific cognition of the universe, which informs the coming to self-cognition of the Logos. Our mind is the very striving of the universal Logos to know itself. It activates the potential self-knowing of the solar systemic being we are. What we come to know about the universe allows the universe to know itself.

IV. The Spiraling Dynamics of the Universe

The mind, however, does stop short of knowing its creator-source, since, by its very nature, it stops short of *being* its source. One can only surmise the greater knowing that dawns when evolutionary maturity leaves behind the duality of mind to realize its "self" as universal being, as being-universe. Such knowing is no longer a knowing of the mind, but the knowing of "being."

For now, and in this existential space of ours, it is essential to the evolution of our system that our faculty of reason proceeds to uncover the divine rationality of the universe in scientific unveilings that propel its gaze and symphonic insights that compel its heart.

※

As the potential kingdom of "souls," humankind is the quintessential manifestation of the Logos. Endowed as it is with the fifth principle of mind, and organized by it, it is the live, individualized expression of the golden mean between the great polarities of Logos and cosmos. Humans are, literally, sons and "thirds" of the fundamental One. Before the human kingdom (evolution wise), or below it (complexity wise), mineral, plant and animal kingdoms harbor no self-principle, hence no individualized *logos*, while beyond it, the angelic hierarchies have no physical body, hence no individualized physical *cosmos* vehiculating them. Emerson writes, "Within man is the soul of the whole; the wise silence; the universal beauty, to which every part and particle is equally related; the eternal ONE."

Teilhard de Chardin describes how the apparition on Earth of humankind, with its potential for self-awareness and reflection, marked the momentous step of a "reflection of life on itself." It marked a turning outside-in of the evolutionary process. The balance of cosmos and Logos shifted, as the Logos began to loom overtly and *consciously* on the horizon of the Earth. From then on, steps in awareness, and their impact on world matters were to become the dominant, if internal, face of evolution.

Even though the meaning of the word *rationality* has been reduced in modern times to the most concrete operations of the mind, the term itself conveys most appropriately the correspondence between mind and golden ratio. The mind is the "mean" that separates and bonds self and world—the spirit and matter of the world being. Its function is to further the harmonious engagement of Logos and cosmos in "spinning" matters and worlds into physical and social structures, arts and sciences, imbued with the wholistic breath of divine reason and the exquisite particulars of human experience.

It takes little effort, based on previous elucidations, to recognize that the human mind operates logarithmically. We saw that perception is threefold, and so is the mind. It has a concrete, analytical, objectifying, empirical-scientific (*differentiating*) cerebral dimension; a whole-intuiting, subjective (*integrative*) soul dimension; and a universal (*at-oning*) spirit dimension, which blends and transcends the duality of objectivity and subjectivity by identifying with what it seeks to "know." It vehiculates information, extracts meaning, and reveals universal truths. Great scientific insights, artistic creations, mystical revelations, and wisdom teachings engage the third dimension. So do experiences of synchronicity—including synchronous discoveries born of a "spirit of the time." They reveal a "mindfulness" that transcends the gap between subjective and objective experienced by the "concrete mind." While mystics are imbued with a feeling of divine oneness, true scientists are "mystics" in the sense that they are moved and motivated by inklings of unity, and devoted to a search for integrative facts and formulas. The threefoldness of the mind as concrete intellect, soulful thinking, and universal mind is akin to the threefoldness of body, soul, and spirit—outside, inside and whole/Identity—which carries the secret dynamics of life in general, and of the generative life of the mind in particular: logarithmicity.

Understanding the threefoldness of the mind restores to the notion of "rationality" its soul core of wisdom and generative power. It affirms the highly rational character of intuitions of a universal order too easily brushed away as "subjective" or irrational by materialistic thinking. It affirms the rationality of analogical thinking—the ability, says Fideler, "to read phenomena hieroglyphically rather than literally, and intuit directly the essence of a thing."[36] It reveals the essential rationality of synchronistic phenomena, which play out the transcendent, nondual "first harmonic" or divine "fundamental" of the mind. Like its Logoic model, the mind is meant to be truly "creative," and in order for its creations and discoveries to be inspiring and vivifying, its three levels must be engaged. Soul and spirit must be part of the equation. Reason, the ratio-nal mean, is by essence unifying and vibrant. For the mind to do a golden work of mediation and interpretation, one that binds the core and peripheries of reality into resonant chords of truth and beauty, it must move from, and be moved by, that which sects and bonds according to the golden mean—love.

Schwaller de Lubicz's lifelong study of ancient Egyptian temples and hieroglyphs led him to a similar threefold view of the mind. He believed that Egyptian consciousness was based on what he termed "functional consciousness"—a way of knowing reality from the inside. What he came to recognize within the language and architecture of Egypt is a wholistic relation to the world, which brought together in symbols the polarities of sense perception and inner knowing—outer objects and inner content. Hieroglyphics did not merely designate a bird; they evoked the cosmic function

---

[36] Fideler, *Jesus-Christ, Sun of God*, p. 20.

IV. The Spiraling Dynamics of the Universe

it exemplified—flight. He contrasted this unitive approach, this "intelligence of the heart" he himself pursued all his life, with the purely cerebral tendency of the modern mind to take apart and analyze a world perceived as foreign and alien, and to "granulate" experience into fragments of time and space. He called poetically to...

> Tumble with the rock which falls from
> the mountain
> ...gather honey with the bee;
> expand in space with the ripening fruit.[37]

In deep resonance with the landscape of evolution we have been contemplating, Schwaller de Lubicz considered that "each individual type in Nature is a stage in the cosmic embryology which culminates in man." Different species, he believed, developed various "functions"—what the Egyptians called *"Neters"* and we translate as "gods"—which have their apotheosis in humankind. Our exploration will continue to lead us to these "gods"-functions of the kingdoms and their relation to the major ratios of musical harmony.

Paradoxically, it is the aspect of the mind most distant from the Logoic core—the world-reifying intellect—on which rests, of necessity, the task to demonstrate how life and matter intertwine in nature's systems. Intellectually satisfying proofs are indispensable for the *whole mind* to be on board with the conception of an intelligent universe-being such as Plato and the majority of humankind before and long after him experienced. The concrete mental faculty, whose analytical skills can flatten and slice world phenomena into sequential quantities and linear reasonings, is the faculty of choice to demonstrate in physico–scientific measurements the Logoic rationality of the universe—the presence of the Logos in the psychophysics of life.

Fechner's discovery is a unique gift in this regard. It reveals the existence of a bridge between psyche and world, a bridge that points to a shared dynamics. The logarithmic modus operandi common to nature and mind signals the Identity they share and make manifest—one as a world without, the other as a world within. An inconspicuous, yet vital ground was thus laid by Fechner's scientific devotion under the Platonic view of unitive world beginnings.

The discovery of the logarithmic character of perception was the first recognition of a mathematical order in a subjective process. Since then, the principle of exponential growth has gained prominence. It has been found at work in cultural trends

---

37 Lachman, "René Schwaller de Lubicz and the Intelligence of the Heart," *Quest Magazine,* Jan./Feb. 2000, pp. 4–11.

and historical processes, including the curve of technological evolution. According to engineer and inventor Ray Kurzweil, the rate of technological, as well as biological evolution is not linear, but accelerates exponentially. The highly publicized 2012 end of the Maya calendar exposed the ancient knowledge, which guided the Mayas' relation to time and history, of an exponential acceleration of our collective evolutionary process—a fact confirmed by detailed research into historical cycles. Teilhard de Chardin's spiral paradigm of evolution presents a similar logarithmic progression from cosmos to Logos–Christos. Cosmosphere, biosphere, noosphere and Christosphere represent exponential cycles of the cosmological spiral. Noosphere, and Christosphere, the "subjective" spheres of conscious humankind, are on a continuum with the "objective" spheres of matter and life, which constitute the first two kingdoms: cosmosphere and biosphere.

Fechner's primary goal was to demonstrate that mind and body, though they appear to be two entirely separate realities, are in fact two sides of one. The main impact of his discovery was to usher psychology into the domain of experimental sciences, now that it showed the potential to become quantified. However, Fechner's law also suggests that psyche is the other side of physicality—being what grants physicality its features in our "perception." As such, it has the potential to usher physicality into the science of consciousness, and lay the ground for a "psychic logic" of cosmos.

As history would have it, Fechner's discovery was swept along the collective wave of fascination for empirical evidence (from e-videre, visible to the eye) ready to take psychology to the lab. A few decades later, however, the advent of quantum physics initiated a turning of the tide. The revelation of an enigmatic rapport between object and subject—world and mind—in particle physics experiments signaled the end of the scientific dream of absolute objectivity. This turn of the tide has the potential to activate the other, complementary, side of Fechner's legacy and entrain a reversal of its orientation, allowing it to feed the new momentum toward a unified science of matter and psyche.

The recognition of one—logarithmic—principle at work in the functioning of the mind, biological growth, and cultural evolution, is a step in the direction of this possibility. Not only does it define a principle of continuity between the two "sides," but it invites the scientific mind to realize that empirical objectivity is but one aspect of rationality, and that a fully rational approach of the reality *we experience* calls for a wholistic blend of sequential objectifying intelligence and participatory identification with the core of what we seek to know. The challenge for the contemporary mind is to allow its well-developed analytical lucidity to make room for a vaster rationality, wholistic and unifying, willing to hold as one questioning the mysteries of matter and consciousness. "What we observe," remarked Werner Heisenberg, "is not nature itself but nature exposed to our method of questioning.... Natural science does not simply describe and explain nature; it is part of the interplay between nature and ourselves."[38] To lift a corner of the veil on the unity of world and psyche is the gift Fechner meant to give. Embedded in his law was a scientific intimation of what he most wished to give his fellow humans: a glimpse of the "angelic being" of the Earth, an inkling of the psychic life of cosmos.

THREEFOLD, TWELVEFOLD AND SEVENFOLD: THE ARCHETYPAL BUNDLE OF SPIRIT, SOUL AND BODY
Where logarithmic dynamics merge with harmonic structure

> *Nature is the realization of the simplest conceivable mathematical ideas.*—A. EINSTEIN

We are going to see that the Pythagorean laws of musical harmony, on which the cosmogony of the *Timaeus* is based, not only affirm the logarithmic character of the universe's unfolding but further illumine it.

The negotiation of differentiation and integration at work in the dynamics of creation is perfectly reflected and expressed in the articulation of the octave series of duplication with the fifth series of mediation. This correspondence brings us back to the laws of harmony as to a luminous source of wholistic understanding, convincing to the mind and grounded in sensory experience, of the cosmogonic deployment into a universe of a primordial sound-vibration—soundless OM of Indian scriptures or big-bang of Western science.

What we are going to find, and this will lead us to the most panoramic vista of this journey, is that logarithmicity sheds a new light on the articulation of the series of seven octaves with the series of twelve fifths, which founds the Pythagorean musical scale. The deeper level of meaning brought forth will imbue with a new glow of "common sense" the Platonic philosophy and the ancient knowledge it conveys about the world construct.

The understanding we arrived at earlier is that the coming together of the octave series with the series of fifths represents the integration of a line of self-duplication (based on the 2:1 ratio) with a line of mediation (based on the the 3:2 ratio or golden "mean")—self-replication of the fundamental on one hand, and bonding mediation of its octave polarities on the other hand.

---

38 Heisenberg, *Physics and Philosophy*.

The mediation happens *within* the self-replication. An obvious fact, perhaps, but a most significant one, archetypally. What we come to discern more distinctly at this point is that the intersection of the line of octave replication with the line of unfolding mediation ("mean"), which founds the scale, determines a seven-octave series of *forward-extensive* replication of the "fundamental" sound of creation and a twelve-note series of self-unfolding mediation (the unfolding of the 3/2 "mean") between the two octavian ends of the fundamental. This 12fold circle of self-bonding (a circle of G, sol and soul) is the *inward-centering* side of the unfurling manifestation of the world being. The 7fold series speaks to the externalizing of Being into dimensions, the 12fold series to its internalizing into selfhood. In the replicating series, the "Same" is brought to distinction; in the mediating series, the distinct is brought into rapport and bonding.

That mediating happens *within* the space of self-replication is a fact, not only for the primary differentiation-mediation of fundamental, octave and fifth, but also for the secondary mediation that the circle of fifths represents within the seven-octave replication of the fundamental. The first one echoes the cosmological differentiation-mediation of Same, Other and Being, which results in the "world soul compound"; the second one parallels the Platonic fractioning of the world soul, which informs the composition of the 12fold–7fold world body.

What the latest context brings to light is that the 12fold circle mediates and binds the extremes defined by the convergence of the two series at the 7th octave. The fundamental vibration creates the *field of self-replication* of the 7-level body of Being–cosmos (the field we call space) and the 12fold circle of mediation constitutes the *inner milieu of bonding* of the world extremes (spirit and matter)—12 tones of differentiated selfhood. This 12-tone selfhood is the archetypal inwardness of being—what we call self, soul, mind—nested at the center of every dimension of Being: DNA in the cell, chlorophyll in the plant, heart chakra in the etheric body, causal body on the mental plane, zodiacal heart in our universe.

✪ The understanding recently gained, that the logarithmicy at work in creation and sense-perception is a conjugation of inside and outside, leads us to realize that the two series refer to the two sides of the world being: self-duplication gives birth to self-similar dimensions or levels of being; self-bonding produces the inner circle of selfhood (logos, son, soul). Logarithmicity illumines in the octave series an addition of dimensions, and in the circle of fifths an unfolding (or multiplication) of "self." It reveals that the mathematics of harmony "fit together" the self-extension of Being with its self-blossoming—external extension into a seven-dimensional body and inner blossoming into a 12fold "selfhood." *They link and unite the outside and inside—the body and soul—of the world Being.* ✪

IV. The Spiraling Dynamics of the Universe

The octave, the second harmonic, is the acoustic experience of the primordial replication of Being (same–other, father–mother)—mathematically, its duplication (1/2); ontologically, its dualization. It speaks to the world creative ("father") impulse of the world being to externalize itself in a cosmos, corporealize itself in a *body* (the "mother"-matter otherness of the same)—an *outside world*.

The fifth interval, the third harmonic, speaks to the primordial mediation and gives us an experience of the tri-unity of the Son, the inner reality of soulful bonding that in-forms nature and cosmos—the *inside* or psyche of cosmos—consciousness.

These two "sides" intertwine in the world reality as they do in music, yet they are also as distinct from each other as world and self, body and consciousness are in our experience. While we readily identify with our psychological processes, our bodily processes remain largely alien to our "selves." How do "I" move my limb? How do "I" breathe and digest? For many lives, the body is an "outside"—until the ego-centered separateness indispensable to the development of selfhood is outgrown by a "self" co-extensive to other selves and elemental nature—until, that is, we know ourselves as lake and bird, and experience the "outside" world as "self" in a realization of Oneness that transcends outside–inside, self and world.

The two series (octave series and series of fifths) are logarithmic in themselves, being at once additive and multiplicative. However, when they are brought together to constitute the scale, each series becomes polarized, relative to the other: the octave series (self-replication) stands out as relatively additive and the series of fifths (self-unfolding) as relatively multiplicative. Octaves are repetitive replications of the fundamental, whereas fifths unfold different qualitative tones within it. When viewed in the musico–cosmological context of their combination-convergence, the octave series becomes the additive, extraverted polarity of world deployment (based on self-replication), and the series of fifths becomes the introverted polarity of self-unfolding (based on internal bonding).

This understanding grants a new depth of significance to the observation made earlier that 12fold and 7fold are defined by the intersection and at-*one*-ment of the 2/1 ratio series with the 3/2 ratio series. The coming together, but for a comma, of 12 and 7, marks the joint fulfillment of each series within a fundamental "One." The self-replicating series of octaves is limited to a 7fold extension by its encounter with the mediating, integrative cycle of fifths, and the mediating series of fifths is limited and defined *as 12fold* by its reciprocal encounter with the octave series. The 12fold and the 7fold define each other and constitute an interdependent duo. What we now see quite clearly is that each series binds the other for the purpose of creating a differentiated fractioning that is also an integrative self-deployment of the fundamental "One"—a unit of life.

Harmony, the Heartbeat of Creation

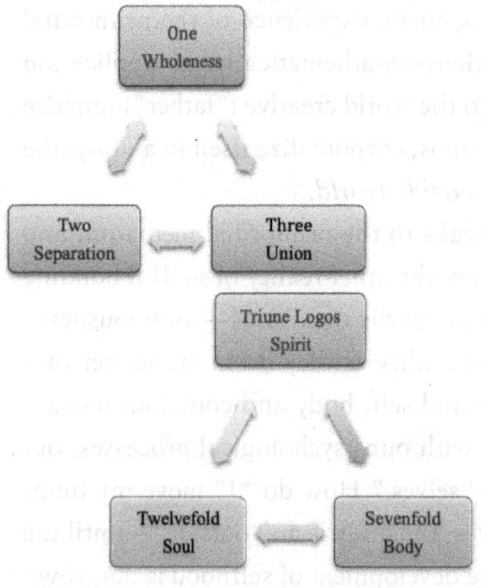

Unmanifest being (upper left); Being manifest in form (lower right)

The new "One," informed jointly by the numbers 3, 2, 1 and their interplay—1, 1/2, 3/2—replicates the primary unmanifest trinity of Same, Other and Being (1, 2, 3) in a second trinity of 3, 7 and 12. This second trinity is the template of the manifestation into "world" of the transcendent trinity. It represents the archetypal structure of cosmic systems and organic forms.

Whereas simple mathematics bring to light the logarithmic principle at work in the intersecting dynamics of the two series of octaves and fifths, it is the unraveling of logarithmicity itself that reveals the nature of these dynamics—*outward replication and inward mediation*—thus lighting up the underpinnings of the 7fold–12fold duo central to cosmological structures and the organization of time and space on Earth.

What Pythagorean harmony captures and presents as a de facto 7fold–12fold dual scale proves to be, in the light of logarithmicity, the musical version of the cosmological co-emergence of the *body and soul* of Being—the co-structuring of the 7-dimensionality of the world (its "seven heavens") and its 12fold center (from chlorophyll molecule to DNA to heart chakra to zodiac). The musical scale—a logarithmic scale—blends the two, and it also keeps them distinct. The seven white and five black keys of the piano clavier display the combination of blending and distinctness characteristic of the seven and twelve pair—diatonic and chromatic scale.

✪ The sovereign simplicity of the fundamental numbers of creation evokes out of its discreet embedding in Plato's *Republic* a mysterious comment. The cryptic remark, quoted below, charges with "awakening" potential considerations that our age might easily dismiss as remote and irrelevant. Instead, Plato alerts the soul to the power concealed in "the little matter of distinguishing one, two, and three"—the power, he says, to usher the soul into "the true day of Being."

> The little matter of distinguishing one, two and three, [which is] something which all arts and sciences and intelligences use in common.... is the sort of knowledge which draws the soul from becoming to being.... [It] has the power of effecting the *turning round of a soul* passing from a day which is little better

IV. The Spiraling Dynamics of the Universe

than night to the true day of being, that is, the ascent from [the world] below [to the gods], which we affirm to be true philosophy.[39]

The image of a turning around of the soul across a threshold evokes the mysterious passing inside out of Dante and Virgile from hell to paradise or the inversion of consciousness described by Fechner as the inner experience of death.[40] It has the tenor of a spiritual scientific hint from an initiate in the mysteries. ☯

※

Once the Platonic paradigm of world manifestation according to the laws of harmony receives this vital insight from the principle of logarithmicity, it stands out as a cosmological model of majestic simplicity. Logarithmicity grants a defining mathematical principle and geometric profile to the metaphysical co-arising and co-evolving of world and soul, matter and consciousness, which result in the geometry of bodies and configuring of souls.

The enigmatic and insistent company that 7 and 12 keep in the structures of time and space (7 days, 12 months, 12 hours, 7 dimensions, 7 colors, 7 planetary orbits), music (7 diatonic and 12 chromatic notes), astrology (12fold zodiac, 7 planets), cosmo-metaphysics (7 pillars of wisdom and 12fold Jerusalem, 7 climates and 12 imams, 7 rays, and 12 creative hierarchies) begins to make sense and reveal its meaningful ground. So does the diptych of body and soul: 7fold body (human chakras, solar systemic planets) and 12fold causal self (individual heart, universal zodiac). So does also the Biblical diptych of the 7 days of Genesis and the 12fold Jerusalem of the Apocalypse. In the 7 days of Genesis, God *separates* Heaven and Earth, light and dark, and then proceeds to differentiate kingdoms. In the Apocalypse, Heaven and Earth, spirit and body, *bond* fully in human consciousness. Whereas the Old Testament has Wisdom building her 7-pillared house (Proverbs 9:1), the 12gated heavenly Jerusalem descends from the future, a "revelation" from the "heavens" within.

The two lines of deployment are interdependent with regard to the whole—the "oneness"—they constitute. This interdependence harbors an essential insight about human evolution, implicitly present, but often distorted, at the core of religious and spiritual worldviews. It suggests that the unfolding of the Egoic lotus, the human soul, to its 12fold Jerusalem-like fullness goes hand in hand with a complete incarnating and penetration of consciousness into the 7-leveled vibrancy of one's bodily constitution.

---

39 Plato, *Republic*, VII, 521–522.
40 Fechner, *The Little Book of Life after Death*, p. 69.

◉ At the source of the 7fold-12fold universe is the *logarithmic logic of the Logos*. In the light of logarithmicity, the laws of harmony reveal how cosmos and soul co-arise as two sides of reality "because" they co-arise as two complementary *motions-motivations* of the world being: spinning out of Logos into 7fold cosmos and spinning in of cosmos into 12fold Logos. A centrifugal motion of differentiation and a centripetal motion of integration characterize the fundamentally dual nature of manifestation: *matter-ialization* and *spirit-ualization*, involvement in body and evolution in consciousness. Seven and Twelve are fundamental numbers in cosmos because they structure the two *sides* of the world being, which, from an inner dynamic perspective, are the complementary *processes* of involution and evolution. ◉

The seemingly linear division of the octave into successive intervals and tones hides a dual up-and-down dynamic of interval formation, which parallels the cosmological dualization of the One into a spirit–matter, psycho–cosmic world. Like the world soul it symbolizes, the musical whole of the octave is the self-resonant space of a polarity: the fundamental and its duplicate, the octave. The 7/12fold musical scale, we saw earlier, emerges from this polarity and its mediating interval, the fifth: an ascent from the fundamental by a fifth, followed by a descent from the upper octave by a fifth creates the interval of a tone from which the scale unfolds. In each direction, the interval of a fourth complements the fifth.

The fifth and the fourth emerge simultaneously, complementarily, from the third harmonic, actualizing in distinct intervals the dual oneness of the octave. The interval of the fifth (up from the lower end of the octave) speaks to essentialization into the triad of spirit—the Logos. The interval of the fourth (down from the upper octave) speaks to corporealization into the quaternary of matter and elements—cosmos. The alternating ascent and descent of fifths and fourths illumines the involutionary and evolutionary breath currents of the world being at the heart of cosmology.

With the sublime sobriety of the tonal world, the Fifth intones the fundamental secret of Logos/cosmos, its Octavian parents and sibling shadow (the Fourth) held in its regal bearing.

◉ What we come to recognize at this narrowing stage of circling around the formative hub of creation is that, while the spiral is indeed the fundamental motion of life, the bundling process of logarithmicity intervenes to contain the potentially unlimited unfurling of this spiral with the imperative to replicate the Logoic trinity into forms that bear the world soul triunity in their proportions (3, 7,

12). What "cuts" the infinite spiral into the limited number of cycles we will be observing in the spiral formations of nature—the details of atoms, shells, plants, human bodies and solar systems—is the conjoining into one "unit" of a process of self-replication with a process of unfolding mediation. The conjugation of the two processes bundles the spiral into a 3–7–12 form, which is the externalized likeness of the Logos (1, 2, 3), a self-same replica of the golden secting–bonding that vessels the triune Logoic "one." This second trinity is the "temple" of the Logos and the universal template of forms. ✪

In the end, music and the laws of harmony unveil the archetypal "enwrapping" of the Logos of life in the spiral-cycling dynamics of living forms and systems. First, the *harmonic series* sheds light on the unfurling of the Logoic vibration into partial wholes (harmonics), which are "sound" fractions of the One. Second, the *musical scale* reveals a bundling process that illumines the formation of units of life and forms of being. While the harmonic series displays a law of sound, which pertains to the cosmic, "divine" ground of reality and sensory perception, the musical scale is a man-made formation, based on the perception and rational ordering of the "natural" laws of harmony.

Using the musical scale and the circle of keys on which composition is traditionally based, humankind exercises the divine art at the source of all forms of life. Music making is emblematic of the integration by a kingdom of nature of the independent capacity to create by "bundling" materials—though not living forms—as the Logos does. The quintessential "likeness" of the Logos, humankind is essentially an apprentice in the divine art of harmonic bundling—of external creations, of its own psyche, and of the world in which it participates. As in music, the secret of the Art is proportionality.

The archetypal logarithmic spiral: The golden spiral

Before directing our steps toward the archetypal core of the logarithmic spiral, we will climb up for a moment to look back at the course of our voyage so far. The initial inquiry into twelve led us to follow Plato along his cosmogonical descent into the harmonic structuring of cosmos. The recognition that behind these harmonic structures is a formative process, which is the involutionary dynamics of Logos in cosmos, brought us to explore the motion prevalent in nature—the spiraling motion by which Life turns matter into forms of being. The marvelous spiral, the ubiquitous motion of life in forms, revealed what appears to be a key principle in the articulation of Logos and cosmos: logarithmicity. Fechner's law about the logarithmic character of sense perception then came into view, revealing that the dynamics at work in the formations of nature is also what governs the subjective process of perception. The logarithmic character of perception brought new insights into the nature of perception itself, which in turn shed light on the function of logarithmicity as a mean to blend outward extension and inward unfolding. The unitive weave of outside-inside we came to recognize in logarithmicity pointed us back to the harmonic operation of world formation, most specifically the 7fold–12fold bundling central to musical scale and cosmological mysteries. Instead of the mathematics of harmonic structuring that our first, *Timaeus*-guided pondering of 3, 12, 7 exposed, what this second round has brought to light is the wholistic bundling of living systems that this "second trinity" presides over. This fundamental threefold bundling phenomenon is now leading us closer to the triune ground of the world.

In summary, this journey has taken us from the primordial, metaphysical "resounding" of Being to the "sound" physical forms of nature, then from these tangible forms to their fundamental formative dynamics. This dynamics is now drawing us to its archetypal ground, where we will discover that the core of the logarithmic spiral is nothing other than Phi, the golden mean. In the end, the core of the harmonic design will prove to be, logically and "marvelously," the core of the spiral cyclic unfolding—a decisive confirmation of the logarithmic envelopment of the Logos in cosmos.

The next step in what has become an adventure to the source and root-principles of the universe will walk us closer to the hub of nature's motions and designs—the archetypal logarithmic spiral—the golden spiral. The simple mathematics involved, while indispensable to show the wondrous fact, may be skimmed over by the reader primarily interested by what they reveal.

Whereas logarithms refer to mathematical progressions of numbers, and logarithmic scales to perceptual progressions in magnitude, logarithmic spirals display the principle of exponential growth in the three-dimensional geometries of organic forms— from nautilus to cochlea to galaxy. As mentioned earlier, the logarithmic spiral follows the rule that, for a given rotation angle (such as one revolution), the radial distance to the pole or origin is multiplied by a fixed amount. In other words, as the angle of rotation increases in an arithmetic progression (a quarter turn or a full revolution for example), the radius of the spiral increases in a geometric progression.

Lawlor echoes our findings when he states that these two numerical progressions—arithmetic and geometric—yield the spiraling forms of nature and cosmos, as well as all the ratios from which the musical scales are constructed. He goes on

IV. The Spiraling Dynamics of the Universe

to say, "We find in the spiralings of gnomonic figures a close association between the temporal laws of sound and the proportional laws of space."[41]

Lawlor does not connect the combination of these two progressions with the principle of logarithmicity itself. However, it has become clear, through the last segments of our exploration, that logarithmicity is the principle at work in the laws of harmony and the spiral cyclic geometry of nature—the "temporal laws of sound and the proportional laws of space." It articulates mathematically the dual principle of *integrative differentiation* at work in the involution/evolution of the Logos in space and time. The "close association" between laws of sound and laws of space points to an identity of principle of which they are complementary expressions. The temporal laws of sound exist only in space—the space that vibration creates—and the proportional laws of spatial forms exist only in time—as the very unfoldment and "becoming" of Being. Space–time is the *one* matrix of involution of Logos into cosmos (space) and evolution (time) of cosmos into self awareness.

With the term *gnomonic,* Lawlor highlights a significant feature of the logarithmic spiral and of logarithmic growth, a feature that grants further insight, this time from the dynamic perspective of formative processes, into the notion of dimensionality explored earlier from the standpoint of cosmic structures. This earlier look focused on identifying the main dimensions or kingdoms of world manifestation—seven "structural" cosmological layers, expressing seven levels of being. What gnomonic growth will shed light upon is the *process of unfolding* of dimensions.

A gnomon is any figure that, when added to an original figure (or subtracted from it), leaves the resultant figure similar, in proportions, to the original. In nature, the shell of the chambered nautilus, a ram's horn, or the trunk of a redwood tree retain their shape as they grow by adding a new growth that is a "gnomon" of the entire organism.

The most perfect figure of gnomonic growth is the golden spiral. The simplest way to draw a golden spiral is to construct a golden rectangle—a rectangle in which the ratio of the larger side to the smaller one is Phi (on the figure, a+b : b+c = Phi). What happens with such a rectangle is that, when squares of equal length to the width of the rectangle (gnomons) are cut from the original rectangle, the sides of the "daughter rectangles" remain in the same proportion. Tracing a curve through the points where this series of "whirling squares" contacts and divides the sides generates a golden spiral. The gnomons are the squares that produce the smaller golden rectangles and the golden spiral.

Figure 37 illustrates at once the swirling in of the spiral and the accompanying phenomenon of gnomonic contraction. The gnomonic construction of the spiral

---

41 Lawlor, *Sacred Geometry,* p. 71.

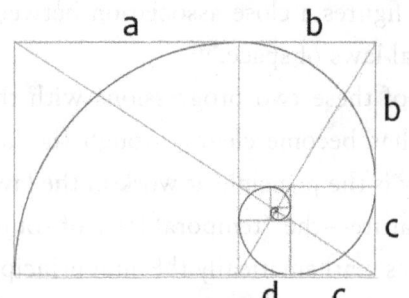

Fig. 37: Gnomonic construction of the golden spiral $a \div b = b \div c = c \div d = Phi$

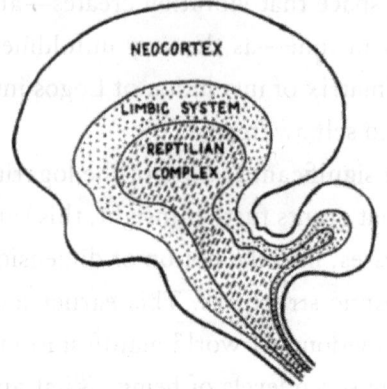

Fig. 38. Gnomonic pattern of brain evolution (Lawlor, Sacred Geometry, p. 71)

highlights the subtraction–addition of dimensions associated with the replication of self-similar figures, displaying a geometric model of the cosmological unfoldment of the chain of Being. Dimensions can be conceived of as cosmological gnomons.

In nature, gnomonic expansion makes visible patterns of successive stages of growth. Old cycles nest into new ones that spiral outward, as may be seen in the bony structures of animals, such as horns and shells. Like the creature within it, the shell grows in size but does not change its shape. It widens and lengthens in the same unvarying proportions. The constant similarity of form rests on a constant relativity of growth. Gnomonic expansion is also the evolving mode of the human brain. The limbic brain, which dominated during the mammalian phase of evolution, unfolds from the reptilian bulb, and is followed by the logarithmic emergence of the cerebral cortex (fig. 38).

Gnomonic spiraling displays in one figure the harmonic principle of self-similar structuring and the logarithmic principle of spiral unfolding. Structuring principles in space and formative principles in time come together in the externalization of a living "one," be it a plant, a shell or a Logos. What we contemplate in forms is a moment of formative time. In time, we participate in the configuring of space.

Interestingly, the word *gnomon* designates the projecting piece, on sundials, which indicates time by the position of its shadow. In other words, it is the piece that turns time into space—quite literally. The spatial marking on the dial allows one to "know" (the meaning of the Greek root: *gno*) what "time" it is in the evolving relation of Sun and Earth. Similarly, dimensional gnomons represent unfolding stages in the externalization of Logos in cosmos. The gnomon implies a divine "knowing" of when-where to "round up" and bundle a phase of growth so that the final form be self-similar—similar to its base. The gnomon represents this knowing of its "self-similarity" inherent to an organic form. It speaks to the self-knowing of

IV. The Spiraling Dynamics of the Universe

a living entity (plant seed, human self, world soul), the self-remembering at work in its replication of a "unit-base" at every stage of its manifestation.

Indeed, the observation of nature begs the question, how does a mollusk "know" the perfect measurement of its next chamber? How does the plant know when-where to start its flowering? Such "natural" knowing of one's "gnomonic" border is a knowing of one's existential proportions. Every instance of such proportional wisdom in nature points to the fundamental triunity of harmony—the logos-proportion itself—which vessels the self-knowing of the soul of the world.

The intelligence of language brings interesting insights again. The term *physiognomy* refers to the unique facial characteristics of a person as a source of insight into their character—"knowing" a person's "nature" by their features. It is hardly a stretch at this point to suggest that the golden spiral, which bundles the second trinity of 3, 7, 12 and its innumerable replicas, is the physiognomy of the Logos. Its key feature—gnomonic expansion/reduction—carries in its geometry and designation the self-knowing constant of the Logos in cosmos. Each particular gnomonic unfolding is also a milieu in which the Logos comes to know a dimension of its being.

"*Gnothi seauton*—Know Thyself," the famous inscription on the temple of Apollo in Delphi, may well be an injunction to know the Logos expressed in the human form—to know the divine "self" present in our subtle physiognomy—for instance our 3-7-12fold composition. In all likelihood, the aphorism came from Ancient Egypt, where an earlier version, engraved in stone in the external temple of Luxor explains, "The human body is the house of God. That is why it is said: Know Thyself!" The Internal Temple offers the next unveiling, "Man, know thyself, and you are going to know the gods."

The human kingdom is a dimension of the god-being of our solar system. The unique Form of the human being represents a gnomonic step in Logoic externalization, an *analogue* of the Logos, informed by the self-knowing of his God. A gnomonic unfolding of the Logos in cosmos, the human constitution deploys a next dimension of divine manifestation—one that includes the faculty to reflect, and to reflect upon the gnomonic signature of the universe inscribed in it. Being the emergence in cosmos of the cosmos' quintessential faculty to know itself, the human being has the potential to "know" its self as "logos"—divinely proportionate in his bodily structure, soulful harmony in progress, monad of love. None of the preceding gnomonic unfoldings (or kingdoms) of the Logos is endowed with this faculty. Deeper than psychological, the gnosis of self to which the "Gnothi Seauton" of Egyptian mysteries and Greek philosophy calls humankind is a psycho-cosmic knowing of our universal selfhood and its archetypal dimensions—the "gods" in us—what Schwaller de Lubicz precisely understood the Egyptian "neters" to represent. "Man, know thyself, and you are going to know the gods."

※

As noted earlier, not all spirals in nature are golden. Even the chambered nautilus, most frequently selected to illustrate the golden spiral, is an example of logarithmic spiral, but not of a golden spiral. It roughly triples in radius with each full turn, whereas a golden-ratio spiral grows by a factor of about 6.85 per full turn, or a factor of Phi with each quarter turn. It remains that the golden spiral is prominently featured in nature, and one more step into the mathematics of the logarithmic series will show that the golden spiral is in fact the archetype of the logarithmic spiral—the divine Idea behind the logarithmic spiralings of nature.

Quite simply, the golden spiral, which is a logarithmic spiral based on the Phi ratio (1.618), is a perfect blend of addition and multiplication processes: each successive term of the geometric progression of the golden series *equals* the arithmetic sum of the previous two terms. No other series presents this ideal feature.

If we look, for instance, at series based on 2 and on 1.25 (slightly above and slightly below Phi respectively) we do not find the same perfection of operative blending: one series produces terms that are greater than the addition of the previous two, the other produces terms that are smaller:

- In the series based on 2:
  $2 (2^1), 4 (2^2), 8 (2^3), 16 (2^4)$—
  8 is greater than 2 + 4, 16 greater than 4+8
- In the series based on 1.25:
  $1.25, 1.56 (1.25^2), 1.95 (1.25^3), 2.43 (1.25^4), 3.04 (1.25^5)$, etc.—
  3.04 is smaller than 1.95 + 2.43, 2.43 is smaller than 1.56 + 1.95.
- By contrast, the golden series based on Phi or ϕ (1.618) yields a perfect blend of multiplication of the base (Phi) by itself and addition of two successive terms. In the series based on Phi:
  $\phi (1.618), \phi^2 (1.618^2 = 2.618), \phi^3 (1.618^3 = 4.236),$
  $\phi^4 (1.618^4 = 6.853)$—
  1.618 + 2.618 = 4.236; 2.618 + 4.236 = 6.853

Each term is both the result of the progressive multiplication of Phi by itself and of the addition of the two preceding terms/dimensions.

$$\phi^3 = \phi \times \phi^2 = \phi + \phi^2;$$
$$\phi^4 = \phi^3 \times \phi = \phi^3 + \phi^2$$

Organic processes of growth and evolution illustrate this logarithmic operation of multiplication-addition. Shelled mollusks build successive coiled chambers as they grow. The human kingdom represents a next addition and unfolding in the manifestation of the world being, marked by the addition of a fourth, then fifth dimension.

IV. The Spiraling Dynamics of the Universe

The converse operation, the fractioning of "one" by the powers of Phi, which captures the reverse, involutionary operation of creation, gives the following golden series:

$1/\phi$ or ϕ^{-1} (0.618), $1/\phi^2$ or ϕ^2 (0.618² = 0.382), $1/\phi^3$ or ϕ^3 (0.236), $1/\phi^4$ or ϕ^4 (0.146)

This series shows the same archetypal property:

$$0.618 - 0.382 = 0.236, \text{ and}$$
$$0.382 - 0.236 = 0.146.$$

As illustrated in the diagram (fig. 39), each term of this "descending" series is at once a product of the division of "one" by the powers of Phi (ϕ^2, ϕ^3) and of the subtraction of a self-similar unit from the one preceding it.

$$1/ = \phi - 1$$
$$1/\phi^2 = 1 - 1/\phi$$
$$1/\phi^3 = 1/\phi - 1/\phi^2 \text{ and } 1/\phi^3 = 1/\phi^2 : \phi$$
$$1/\phi^4 = 1/\phi^2 - 1/\phi^3 \text{ and } 1/\phi^4 = 1/\phi^3 : \phi$$

It is the unique property of the Phi based, "golden" spiral that each successive quarter turn is based on a fractioning/multiplication of the base that is also a subtraction/addition of a self-similar dimension. In other words, each self-replication in the golden series is at once a self-fractioning *and* a self-subtraction, a *self*-multiplication and a *self*-addition. The unit-base Phi points to the archetypal core "self" of every addition-and-multiplication process of world manifestation and natural growth—and every dimension that results.

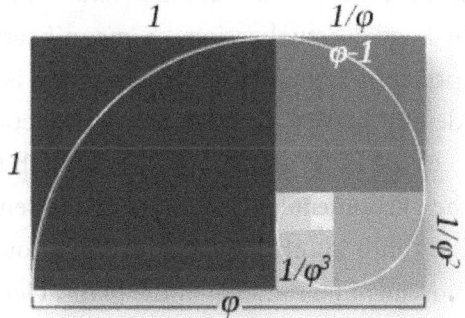

Fig. 39. Golden spiral: The length of the side of a larger square to the next smaller square is in the golden ratio

Continuing to draw the golden spiral displayed in the diagram in the direction of its magnification would bring as the next gnomon a square with side ϕ (ϕ^2, 1 × ϕ^2), thus completing the next golden rectangle with side $\phi + 1$ (again, $\phi^2 = \phi + 1$). Central to the marvelous spiral is the unique relation and articulation of 1 and Phi—the "one" and Phi—*the unique mediation that Phi provides for "the one" to literally "turn" itself into many that remain "same"*—many that are dimensions of its self. Phi is the archetypal mean by which the great circle of the One turns into "squares" of self-manifestation.

This figure of harmonic unfolding of "one" into self-similar figures through a process that is at once division (of itself) and subtraction (of successive self-similars) is the geometric model of the spiraling that curves the forms of nature and the revolvings of cosmos—the archetypal spiraling that points to the metaphysical unfolding of the Logos into a physical cosmos *through seven dimensions of Being— Logoic, monadic, atmic, buddhic, mental, astral, etheric-physical*. The golden logarithmic fractioning—$1/\phi$ (0.618), $1/\phi^2$ (0.382), $1/\phi^3$ (0.236), and $1/\phi^4$ (0.146)—illumines the Platonic process of creation, the rhythmic descent of Logos into cosmos.

The golden spiral reveals the mathematical idea behind the logarithmics of nature and cosmos. Underlying the phenomenon of gnomonic growth is the convergence into an unfolding series of self-similar steps, of an addition of dimension with a whole multiplication of "self"—a perfect blend of quantitative increase and qualitative, proportional magnification. This constant proportionate blend of external physical addition and internal growth characterizes the form-making process of life in any domain, physical and psychic. Every dimension of world and consciousness is a partial expression (a fraction), and also an expression *à part entiere* (a "whole part") of the Logos. The emergence of the human kingdom, for instance, represents the "addition" of a dimension—the self-reflective, mental dimension—to the physical, etheric, and astral manifestation of the world soul, and it heralds a major step in Logoic actualization—a magnification of Logoic expression in cosmos. Every formation of nature or consciousness is the manifestation of a living "entity" whose identity persists through states of metamorphosis by dint of the unitive process of external build up and internal growth that logarithmicity ensures. Metamorphosis, the phenomenologist Ronald Brady remarks, "is not the outward alteration of one form into another but the differing outward expressions of an inward idea." The movement up the stem of the plant, Henri Bortoft explains, is not "made out of" the sequence of organs [that is, leaves], but the organs are "made out of" the movement. "Suddenly," he describes, "there is a movement, a dynamic movement, as you begin to see not the individual leaf but the dynamic movement. The plant is the dynamical movement. That is the reality."[42] The logarithmic constancy of quantitative addition and proportionate self-multiplication is what allows entities to exist as them *selves* in partial, yet "whole" forms.

The "quanta" of logarithmic growth are both similar, additive replicas of a singular core, and different phases of its unfoldment. Logarithmicity perpetuates the primordial harmony of inside and outside—soul and body. The being-ness of every form of life, which includes an inherent knowing of its own wholeness, a *soul*-like knowing of itself, curbs quantitative accumulation to bundle up a "whole" new

---

42 Bortoft, *Imagination Becomes an Organ of Perception;* conversation with Claus Otto Scharmer, London, July 14, 1999.

IV. The Spiraling Dynamics of the Universe

dimension, qualitatively different, but analogous in proportions to its fundamental signature. "Wholeness" concludes a stage of growth and induces the beginning of a new dimension. In the plant, the grown stem gives way to the unfolding of a flower. In our brains, an additive accumulation of data suddenly blossoms into a genuine understanding. The corresponding step in human psychospiritual development is called initiation—a threshold of completion, which marks a quantum leap of new beginning. The footprint of the Logos in the forms of cosmos, logarithmicity speaks to the quantum, initiatory model of cosmic and psychic evolution.

In modern terms, we would say that the cosmological process of involution-evolution is a "fractal" descent–ascent of "one" into many self-similar fractions of itself by mean of Phi—the vehicle of the triune Logos, the world soul. A fractal is a geometric motif or pattern that repeats itself at different scales of magnification/reduction. Like the logarithmic spiral, fractality is a phenomenon of infinite reduction and infinite expansion. Fractalization (self-similarity) is increasingly recognized as the basic principle behind nature's self-organization. This principle of scale invariance governs the geometry of the golden spiral and the harmonic cosmology it illumines. *Phi* is simply the ideal scaling mean—the very archetype of fractality; the golden spiral is the archetypal fractal process of nature. It replicates the transphysical process of creation—the vortical motion that turns a Logos into cosmos while winding up matter into increasingly complex organisms.

Being, the "one," wraps itself in the 3fold oneness of Logoity to descend through a logo-rhythmic stairwell into partial replicas of itself—world dimensions—through which matter spirals up into new forms of being. Evolution is simply the other side of the creational descent—the centripetal ascent up the stairwell of creation.

The Slavonic liturgy, which has kept in its words, melodies, and rituals the qualitative radiance and warmth that imbued the inner sight and senses of the early fathers of the church, brings to bear upon the death and resurrection of the Logos-Christos of the New Testament the Old Testament's Psalm of creation: "*You wrap yourself in light as with a garment...*" In so doing, it transposes the involutionary wrapping of the Logos in the objective light of creation, praised by the Psalmist, into the evolutionary leap of "resurrection," when, at the apotheosis of individuation, Jesus-"Christos" wraps himself in an etheric body of light, his flesh galvanized beyond molecules, quintessentialized beyond elements, by the transmutative fire of monadic being.

The sweeping transposition lights up the complementarity of creation and resurrection, the two fundamental gestures of the universe—existentializing submergence of the Logos in creation and essentializing emergence of the Logos in humankind. Outer manifestation. Inner emergence.

The event of resurrection points to the supreme horizon of humankind—the consummate wedding of material physicality and solar vibration. The more evolved the individuality, the more fully and deeply the monadic consciousness penetrates into and takes hold of the vehicles. Resurrection suggests the attainment of a vibrational presence of being apt to transfigure the body and galvanize matter. In the words of Master Morya, a Tibetan Master of wisdom, "The suffering and resurrection—or the transformation of matter—by Christ provided the attainment of the supreme earthly achievement. But no one knew about the disintegration of the body into the atomic state. People thought that his body had been stolen away by his disciples."[43]

THE FIBONACCI SERIES: THE GOLDEN SPIRAL ADAPTED TO FINITE ENDS

There is a close proximity between the golden spiral and the Fibonacci series, a number series so prevalent in patterns of nature and rhythms of living growth that it is regarded as a law in biological growth and reproduction.

The Fibonacci series is a sequence of numbers, in which each term is the sum of the two previous terms: 0, 1, 1, 2, 3, 5, 8, 13, 21, etc. Known in ancient India, the series was introduced to Western mathematics in 1202 by Leonardo da Pisa. Traveling as a boy with his merchant father, he learned about it from Moorish scholars and brought the information back to Europe. What characterizes the Fibonacci series is that the ratio of any two successive Fibonacci numbers after number 3 approximates the golden ratio Phi as n approaches infinity. In other words, the further one advances in the sequence of numbers, the closer does the ratio of a term to the one before it gets to the golden ratio: 1.6180339: $1/1 = 1$, $2/1=2$, $3/2 = 1.5$, $5/3 = 1.66$, $13/8 = 1.625$, $21/13 = 1.615$, $34/21 = 1.619$, $55/34 = 1.617$, $89/55 = 1.6181$.

Like the golden spiral, the Fibonacci series is an expression of the fractioning-fractaling principle at the core of our "turning worlds." The diagram (fig. 40) shows how the Fibonacci number-series unfolds around the golden spiral as its framework. Each gnomon is a square whose side is a Fibonacci number, and also the addition of the two previous Fibonacci numbers.

The Fibonacci series offers another avenue of insight into a few facts that the Pythagorean harmonic cosmogony has brought us to recognize. Fibonacci numbers highlight, for instance, the unique nesting of Phi, the divine proportion, at the center

---

43 Master Morya, *Agni Yoga*, sloka 8.

IV. The Spiraling Dynamics of the Universe

of musical harmony. The "dominant" of the diatonic scale, the fifth/5th, is also the 8th of the 13 notes that comprise the octave. 8/13 is a Fibonacci ratio (.61538), which approximates Phi. This 8th note, the fifth, is the central Fibonacci-Phi number within the frame of the 13fold octave whole. On the keyboard, every octave from C to C above, displays 8 white keys and 5 black keys, split into groups of 3 and 2. A scale is composed of 8 notes, of which the 5th and 3rd create the basic foundation of all chords. All these number-relations are part of the Fibonacci series.

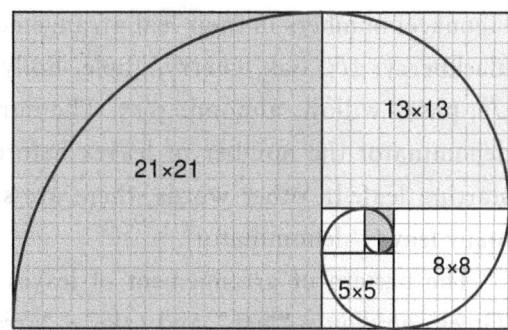

Fig. 40. Fibonacci underpinnings of the golden spiral

Plato's harmonic cosmogony exposed the key significance of number 12. 12 also finds itself highlighted in the Fibonacci series. The first perfect square in the series, 144 is also the 12th number and its square root is 12. The first perfect square along the spiraling motion is thus associated with number 12. This perfect "squaring" of life in form evokes the completion of human evolution suggested in the Apocalypse—the full penetration of fourfold matter by triune spirit, mirroring from the end of time the 12fold deployment of the world soul in the beginning.

The Fibonacci series is at work in the branching of trees and the arrangement of leaves on a stem: 3 leaves in 5 turns, 5 leaves in 8 turns. It can be observed in the fruitlets of a pineapple and the flowering of an artichoke, the pattern of an uncurling fern and that of a pinecone. Flowers generally exhibit a Fibonacci number of petals: 3 petals: lily, iris; 5 petals: buttercup, wild rose; 8 petals: delphiniums; 13 petals: ragwort, corn, marigold; 21 petals: aster, black-eyed susan, chicory; 34 petals: plantain, Pyrethrum; 55 or 89 petals: Michaelmas daisies, the Asteraceae family. The sunflower has 21 spirals on its head in one direction and 34 going in the other—or 34 and 55, or 55 and 89, even as high as 89 and 144, depending on the species—all consecutive Fibonacci numbers. The same is true with daisies.

The outside of a pinecone has spirals that run clockwise and counterclockwise, and the respective numbers of these two sets of spirals correspond to a sequence of Fibonacci values. For example, the petals of a Monterey pinecone show 8 spirals in one direction crossing 13 spirals in the other. Similar patterns arise in other plants whose leaves grow in a spiral around a central stem; each successive leaf may be 1/2 way around (opposite) or 2/3 of the way around, or 3/5, etc. The estimation is that 90 percent of all plants exhibit Fibonacci numbers in their leaf patterns. In flowers, spiraling and flowering fuse. Common trees show the following

Fibonacci numbers in their leaf arrangements: elm, linden, lime: 1/2; beech, hazel, blackberry: 1/3; oak, cherry, apple, holly, plum: 2/5; poplar, rose, pear, willow: 3/8; pussy willow, almond: 5/13. The numerator is the number of turns and the denominator the number of leaves before one reaches a leaf directly above the starting leaf. In other words, there are so many turns (numerator) for every so many leaves (denominator).

The pattern of arrangement of leaves on a plant stem is called *"phyllotaxis"* (Greek: *phyllon* = "leaf," and *taxis* = "arrangement"). The most common angle of divergence between successive leaves is 137.5°, the golden angle. It is the angle that divides a full 360° angle in a golden ratio—360/Phi—but measured in the opposite direction. If we look at a plant from above, we see how the leaves are arranged in such a way that leaves above do not hide leaves below. The fact that the golden angle ensures maximum sunlight and rainfall for all the leaves is often advanced as the "reason" or explanation for such a perfect arrangement. Adaptation of inside to outside, here of Earth formation to light, is the way the scientific mind "rationalizes" the perfect creations of nature—the way it justifies its perfect ratios. If adaptation is the key factor in the ordering of plant growth, how intriguing that there could arise such marvelous reason in the plant kingdom to orient and construct itself just so. Where in the plant would such adaptive intelligence be found? Both explanations—reasonable rationale of adaptation or golden *ratio-nal* explanation—acknowledge an unseen intelligence at work in orienting the plant motion of growth in a way that ensures the most perfect coming together of earthly leaves and sunlight—the most perfect conjugation of energy and matter; an intelligence that shoots and turns the "stemming forth" of plant at such intervals that there can be fullest interplay between earthly matter and Sun—the very definition of the golden mean. Both explanations are in agreement as to the exquisite *adaptive mediation* at work in nature's engagement of Sun and Earth, light and substance.

The only factor that makes the adaptive argument more acceptable scientifically is that it does not challenge empirical "rationality." Yet, however reasonable and "rational" (irrationality-proof) the adaptive rationale may be, it lacks a genuine "intelligence" of the matter—the vibrant rationality that is experienced when, instead of invoking adaptation as a rationale for a remarkably "functional" world, one begins to regard the mediation provided by proportionality as the very principle of the universe's structures and motions, the dynamics of harmony at the source of all its operations, the highest "reason" that presides to its ongoing construction.

The ideal mean to actualize such reason in a physical cosmos, the golden ratio is precisely *the* key to the adaptive engagement of every polarity in the universe

IV. The Spiraling Dynamics of the Universe

(such as sunlight and earth particles). The highly rational (ratio-based) constructions of nature bear witness to the adaptive intelligence of harmony-love. The golden ratio is both the mean by which life adapts itself to a diversity of forms as it swirls its oneness into many similar "ones," and the reason–ratio behind the (logo-rhythmic) adaptation of matter to the geometries of harmony that arrange particles into complex vessels of life and mark the intervals of leaves spiraling up toward the Sun—into sun-like flower and fruit.

A purely mechanistic consideration of adaptation misses the essential dimension of nature's adaptability—the transmission of life. It fails to see the generative intercourse at work in adaptation. It fails to recognize in mediation and harmonization the mean for the wholeness of life to pass and exist in forms. Beyond the intelligent adaptability of nature, the golden mean accounts for the presence of life—Being—in nature. Harmony is the rationality of the "living god."

Indeed, says the Emerald tablet, "as all things have been and arose from one by the mediation of one [logos]: so all things have their birth from this one thing by *adaptation*"—in other words, harmonization.

How fascinating that the specter of irrationality, an "irrationality" that only speaks to the lack of a reasoning apt to embrace the laws of life, manages to obliterate the golden ratio that rules universal manifestation—the proportional rationality of planetary order and plant growth, cosmic and organic structures. How ironic that in the name of rationality we would fail to recognize the key to the harmonic unity in diversity that makes the universe highly rational—reasonable. And how remarkable that the utmost rationality of our world should lie in an irrational, or transrational number—*Phi*—granted that this "irrationality" is in reference to the attempt to translate or "figure out" proportion in a number. Proportion defies this attempt magnificently by affirming a sovereign freedom from the limitation of numbers. Proportion is essentially a ratio, not a number—at least not a human number. It transcends the quantitative integrity and separate multiplicity of "integers." The triune "divine proportion" is a *logoic* number—*the "logarithmos"*—the divine "number" of bondedness, harmony, and "friendship." Is it not reasonable to expect that only a measure that transcends the mathematical framework of our divisible worlds would be suitable for the Logos-mediator of the nondual, indivisible power of life? Our own experience tells us how the sense of proportion transcends the intellect, drawing instead on the wholistic intelligence of feeling.

In the end, we see how deeply true the adaptive reasoning is, and how congruent the functional and ontological explanations really are. Both ends of the spectrum of "explanation"—modern rationale and Platonic ratio—agree in pointing to the Intelligence of nature as a wisdom of mediation, an inherent *common sense* of how to engage and negotiate the powers of Earth and Sun, the interplay and conjugation of matter and energy.

✺

✪ Even the golden mean has to adapt itself to its *own* end, which is the manifestation of infinite being into finite forms. The process of landing into forms that have beginnings and ends requires a "trimming" of the infinite spiral to a finite point, a "rationalizing" into whole numbers of the irrational number of the Logos. If it is to shape separate forms and guide their growth, the infinite spiral must find a way to end and to begin. As it approaches the physical ground in the forms of nature, the irrational number Phi literally "turns" into whole number fractions such as 34/21, 13/8 (found regularly in flower/plant formations), the quotients of which diverge further and further from the divine, irrational value, to reach 3/2, the interval of the musical fifth, very close indeed to the golden mean, then 2/1, the unison of the octave, then 1, the ground "number" of any "whole" living form. This recognition answers the question raised earlier, of why the profound analogy of principle between golden mean and musical fifth (0.618 and 0.666) does not translate into the same exact numerical value. ✪

To embody itself in separate forms, the world soul must clip the wings and curb the momentum of its golden curve of love—its "mean" without ends—to fit the "ends" of an earthly world. To mold the soul into bodies is the grand issue of divinity. As it comes close to the ground, the logarithmic breath literally bends itself "out of shape" in order to "fit" into finite physical forms. The Word articulates itself into words.

The "discrepancy" between Fibonacci and golden spiral is the gap between the quantitative constraint of whole numbers and the qualitative principle of proportionality. Their convergence in the natural world illustrates how Nature astutely and elegantly catches up with divine motion and turns earthly components into divine gears—how it raises geo-metry into harmony and lands harmony in geometry. The gap speaks to the milieu in which the soul willy-nilly unfolds its wings, and spins infinity out of the finite constraints of physical existence—through them. To radiate infinity in finitude is the divine mystery of beauty. To raise finitude into the transcendent fabric of the soul is the beauty of human mastery.

Negotiating between Phi proportion and Fibonacci series is how infinity surrenders enough to finiteness to arise from it, and spin to its vibrant stillness the tumbling and balancing of the elements. Wobbling between Sun and Moon cycles, solar and lunar forces, is how the Logoic spiral bundles itself into earthly rounds of seasonal flowers and calendar years of history. The yearly comma of 11 days and the grand precessional comma of a degree per 72 years speak to the evolutionary tension between the gravitational pull of the Moon and the levitational pull of the Sun. Steiner points out how one degree of precessional motion (25920:360 = 72 years) is the average of a human life. Indeed, quite similarly, does the human psyche wobble between the lunar tides of the personality and the solar stillness of the soul, who, Phi-like, wisely paces its emergence from the coarse spirations of desires and the limited constructs of the concrete mind.

For the quality of divine proportionality to descend and dwell in quantity and number, the sphere of oneness must accomodate itself to the squareness of matter, and the threefoldness of divinity to the fourfoldness of physicality. Phi must compromise its integrity to fit the "integers" of quantity, the quanta of form. And so Phi curbs its transrational nature to fit the 2/3 ratio of the musical fifth (the third harmonic) and give the finite human ear an acoustic impression of transcendent medianity. In turn, the squareness of the earthly bends its neck to the inspiring embrace of the spiraling breath of life, so the grace of the Three, the very soul of life, may whisk and round out "whole" organic forms out of its warring elements.

Kepler was aware of these converging ratios of Phi and Fibonacci series standing like a pair of immortal and mortal twins behind the forms of nature. He spoke of them as "the divine and its shadow":

> There is the ratio which is never fully expressed in numbers and cannot be demonstrated by numbers in any other way, except by a long series of numbers gradually approaching it; this ratio is called divine, when it is perfect, and rules in various ways throughout the dodecahedral wedding. Accordingly, the following consonances begin to *shadow forth* that ratio: 1 : 2 and 2 : 3 and 2 : 3 and 5 : 8. For it exists most imperfectly in 1 : 2, more perfectly in 5 : 8, and still more perfectly if we add 5 and 8 to make 13 and take 8 as the numerator.[44]

The gradual accommodating of the transrationality of Phi to the whole number fractions of "corporeity," which the Fibonacci series illumines, mirrors the primordial deployment of the golden mean in world creation. To create a world required from the One cosmic being a beginning operation analogous to the microcosmic conception of any organic form. In both cases, as the identical beginnings (1, 2, 3) of Platonic cosmology and Fibonacci series make clear, what opens the way for forms of matter-spirit to unfurl through the Logoic spiraling of life are the first three numbers. One, two and three enact the threefold dynamics of 1) creative being (life), 2) duplication-externalization (matter), and 3) integrative consciousness (form). They register how 1, the One, has to duplicate itself for the vibrant relation with its self-same "other" (2/1) to generate the archetypal mediator of duality, which is the "mean" for life to move as one triune harmony (3/2) through

---

44 See http://www.spirasolaris.ca/sbb4e.html.

the multiplicity of forms. The first three numbers reappear in the world of forms as "one" spermatozoid uniting with an "other" egg (2) and conceiving the germ of a whole new organism—a zygote (3). In the flowering plant, the "union" of stamen and pistil by pollination generates the seed, which contains the germ of a new plant. From two, a new "one" is born, which transcends the distinct units from which it proceeds as the golden mean transcends the quantitative distinctness of the first two numbers.

The golden proportion is a 3fold dynamics and can only come to be with the existence of three terms. In the beginning, before any such "term" exists, we could say that the world formative spiral-to-be uses the only mean possible (division–multiplication in 2, *the 1/2 ratio*) to emerge as the proportionate bonding of the two, which is the Fibonacci approximation of the golden mean: 3/2. The cosmological underpinnings illumine the "deviation" of the Fibonacci series from Phi in the germination of forms.

To live in fourfold physicality, the trinity must accommodate its transrational "value" to number-units. The incarnation of Being in forms of matter exacts a squaring of the world soul. Plato's pronouncement that the soul of the world is crucified on the cross of the world is a spiritual scientific insight. The world soul is literally crucified (squared) to fit the rational numbers and finite geometries of the fourfold physical world.

To penetrate the depths of the Earth and turn it into his body of light ("this is my body"), Christ goes through crucifixion. In profound analogy, or self-similarity, with the scientific logic of the universe, sacred history reveals how humankind is meant to take in and experience the soul life of creation and evolution. The Christ epitomizes the deeds and sufferings of the golden mean, whose function is to "cross" and mediate the gap between spirit and matter and to unfold the soul harmonics in which Heaven and Earth, sameness and otherness unite creatively.

The quadrature of the circle is the involutionary problem of god. Our problem is the circling of the square into the 12fold "City-bride" of the Logos. If molding the soul into bodies is the grand issue of divinity, to incorporate the cosmic breath in time–space experience is its human counterpart. Symbolically, musical man tempers gaping commas into harmonious sounding musical scales and cultural man adjusts soli-lunar cyclical gaps into rhythmic calendars.

✪ The mysterious, graceful shape of the treble clef (G clef, or *clef de sol*) harbors a signature of the essential identity of golden mean and musical fifth. Wrapped in its unique flow, the waveforms of the fifth and the fourth may be seen curving

IV. The Spiraling Dynamics of the Universe

around the axis they cross at the golden section point—their geometric expression (fig. 41). The whole movement unfurls out of the spiraling gesture emerging from the center, the primordial "speira"-germ, which points to the origin of the music of creation, the transcendent Sol—the Logos of harmony and divine proportion. ✪

The first three harmonics, expressed geometrically in the golden mean, are the vehicle of the trinity. Encrypted in the G clef—the golden "key" of sol—is the keynote of the triune Being of the universe: harmony and grace.

Andrej Roublev's icon of the Trinity (fig. 42) carries in the fabric of its exquisite representation of the Three the "key" to their spiritual grace and harmony: the gesture of the golden spiral and the countenance of the clef de sol. Also called "The hospitality of Abraham," this painting of the three angels who, the Bible recounts, visited Abraham at Mambre, has been widely interpreted as a rendering of the Holy Trinity.

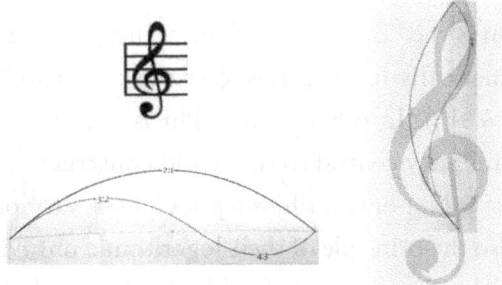

Fig. 41: (above) treble clef or clef de sol; (below left) waveforms associated with the octave (2:1), the fifth (3:2) and the fourth (4:3); (right) the first three harmonics superimposed on a treble clef

The heads-halos of the three beings, as well as their rods, are visibly in a rapport of "divine proportion" to each other (each set of three in a reverse golden order), and the "Log-arithmos" is perceptible in the counter-clockwise progressive increase in distance between third (on the right) and second, second and first, then first and third rod–being. The great mediator, the second, is the threefold "Logos"—the golden mean between the primordial polarities who occupy the octave position referred to in Christianity as "Father" and "Holy spirit." These discreet geometries contribute to the impression of "divine" harmony emanating from the painting.

Fig. 42. Andrej Rublev's Icon of the Trinity. fiftheenth century, Tretyakov Gallery, Moscow

Phi, the "Figure" and Profile of the Logos

> *The final truth about a phenomenon resides in the mathematical description of it; so long as there is no imperfection in this, our knowledge of the phenomenon is complete.*—James Jeans, *The Mysterious Universe*

Phi is the "base unit" of the archetypal logarithmic spiral, which is expressed mathematically in the series: ϕ (1.618), ϕ^2 (1.618² = 2.618), ϕ^3 (1.618³ = 4.236), ϕ^4 (1.618⁴ = 6.853). In other words, Phi is the core of the archetypal pattern of fractal self-similarity central to the world construct.

⊙ Not only is Phi the principle of harmonic structuring of nature's forms, but it is also the principle of their logarithmic unfolding. What we behold in Phi is the essence of the phenomenon of mediation key to the variegated self-sameness of creation. ⊙

The differentiating-integrative impulse at work through Phi—as Phi—is the pulse of the world. The logarithmic spiral—the unfolding of Phi in time and space—is the dynamic signature of the universe, the profile of its identity, the trace and expression of its "self." The soul of the universe is encrypted in the golden mean, the bonding intelligence that generates whole, harmonious forms of being. Its inner reality is the world creative vibrancy of love. The essence of the universe is at once hidden and revealed in the omnipresent love–rhythmic, separating-bonding intercourse of its polarities, which begets the multi-beauteous forms and dimensions, sounds and colors, bodies and psyches we call the universe.

Wrapped in the golden bonds of nature's geometries and dynamics is the jubilant, creative grace of love—*our* universal identity. All forms of nature are forms of love. We sense this in moments of inspiration. However, to know as a rational fact that love is the essence of this entire creation, the life transmitting source and form shaping agent of all forms of existence is a stunning and moving realization.

Sexual intercourse, which bonds the male and female "sections" of living species in procreative vibrancy, is one form of enactment of the golden mean. We may only imagine what subtle deployments parallel the conception of bodies in the loving intercourse of souls. Jesus' words to his mother, as he is about, with her help, to transmute water into wine, open a window into this mystery, "Pay heed, O woman, to the power which flows between me and you."[45] This power, which fires water into wine, is the power that solarizes the Earth, enlivens matter and spirals up the love between souls to its divine potential: to kindle life.

Wisdom traditions and religions intimate, in one form or another, that "God," the Lord and Logos of the universe, is a unitive trinity, and that the

---

45 John 2:4; in Madsen, *New Testament: A Rendering*, p. 215.

IV. The Spiraling Dynamics of the Universe

involutionary-evolutionary process is a process of integrative dualizing. The Bible, for instance, narrates the Genesis of the world as a process of successive separations-gathering: First, God polarizes the void into Heaven and Earth and bonds the two with light; he then separates the waters above from the waters below and bonds them with sky; he separates land and water (form and energy) and bonds them with vegetation.

The separating-bonding dynamics of love that intones and lights up the universe pulsates through it with its formative gesture—from the separating-binding of matter-energy that spins atoms, shells, and trees, evolves crystals and revolves galaxies, to the alchemical harmonization of self and world which forges the blossoming of human souls. The self-resounding that unfurls Being into wave fields of "existence," courses through all its parts and spirals into its multilevel blossoming, is the vibrant continuo that "minds" matter, and reminds it of its Being, swirls it into living forms, and turns lesser forms of consciousness into whole self-aware individualities. It is Nature's presence of mind in everything she does.

Goethe experienced this mind of nature and felt taken up, "unasked and unwarned, into the whirl of her dance."

> She has thought, and she ponders unceasingly; not as a man, but as Nature.... Each of her works has an essence of its own; every shape that she takes is in idea utterly isolated; and yet all forms one.... The meaning of the whole she keeps to herself, and no one can learn it of her.... She is whole and yet never finished.... She hides herself in a thousand names and terms, and is always the same.[46]

The differentiating–integrative dynamics that conspires in the spiraling forms of creation is also visible in the workings of the mind—most transparently in the structure of its primary instrument—language. "Speech or language, she has none," Goethe further muses about nature, "but she creates tongues and hearts through which she feels and speaks." The core structure of language is a 3fold binding—of subject and object with a verb. The verb bonds subject and object into a creative dynamics, which engages a next spin of communication, world understanding, or world enacting. The archetypal sentence points to the basic "operation" of language, which is to *create new bonds* between subject and object, self and other, self and world, world "portion" and world "portion," and through such bonds, generate new forms of awareness and experience. The archetypal verb is: "to bond." Most profoundly and exactly is the Logos called the "verb" (*in principio erat verbum*). The verb of creation is "to love." The Logos (metaphysically), love (experientially), ratio (mathematically), and verb (grammatically) all point to the same instance and core dynamics of bonding, which gives life to the world and keeps world and language alive.

---

46 Goethe, "Nature: Aphorisms," *Maxims and Reflections,* pp. 52–54.

The time–space dimension, the dimension of the human mind, is where all these considerations have their reality—as well as their limited value. This time–space-mind dimension, which is at once the matrix of the spiraling out of Logos into cosmos, and the psychological context of the spiraling in of cosmos into Self, is the milieu of "rationality" where humankind, the organism of self-awareness of the Logos, attempts to figure out the world, in ways and steps that bear witness to its own unfolding. The time–space dimension stops short, however, as does the dimension of mind, of comprehending the greater dimension within which it exists—precisely because it exists *within* it. It cannot comprehend the nondual universality of being that dawns when the human "monad" outgrows his fully blossomed Egoity and lets go of the beautiful causal form of his being to identify with a more essential dimension, which transcends individuality, mind, space, and time. The rational grasp of the universe, such as we are attempting in this work, speaks to the time–space reality of individuality (soul), the zodiacal reality of the individuation process. Beyond the time–space–mind womb of zodiac and soul, there dawns a greater order of consciousness, which transcends time, space, and the rational language of the mind.

THE GOLDEN DYNAMICS OF THE LOGOS: THE PULSE OF THE WORLD

The Lord answered Job out of the whirlwind. (Job 38:1)

And Elijah went up by a whirlwind to heaven. (2 Kings 2:11)

We now see how the golden spiral is the deployment in "form" of the golden mean, the mean to *turn* life into forms, as the Logos turns Being into an evolving milieu of self-awareness. What we are about to discover next is that the dynamics of exponential self-replication, which defines logarithmicity, is not only the pattern of unfolding of world forms, but it is the dynamics *intrinsic to the golden mean itself.*

Similarities between the defining features of logarithmicity and the underpinnings of the golden mean have already suggested kinship. Addition is a sequencing process that affirms differentiation, and self-multiplication a "bundling" process that affirms sameness.

✪ Mathematical facts will show unequivocally that *logarithmicity is the active principle of the "mean" that weaves together the Inside and Outside of our universe*—consciousness and cosmos—in other words, that it is the dynamic core of the golden mean and of its form-generative potential. *Logarithmicity* is not only the dominant feature of the deployment of the Logos into cosmos, *it is the dynamics of his triune identity.* ✪

IV. The Spiraling Dynamics of the Universe

That logarithmicity is at work in the golden proportion can be demonstrated, Lawlor makes clear, by the fact that "Phi is the unique division that fulfills the characteristic: $1/\phi + 1/\phi^2 = 1$."[47] Put in slightly different terms, so as to make this cryptic content actually visible in the diagram of the Phi-based logarithmic spiral presented earlier (see fig. 39, page 217), the same equation reads: $1/\phi = 1 - 1/\phi^2$.

Indeed, that $1/\phi = 1 - 1/\phi^2$ is simply what we see when we look at this diagram. In other words, we readily *see* the equation that demonstrates that logarithmicity is intrinsic to Phi by simply looking at what the diagram displays, which is that Phi is the unique partitioning of One with the potential to *generate a series of self-partitionings,* in which each term of the division equals the subtraction of the two previous terms from each other. It is the only fractioning of "one" that fuses self-division and self-subtraction, the only self-division that propagates itself in self-similar wholes that are also graded subtractions of each other—thus generating a spiraling stairwell of dimensions that replicate the fundamental proportion of world-creative harmony, the golden physiognomy of the Logos.

Another way to sight the logarithmic principle lodged in the golden mean, as its world begetting potential, is to shift point of view as to what constitutes the "unit"—the "one"—in the geometric formula of the golden mean. Instead of associating "1" with the whole string as we did when we first introduced the golden mean, we are going to attribute "1" (i.e., the character of a "unit") to the mean term.[48]

A simple diagram will help to grasp this shift and what it reveals. As the three segments display, the golden section is the unique division of a string such that the large segment is to the small segment as the whole string is to the large segment: B is to C as A is to B. For a string of length "1," the value of this unique proportionate sectioning is Phi. If we designate A as the unit, in other words, attribute "1" to the entire string A, the value of Phi, 0.618, is associated with the mean term B:

```
A = 1.000 = B + C
B = 0.618                          A/B = B/C = 1.618 = Φ
C = 0.382
```

If, on the other hand, we imaginatively designate B as the unit, and attribute "1" to the mean term B, calculations show that the value for A, the whole string, is 1.6180339 (the greater golden value) and the value for C, the smaller segment, is 0.6180339 (the lesser golden value). These are referred to as φ and Φ.

These two numbers are unique "golden" values, in that their multiplication equals their subtraction equals one: $\phi \times \Phi = \phi - \Phi = 1$. Not only does the difference between

---

47 Lawlor, *Sacred Geometry*, p. 46.
48 Ibid., p. 49.

the two (ϕ and Φ) equal "1," as could be expected given the simple division that generated it, but their product as well equals "1." What is significant about this equating ($\phi \times \Phi = \phi - \Phi = 1$) is that the *multiplication and subtraction of two numbers* amounts to "one"—to one and the same unit. Such "unity" of the *radically* different processes of multiplication and subtraction happens only with the golden values of Φ and ϕ and their multiples, and it points to what lies at the source of these unique values: the golden mean proportion: a is to b as b is to c. The only possible convergence of subtraction and multiplication processes lies in the value behind our second unit (B = 1). This value is Phi, B having been obtained by golden secting the original "unit" A.

In other words, Phi is the only ratio for which the subtraction and multiplication of the sections it creates come to one. It is the only way or mean to make the radically different operations of subtraction and multiplication, arithmetic and geometric progression—one. By the same token, it shows that the two processes do have a hub of convergence, and that there exists a value for which they are "one." This value is Phi.

What makes the convergence of subtraction and multiplication meaningful is that it represents a convergence of the world of quantity and the world of quality—"how much" and "what kind." The quantitative world deals with the perception and subtraction/addition of distinct components, the qualitative world with the intuitive apprehension of an internal "kind" of reality or "species" of life. The "how much" of an organic formation is its number of self-same levels or dimensions (its constitution). The "kind" of reality present in it is the "sameness" of its form, which points to its "unit-base," or core entity (its species). One concerns itself with physical values, the other with soul values. Body and soul converge in—because they emerge from—the one entity that expresses itself in these two facets of being. By *at-oneing* these two processes and their associated worlds, Phi heralds the oneness of *Life* that transcends the apparent dichotomy of body and psyche, world and consciousness.

Dualization and mediation, separation and bonding, the primordial operations associated with the two movements of world construction and self-unfolding, do indeed have a common source: the Logos that precedes and transcends the division between cosmos and self, outside and inside. The common root of the dual unfolding of cosmos and self, Phi is the key to their unfolding as one.

The mathematical move to mark the mean term as "1," and have it be the central "one," parallels the metaphysical transition that marks the first step of creation: the transition from the primordial One (unmanifested Being) to the triune One (the Logos). In the ontological unfolding, the threefold "one," the Logos, is the golden

IV. The Spiraling Dynamics of the Universe

mean between the fundamental "Two" of Same and Other. The unifier of the two, the Logos becomes the new "one," the Phi "unit" base of the cosmic stairwell.

Giving the golden mean term B the value "one" in our diagram, making it the central "unit," reveals its unique capacity to "unify" or at-one the two lines of arithmetic and geometric progression—duplication and mediation—which speak respectively to the involutionary process of dimensions-creating in space and the evolutionary process of self-unfolding in time. In Phi, the duplication and mediation processes, which we have come to recognize as the principles of externalization of the Logos into cosmos and its internalization into soul, are one. Through Phi, they remain secretly one. Phi is indeed the mean to unfold a cosmos in which outside and inside, world and soul are one, continuously and throughout. This perfect happening is embedded in Phi itself—it is the very secret of the "golden number" Phi, the secret of the Logos.

A similar shift in the attribution of the value "1" to the middle term in the geometric representation of the golden mean's logarithmic unfolding offers equally rewarding insights into what is *implied*—literally, "folded in"—the phenomenon.

A Phi based logarithmic spiral (with Phi as the base unit, and 1 as the mean length, as shown in fig. 39) shows that what is carried out through the magic number Phi is a continuity of proportional reductions of a fundamental "one": a continuous oneness, a chain of smaller and smaller self-similar whole fractions of "one." The archetypal unit of proportion, Phi, is what goes down the spiraling stairwell of manifestation as the identity–integrity of the whole in every part.

The vessel of the universe's identity is a *bonding constant—a constancy of harmonious bonding.* Phi is the divine unit base of an exponential world-unfolding based on perfectly harmonious *rapport*. The intuitive perception of such a fundamental harmony breathing through all forms of life is an awesome source of wonderment.

What the two permutations between *"Phi"* and "one" do is shift our perspective. Standing on the Phi ground (the new "1") rather than the larger ground from which the golden mean emerged (the primordial "one"), we see its reality unfold, literally, inside out. We watch the world dimensions unfold from the perspective of the Logos. This shift turns our gaze around to what the Logos "does" from within his golden, Logoic ground. It gives us to contemplate how his golden mean triunity deploys a (logo-)rhythmic chain of whole, triune worlds, which are self-similar fractals of the Logoity held in Phi. "Having pervaded the universe with a fragment of Myself, I remain," says Krishna in the Bhagavad Gita.

✪ Logarithmicity is the essence and the gesture of Phi, and Phi is the "proportion-number" of the Logos—the "arithmos" of the Logos. ✪

✪ Discovering how the logarithmic principle is in germ in the golden ratio opens our eyes to the numinous logo-rhythmic identity of the Logos of the world behind his self-same gesture in the geometries of nature. Looming behind the Platonic apportioning, secting, and binding is a primordial triune dynamic of vibrant separating and bonding.

In the end, the dynamic core of Phi discloses the heartbeat of all living forms. The spiraling motion of life in nature points to the milieu of love-harmony of the trinitarian Logos as the source that instructs matter—"in-structures" it—in the harmonic intelligence of creation, and shapes organic forms to its pulsation. Nature, the great cosmos outside, points to the great inside—the world soul—and the dynamics of love that makes them one. All organic forms are forms of bonds (from the objective perspective of science) and forms of love (from the inner perspective of soul) between the "inside and outside"—the logos and cosmos of Being—self and world. ✪

Filled with logarithmic momentum, the golden ratio is the geometric formula of the Logos, the perfect vehicle for the world soul to indwell nature and cosmos. Vesseled in the proportional fractioning of "one" (as Phi) is the love-harmony principle of the Logos. Sealed in its logarithmic unfurling is the potential for the world being to articulate—"word"—itself by swirling-spiraling-spiriting matter into innumerable forms of "itself," infinitely diverse, yet infinitely similar, infinitely large and infinitely small.

The logarithmic cosmos is the *body* of the world-creative Logos; the threefold mind—his gradually unfolding *self*-awareness; logarithmicity is his *life* pulse—the heartbeat of harmonic bonding, which quickens the relation between world and being. The trio of Same, Other, and Being reappears in human nature as the trinity of spirit, body, and soul. It reappears in consciousness as the faculties of willing, thinking, and feeling.

Love and golden mean speak respectively to the inner reality and outer geometry of creation. It is equally true to say that love propagates life, as to say that the golden mean unfurls the process of integrative differentiation of the "one" (life) that produces its many forms. It is equally true to say that god spins the stars, as to say that the logarithmic dynamics of the golden mean "turns" cosmic matter into the Logoic organisms of seed and self, star and galaxy.

Ecstatic with the inner knowing that Love moves the planets to revolve around the Sun, whirling dervishes tilt their head at the golden angle of creative intercourse between above and below, to enter the self-spinning deployment of Logos into cosmos and the self-gathering of cosmos into Logos.

IV. The Spiraling Dynamics of the Universe

THE SPIRALING UNIVERSE:
PHYSICAL FACTS OF NATURE, BODY AND COSMOS

> *Taking nature as exhibiting thought for my guide, it appears to me that while human thought is consecutive, Divine thought is simultaneous, embracing at the same time and forever, in the past, the present and the future, the most diversified relations among hundreds of thousands of organized beings...*—LOUIS AGASSIZ[49]

We are now ready to look at the formations and organisms of nature and cosmos with an eye for the patterns whose principles we have been uncovering. Spiral cyclic motions and patterns are what nature and cosmos display all around us, visibly and less visibly: from plant growth to planetary orbiting and solar systemic spiraling, from insects circling around light to tropical cyclones; from the shells of mollusks to the wings of birds, from the hairlines of the human crown to the spiral formations of ear and nose, from chakric vortices to the arms of spiral galaxies, from the pattern of seeds in a pineapple and a pine cone to the DNA molecule.

Our exploration has led us to see in the spiral the geometric profile of the universe and its Logoic "germ"—a fact that the very etymology of the word *spiral* seems to intimate. Spiral comes from the Greek word *speira,* which means germ.

The prevalence of spiraling patterns in ancient sites across the globe suggests a deeply seated awareness of the cosmogonical significance of the spiral. A central motif in many Goddess-worshipping pre-historic centers in Europe, it symbolizes the great mother's creative womb. Indeed. Ancient Crete, where the mother-centered Minoan civilization flourished, shows an abundance of such labyrinth-like patterns. The Chora church in Istambul displays at the center of one of its vaulted ceilings a nautilus like scroll, borne downward on the wings of an angel, as if to point to the central secret of creation: the unrolling, from a transcendent heavenly source, of a

Fig. 43. Ceiling of the Byzantine Chora Church in Istanbul

---

49   Agassiz, *Essay on Classification*, p. 131.

universe at once dual (Sun and Moon) and "marvelously" one, apt to mediate between indivisible and divided worlds and also roll the two into one (fig. 43).

The fundamental seed-pattern of the universe blends the two[fold] dynamics of "extraverting spirit and centroverting matter," which underlies the enfolding of spirit in forms of matter and the unfolding of matter into forms of being—into the "universing," universe-making, gesture of a dual spiral (Latin: *vertere* = to turn)—two currents unfolding and refolding One turning world.

The spiral is the signature of *"spirit"* involved in the downward stairwell of world dimensions and the evolving growth of every living form: the centrifugal *(ex)spiration* of life into elemental peripheries, and the centripetal *(in)spiration* of elements into an ascending chain of intelligent organisms, self-aware beings, and cosmic systems. The spirit of the spiral abides discreetly in such words as spring, sprout, and spora (seed, sowing), spire and re-spiration, sperma, and spirituality.

As "rational" Voltaire intuited, what produces our mind and thoughts is indeed a spiraling principle of inspiration and respiration. "There must be something within us that produces our thoughts. Something very subtle; it is a breath; it is fire; it is ether; it is quintessence; it is a number; it is harmony." The Greek word for *spiritus* is *pneuma,* "that which is breathed or blown"—the breath of life.

Spiraling is how the vibrant Oneness of life negotiates the polarities of spirit and matter, center and periphery, movement and form, large and small, inside and outside. The inside seeks to unfold and the outside to reveal its innermost. The world soul unfurls and appears, Phanes like, out of the divine beginnings, as the universe of minerals, plants, and humans arises from the obscure ends of matter, rolling matter up into forms of life and turning flesh into organs of light for its light—eyes for the "I"-ness of the world soul to appear.

The spiral is the dynamic mean of creative engagement of the two currents of world becoming—matter differentiation and spirit integration, centrifugal and centripetal—whose conjugation conceives and produces the germs of all that exists. So is the intercourse of yin and yang, said to produce the 10,000 beings. In the words of George Oshawa, the founder of the macrobiotic diet and reviver of a worldview based on ancient yin–yang Taoist concepts:

> From the largest stars and planets to the microscopic basic elements, all things in this universe conform to cyclical spiralic form...this form is established by the interaction of two circular energies which are antagonistic to each other.... All [these] bodies are moving in logarithmic spirals.... All things grow through a union of complementary opposites. Structures in nature are formed by spirals that move in opposite directions—clockwise and counterclockwise.[50]

---

50 Oshawa, *The Art of Peace,* p. 65.

IV. The Spiraling Dynamics of the Universe

Randolph Stone shared the same archetypal view:

> All energy in creation vibrates as a logarithmic spiral, a three-dimensional spiraling vortex, undulating in both directions, centrifugally and centripetally, simultaneously.... Every level of macrocosm and microcosm reflects the polarity dynamic of energy radiating outward from a nucleus and being received again into that neutral center... All creation is an undulating spiral helix.[51]

This bidirectional motion induces spiral–vortical processes of self-similar expansion-compression in space, involution-evolution in time. Levels of being and forms of nature unfold along spiral cyclic, helical patterns.

Etymologically, the term *universe* means "one coil"—keeping alive, in the muted eloquence of a word, the ancient understanding that all creation and all subsystems of nature are part of one vast spiraling harmony.[52]

To explore some of the layers of this spiraling universe, we will follow the ascending course of matter-spirit through the kingdoms of nature, starting from the mineral, continuing with the plant kingdom, the human kingdom, and the macrocosm.

The mineral world, the realm of atomic elements, chemical compounds and their aggregation into crystals, gems and rocks, represents the most "elementary" organization of matter. At the end of the nineteenth century, Mendeleev discovered that classifying the elements according to their atomic weight or mass was revealing an interesting fact—a "periodic" recurrence of similar physical and chemical properties.[53] This hint at an underlying periodic order led Mendeleev to display the elements in rows of columns reflecting these similarities. Self-sameness was emerging as a key feature of the atomic world.

A different charting of the elements, published in 1900, involved a spiral. Classes or families of elements with similar properties lined up on the same radii, revealing stages of cyclic repetition (for instance: H, CL, Br, I). Originally found in Erdmann's *Lehrbuch* (1900), this spiraling display of the chemical elements was reintroduced by Leadbeater and Annie Besant in *Occult Chemistry,* as they saw that the order and classes resulting from their own investigation concurred with Erdmann's arrangement on a curved line (fig. 44). They were intrigued by this curved arrangement and

---

51 Randolph Stone, quoted in Burger, *Esoteric Anatomy,* pp. 145, 148, 178. The *Webster* dictionary defines *helix* as a three-dimensional spiraling form (Greek *eilyein* = roll, wrap).

52 Ibid., p. 170.

53 Chemical elements consist of a single type of atom. Atoms differ in charge, atomic weight, number of protons, and electrons. The chemistry of an atom is tied to its atomic weight, which is related to its atomic number—the number of protons in the nucleus, which is also the number of orbiting electrons.

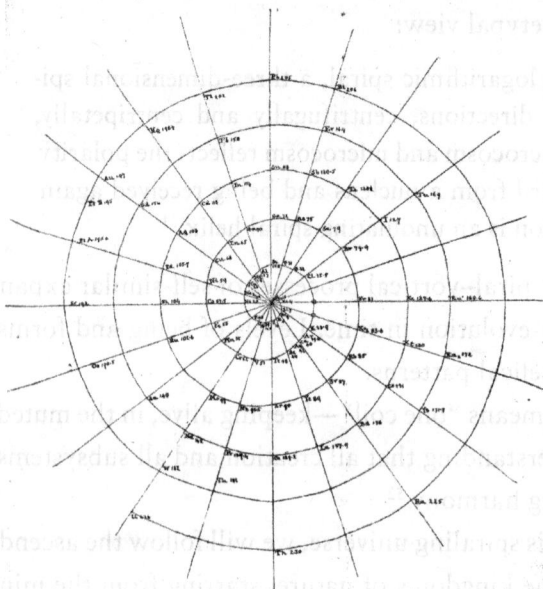

Fig. 44. *Periodic table from Erdmann's Lehrbuch, reproduced in Leabeater and Besant, Occult Chemistry*

struck by its "curious resemblance with the curves within the shell of a nautilus."[54] This observation highlights the striking parallel between the geometric unfolding of the elements by order of complexity and the spiraling growth of biological systems. The discovery of a spiraling self-sameness at work at this most invisible level of the mineral order is an exquisite example of the marvelous facts brought to light by modern science.

The spiraling arrangement reveals that, on the basis of their atomic weight, the elements follow a logarithmic order. The discovery was corroborated by the findings of Dr. Stoney presented in Theodore Andrea Cook's *The Curves of Life* in 1914. Clearly, minds were converging to sight a fundamental ordering of matter. Cook writes:

> In 1888, Dr. Johnstone Stoney submitted to the Royal Society a memoir on the "logarithmic law of atomic weights," which, however, was not published in full.[55] After many fruitless efforts to extract information from the curves obtained by plotting the atomic weights, it happily occurred to Dr. Stoney to employ the volumes proportioned to the atomic weights. When this was done, the resulting figure at once suggested a well-known logarithmic spiral. Close scrutiny justified this suspicion. In other words, the *relations of all the known elements to each other could almost exactly be expressed by the logarithmic spiral.*

Cook's additional remarks are worth quoting in full as they illustrate what frequently happens with comprehensive scientific insights such as Stoney's. Later discoveries fill in initial gaps in the archetypal sequence and end up confirming the larger intuition. We saw a similar development in the case of Bode's law. Cook writes:

> If this held true of what was known already, it became apparent that it would also hold true of what was to be discovered later on; and that if new elements were discovered after 1888, they would find their right places in the gaps indicated

---

54   Besant and Leadbeater, *Occult Chemistry*, p. 28.

55   "Lord Raleigh consulted the original manuscript and gave some extracts from it and remarks upon it"; *Proceedings of the Royal Society*, series A, vol. 85, 1911, p. 471.

in Dr. Stoney's spiral diagram. This remarkable process had already occurred in Mendeléev's Periodic System since the year of its publication in 1869: and the fact that it has also occurred in the spiral system is one of the most convincing proofs that the spiral system is not merely a correct hypothesis, but a fundamental law. The total of the elements known in 1912 was about eighty-three. Six elements were missing in 1888 in Dr. Stoney's diagram, between hydrogen and lithium; Sir William Ramsay discovered helium in 1895, which fills one of the gaps, though the position is not mathematically exact. But on the sixteenth radius an even more remarkable corroboration was effected, in what had hitherto been a gap between the most intensely electro–negative elements (such as fluorine, chlorine, bromine, and iodine) and the most electro–positive elements (such as lithium, sodium, potassium, etc.). This gap was filled with absolute appropriateness by the series of inert gases: argon, discovered by Lord Raleigh and Sir William Ramsay in 1894, and helium, neon, krypton, and xenon, discovered by Sir William Ramsay between 1895 and 1898, five new elements which occupy places foretold to be necessary to the Mendeléef series as well.[56]

Other spiraling representations of the periodic system of the elements have been proposed since the time of Dr. Stoney. Philip Stewart, for instance, designed a model he called "the chemical galaxy," pointing to the self-sameness of macrocosmic and microcosmic systems.[57]

The entire periodic table spans seven octaves. Like every eighth white key on the piano, each eighth atomic group marks a new turn of the spiral, with a repetition of chemical properties at each rung. The phenomenon of periodic recurrence is analogous to the musical phenomenon of "resonance" between the same notes one, two, three octaves apart. The fields of resonance exposed in the periodic spiral of elements mirror the spiraling deployment of world dimensions that, like the musical scale of the periodic "groups," involve analogous sequences (or groups) of components.

Gary Peacock, a musician who undertook studies in molecular biology in the midst of his musical career, found that the relationships in the periodic table of elements mirror the overtone structure of harmony. "It becomes clearer and clearer to me," he writes, "that the actual structure of tone in music and the actual structure of matter are the same."[58]

The *Timaeus* pointed to the relation between the laws of harmony and the formation of the Platonic-solid–like structures of the elements, and *Occult Chemistry* advanced that the Platonic solids inform the shapes of chemical atoms and actually line up behind their valency groupings. Bruce Burger synthesizes this view in one

---

56   Cook, *The Curves of Life*.
57   See http://spiralperiodictable.com.
58   Tame, *The Secret Power of Music*, p. 228.

concise statement: "The ancient wisdom describes the fields of resonance emanating from the center as elements," which are "defined by specific harmonics."[59]

Plato's cosmology led us indeed to view the five elements associated with the five Platonic solids as harmonic, wholeness carrying fractions of the whole, embodying phases of greatest consonance in the relationship between spirit and matter—between the opposite currents of spirit's involvement in forms of matter and matter's evolving organization of forms of being. "All creation," Burger concludes, "entrains through sympathetic vibration with the archetypal keynotes of the universal harmonics of the elements"—which, as we shall see, are the keynotes of the universal octave.[60]

According to Leadbeater and Besant's clairvoyant research, the smallest unit of elemental matter, the atom, is itself composed entirely of spirals. Each spiral is composed of spirillæ, which are in turn comprised of minute spirillæ. This is how the authors describe this multidimensional multilayered spiral "structure":

> Three whorls surrounding [a central] "hole" with their *triple* spiral of two and a half coils, and returning to their origin by a spiral within the atom; these are at once followed by *seven* finer whorls, which following the spiral of the first three on the outer surface, and returning to their origin by a spiral within that, flowing in the opposite direction—form a caduceus with the first three... In the three whorls flow currents of different electricities; the seven vibrate in response to etheric waves of all kinds—to sound, light, heat, etc.; they show the seven colours of the spectrum; give out the seven sounds of the natural scale; respond in a variety of ways to physical vibration—flashing, singing, pulsing bodies, they move incessantly, inconceivably beautiful and brilliant.[61]

The threefold and sevenfold spirals are emblematic of what we have found to be key features and numbers associated with the cosmological beginnings and peripheral constructs of the universe.

The clairvoyant researchers discern two types of atoms:

> ...in this ultimate state of physical matter...alike in everything, save the direction of their whorls and of the force which pours through them. In the one case, force pours in from the "outside," from fourth-dimensional space, and passing through the atom, pours into the physical world. In the second, it pours in from the physical world, and out through the atom into the "outside" again, i.e., vanishes from the physical world. The one is like a spring, from which water bubbles out; the other is like a hole, into which water disappears. We call the atoms from which force comes out positive or male; those through which it

---

59   Burger, *Esoteric Anatomy*, pp. 148, 163.

60   Ibid., p. 254.

61   Besant and Leadbeater, *Occult Chemistry*, p. 52.

disappears, negative or female. All atoms, so far as observed, are of one or other of these two forms.[62]

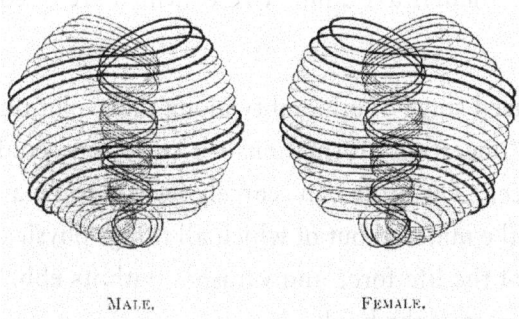

MALE. FEMALE.

Fig. 45. The "ultimate" atoms, as seen clairvoyantly by Leadbeater and Besant

These two types of atoms illustrate the dual motion of centrifugality and centripetality—involution-evolution—characteristic of the holodynamics of the universe. How remarkably congruent, that the caduceus like breath of life that informs manifestation would show up so precisely at this micro level—first in the dual spiraling-out and spiraling-in dynamics of every atom, then in the polarity of "male and female atoms" formed by the opposite currents of universal flow.

Stephen Phillips, who, as a graduate student in physics in the late 1970s became interested in Besant and Leabeater's work, confirmed and reinterpreted their clairvoyant descriptions of the chemical elements as "completely consistent with the Quark, Quantum Chromodynamic, and Super-String theories of modern subatomic physics."[63] His own research led him to conclude that their observation of the ultimate atom was not of atoms, but of subquarks and quarks.

Crystals have also been found to enter a classification into seven types: cubic, rhombohedal, hexagonal, triclinic, monoclinic, tetragonal, and orthorhombic. In the same way that the seven finer spirals of the atom *"give out the seven sounds of the natural scale,"* the seven types of crystals may be viewed as "the crystallization, at the mineral level of being, of the seven elemental harmonics of creation."[64] These seven major groups differ in the angular and length relationships of their respective axes, intimating again how specific geometric fractionings, evidenced in the angular relations of the parts of a crystal-"unit" to its virtual sphere, translate into forms

---

62 Ibid., pp. 48-50 (plate 2).
63 See http://blog.world-mysteries.com/science/micro-psi-and-string-theory-how-occultists-beat-physicists-to-the-punch.
64 Burger, *Esoteric Anatomy*, p. 147.

what is at play behind forms: harmonic intervals. The cube, for instance, features three axes of identical length perpendicular to each other; the tetragonal system, three mutually perpendicular axes, only two of which are of equal length.

Preceding the organized stage of minerality along the evolutionary path of forms is a quantum field of "uncertain" formations—a stage, one could say, of unformed, "unworded" substance. "The atom can scarcely be said to be a 'thing,'" notes Leadbeater, "though it is the material out of which all things physical are composed. It is formed by the flow of the life-force and vanishes with its ebb."[65] We will return to this field in the next part of the book.

The plant kingdom exhibits most conspicuously the fundamental spiraling dynamics of the universe. Visible in the growth patterns of plants and trees, and the structures of many vegetables, fruits and flowers, are dual spirals of growth, which precipitate in physical formations the swirling conjugation—in and out, downward and upward—of Earth substance and Sun forces. This dynamic interplay organizes elemental matter along its vibrational ordering. Burger concisely articulates, "All systems of vibration in nature are subsystems within the circulation of life energy between the Sun, as the nucleus of the centrifugal field, and the Earth, as the nucleus of the centripetal field."[66]

Fig. 46: (top left) Romanesque Broccoli; each floret represents an identical but smaller version of the whole; (top right) Pinecone shows Fibonacci spirals in clockwise and counterclockwise directions; (bottom left) wildflower; (bottom right) fern

The 7fold structure of the plant reproduces the 7fold octave-like spine of the universe, and the 12fold microscopic structure of the chlorophyll bears the seal, in the plant kingdom, of the archetypal bonding of world polarities—in this case, mineral geometries and solar light. Growth plays out the ascending-descending interpenetration of Earth and Sun—into the formations of stem, leaf, and flower. What plant growth gives us to witness is the concomitant climbing up of matter and

---

65 Besant and Leadbeater, *Occult Chemistry*, p.16.
66 Burger, *Esoteric Anatomy*, p. 174.

climbing down of light—along the *proportionate, harmonic means* where they jointly emit the phases-tones of nature's unfolding: F of the leaf, perhaps, and G of the flower.

Goethe, the enlightened naturalist who discerned in plants this sevenfold metamorphic climb, is also the poet gazing, through the contemplative eye of Faust, at the rising and descending Powers of the world intercourse, exchanging "golden urns" in "melodious ringing."

An analogous interplay of polarities forms the core of the human energy system, the etheric organism of energy currents, which literally in-forms the physical body, weaves its organization and sustains its life. Another designation for this energy is *prana*, a Sanskrit term that means life or life force. In Chinese culture, it is called *Chi*. "Prana is, on the subtle level, what the breath is on the physical plane," Georg Feurstein explains.[67] According to the Rig Veda, *prana stands for the breath of the Macranthropos, or Cosmic Person*, and elsewhere for the breath of life in general.

Contrasting with the common association of the breath with axis, verticality or column of air, Prana is associated with envelope, sphere, circularity, aura, mandorla—a dual perspective that, when we reach a more distinct picture of the archetypal form of life, will prove to be, beyond the apparent contradiction, an actual insight into its dual dynamics.

"The life force clothes a person as a father would clothe his dear son," says the Atharva-Veda.[68] This image echoes Rudolf Steiner's clairvoyant observation of the radiant cosmic aura "complete with Sun and zodiac," which, he describes, clothes every child upon his arrival-birth on Earth. It also evokes the traditional representation of "sons of God," such as Mithras and Christ, enveloped in the twelvefold auric mandorla of the matrix-patrix of the world.

The energy practitioner is aware that Prana, or Chi, does not circulate in straight lines, but in arcs, circles and spirals—small, tight spirals along acupuncture meridians, as well as large circles and spirals swirling the auric field. Disciplines such as Tai Chi or eurythmy are based on a wisdom of this etheric field and its vital significance for the harmony of body and psyche, self, and cosmos.

Modern research based on MRI and color Doppler flow imaging techniques has demonstrated the prevalence of spiral flow in the circulatory system, showing how this motion of life continues on the physical plane. "There is a spiral-helical property to laminar blood flow.... A clockwise rotation has been demonstrated

---

67 Feurstein, *Tantra*, p. 148.

68 *Atharva-Veda*, 11.4.11.

during systole...a counterclockwise rotation has been shown in diastole in the descending aorta."[69] Under the name "Spiral Laminar Flow," a technology of prosthetic grafts has been developed, on the basis of these findings, to "restore the blood flow pattern to the body's natural state—spiral flow." The researchers point to drawings dating back to Leonardo da Vinci, which already convey the spiraling character of the blood flow in the body's vessels. Leonardo sketched what appears to be spiral cardiac emptying—spiral flow generated within the left ventricle of the heart.

Traditional Hindu wisdom describes how this fundamental holomovement of energy is centered on two major spiraling currents of "pranic" energy, called *Ida* and *Pingala,* intertwining around a central channel called *Sushumna,* to form the threefold axis of our energetic being. The central and primary *nadi*—the Sanskrit term for a channel of *prana*—called Sushumna, connects the base chakra to the crown chakra. On either side of the Sushumna, the two main secondary channels of Ida and Pingala interweave their opposite and complementary currents. On the right side, Pingala, the extroverted, solar nadi, is associated with masculine attributes, the color red, the positive polarity, and the Yang element of Chinese philosophy. It corresponds to the left hand side of the brain. On the left side, Ida is the introverted, lunar nadi associated with feminine attributes, the negative polarity, a pale color, and the Yin element of Chinese philosophy. It refers to the right-hand side of the brain. Ida contains the descending or centrifugal vitality (apana), and Pingala the ascending or centripetal vitality (prana). Polar opposites of each other, Pingala and Ida pursue a serpentine pattern, a double-helical pathway around a neutral Sushumna. "Together, John Woodroffe points out, the three nadis form the figure of the Caduceus of the God Mercury, which, some say, is their representation."[70]

This threefold channel steps down "the universal creative energy" (*prakriti*) into the human form. What Indian philosophy refers to as the three *gunas*—three primordial qualities or modes of motion called *rajas, tamas,* and *sattva*—are the qualities of consciousness associated with the three energies of Pingala, Ida, and Sushumna. Analogous to the yin yang and Taichi of Chinese cosmology, this fundamental threefold is an expression of the 1, 2, 3 (same, other, and Being) of Platonic cosmogony. Each formula is a particular expression—energetic, philosophical, and mathematical—of the trans-physical triune "selfhood" of the world.

Strung "like jewels" along the brilliant thread of the Sushumna are the seven chakras—seven pulsating whirls of energy generated by the vibrational interference of the intertwining currents. Clairvoyant observers describe chakras as energy fields

---

69 See http://vascular-flow.com/spiral-laminar-flow-technology.

70 Woodroffe, *Shakti and Shakta*, ch. 29.

IV. The Spiraling Dynamics of the Universe

that resemble rotating color wheels, somewhat funnel or vortex-like. Each chakra represents a particular phase of harmonic conjugation between the vibrational currents of "Heaven and Earth"—descending life energy and rising responsive matter—"exchanging their golden urns" between base and crown.

When Leadbeater compares the general appearance of a chakra to "the bell of a flower of the *con-volvulus* type," he captures the dynamics of the "marvelous spiral" at work in both chakra and flower: the harmonic co-spiraling involvement of Earth and Sun, which replicates the primordial conspiring of matter and spirit into a world vortex. He adds, "The stalk of the flower in each springs from a point in the spine, so another view might show the spine as a central stem from which flowers shoot forth at intervals, showing the opening of their bells at the surface of the etheric body."[71]

Stone calls "ultrasonic core" the central pathway from which the chakras radiate and converge as centrifugal and centripetal vortices. This ultrasonic energy flow, he explains, is actually "the soul of man in this mortal body, the primary energy which builds and sustains all others."[72] The chakras constitute a step-down system from cosmic to solar systemic, to planetary, to human structuring. They step down the seven dimensions of the world being into the psychophysiological vehicles of human beings. Like the kingdoms, chakras are outposts of these primordial levels: physical, etheric, astral, mental, buddhic, monadic, logoic. The chakric system as a whole is a micro-replica of the prototypical body–soul of the world being—7 dimensions of Being and 12fold soul core—solar system and zodiac.

Through sympathetic vibration, the first four chakras resonate with the four elements, the fifth with their quintessence, the sixth with the primordial dual field, and the seventh with the creative source. The base chakra of physical survival is entrained by the earth element, the sacral center of etheric vitality resonates with water, the solar plexus of emotional activity with fire, the throat of verbal communication with air, the heart with ether (universal harmony–love), the Ajna center with the discerning-unifying light of the Logos (universal mind), and the crown with the primordial vibration of the creative sound (universal will).

Every chakra is a fractal of one of the major sub-vortices of the "one coil" (universe). Each resonates with a major harmonic fraction of the fundamental sound of the world being. This harmonic interval or ratio is the qualitative reality underlying a chakra and the corresponding kingdom of nature. Measurements have shown a coincidence between the chakras and the nodes of a vibrating string stretched between top of the head and feet: the half-way point between head and feet coincides

---

71 Leadbeater, *The chakras*, p. 3.

72 Randolph Stone, quoted in Burger, *Esoteric Anatomy*, p. 218.

with the genitals, the half-point of the upper half with the heart, while further divisions (twelfth harmonic) coincide with sacral, solar plexus, throat, base of the nose. The forehead (*ajna*) corresponds to the twenty-fourth harmonic. The chakras may thus be viewed as lining up in a micro-"octave" between crown and base—an instrument to be tuned and refined into a unique epiphany of universal harmony.

The three currents of Ida, Pingala, and Sushumna represent the physio–psycho-spiritual trinity that defines a human being: involutionary, lunar energy of personality; evolving, solar energy of soul; axial energy of the monad, the universal self. Such a view of body, soul and spirit as a threefold braiding of energy currents engaged in a synergetic process—as opposed to the more traditional layered representation—illumines the dual opus of incarnation and spiritualization that defines the labor of individual growth. The descending–ascending levels of our subtle physiopsychology correspond to specific phases (and proportions) of harmonic intercourse between materializing–involutionary and spiritualizing-evolutionary currents: self-survival instincts (base chakra), reproductive biological forces (sacral center), emotional drives (solar plexus), individuating-universalizing power (heart), intelligent creative expression (throat), integrative intuitive mind (*ajna*), and spiritual identity (crown). We see how the human constitution replicates the primordial differentiation of the one(-ness) into a 3fold, and the deployment of the 3fold One into a 7fold form centered on a 12fold nucleus.

In the course of evolution, as natural urges and personal drives become imbued with soul quality, the vehiculating substance becomes more organized and vibrant. The chakric dimensions increasingly reverberate the harmonic tones, geometries and colors of the world octave, and the great soundness of cosmos begins to play itself out through a human being—as a "different," unique sameness.

Yoga aims at facilitating this natural evolutionary process. Tantra yoga focuses specifically on directing the subtle airs from the two side channels into the central Sushumna, in order to activate the integrative life force, the Kundalini energy coiled at the base of the spine. Eventually, the kundalini uncoils and ascends through every chakra. When it reaches the top of the head, a great sweep of unification-integration takes place. Reaching "union"—the very meaning of "yoga"—the yogi attains liberation from the dual worlds and enters a new octave-dimension of consciousness. A unique harmony of Heaven and Earth has been achieved—self-realized—realized in a "self."

✺

If harmonic fields underlie the formation of anatomical structures and the regulation of organic functions, such harmonies ought to be found in the geometries and functions of our physical bodies. Among the most eloquent and conclusive works to

IV. The Spiraling Dynamics of the Universe

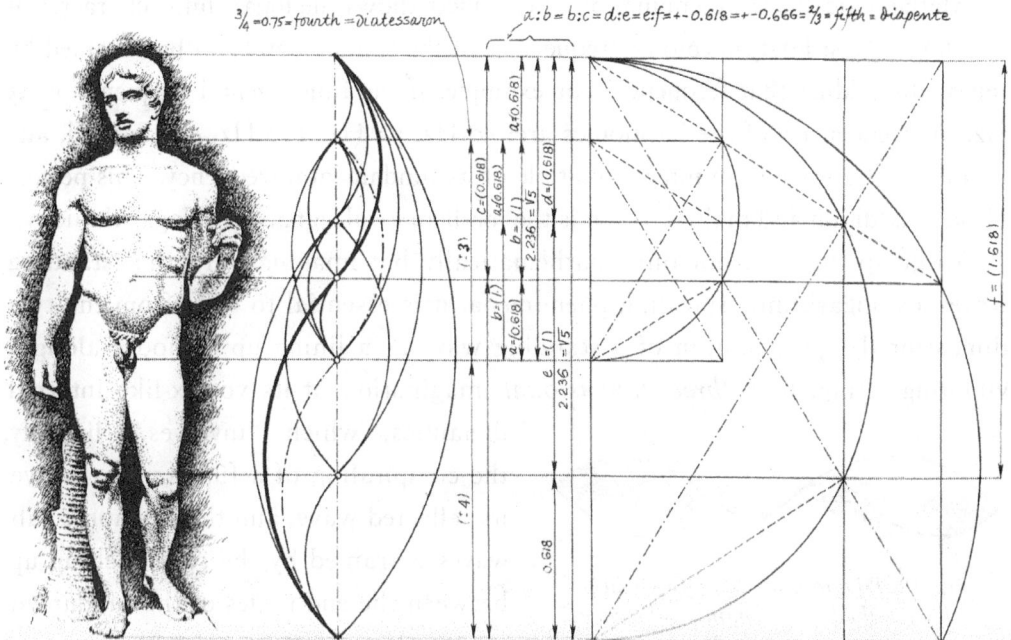

Fig. 47. The spear bearer or Doryphoros by Polycleitos
(Doczi, The Power of Limits, p. 104)

demonstrate, indeed, this harmonic structure of human and animal bodies are those of the Hungarian architect Gyorgy Doczi. In *The Power of Limits,* he presents a number of careful measurements that show the predominance of harmonic proportions in the structure of organic forms—from leaf to insect, from fish to butterfly, from animal to human body.

A sculpture by the fifth-century-BC sculptor Polycleitos, also credited for a lost treatise about the proportions of the human body, as Doczi points out, illustrates the overall harmonic composition of the human body, shaped by standing waves that sustain a double helix centered in the navel (fig. 47).

The dynamics of a standing wave is a dynamics of logarithmic spiraling. We remember that a standing wave pattern is created in a self-contained space (object, metal plate, instrument, body of water) when its medium (air, powder, water, metal, etc.) vibrates at frequencies/wavelengths that allow back and forth reflective waves between the extremities to interfere constructively, so that specific areas along the medium appear to be standing still. Such patterns happen only with frequencies that represent whole multiples of the fundamental frequency of the vibrating string, air column or the natural frequency of the object—in other words, with half-wavelengths that represent a whole fraction of the fundamental. Any other frequency will result in irregular disturbance.

Mathematically, the combination of two facts shows the logarithmic character of standing waves: First, harmonic frequencies of the fundamental can be obtained by repeatedly adding that frequency. For example, if the fundamental frequency is 25 Hz, the frequencies of the harmonics are: 50 Hz, 75 Hz, 100 Hz, etc. Second, any harmonic of a wave is an integer multiple of its fundamental frequency. This perfect blend of addition and multiplication is the mathematical signature of logarithmicity.

To grasp the fact, not only mathematically but imaginatively, that standing waves are logarithmic spiraling phenomena, it is essential to shift from the two dimensional representation of a standing wave as a lining up of nodes along a vibrating string, to a *three dimensional* imagination of its vortex-like internal dynamics, which "involves," literally, the co-spiraling of a fundamental wave, its reflected wave, and the harmonic sub-waves entrained by the two. Welling up between the end nodes of the fundamental wavelength associated with a particular standing wave are harmonic waves of smaller and smaller wavelengths, swirling in together constructively. The standing wave is this bundle of logarithmic spiraling-in sub-waves.

Fig. 48. Diagram of a standing wave with its embedded harmonic sub-waves

As noted earlier, the harmonic series is a logarithmic series. The fact that standing waves represent by definition a set of *harmonic* waves—in other words, waves whose half-wavelengths are whole number fractions of the fundamental wavelength—implies that they form three-dimensional logarithmic spirals. A direct way to perceive this logarithmic spiraling is to look at forms of nature that exhibit most tangibly the harmonic scheme of a standing wave. The nautilus is a good example.

Figure 49 shows the similarity of shape between a shell and a simple graph of the nested harmonic sub-waves entrained by a fundamental wave of vibration. Figure 50, an outside view of a seashell, shows unequivocally that logarithmic spiraling is the internal dynamic of a standing wave.

Conversely, the logarithmic spiral can be conceived as a series of waves of proportionately increasing wavelengths and decreasing frequencies.

Doczi's findings show how logarithmic spiraling is the fundamental dynamics underlying the structure of human and animal bodies. The logarithmic spiral can be superimposed on the fetus of animals and humans. Every living organism is informed by the holomovement of nature, which is the logarithmic dynamics of the Logos. The world arises out of resonant harmonic fields, and harmonic proportionality is the fundamental organizing principle of macrocosmic systems and microcosmic bodies.

IV. The Spiraling Dynamics of the Universe

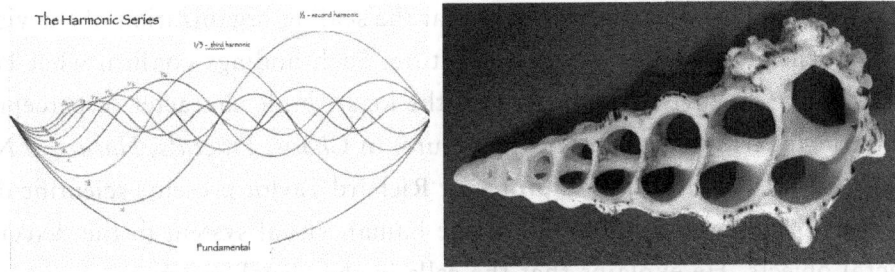

Fig. 49. The longitudinal cut of a shell displays a shape evocative of the harmonic series

 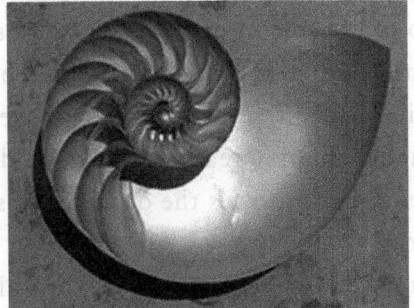

Fig. 50. Side view of a seashell (left) evokes harmonic diagram above; Longitudinal cut of a nautilus shell (right) shows a logarithmic spiral

To Doczi's demonstration of the harmonic structure of bodies, Lawlor adds insights into the harmonic organization of the sense organs. The sense organs, he says, step down the universal harmonic field: "Our different perceptual faculties such as sight, hearing, touch and smell are a result of *various proportioned reductions* of the one vast spectrum of vibratory frequencies.... Our visual sense differs from our sense of touch only because the nerves of the retina are not tuned to the same range of frequencies as are the nerves embedded in our skin."[73]

Further research in the science of harmony may show one day that the five senses correspond to the elemental fields of consonance generated by the major Pythagorean intervals. They exemplify the role played by the major harmonic ratios in conveying a whole and "sound," if fractional, experience of the world—be it acoustic, visual, olfactory, be it a sense of touch or taste.

Lawlor suggests that our sense of smell is organized to match geometric molecular constructs and interact with them. The experience we have of fragrance, he says, is the experience of a certain harmony of internal rapport, a certain bonding. "Any chemical substance which is *bonded together* in the same geometry as that of the rose will smell as sweet." Fragrance hints at a unique form of harmonious bonding, and harmonious bonding is fragrant. Great souls have been known to emit rose petal like fragrance.

---

73 Lawlor, *Sacred Geometry*, p. 5.

Similarly, scientific research reveals that the scaling organization of the visual system matches the fractal patterns of nature. Such findings confirm what Fechner's law led us to envisage, namely that the logarithmic character of perception matches the logarithmic geometries of nature. In *Chaos, Fractals, Nature: a New look at Jackson Pollock*, physics professor Richard Taylor presents scientific data that point to the perfect suitability of the human visual system to the detection of fractal objects. He explains that the cells in the visual cortex are grouped in channels "distributed in spatial frequency in a way that parallels the scaling relationship of the fractal patterns in the observed scenery."[74] The speculation is, he comments, that "this is not coincidental but the consequence of the visual system's adaptation to the fractal character of the natural environment during evolution." Is gradual adaptation really the key? Or does this similitude speak to the ontological ground that precedes the dualities of subject and object, seeing and seen, soul and world?

Taylor's fractal research laboratory also discovered that exposure to fractals automatically relaxes people, as stress reduction is triggered by the physiological resonance that occurs when the fractal structure of the visual system matches that of the fractal image being viewed.[75] Interestingly, he identified consistent regularities, recurring on finer and finer magnification in several authentic paintings by Jackson Pollock, and suggested this presence of fractal patterns might explain their popularity.

Steven Lehar, a Research Fellow in Ophthalmology at Harvard University, complements Taylor's view, when he advances, in his harmonic resonance theory, that "spatial patterns in perception and behavior are mediated by spatial standing waves in neural tissue."[76]

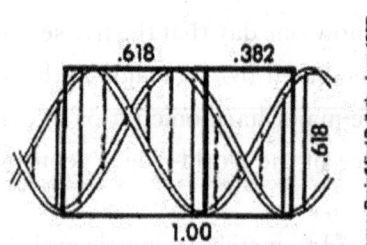

Fig. 51. Golden proportion in the DNA helix

Harmonic bonding and spiral-cycling also informs the structure of the DNA. Lawlor advances that the DNA carries out its replicating power not only through its molecular composition, but through its *helix* form.[77] One revolution of the double helix measures 34 angstroms and its width is 21 angstroms. The ratio 34/21 is a Fibonacci ratio, which points

---

74   Taylor, *Chaos, Fractals, Nature*.

75   See https://blogs.uoregon.edu/richardtaylor/2016/02/03/human-physiological-responses-to-fractals-in-nature-and-art.

76   See http://www.tokenrock.com/harmonicnature/resonance.

77   Lawlor, *Sacred Geometry*, p. 4.

IV. The Spiraling Dynamics of the Universe

to Phi; 34 divided by 21 equals 1.619, a close approximation of 1.618—Phi. As shown on the figure (fig. 51), the relative dimensions of the helix reflect the golden proportion. We also noted earlier that the DNA spiral reproduces the structure of a ratcheted dodecahedron, a dodecahedron spun in the pattern of a concertina with 12 turns of 30 degrees for each stage. In other words, the informative molecular structure at the heart of every cell is a dodecahedric helix. A cellular analog of the 12-petal lotus, it is a minute "causal" structure, a musical instrument that plays out uniquely individual variations on the theme of the human constitution.

Spiraling shapes are also visible in the structure of bones and muscles. Two interacting sheaths of muscles spiral from toe to head, resisting, complementing and balancing each other. Complementary muscle pairs never act in isolation. Every physical action engages the spiral dynamics of creatively engaged pairs. The ancient practices of yoga and Tai chi are based on an understanding of these underlying formative processes. Tai chi cultivates movement based on the spiraling dynamics of etheric energy, and yoga operates from an intelligence of the intertwined spiraling of muscles and the balancing of polarities.

A number of organs show evidence of the same signature. Sense organs, we saw, follow logarithmic spiraling shapes. The pineal gland, also known as the pinecone gland, is a spiraling structure akin to a vegetal form. The convolutions of the intestines and the contractions of the heart exhibit vortical dynamics.

From skeletal constitution to sensory periphery to DNA quintessence, the human body shows the same fundamental harmonic architecture as any living form in nature. In Bruce Burger's words, "The body, as all of nature, is a fabric of attunement. Every muscle and bone is a field of energy vibrating as an expression of attunement with the standing waves of universal law."[78]

Theodor Schwenck, a Swedish pioneer in water research, noted how the spiraling forms of nature point to the mediating factor of movement, as well as the mediating elements of water and air, and how through them both, energy, the energy of a spiritual being, works its way into matter.

> Forms arise out of the gaseous or liquid elements, but above all it is to movement that we must ascribe these forms. They are the movements of a spiritual being descending from the world of cosmic laws to take on bodily form through the circulation of air and water.[79]

---

78 Burger, *Esoteric Anatomy*, p. 236.
79 Schwenk, *Sensitive Chaos*, p. 126.

Schwenck's experiments with water and air phenomena capture dynamics evocative of organic formations. For instance, pictures taken when air or water is stirred by a secondary current of the same medium (sound vibration in the air or current created in water) show vortical patterns reminiscent of the shapes of living systems—mushroom, cauliflower, brain (fig. 52, 3 & 4), spinal column (fig. 52, 2), tree-like shapes—intimating how wave dynamics underlie the formative processes of nature. The interplay of water currents clearly replicates and illustrates, in a slow motion fit for human perception, the cosmogonical processes engaged by primordial waves of vibration. Schwenck points out how many unicellular water animals have incorporated the archetypal spiraling movement of water not only in their shapes, but in the way they propel themselves along with a screw-like movement. "The forces of the flowing water simply demand the spiraling form," he observes.[80] Schwenck brings to his observations and his very capacity to observe the cosmological understanding that these vortical currents, which come to rest in the forms of nature, result from a combination of the spherical tendencies of water with the gravitational forces playing on it.

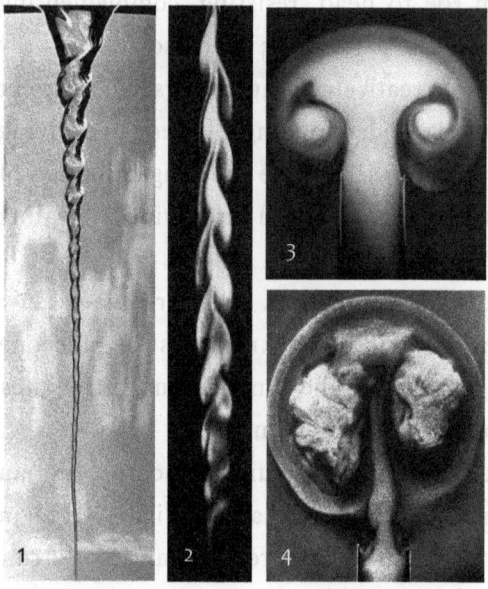

Fig. 52: (1) simple vortex produced in water; (2) vortex of air produced by a flute sound; (3) An underwater stream of liquid flowing into still water creates a mushroom like form; (4) A slight variation in the previous experiment produces a new form evocative of a brain or a walnut (Source: Theodor Schwenk, Sensitive Chaos, illustrations 41, 69, 46, 48)

Schwenck's experiments show in the workings of nature what mathematical and geometric abstractions explain about standing wave dynamics. They show how the countercurrent induced by a sound vibration pulsed in an air medium, or a stream pulsed in still water, generate stable, "standing" vortices.

"The vortex in the water," Schwenck concludes, "is like an archetypal organ—an Ur-organ having within it all the potentialities of differentiation and formation." Schwenck's insight strikes a resonance, in our context, with the notion we arrived at earlier, of a logarithmic (vortical) "common sense" and common sense organ. In nature, Schwenck adds, the vortex reveals different stages of itself—"ranging from

---

80   Ibid., pp. 21–23.

IV. The Spiraling Dynamics of the Universe

an invisible creative dynamic principle to an organ in all its finest detail. All the intervening stages and manifold variations of the vortex formation can be discovered everywhere in nature."[81]

The macrocosmic forms, the physical bodies of cosmic Logoi, reveal the same dual spiral dynamics as observed in plants and flowers (fig. 46, page 242). While evidence of this fundamental pattern of galaxies had to wait for modern telescopes and space explorations, the mystical consciousness has for long perceived the subtle gesture of the world and its beings (fig. 53).

Fig. 53: (left) The barred spiral galaxy NGC 1365 (NASA/JPL-Caltech); (right) Angel (from "The Shahnameh," a tenth-century Persian poem by Firdowsi)

From the perspective of astronomy, our solar system orbits around the center of the Milky Way galaxy, whose dual spiral shape appears clearly on figure 58.6 below. Blavatsky states that it spirals forward around a central point located in the Pleiades, in the neck of the constellation of Taurus—a meta-center that esoteric sources refer to as the central spiritual sun.

Schwenck's contemplation of the vortex as the archetypal formative motion led him to significant insights into the fundamental dynamics of the solar system:

> With its different speeds—slow outside, fast inside—the vortex is closely akin to the great movements of the planetary system... [and] depicts in miniature the great starry universe. The Sun corresponds to the suction center of the vortex, where the speed is theoretically infinitely great.... Apart from minor details, it follows Kepler's second law of planetary movement: a given planet circles around the Sun as though in a vortex in as much as it moves fast when near the Sun and slowly when further away... a law which applies to all the planets.[82]

---

81   Ibid., pp. 47–49.

82   Ibid., pp. 44–45.

The spiral cyclic motion of cosmic systems displays a blend of "spherical and gravitational forces" that speak to the creative interpenetration of the enfolding involutionary dynamics of being and the unfolding evolutionary dynamics of cosmos.

Rudolf Steiner takes us one step further "into" the center, suggesting that we "imagine the space taken up by the Sun as a kind of sucked out space—a space that is not only empty but just the opposite of being filled with substance. With the Sun, there is suction—not pressure, as in a gas-filled space, but suction. The Sun is filled with negative materiality."[83]

With regard to the structure of the solar system per se, we owe a remarkable insight to the nineteenth-century Harvard professor of mathematics and astronomy Benjamin Peirce. In his *Essay on Classification*, Peirce's colleague and friend Louis Agassiz, himself a renowned biologist and geologist, explains that Peirce discovered a "most perfect identity" between the fundamental laws that regulate the arrangement of the leaves in plants and the revolutions of our solar system. Planetary orbits form a pattern analogous to the phyllotactic pattern of leaves on a stem. Considered from the perspective of time rather than distance as in Bode's law, planetary revolutions show a pattern of progression analogous to the Fibonacci pattern observed in the unfolding of leaves around the stem of a plant.

> It is well known that the arrangement of the leaves in plants may be expressed by very simple series of fractions, all of which are gradual approximations to, or the natural means between, 1/2 or 1/3, which two fractions are themselves the maximum and the minimum divergence between two single successive leaves. The normal series of fractions which expresses the various combinations most frequently observed among the leaves of plants is as follows: 1/2, 1/3, 2/5, 3/8, 5/13, 8/21, 13/34, 21/55, etc. Now, upon comparing this arrangement of the leaves in plants with the revolutions of the members of our solar system, Peirce has discovered the most perfect identity between the fundamental laws which regulate both, as may be at once seen by the following diagram, in which the first column gives the names of the planets; the second column indicates the actual time of revolution of the successive planets, expressed in days; the third column, the successive times of revolution of the planets, which are derived from the hypothesis that each time of revolution should have a ratio to those upon each side of it, which shall be one of the ratios of the law of phyllotaxis; and the fourth column, finally, gives the normal series of fractions expressing the law of the phyllotaxis.[84]

What Peirce demonstrated is that the golden mean, which the Fibonacci series quite literally articulates, is the pulse of the solar system and its organizing principle.

---

83 Steiner, lecture, June 24 1921.

84 Agassiz, *Essay in Classification* (1857); see http://www.spirasolaris.ca/sbb4d2bs.html.

IV. The Spiraling Dynamics of the Universe

Planet	T1(Days)	T2(Days)	Fraction
Neptune	60,129	62,000	
Uranus	30,687	31,000	1/2
Saturn	10,759	10,333	1/3
Jupiter	4,333	4,133	2/5
Asteriods	1,200 - 2,000	1,550	3/8
Mars	687	596	5/13
Earth	365	366	8/13 ⎫
Venus	225	227	13/21 ⎬ 8/21
Mercury	88	87	13/34

Fig. 54. Peirce-Agassiz table (1857)

The table (fig. 54) shows the remarkable proximity between the first and second columns—T1, the actual number of days per revolution, and T2, the hypothesized number of days. This phyllotactic pattern of planetary revolutions evokes the image of Yggdrasil, the mythical world tree, which connects the nine worlds in Norse cosmology.

Based on these results, Peirce and Agassiz extrapolated that "under this law there can be no planet exterior to Neptune, but there may be one interior to Mercury"— an intriguing conclusion in the light of Pluto's recent demotion on one hand, and, on the other hand, theosophical hints about a yet-to-be discovered planet between Mercury and the Sun—the planet Vulcan—referred to as "the heart of the Sun." Interestingly, mythology's portrayal of Vulcan as a forging, demiurgic like power suggests that Vulcan is very "close" to the Sun.

Ovid tells the story of how Venus and Mars were caught making love in the net fabricated by Vulcan, the blacksmith of the gods. What Vulcan's net exposes is the "fabricating" secret of creation—the secting-bonding of masculine and feminine polarities from which world harmony is conceived. Harmonia was indeed the mythological daughter of Mars and Venus. Being the god who fashions the web of bonds that carries the generating secret of everything in nature, Vulcan is as close to the Sun God, indeed, as his "forging" hand is to the Platonic demiurge.

In his *Commentary on the Timaeus of Plato*, Proclus alludes to the same mythological episode to suggest in Plato's demiurgic operation of bonding a Vulcanian feat, thereby confirming the proximity (functional and topological) of Vulcan to the Sun.

> Through analogy, the universe is completely rendered one, this having the power of making things that are divided to be one, of congregating things that are multiplied, and connecting things that are dissipated. Hence theologians surveying the cause of these things in the Gods, enclose Venus with Mars, and surround them with Vulcanian bonds, the difference which is in the world being connected through harmony and friendship. All this complication and connection likewise

has Vulcan for its cause, who through demiurgic bonds connects sameness with difference, harmony with discord, and communion with contrariety.[85]

A different line of measurement, based on the average distance between each successive planetary orbit, leads to the same conclusion. The average of the mean orbital distance between each planet and the one before it approximates Phi. Following is a list of the actual mean distances in million kilometers, per NASA. In parenthesis are the relative mean distances, taking Mercury as 1: Mercury 57.91 (1.00000); Venus 108.21 (1.86859); Earth 149.60 (1.38250); Mars 227.92 (1.52353); Ceres (the largest asteroid); 413.79 (1.81552); Jupiter 778.57 (1.88154); Saturn 1,433.53 (1.84123); Uranus 2,872.46 (2.00377); Neptune 4,495.06 (1.56488); and Pluto 5,869.66 (1.30580). The total is 16.18736, which, when divided by 10, gives the average of 1.61874. Phi is 1.61803.[86]

On retiring from his duties as President of the American Association for the Advancement of Science in 1853, Benjamin Peirce invited his audience to marvel at the laws he had uncovered and at the divinity of Pythagoras and Plato's "spiritual instincts," now confirmed by modern science:

> Modern science has realized some of the most fanciful of the Pythagorean and Platonic doctrines, and thereby justified the divinity of their spiritual instincts. Is it not significant of the nature of the creative intellect and the simplicity of the great laws of force that the same curious series of numbers is developed by the growing plant which assisted in marshalling the order of the planets? And that the marriage of the elements cannot be consummated except in strict accordance with the laws of definite proportion?[87]

If we extend this exploration of the spiraling universe to the causal body on the mental plane, a simple quantum leap in "reasoning" along the logarithmic, analogical order suggests the logic of successive lives. The notion of incarnating cycles is simply a perfectly "logical" replica of the vortical architecture and evolutionary cyclicity of the logarithmic universe we partake of—the deployment, on a microcosmic scale, of the whirling order of incarnating spheres and cycles of the Logos—as well as a replica, on a larger scale and greater order of reality, of the seasonal cycles of plants. The causal vortex is but a higher version, in "mental substance," of the vortex of a plant and a lesser version of a solar system. As a seed unfolds into a 7fold plant, as a solar Logos manifests in a 7fold world, a causal self incarnates again and again into 7fold bodies.

---

85 Proclus, *Commentary on the Timaeus of Plato*, book 3, 2.27.
86 See http://www.british-israel.us/34.html.
87 Peirce, address, in *Proceedings of the American Association for the Advancement of Science*.

IV. The Spiraling Dynamics of the Universe

One begins to sense the grand and simple *rationality* of ancient teachings about the soul's journey in and out of incarnation. Birth and death point to the event-processes by which the soul involves and dis-involves itself from the planetary spheres that provide it with enveloping vehicles, in the same way that a solar Logos involves itself into the revolving spheres of its cosmic system and rises to self-awareness through their evolution in time. As the soul descends into incarnation once again, it gathers its vehicles (from mental to physical body) from the substance of the planetary spheres that make up the body of our solar system—which is, indeed, *our* universal body—by attracting substrata that resonate with the frequencies of its evolving consciousness. The planetary spheres, the partial harmonic fields of the solar Logos, represent different functions (and vehicles) of the psyche. To the descending soul, they offer the fields of vibration out of which the soul crafts, informed by its own past and stage of unfolding, the vehicles in which it clothes itself for its upcoming existence.

The incarnational descent of the soul through vortices of mental, emotional, etheric-physical substance tuned to planetary tones, down through the birth canal, has its counterpart at death in the withdrawal of the soul from vehicles potentially reorganized and refined by a lifetime of growth. This passing away is a spiraling and essentializing of the consciousness out of the *corresponding* planetary spheres. In between lives, the qualities and patterns developed in a preceding incarnation—mentally, astrally, and etherically—are held around the causal body in a seed-form (not unlike electrons around the nucleus of an atom), ready to inform the vehicles of the next incarnation, and land the self in a new earthly personality, which will be its next laboratory of growth.

Jung understood the ancient and fundamental connection of astrology with this worldview:

> To the alchemists, the connection between individual temperament and the positions of the planets was self-evident, for these elementary astrological considerations were the common property of any educated person in the Middle Ages as well as in Antiquity. The ascent through the planetary spheres therefore meant something like the shedding of the characterological qualities indicated by the horoscope, a retrogressive liberation from the character imprinted by the archons.[88]

The incarnating–excarnating pulse of the soul parallels the pulse of the Logos spiraling down and up shells and plants, animal and human organisms. As kingdoms and species are parts and parcels of the evolutionary life of the Logos, successive physical lives are parts of the evolutionary life of the human soul. At work in the

---

88 Jung, *Mysterium conjunctionis,* par. 308.

grand and minute spirals of the universal scheme, physical and transphysical, is the opus of manifestation and revelation of cosmic and individual "logoi," who speak to the fundamental individuating process of cosmos.

The maturing work of numerous incarnations raises the vibrational quality of the vehicles and eventually brings their central, "causal" quintessence to full blossoming, as the consciousness of the indwelling self expands to the loving radiance, all-consonant understanding, and vibrant creative power of a universal self, who realizes himself as one with the elements and kingdoms of nature. At this point, the (micro)-logos of such a microcosm surrenders the beautiful luminous form he has outgrown—the 12petal Egoic lotus that his expanding consciousness has brought to full blossom. The causal body disintegrates, releasing the monadic entity to a next octave of experience—an event referred to in the Ageless Wisdom as the fourth initiation. The death of a star is the macrocosmic correspondence of this destruction of the microcosmic causal body. A star, which is the body of a solar Logos, collapses-explodes when its indwelling Logos has completed this "star" cycle of experience and outgrown this form of being.

Instead of disconnected circles, orbits, and dimensions, zodiac and planetary spheres without, soul and vehicles within, reveal themselves as grand vortices of incarnating descent and ascending growth, incubators of forms of harmony (in nature) and entities of love (in human beings). A trans-dimensional *spira mirabilis* unfurls, in successive harmonic fields, different proportions of soul to matter—different dimensions of being—which guide along their internal helix the evolution of a consciousness increasingly "fit" for Logoic expression.

In essence, the archetypal vortex of spirit-matter intercourse—of which zodiac, solar system, and causal body are universal, systemic, and individual expressions—deploys the major harmonic dimensions of Being: mineral–physical (from carbon to diamond), plant–etheric (from stem to flower), animal–astral (from desire to devotion), human–mental (from personal to universal love, intelligence, and will).

This grand vortex of the world's becoming is the real setting of the divine "Comedy" of evolution—the spiral cycling down of heavens (world soul) into hells (limiting bodies), which "turns" chaos into cosmos, and cosmos into a glorious rose-revelation of being.[89] The spiraling journey of Virgil and Dante down the circles of hell, then up the circles of purgatory and paradise, deploys a picture of the inner worlds of soul experience, which mirrors the vortical landscape of world creation and evolution—indeed the weft and warp of the weave of goddess Natura.

The "marvelous spiral" is the movement of the marvelous germ of love, the Logoic principle, which engages matter and spirit, self and world, into whirling

---

89 The Jewish tradition speaks of the descent into the abyss-goufre of the body.

unit-forms of being. In this fundamental gesture lies the ubiquitous advent of love that is the universe, the dynamic rhythm of expansion and binding, dispersal and gathering that compels essence to existence and attracts existence to essence, multiplicity to oneness. Referred to as love by the subjective soul, harmonic proportion by the objective mind, experienced as attraction-repulsion by the emotional-instinctual self, grasped as electromagnetism by science, the Logos is the dynamic heart of the universe and the logo-rhythmic spiral the geometry of its pulsation at work and at play throughout the physics of nature.

Central to the "mindedness" of cosmos and nature—and the human "reason" able to reflect and reveal it—is the golden ratio. When we realize that the golden mean is at once the object (world) and subject (mind, soul), the ratio and reason, of the one opus of world-and-soul deployment, we also see that individual consciousness is a potential replica of the zodiac and zodiac the grand model of selfhood.

The Grand Self-Sameness of the Forms of Nature and Cosmos: The Spherical Vortex

> *Indra's form is to be seen everywhere*
> *For of every form He is the Model.* —Rig Veda, 6.47.18

> *The demiurgic bond makes all things to be in all, and exhibits the same things in each other, according to all possible modes, empyreally, aerially, aquatically, and terrestrially.* —Proclus, on Plato's Timaeus

This is how Walter Russell, sculptor and painter for the White House, and a national figure skating champion, relayed a thirty-nine day spontaneous experience of ecstatic knowing, which revealed to him how energy moves in the universe.

> I was made to see the universe as a whole and its simple principle of creation as one unit, repeated over and over, endlessly and without variation, as evidenced in the universal heartbeat to which every pulsing thing in the light-wave universe is geared to act as one unit of the whole.

To perceive logarithmicity as the fundamental dynamics it represents in world creation, we have to contemplate it in its "natural" setting, which is the phenomenon of harmonic vibration and resonance. When a tone resounds, when a Big Bang launches a universe, a vibrational wave has conjugated itself with a reflective wave constructively and harmoniously. The secondary harmonic waves generated form a logarithmic progression of nested vortices, an "addition" of self-similar spiraling formations, which are whole fractions of the fundamental wavelength.

Only when we reset logarithmicity in its phenomenological context, can it descend from the abstract realm of mathematics or geometry to regain its live, world-formative countenance. While mathematics and geometry lay out the mechanisms of logarithmicity, they do not account for its involvement in the spiraling shapes and motions of nature. What accounts for this involvement are the laws of harmonic vibration—the laws of the world's beginning.

If, mathematically and geometrically, a logarithmic spiral is unlimited, such is not the case for the spiraling forms of nature. The formations of nature are finite. Up to this point, we have been looking at the spiraling gesture at work in organic forms, but not at how a finite *whole,* an organism, is formed out of such spiraling. What is the nature of the mysterious time–space limit placed on the logarithmic spiraling of life by a form, an organism or a cosmic system? Why does the ammonite or the nautilus stop growing new chambers after a few cycles?

The mystery of the boundaries that define organic forms is the mystery Doczi approaches in his book, "*The Power of Limits.*" Essentially, the book illumines the fundamental "wrapping up" of the spiraling flow of life that results in the formation of any living organism or system—from a butterfly to a human body to a galactic system. Every form limits the potentially infinite course of waves of vibration to "involve" them—literally "turn" them—into relatively autonomous systems, which are sub-organized by smaller harmonic enclosures. Doczi's measurements capture the harmonic fractioning that underlies the self-bundling formative order of nature, its quantic mode of organization and organ-making.

The question is, how does this phenomenon actually happen? What is the law at work in this "limiting" operation? What we have come to understand so far is that in organic forms, the infinite exponential expansion/reduction of the spiral is limited by the necessity to generate a self-same replica of the Logoic "unit" of the universe. Behind the finite spiralings of atom, shell, plant, bodily form, is the operation that "turns" the life-giving harmony of the triune Logos into finite *forms of itself*—minute and galactic. Every form seeks to deploy and achieve, if only in part, a likeness of the physiognomy of its Lord.

We saw that organic forms involve standing waves, but a standing wave per se does not constitute a form. So, how does a whole organism form itself out of spiraling waves—*physically*? How do organisms and systems turn standing wave patterns into standing wave structures? What does an organic form ultimately—archetypally—look like?

These questions begin to resolve into answers when we realize that the understanding of standing waves is largely based on *longitudinal* waves. However, in nature, vibration does not travel longitudinally, it spreads *spherically.* Almost all

IV. The Spiraling Dynamics of the Universe

audible sounds are bubble-like, not wave-like as commonly believed. Sound propagates spherically, and the vortical formations of nature are roughly but essentially spherical.

The oneness of life, expressed geometrically in the sphere, is the fundamental reality of any existence, and existence happens in the context of this fundamental wholeness of life. A "whole" life is what nests itself in an "organic" form. The expression in earthly measurements—in "geo-metry"—of this wholeness of life is the sphere. The sphere is the archetypal referent of any organic form, and every organic form is, essentially, a sphere—a whole of life. Every form unfolds between a source-point-seed of life and an ideal spherical manifestation.

As we seek to grasp the archetypal model or gesture of living forms—a gesture that would encompass and blend into an autonomous form the 7/12fold differentiating pattern and the spiraling motion of the cosmic logic of harmony—it becomes essential to realize that the spiralings at work in natural forms are conditioned by their ideal spherical finiteness. In the same way that standing waves can only exist within the octave-like whole of a fundamental half-wavelength of vibration, the finite wholeness of any organism, regardless of its species, exists within the geometrical "octave" that the sphere represents for the point. This archetypal unit of the sphere draws the spiraling out vibration of life to curl around its periphery (in the same way that the standing wave bounces back from the end of its medium) and reverse course into a centripetal vortical motion. As it turns the vibratory flow around and back to its source, the second vortex interferes with the first constructively and harmonically. The harmonic interferences of centripetal and centrifugal currents create still nodes and "standing" forms. The result is the formation of a *standing wave structure*—a dual vortex (at once centripetal and centrifugal), a form called spherical vortex or torus (fig. 55).

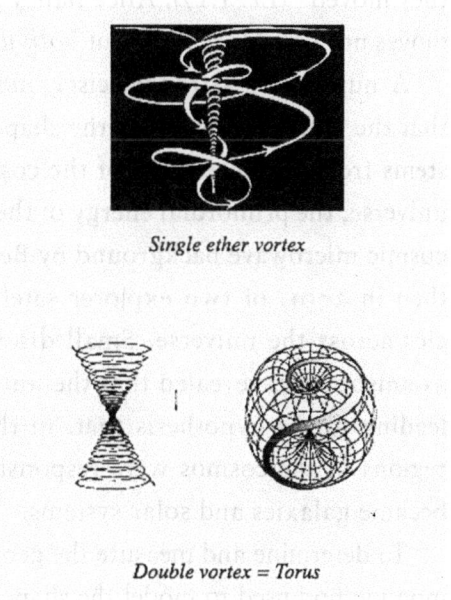

Single ether vortex

Double vortex = Torus

Fig. 55. Single and dual vortex (D. Winter: http://www.soulsofdistortion.nl)

The torus is a vortex that folds back on itself. The energy flows inward through one end, circulates around the center and exits through the other end, to turn its vorticing into a ring, which is then drawn back to the center. Plants and trees display this energy flow dynamics. So do hurricanes and tornadoes, and so do magnetic fields around planets and stars. In hydrodynamics, the torus is a unique flow form, which allows fluids to spiral inward and outward *on one and the same* surface. Its seamless extraverting and introverting dynamics creates what could be perfectly called, indeed, a "*Uni-verse.*" A singular bidirectional whole. Unity in *di-versity*—indeed.

The cosmological notion of a Being manifesting into a universe that expands out and returns-contracts to the center suggests that such may be the form of the cosmos as a whole. The outpouring and return to itself of the fundamental breath wave, the exhale and inhale of the divine breath, emits the primordial toroidal form of a "unicellular" world, whole and sound in its bidirectionality, and, like any form, finite.

This wheel turning upon itself echoes Plato's description of the world soul as a dynamic form "revolving back upon herself." The world soul "compound is then proportionately divided and bound together and *revolves back upon herself.*"[90] It is also reminiscent of Ezechiel's vision of the four living wheels under the throne of the Lord, each with "a second wheel turning crosswise within it.... Whenever they moved, they moved in any of their four directions without turning as they moved" (Ez. 1:17). Alice Bailey similarly points out how "the zodiacal wheel moves not only clockwise, but *both ways at once*, and *also at right angles to itself.*"

A number of astrophysicists, including Joseph Silk, have launched the idea that the entire universe has the shape of a giant doughnut torus. This hypothesis stems from measurements of the cosmic microwave background radiation of the universe, the primordial energy of the Big Bang. In the mid-1960s, the discovery of cosmic microwave background by Bell Labs led to the lauching by NASA, in 1989 then in 2001, of two explorer satellites intended to measure radiation frequencies across the universe. Small discrepancies in temperature fluctuation known as anisotropies revealed that the universe consists of regions of varying densities, leading to the hypothesis that, in the early stages of the universe, these denser regions of the cosmos were responsible for attracting the matter that ultimately became galaxies and solar systems.

To determine and measure the geometry of this background radiation is the common method used to model the shape of the universe. In 2004, NASA released new information about discoveries by the European Space Agency observatories that so-called black holes in our universe are doughnut-shaped torus formations (fig. 56). Astrophysicists studying a black hole observed a torus like structure "spouting out both sides of the doughnut-shaped galaxy's center." At the time of this book's

---

90 Plato, *Timaeus*, 37A.

IV. The Spiraling Dynamics of the Universe

publication, the first actual image of a black hole has been captured (April 2019), deep in the heart of the galaxy, known as Messier 87, in Virgo (fig. 56).

Itzak Bentov's book, *A Brief Tour of Higher Consciousness,* is dedicated to the idea that all reality, including consciousness, can be modeled on the torus. He views galaxies as toroidal forms with "white holes" putting out energy and "black holes" on the opposite side taking it back in.[91] In *The Reflexive Universe,* Arthur Young also explores toroids as models of the primary pattern that nature uses for life at every scale.

Babbitt, Leadbeater, and Besant depicted the basic building form of matter, the atom, as torus shaped ether, and called this primary flow form the *"Anu"* (Sanskrit for "atom"). "The atom has—as observed so far—three proper motions, i.e., motions of its own, independent of any imposed upon it from outside. It turns incessantly upon its own axis, spinning like a top; it describes a small circle with its axis, as though the axis of the spinning top moved in a small circle; it has a regular pulsation, a contraction and expansion, like the pulsation of the heart."[92]

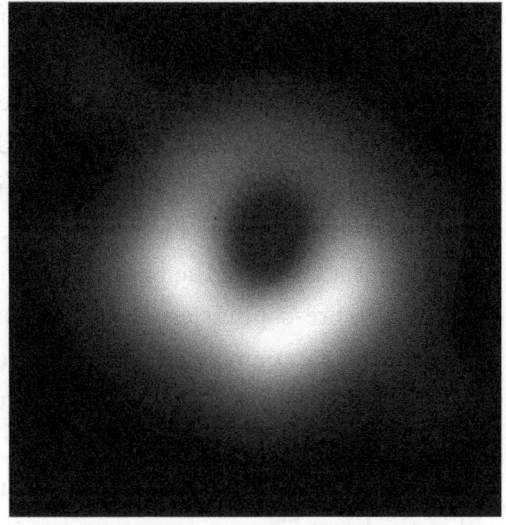

Fig. 56: The first image of a black hole, using Event Horizon Telescope observations of the center of the galaxy M87. Source: Event Horizon Telescope Collaboration

Preceding them in the eighteenth century, Emmanuel Swedenborg taught that matter is made up of a series of particles in ascending order of size, each of which composed of a closed vortex of energy, which spiraled at infinite speeds, giving the appearance of solidity.

The physical heart, unsurprisingly, is a powerful illustration of this fundamental dynamic form. It is described as pulsing from top to bottom and back up in a twisting fashion. This vortical pulsation has a specific rhythm of diastole and systole known as the Cardiac Cycle. First it contracts, stretching itself downward, then it expands in breadth, drawing up the inner layers. A new contraction follows, with an extension downward, which again withdraws to spread out in breadth and

---

91 See http://harmonicresolution.com/Toroidal%20Space.htm.
92 Leadbeater and Besant, *Occult Chemistry,* p. 52.

so on. Schwenck's schematic drawing of this motion, reproduced below, evokes Babbitt and Leadbeater's representation of the atom. The heart illustrates most eloquently Schwenck's remark that "the organ, a form in space, is at the same time a movement in time. Form and movement are interrelated, uniting in a rhythmical process in space and time."[93] Interpenetrating rhythms and movements in the pulsating vortex of the heart result in the formation of the separate chambers.

Although the heartbeat has been studied in great detail, there is no scientific explanation for why it evolved to beat as it does. It has been proposed that it beats according to the recursive Phi proportion.

"The musculature of the heart and arteries," Schwenck elaborates, "all the way down to the pre-capillaries, is spirally oriented, and both the heart and arteries move spirally to augment the momenta of the blood. The fibers of the heart are a physical echo of the creative movements by which it was begotten. In spiraling paths they swing down to its apex and then rise again to its base. They make the same movements and emphasize the revolving vortical streaming of the fluids within the heart."[94]

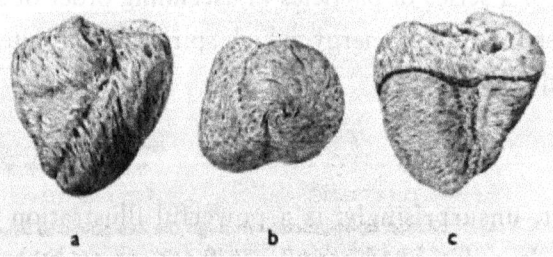

Fig. 57: (top) schematic representation of the spiraling fibers in the heart (natureinstitute.org); (bottom) heart muscle fibers in the ventricles

---

93   Schwenk, *Sensitive Chaos*, p. 91.
94   Ibid.

IV. The Spiraling Dynamics of the Universe

> *There is nothing in my life that equals the sense of ecstasy I have felt in discovering nature's beautiful agreement.*—BUCKMINSTER FULLER

The dual vortex phenomenon lights up a stunning pattern of identity among the variegated forms of nature. Indeed, says the *Bhagavad Gita* in most remarkably precise terms, "O Arjuna! The Lord abides in the heart region of all beings, whirling all beings by his power, [as if they were] mounted on a machine."[95]

Through the lens of this universal template of the torus, we come to see the sameness that in-spirals and inspires all living forms—from atom to fruit, from shell to plant, from butterfly to human being to galaxy. Cut in half, mushroom and tomato, cherry and kiwi disclose the universal heart—the archetypal, toroidal "organ" of life. Tree and galaxy display the same embracing curve of the toroidal auric field that bundles nature's forms. The "world fruit" arises, as does the figure of Mithras–Phanes (fig. 3, page 28), whose wings bridge the flaming halves of the world egg, forming what we now recognize as a dual vortex of rising kundalini and wing-like, descending auric currents. The 12fold mandorla surrounding the (7fold) spiraling kundalini intimates how the swirling harmonic ground of the world comes to full expression in this being, who fleshes out the Platonic unfurling of space and time, world and self, into a soulful form. The torus also lights up in the caduceus a dual vortex, sustained by the interplay of a centrifugal, incarnating, earthward current of energy, and a centripetal, rising current. "The torus is literally around all life forms, all atoms and all cosmic bodies such as planets, stars, galaxies. It is the primary shape of existence," affirms Drunvalo Melchizedek.[96]

Based on his clairvoyant investigations, Rudolf Steiner describes as follows the first outline of the human body, formed as a condensate of the spiritual during the first condition of the Earth, which he refers to as old Saturn. "When the Earth existed as Saturn, only the first germs of the human kingdom dwelled on it.... The human form was like a kind of auric egg, and within it was a remarkable scaly structure, a sort of vortex, shaped like a small pear and as though made of oyster shells."[97]

This archetypal form appears also in works of art. Lorenzetti's painting of Michael (fig. 59.1) offers a spiritual imagination of the unified spiraling gesture and dynamic countenance of a "living one" who blends into one luminous toroidal field the flaming spirit descending from above and the fiery dragon rising from below. No longer is this universalized being "tergiversating" between the pull of above

---

95 *Bhagavad Gita*, 18.61.
96 Melchizedek, *The Ancient Secret of the Flower of Life*, p. 156.
97 Steiner, *Founding a Science of the Spirit*, p. 73.

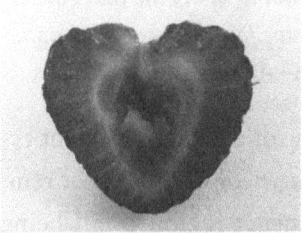 

Fig. 58.1–2: Longitudinal and transversal cuts of a strawberry, showing its vortical and toroidal dynamics

Fig. 58.3–4. Pinecone: view from above (left) highlights the toroidal form; side view (right) emphasizes the vortical shape.

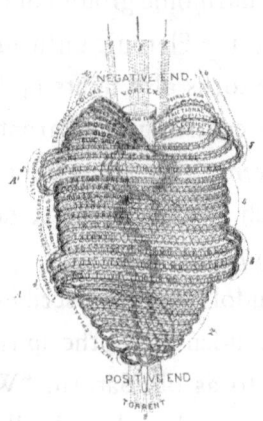 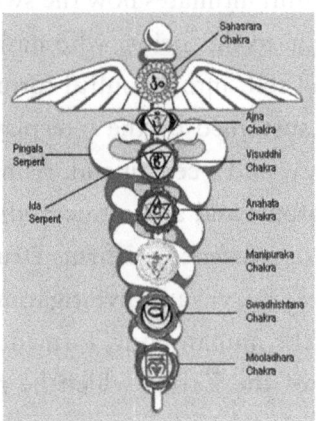

Fig. 58.5–6: (left) Babbitt's atom; (right) diagram of the caduceus, showing the three currents of Ida, Pingala, and Sushumna crossing at the seven chakras

 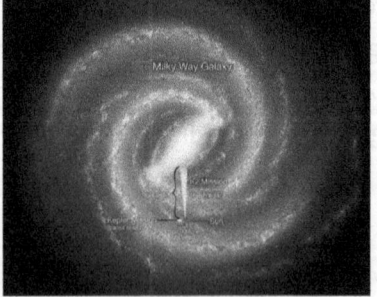

Fig. 58.7–8: (left) red cabbage head cut in half (reddit.com); (right) Milky Way Galaxy (NASA/Ames/JPL-Caltech)

IV. The Spiraling Dynamics of the Universe

Fig. 59.1: (left) Archangel Michael and the dragon (Ambrogio Lorenzetti);
Fig. 59.2: (right) The Hindu God Shiva Natajara, dancing creation and destruction

and below, right and left, past and future, but circulating them consciously into one "univer-ting" wheel. Evocatively close in form and meaning is the representation of the Hindu God Shiva dancing the fiery wheel of life and death—the uni-vortex of creation and destruction that is the universe (fig. 59.2). Both the transfigured man-Archangel of the West, and the embodied god-universe of the East deploy their dance from the animal power mastered in their feet.

This subtle reality of dance and flow is how Richard Rohr, a contemporary Franciscan teacher, describes and experiences the Christian Trinity. He notes how the best metaphor the early Christian Cappadocian Fathers could offer after three centuries of highly sophisticated thinking about the Trinity was, "Whatever is going on in God is a flow, a radical relatedness, a perfect communion between Three—a *circle dance of Love*. God is not just a dancer," he adds. "God is the dance itself."[98]

Every living thing is a minute choreography of the divine dance, a fractal of the toroidal cosmos. Like its Logos, every entity exhales–unfolds–externalizes itself into a form, which is inspired and spiraled-in by its self-organizing principle. Each is drawn to turn itself into a "world" that also aspires to its source, and in this dual motion discovers itself. If everything alive—tree, animal, human—rises "across or amidst" gravity, it is because every creature breathes between the pull of gravity and the call of levity. Every living thing is involved at once in cosmos making and Logos becoming. Every soul envelops itself in layers of subtle and dense matter to develop

---

98 Rohr, *The Divine Dance*, p. 27.

awareness of itself. As the intertwining currents of the etheric body illustrate, and Doczi's measurements intimate, all living forms are woven upon a dual spiraling loom. Every form of existence is a toroidal gesture of ascending-descending, spiraling out and spiraling in, unfolding and folding back. To imagine oneself at the center of the toroidal form we are is the simplest way to experience how the axial transfer from centripetal to centrifugal verticality—which occurs at the heart center—spirals out to expand, ring-like, quasi-horizontally around us, then spirals-contracts down to reverse its course inward and upward again.

☉ This dual vortical flow, and the nested self-similar toroidal flow-forms it sustains, is the universal Form—the archetypal body of the Logos. To see this dynamic gesture in the many physionomies of nature is to divine in all forms—literally—one pulsating *heart*. ☉

How the 7/12foldness of forms might fit with their overall toroidal gesture is the next question raised by the sighting of the universal spherical vortex. To explore this question, we turn to cymatic research and its investigations of wave conjugation phenomena within a three-dimensional spherical medium. They will prove to lift the veil on how these two major insights into the forms of nature are indeed intrinsic to each other—how 7foldness is intrinsic to the torus, and how the torus is the integrative dynamics of the 7/12foldness of forms.

As briefly touched upon earlier, cymatics explores the formative effects of sound waves. Beautifully photographed experiments show how sound waves pulsated in a milieu such as a metal plate of sand, a petri dish of water or a bubble of gas cause the formation of nature like patterns and designs, both inorganic and organic.[99] Most stunningly, it reveals how specific frequencies generate structures akin to the fundamental Platonic shapes.

Buckminster Fuller was the first in modern times to explore the relation between sound and geometric forms and to discover how Platonic shapes would form on the surface of a sphere pulsed with specific frequencies. "Angle and frequency modulation," he asserts, "exclusively define all experiences, which events altogether constitute the Universe."[100]

Using a balloon submerged in blue dye, he pulsed it with diatonic sound frequencies (those in the ratios of diatonic musical intervals). A small number of evenly distanced nodes formed across the surface of the sphere, along with thin lines connecting them to each other (fig. 60, a photo by Hans Jenny illustrating Fuller's

---

99   See Jenny, *Cymatics: A Study of Wave Phenomena and Vibration*, 2001.
100 Quoted in Edmonson, *A Fuller Explanation*, p. 67.

experiment). The "nodes" represent areas of stillness where standing waves touch the surface of the sphere and reverse their course. The thin lines connecting them form the vertices of polygonal shapes. The number of nodes, and the consequent nature of the emerging polyhedron depend on the frequency of the vibration. The higher the vibration, the more numerous the nodes. Buckminster Fuller documented the following:

- Four evenly spaced nodes form a tetrahedron.
- Six evenly spaced nodes—an octahedron.
- Eight evenly spaced nodes—a cube.
- Twelve evenly spaced nodes—an icosahedron
- Twenty evenly spaced nodes—a dodecahedron.

Hans Jenny, the Swiss physician who continued Buckminster Fuller's work and founded Cymatics, conducted similar experiments to test standing wave vibrations in volumes of fluid. When he vibrated a droplet of water containing a very fine suspension of light-colored particles with frequencies in diatonic ratios, he saw the Platonic forms appear, surrounded by elliptical curving lines connecting the nodes.

Fig. 60: Hans Jenny's Platonic Solid formation in spherical vibrating fluid (Jenny, Cymatics: A Study of Wave Phenomena and Vibration; used by permission)

While we already know from the basic rules of constructive interference that if the half wavelength of the vibration emitted is not in a whole number ratio of the volume's radius, chaos will result, these experiments reveal a next level of selection: in order for Platonic forms to appear, the half wavelengths of the sound waves pulsated must be in a diatonic, musical ratio of the radius of the plate or sphere.

Why are diatonic ratios uniquely apt to produce Platonic shapes? As we approach this question, a brief return to Leadbeater and Besant's micro-observations of the elements and their insights into Crookes' periodic table will provide a pivotal insight into the internal dynamics of the Platonic volumes, and thus a vital link between elemental forms of nature, vortical formations and cymatic experiments. What Leadbeater and Besant's investigations revealed is that nested *vortex cones* were the formative dynamics of the Platonic solids—which "solids" stood behind the symmetry groups of chemical elements identified by Crookes in reference to their valency. As illustrated in the figures below, different combinations of dual vortices were behind the different solids. If Plato taught that the Platonic shapes informed the elements of matter, the question remained of how such geometries actually arise from a vibratory field. The vortical dynamics that the two occultists observed as constitutive of the Platonic shapes converge with the logic of formative harmonies that has brought us to the torus, to provide what may be, to date, the most lucid tracing of the vibratory "involvement" and envelopment of energy into particles and forms of matter.

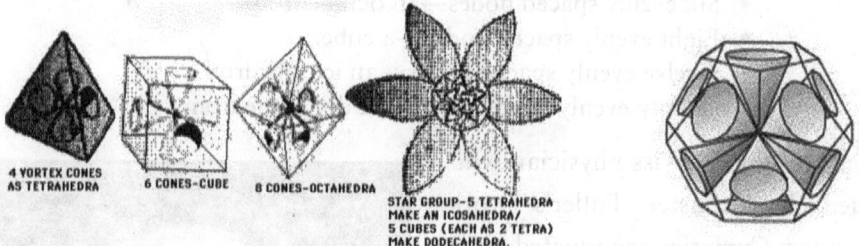

Fig. 61. 1: From William Crookes' geometric table of the elements (left); (right) 12 cones. Dodecahedron (divinecosmos.com)

Differing slightly from fig. 60, the next picture reveals how the "lines" of the polyhedra are indeed vortex-cones.

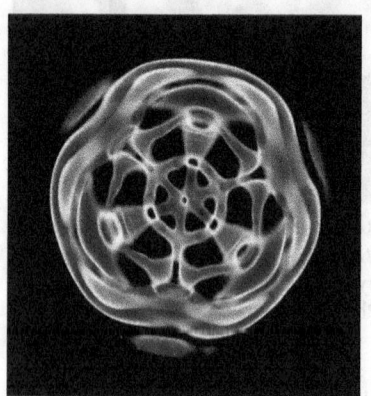

Fig. 61. 2: The lines forming the polyhedron are vortex-cones (decibeljournal.com)

Each of the vortex cones through which the energy of life shapes matter represents one half of a dual vortex or torus. Different sets of cone-pairs or toruses organize the energy flow into different Platonic patterns. Each is shown to result from the combination of a specific number of dual vortices: the tetrahedron contains 2 vortex pairs or toruses, the cube 3, the octahedron 4, the dodecahedron 6, the icosahedron 10 vortex pairs (fig. 61.1).

What Leadbeater and Besant's observations suggest is that Buckminster Fuller's Platonic volumes are made of dual vortices—and dual vortices harmonically related. The beginning of an inkling arises, as to why diatonic frequency ratios would be needed to generate these harmonious sets of dual vortices, which are the informing patterns of Platonic solids.

Connecting Fuller and Leadbeater–Besant's experiments engages the core of the leap we took a short while ago, as we sought to grasp the archetypal gesture of organic forms: the leap from longitudinal standing waves to spherical standing wave structures or toruses. The corresponding leap in nature is a cosmological one—from the formation of a harmonic field to the formation of Platonic elements and organic forms. What is at stake in a simple standing wave is the basic strand or "weave" of

IV. The Spiraling Dynamics of the Universe

a wholistically vibrant *field*. What is at stake in the torus or standing wave structure is the wholeness of a *form*.

Standing waves simply select for their harmonious interferences wavelengths that are in a whole number ratio of their fundamental. Platonic standing wave structures, on the other hand, rest on a next order of selection. They involve a combination of harmonically related standing waves.

When we try to understand why only diatonic ratios generate Platonic toroidal sets, we are led again to ponder the radical imperative of *wholeness*, which guides the formation of all systems in nature. As long as we deal with objects, musical instruments, strings, the simple rule of whole number ratios suffices to generate harmonious sound. The generation of organic forms implies a next order of necessity. Organic forms are essentially "wholes" of life. Like the forms of nature they in-form, the Platonic solids are defined in reference to the wholeness of the sphere, which is the ideal volume of three-dimensional forms. Plato's "beautiful" intermediaries between soul and body, they are the descending steps of life into form—the geometric intermediaries between point and sphere. They resonate with the planetary intermediaries of Kepler, and they echo the diatonic intervals of Pythagoras. Indeed, a living system is essentially a (creative) duplication—a likeness or "double" of the primordial "one." It is octave-like.

To be alive, a form must fit together the extremes of life and matter through its "mediating" organization, the secret of which is harmony. The Platonic forms are the structural expression of this necessity. They "fit together" the extremes of point and sphere, center and periphery, via the secret mean seeded at their core: the golden mean and its musical equivalent, the fifth, followed by its court of major musical intervals. They stand between Heaven and Earth as the five whole, harmonic steps of the corporealizing process.

The archetypal model of this intermediation is the musical scale of major intervals between fundamental and octave. The diatonic ratios are the expression of the musical proportion—6:8::9:12—which fits together the one (life) and its duplication (into a *whole*, living form).

To understand why diatonic ratios are essential to the formation of the Platonic solids is to realize that they are the means to create perfectly wholistic geometric intermediaries between the center and periphery of the sphere, as they do musical tones between fundamental and octave. Similar to the diatonic intervals, the Platonic solids are the only "sound" polyhedra to bind and bridge in harmonious solid forms the extremes of a sphere—they are the geometric "tonalities" and intervals of "wholeness." To grasp this internal harmonic bonding of the sphere is to hear the fundamental music of living forms.

Diatonic intervals are the intrinsic harmonies of a sphere—its "natural frequencies"—in the same way that the frequencies at which standing wave patterns occur within an object define its natural frequencies. There are five Platonic solids, and there are, archetypally, five frequency ratios that can generate Platonic shapes within a sphere. With the fundamental and the octave (the point and the sphere, geometrically), they stand behind the basic sevenfoldness of forms.

✺

The leap from standing wave to spherical vortex, then to a Platonic composition of spherical vortices speaks to the cosmological step from the triune ground of harmony to the formation of elements and organisms—from harmonic field to harmonic form. It represents a stepping up in harmonic complexity, which can be defined as a leap from resonance to consonance, from resonant waves in whole number ratios to consonant standing waves in musical proportions.

The torus heralds the leap from the spiral-weave activity of the ground of cosmos and fabric of nature to the bundling up of this spiraling into organic forms. It represents the next installment of the principle of polarization (in this case: a dual vortex) and bonding at the foundation of harmonic cosmology, the principle epitomized in the golden section, and enacted in the logarithmic spiral. The torus is the expression of this principle *in the forms* of nature and cosmos. If the torus is the archetypal structural dynamics of living forms, it is because it provides the condition of wholistic proportionality indispensable for life to abide in forms.

❂ The first level of logarithmicity engages a primordial harmonic fractioning-spiraling *field* of standing waves, based on *resonance*. For basic standing wave patterns to be generated in a medium such as air or water or along a string bound by a fret, the only necessity is for the half-wavelengths of the waves involved to be whole number fractions of the fundamental.

The second level, at work in toroidal *forms,* involves standing waves whose fundamental wavelengths are in harmonic proportion with the implicit sphericity of the form. When a spherical balloon of gas or fluid is vibrated with a tone, as in Buckminster Fuller's experiment, the formation of spherical standing wave structures—dual vortices—involves more than the standing waves that sustain a tone or a field, it involves a particular combination of standing waves apt to form a harmonious "whole." More than just bonding the extremes of a string or an object, the imperative, for such a standing wave complex, is to bond the "extremes" of source and spherical periphery—life and form—harmonically. To constitute a unitive whole, the form must be an octave-like, spherical vortex, in which spherical standing waves or sub-vortices are in *musical proportion or consonance* with the whole. As whole fractals of the universe, living systems are organized on the basis of consonant intervals. ✪

The major consonant intervals are the octave, the fifth, and the fourth, also referred to as the super particular ratios 2/1, 3/2, and 4/3. 2/1 is the octave, or *diapason* (Greek for "across all"). 3/2 is the perfect fifth or diapente ("across five"); 4/3 is the perfect fourth or diatessaron ("across four"). These three intervals and their octave equivalents are the only absolute consonances of the Pythagorean system. All other intervals represent lesser degrees of consonance and varying degrees of dissonance.

Based on the tone correspondences noted earlier for the angles (or fractioning ratios) of the Platonic polyhedra, there does not seem to be a direct equivalence between a specific musical interval and a Platonic solid. What we find, however, is a definite display of major harmonic intervals—1/2, 3/2, 5/4.

Solid	Maraldi angle	Tone
Tetrahedron	4x180 = 720	F#
Octahedron	6x240 = 1440	F# (octave, 1/2)
Cube	8x270 = 2160	C# (5th from F#, 3/2)
Icosahedron	12x300 = 3600	A# (3rd from F#, 5/4)
Dodeca	12x540 = 6480	G# (5th from C#, 3/2)

Whether we come from the general perspective that whole forms are archetypal octaves (hence sevenfolds) or from the Platonic-Keplerian perspective that the planets are strung between Heaven and Earth like the Platonic solids-elements, a 7fold is implied: 7 diatonic intervals; five Platonic waveforms + point-source (Phi) + surrounding sphere of wholeness = 7. "There you have the explanation of the number of planets," said Kepler when he discovered how the solar system could be viewed as a harmonious nesting of Platonic solids.

Similarly to how yoga philosophy explains that the 7 chakras are the major but not the only chakras, cymatics shows that numerous complex formations can be derived from different vibrational frequencies (other than the major ones) and reproduce designs of nature—from a tortoise shell to the rudiments of a human face. Seven is simply the archetypal number of the world body.

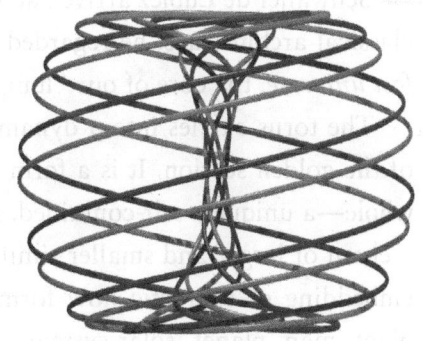

Fig. 62. A torus knot of twelve rotations around the toroid, and seven helical twists through the toroid hole. (Don E. Mitchell, see www.youtube.com/channel/ UCnyhur57xuMpGMii7fHg0KA)

Discreet signs of 7foldness are associated with the torus itself. The so-called 7-color theorem formulates a unique topological property of the torus—"that a map drawn on its surface requires seven colors in order for all bordering countries to be distinguished by differences in color."[101] In other words, the number of colors sufficient to color map a torus divided into regions so that neighboring regions have different colors is seven. The toroidal heart has seven muscles, which pump blood in seven regions, which correspond to the color map of the torus.

Furthermore, recent studies of "torus knots," which describe the topology of a mathematical relationship of two spin momenta, have shown that "dual rotations with independent but coupled centers harmonize at certain integer ratios to form a torus knot." Don E. Mitchell illustrates his point with a torus knot in a ratio of 7 twists per 12 rotations—a dynamic geometry remarkably evocative of the 7fold human figure nested in the 12fold Egoic-zodiacal lotus.

The key factor in toroidal embeddedness is the logarithmic principle of Phi, and Phi is indeed the archetype of consonance that governs the entire creation. As we saw when we explored the Platonic solids, the five solids "grow" from the one "divine seed"—Phi—and each is related to the preceding and succeeding one by Phi or a Fibonacci ratio. The Fibonacci rapport between the number of opposite spiraling currents visible in fruit and vegetables is, in essence, Phi, and the ratio 3/2 is also the source of the 7/12 diatonic notes of the scale. Everything unfolds from this divine proportion, which, embedded in everything, carries into every form the all-consonant soul of life.

Schwaller de Lubicz arrived at a similar view of Phi. More than a central item in classical architecture, he regarded Phi as *the mathematical archetype of the manifest universe*, the core of our "lumpy" world of galaxies and planets.[102]

The torus carries into a dynamic three-dimensional form the unique character of the golden section. It is a form that *separates and unites*—a *relatively* separate whole—a uniquely self-contained, self-sustaining wave field that is also a fractal in a chain of larger and smaller similar forms. Toroidal dynamics enables a seamless embedding of energetic flow forms from micro-atomic to macro-galactic—atom, plant, man, planet, solar system, galaxy. Each entity has its unique identity while also being connected with the whole universe. Arthur Young remarked: "The self in a toroidal universe can be both separate and connected with everything else." Indeed, such is the secret marvel of Phi: separation and union.

---

101 Young, *The Reflexive Universe*, p. xxv.
102 West, *The Serpent in the Sky*.

IV. The Spiraling Dynamics of the Universe

"It is amazing," Doczi writes, "to behold such unity in the manifold diversities of nature, each species developing freely its uniqueness, yet all united by sharing the same simple, dinergic and harmonious *proportional limitations*."[103] Doczi is captivated by the intention embedded in this principle of proportional limitation, an intention he also finds expressed in Leonardo da Vinci's studies of good proportions: "...every part is disposed to unite with the whole, that it may thereby escape from its incompleteness." In the proportionate character of every portion, lies the disposition to union of an infinitely diversified world, a world at once kaleidoscopic and wholistic by virtue of its all pervasive unitive bonds—a wildly variegated nature rooted in a subtle field of harmonic rapport, barely visible to the physical eye, yet perceptible in its grace to the eye of the soul. The "divine" proportionality, which is the quiet breath of nature and the grace of its subtle rhythmic continuo is, quite literally, the pulse of the love principle at the heart of creation.

❋

According to the harmonic model, the toroidal triuning of the primordial polarity of Being into a universe-bundle of harmony is a 7fold body/12fold soul composition. 3, 7, 12 is the formula of deployment into form of the Logoic triunity.

Embryology offers a complex, multifolded illustration of the formative process of a toroidal living organism, a process, Rumi affirms, which *we* conducted—"We made the body, cell by cell we made it."[104] Indeed, an attentive look into the initial phases of embryological development reveals the signature of the Logoic "I" initiating its manifestation in the cosmic womb of a mother: the now familiar course of a 3fold emergence, followed by a 7fold/12fold differentiation.

The conception event reenacts the cosmological emergence, out of the coming together of the two—female ovum and male spermatozoid—of a third and unique proto-organism: the zygote. This triunion of the two into a new, unicellular organism, mirrors the emergence of the Logos from the interplay of Same and Other, cosmic

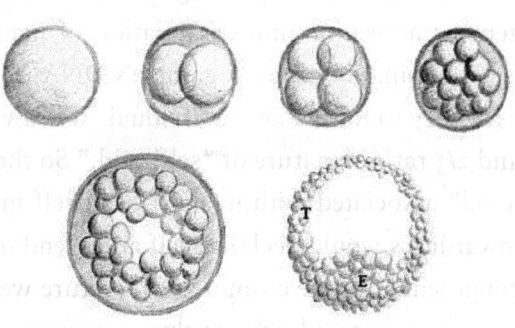

Fig. 63. *The first week of embryonic development: zygote (top left) to blastula (bottom right); E = embryoblast. T = trophoblast (J. Van Der Wal)*

---

103 Doczi, *The Power of Limits*, p. 92

104 Following is Rumi's entire short poem: "*Wine got drunk with us, not the other way. / The body developed out of us not we from it. / We are bees and our body is a honeycomb. / We made the body; cell by cell we made it.*" Quoted in Jaap van der Wal, "The Embryo in Us."

Father and Mother. A new entity begins to exist and unfold through a process of self-division.

The first week of development (fig. 63) is marked by the rapid division–multiplication of the zygote and virtually no growth. This 7-day process of division, not unlike the Biblical 7 days of genesis, results in the formation of the morula. After the 64-cell phase, the morula, until then a solid ball, folds inward into a hollow sphere, the blastula (also *blastosphere,* from *blastos,* or sprout). Commenting on pictures of this stage, Drunvalo Melchizedek elaborates, "the north pole starts dropping through the space inside, going down toward the south pole, and the south pole comes up through the space to meet the north pole.... The hollow sphere then becomes a torus."[105]

Embryologist Jaap van der Wal highlights the phenomenon of dualization between "a center (embryoblast) subjected to centripetal forces and tendencies and an opposite pole with a centrifugal tendency (trophoblast), a periphery that has a completely different relationship to the environment, because it connects with it and interacts with the maternal environment in a metabolizing way."[106]

The blastula consists of a spherical layer of about 128 cells surrounding a central fluid-filled cavity called the blastocoel, bound by a single layer of cells. "128" corresponds to 7 "octaves" of cell duplication: 1, 2, 4, 8, 16, 32, 64, 128. Remarkably, it is at this point that the organism "bundles" its duplicating process to enter a next stage.

The hollowing process marks a dualization into inside and outside, centripetal and centrifugal. Concomitant with this morphological emergence of inwardness, and paralleling it, the DNA becomes activated. The end of cellular division coincides with the beginning of zygotic transcription, which involves both zygotic genome activation and degradation of maternal transcripts. The zygote's genome is a combination of each gamete's DNA. It contains all of the genetic information necessary to form a new individual. We saw earlier how the DNA carries the 12fold and 2/3 ratio signature of "selfhood." So the 12fold DNA, and the principle of "selfhood" associated with it, inscribes itself in the 7fold cellular body. That a 12fold inwardness would declare itself at the end of the 7fold cellular division is strikingly congruent with the cosmological picture we have seen emerge. From then on, development comes under the exclusive control of the zygotic genome.

Following the primordial 1, 2, 3 process of tri-union (ovum, spermatozoid, zygote), a secondary tri-union thus takes place, which is a 3, 7, 12 unfoldment. The triune "one," the zygote, deploys itself into a 7fold multiplication of cells,

---

[105] Melchizedek, *The Ancient Secret of the Flower of Life,* pp.192–193.

[106] Van der Wal, "The incarnating embryo—The embryo in us," in *Foundations of Morphodynamics in Osteopathy.*

which is then personalized by a 12fold zygotic DNA. With the 7fold division, the hollowing into a torus, and the beginning of 12fold "self" transcription, a being externalizes its self in a 3/7/12fold spirit-body-soul tri-unit.

During the next phase, the single-layered blastula is reorganized into a trilaminar ("three-layered") structure known as the gastrula. The three germ layers, called ectoderm, mesoderm, and endoderm—outside, inside, middle—herald the tripartition of tissues and organs characteristic of the upcoming organogenesis. The germ layers are not anatomy, "they are functional principles," Jaap van der Wal passionately points out, as he shows, step by step, how embryonic development proceeds through operations of dualization and threefolding, and how the growth of the human organism is essentially a dynamics of polarization and mediation, a gesture of separation and union.

THE DUAL VORTEX NATURE OF THE HUMAN EXPERIENCE

> *When we try to pick out anything by itself, we find it hitched to everything else in the universe. One fancies a heart like our own must be beating in every crystal and cell.*—JOHN MUIR

When we begin to see in every organic and cosmic form the vortical dynamics of the heart, we begin to sense within their still shapes a pulsating gesture. From cabbage to galaxy, from the "pulsing bodies" (Leadbeater) of atoms to the vital bodies of human beings, forms reveal a fundamental dual spiraling pattern of systolic contraction and diastolic expansion around a central milieu or "light hole," which hosts the suspense of reversal between the two. Streaming in and out of the center is the pulse or breath of the entity (vegetal, human, or cosmic) manifesting in this form.

As the heart illustrates, the dual vortex turns into one "uniflow"—one re-spiration—the exhaling, outspiraling breath of an entity into a form of itself—be it galaxy, fruit, shell, or heart—and its inhaling, spiraling back from periphery to source. The reversal of direction is how life, quite literally, *turns into* a form of itself—cosmic, vegetal, animal, human—which harbors the center, seed, or "heart" of a smaller world. The (seeded) fruit at the edge of the tree, the heart of a human being at the periphery of the solar system, the galaxy at the reversal of a cosmic Breath—all epitomize at different scales the peripheral *return to itself* of a world entity, which in this return externalizes itself into a smaller form. Fruit, heart, star, galaxy are epiphanic forms of Life. Each externalizes an entity into a microcosm through which the entity becomes known, seen, tasted, and ushered in turn into reproducing itself or creating a world. When the fruit is tasted, its uniqueness (as a kind) is tasted and known by a self who is itself a taste of the Logos.

The organ of the heart is at once a dual vortex formation at the periphery of the solar system (pulmonary circulation) and a Sun-like center for the human body (systemic circulation). It receives and it gives. A densified, physicalized pulsation of etheric solar life returning earthly, venous blood to the Sun (pulmonary circulation), the heart is also an active source of solar, oxygenated blood flow to the peripheries of a body (systemic circulation).

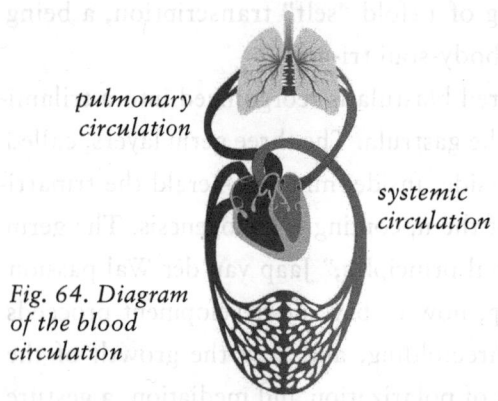

pulmonary circulation

systemic circulation

Fig. 64. Diagram of the blood circulation

The notion, present in various traditions, that the blood is the carrier of the human "I" opens a next level of contemplation of the dual vortex at the heart of our being. We see, indeed, one octave higher, how the causal or Egoic body is at once the solar/soul *center* of a human existence and the *peripheral* (in reference to the solar being) lotus-vortex of a human "I" consciously returning creation to its Logoic center.

The self experiences this vortical "return" to its Logoic source in the "just" choices and decisions it gradually brings itself to make—"just" meaning proportionate with regard to the center/periphery, spirit/matter, self/other polarities of one's being. The "turn" hinges upon the willingness to muster enough resistance to elemental, emotional, mental pulls and patterns, to initiate responses, actions, thoughts, creations that engage and harmonize the spiritualizing and incarnating currents of one's being. The turn lies in awakening to the incontrovertible necessity (for *real* well-being) to shift the self-centered course of instinctual nature in response to the decentering–recentering call of the soul, and row upstream to intervals of balanced interplay such as the fifth epitomizes in relation to the octave. As Lorenzetti's painting (fig. 59, page 267) suggests, to master the dragon is to engage it into a dance and so return the *whole* to itself—turn the whole and its polarities of light and dark into the unique beauty of a "self."

At the archetypal turn of worlds, the presence of "the whole"—whole of one's being or wholes of which we participate (family, group, country, planet)—challenges the insistence of the part. Similar to how the musical scale hinges on the fifth, Earth evolution hinges on this internal—individual and collective—reversal of tide, a reversal vividly captured in Virgil and Dante's intricate, arduous and "panting" turn from hell to paradise in the *Divine Comedy*. To reverse ourselves from hells of separation to a paradise of wholeness, one must indeed, in the language of the alchemists, go *"contra naturam"*—against nature. One must, against the passive currents of self-seeking inclinations, redirect one's will upstream toward the vibrant center of the universe within. Dante recounts:

> When we had reached the point at which the thigh revolves, just at the swelling of the hip, my guide, with heavy strain and rugged work, reversed his head to where his legs had been, and grappled on the hair, as one who climbs—I thought that we were going back to Hell. "Hold tight," my master said—he panted like a man exhausted—"it is by such stairs that we must take our leave of so much evil."... "Before I free myself from this abyss, master," I said when I had stood up straight, "tell me, how is he so placed head downward? Tell me, too, how has the Sun in so few hours gone from night to morning?" And he to me: "you still believe you are north of the center, where I grasped the hair of the dammed worm who pierces through the world. And you were as long as I descended; but when I turned, that's when you passed the point to which, from every part, all weights are drawn."[107]

It is through such turns that the vortex-"lotus" of causal Egoity begins to pulsate and, like fruit, star, or galaxy, emit the juice and light of quintessential selfhood. The Egoic lotus hosts the fifth of "I" consciousness—the solar essence that invests humankind with the heart-like capacity and responsibility to turn itself *consciously* toward its Logos source—and so evolve at the heart of creation, and out of the human heart, the harmonious whirling sphere, Michael- and Shiva-like, of a "kingdom of soul."

A latent outpost and octave of the Logos in the midst of creation, the Egoic lotus is suspended between the world-being it emanates from and the human being emanating from it. Evocative of the heart at the center of the blood circulation, this imagination of the Egoic lotus is also reminiscent of Steiner's grandiose picture of the heart as a lifetime mediating chamber between macrocosmic and microcosmic being.

As we saw earlier, Steiner describes how a gradual drawing in of the radiant star-ether body in which the human being descends into the earthly world concentrates around puberty into a distinct structure—the etheric heart—in the midst of which the physical heart is suspended. At the time of death, he says, the heart releases into the cosmos the qualitative content of the life just completed. Coming to existence as an organ through a systole-like contraction of cosmos, and ebbing out of existence through a diastole-like expansion back to cosmos, the heart is at once an incarnation of cosmos and the fruit of a human life—a concentrated microcosmos, which becomes a milieu of creative input into the universe. Fruit for the gods we are. Not unlike humankind plucking nature's fruit from the tree, the universe takes into its evolution the unique fruit, gorged with deeds and seeds, of the human heart.

---

107 Dante, *The Divine Comedy: Inferno*, last verses.

The dual spiraling dynamics of our Milky Way galaxy (fig. 58.6) displays the same concentrating and expanding (spiraling out) pattern-gesture. Similar to how, with his last breath, a human being releases the astral-mental contents of his heart into the cosmos, the galactic world eventually releases its evolutionary contents, as would a fruit its seeds or a human being its deeds, when "death" expands it back in the continuum of world creation.

With the last images—of blood circulation and lifecycle of the heart—we see the dual vortex deploy itself in the alternative form of a lemniscate (a figure-eight or infinity symbol), one half of which moves into a physical form, then out of it (similar to the internal vortex of the torus), while the other half moves out into a vaster or subtler dimension then comes back into the physical microcosm (similar to the external phase of the torus).

The toroidal inside–outside wholeness of fruit, shell, heart organ, Egoic lotus, and the corresponding lemniscate dynamics of etheric flow, blood, human Egoity between a smaller form or system of life and a greater one (physical or nonphysical) in which the first is nested are displayed together in the heart. The heart is at once a toroidal organ and the hub of a lemniscate between pulmonary and systemic circulation—between the microcosmic peripheries of a human body and the macrocosmic periphery of the Earth's atmosphere—between form and life.

The archetypal perception of the mediating dynamics of the heart in the patterning of organic forms and cosmic systems brings back to mind the great X (*Chi*) of Plato's cosmology (fig. 23, page 73). What the contemplation of the fundamental gesture of forms—macrocosmic contraction and microcosmic expansion—suddenly does is deploy this abstract X into a dynamic, spherical dual vortex. Welling up from within the dry sober lines of the X is the flow form of a dual vortex centered on the Sun (bringing together figures 23 and 62 will make this clear to the reader).

Plato's X "depicted" the demiurge's division-union of the soul compound into two circles crossing at an angle and revolving in opposite directions—one, zodiac like, revolving as one "Same" whole, the other revolving as a sevenfold "Other" (planetary) circle "across" the first. The two, says Plato, crossed at intersecting points. Viewed through the lens of the dual vortex dynamics of the heart, these points now suggest the exit and entrance of a toroidal universe, turning inside out, then outside in.

Initially, Plato's X led us to see the central place of the Sun, in the eyes of the ancients, at the "crossing" of zodiacal circle and sevenfold planetary circle. Belonging to both, intermediating between the two, the Sun was regarded as a threshold

IV. The Spiraling Dynamics of the Universe

between the Sameness of the spirit world and the otherness of corporeal, divisible worlds—a threshold of soul and selfhood.

What is happening now, as Plato's X and the lemniscating circulation of blood, solar life, and self merge into one picture, is that what stood as distinct and relatively abstract facts—that the Sun is the heart chakra of a cosmic being (upper half of the X), and also the center of its own (solar) system, generating it as well as attracting it back to its source (lower half of the X)—now arises vividly and movingly before us as *one dynamic "circulation."* The planetary 7fold is no longer statically and spatially "within" the zodiacal 12fold (as we traditionally see it), but the two "circles" are *moving indeed "across" each other*, in torus-like fashion.

One might object that the planetary 7fold does not move out and around to "turn into" a zodiacal whole, then back again into a 7fold spiraling. However, quite similarly, one does not see the 7fold etheric human being flow out and around its aura, nor do we see the human "I" slip out and around its universal "outside"—although death and sleep both point in the direction of human consciousness cycling out and around, and then in again. The pace of this flow, in and out of vortical formations, depends indeed on the time frame of a particular entity's life cycle—from the season of a plant to the decades of a human being to the eons of a solar system. In the same way that we experience ourselves, at least during the day, as an "inside consciousness," so does the solar system exist and experience its existence as a "within" of planetary and human evolution.

While the internal and external vortices of a shell are visible to the naked eye (see figs. 49, 50, page 249), the nonphysical, "outside" swirling powers of Life that shape our being (human or solar systemic) are invisible to our "inside" perspective. The other face of the dual vortex we are is essentially cosmic and "spiritual"—being that in which we have our being. In essence, the X represents the crossing from spiritual essence to physical manifestation and back, and the dual vortex we come to fathom behind the X brings forth a view of the spiral cycling of life and all its beings across a ubiquitous archetypal threshold between essence and existence.

This perhaps surprising view of reality finds corroboration in Rudolf Steiner's spiritual-scientific insights regarding the mysteries of night and sleep:

> As our life of soul by day has its dwelling place in the circulation of the blood, so our life of soul by night is inserted into something that is a copy of the planetary movements of our solar system.... Inspired Knowledge leads us to see how the inner life of nighttime is...in fact a copy of the planetary movements of our system.... Having lived during the second stage of sleep in the copy of the planetary movements, [humankind] now lives in the constellations, or rather in copies of the constellations, of the fixed stars of the zodiac.[108]

---

108 Steiner, *Spirit as Sculptor of the Human Organism*, pp. 9–11 (earlier translation used).

As the external swirl of a shell mirrors or "copies" cosmic etheric motions, it is simply in accord with the noble logic of the universe that the "outside" of our being, the cosmic spiritual side of our "self," would present itself to the inspired knowledge of an initiate as a "copy" of the world-being surrounding us and in whom we exist. Recognizing the toroidal dynamics intrinsic to all forms of nature and cosmos sparks new insights into the day–night rhythm of our being—insights that echo Steiner's indications. Indeed, when we allow ourselves to (spiritually) imagine the subtle form of a human being as a dynamic spherical vortex, we begin to see the cosmic logic of the daily rhythm of human experience. It becomes simply a "reasonable" manifestation of the *whole* of our consciousness life that we should pass out of awareness of "self" at night and move–reverse ourselves through our "other side," as our being "turns around" itself to partake in deep sleep of the streams of cosmic Lives that form-inform us and are our (unconscious) greater Self—"unconscious" indeed, since it pertains to a consciousness that transcends our mind.

It would be closer to the reality of this night journey to the other side to say that we *become* the other side of our "selves," and that in this reversal we unconsciously experience ourselves as impressions, or imprints (copies), of the swirling breaths of the cosmic Lives in whom we have our being. Indeed, Steiner says:

> Our soul is permeated from going to sleep to waking by the planetary forces and by the forces that reveal themselves in the constellations of the fixed stars.... We must say for the nighttime: there revolves in us a copy of the movement of Mercury, of the movement of Venus, of the movement of Jupiter.... From being personal and human our life becomes cosmic during sleep.[109]

We come to see in a new light the mysteries of sleep and death. While we (our "I" awareness) go around the other side of our etheric being at night, at the time of death, we are altogether released to the other side—the cosmic "outside"—of the (solar) system that clothed us with planetary garments, differentiated psyches and bodies, chakric dimensions and faculties.

This view of a cross-dimensional toroidal and lemniscating dynamics of living forms also sheds light on Steiner's bemusing insight about the "return" and reversal of the long bones of a previous incarnation into the round cranial bones of the next.

※

Although the turning points of the dual vortex appear to coincide with the experience of going to sleep and waking up—hence with dawn and sunset, when the ecliptic circling of the Sun crosses the horizon circle of the Earth—the actual reversals of direction take place at the heart of day and night: noon and midnight. The

---

109 Ibid., pp. 12, 9 (earlier translation used).

IV. The Spiraling Dynamics of the Universe

Sun starts its re-ascent toward dawn at midnight and begins to descend toward setting at noon. In other words, within the dual vortex of human experience, noon and midnight mark critical thresholds of conversing and reversing ("universing") between the consciousness awakening forces of the Sun and the form-tending forces of the Moon.

Every new day begins at midnight. In the human experience, the middle of the night is the time we literally pass out of a day cycle of the cosmic system of which we participate—into a new one. The conscious forces of the "self" sink into the meta-conscious, renewing forces of cosmic life. However ubiquitous this midnight is from an objective perspective, it marks the universal time of transition to a new day, the universal experience we have, as earthly beings in cosmos—cosmic beings—of our human system being mysteriously restored and re-enlivened by the pulse of a new Sun–Earth cycle. Midnight points, archetypally, to the pulsing "heart" of our daily life on Earth.

Following the high and low of the daily course of the Sun into their yearly correspondence, we begin to understand *organically* (i.e., from the perspective of our whole cosmic organism) the ancient association of the solstices of Cancer and Capricorn with the "Gates" of incarnation and initiation. Cancer, states Alice Bailey, has been recognized down the ages as "the doorway into life of those who must know death," just as the constellation Capricorn is called esoterically the "doorway into life of those who know not death."[110] In Cancer, the Sun begins to descend toward the Earth and the days become shorter. In Capricorn, the Sun begins to re-ascend in the sky, a new Sun cycle begins and daylight increases.

The dual vortex character of organic systems reveals in Cancer and Capricorn the major turning points *in the yearly cycle of our Sun-Earth-Moon dual vortex*—the places where the expanding forces of solarization negotiate their world creating, harmonic "turn around" with the contracting, in-turning, lunar forces of incarnation to "beat" the golden pulse of evolution.

Brought to bear upon the ancient knowing about the Gates, the dual vortex view of our system and its alternating consciousness illumines in the solstices thresholds

---

[110] Bailey, *Esoteric Astrology*, p. 83; also; "Cancer and Capricorn are the great gates of the zodiac—one opening the door into incarnation, into mass life and human experience, whilst the other opens the door to the life of the spirit, into the life of the kingdom of God, the life and purposes of the hierarchy of our planet. Cancer admits the soul into the world center which we call Humanity. Capricorn admits the soul into conscious participation in the life of that world center which we call the Hierarchy.... Man continuously and consciously enters into life in Cancer, the constellation under which the law of rebirth is applied and administered. But it is only on the reversed zodiac that the man learns to pass with equally conscious purpose through the gate of Capricorn." And: Cancer is "the sign in which humanity as a whole integrated unit is born" (p. 102).

of reversal not only in the cycle of the year but in the cycle of the soul—be it via sleep and awakening in our day/night experience, light/dark in our experience of the year, birth and death in the incarnation/excarnation experience of the soul, dual and nondual consciousness in the spiritual experience of initiation. In this light, the solstitial Gates shine anew as portals of passage from zodiacal, soul-spirit life to planetary existence, and from the existential within to the great cosmic without—places of crossing between cosmic and earthly experience.

The solstices point to a next level of interpretation of the crossing of circles and dimensions symbolized in Plato's X. We first read in this X the intersection of the ecliptic path of the Sun with the celestial equator of the Earth at the equinoxes. The crossing associated with the solstices, on the other hand, points to the intersection of two spherical vortices: the smaller vortex of Earth–Sun–Moon and the greater zodiac–Sun–Earth vortex in which it is nested. One engages the dual dynamics intrinsic to the Earth system, the second its relation with the larger cosmos. Not only can Plato's X be understood in these two different ways, but in both cases the crossing points (equinoxial or solstitial) can be justly regarded as intersections of zodiacal life and planetary existence.

The solstices, which display to the naked eye the "turning around" standstills of the Sun—from darkness to light in Capricorn, from light to darkness in Cancer—thus point to a deeper, soul-spiritual ground of reality than the purely physical clockwork-like "mechanism" of orbital planes crossing each other. They point to the lemniscating flow of the soul of the universe in and out of spiritual and existential worlds, individual and cosmic consciousness, and the reorienting shifts that accompany such cyclical turns of our solar, planetary, and human being.

Analogous to the midnight of the daily cycle, the solstice of Capricorn initiates a new heartbeat of our Sun–Earth–Moon cosmos. While the daily midnight restores and "re-sources" our human system and self, the global midnight of the year delivers the next pulse of our cosmic system, which means that it literally "re-heartens" our planetary consciousness and self. The planetary self is quickened within the individual self, the cosmic heart within the human heart.

The solstice of Capricorn has been since ancient times associated with cultural festivals celebrating the return and "resurrection" of the Sun, as well as mystery rituals aimed at activating the initiatory opportunity this turning point represents for human consciousness. Steiner describes how everywhere—from the Eleusinian to the Egyptian mysteries, from the mysteries of the Near East to the Babylonian-Chaldaic mysteries, the mysteries of the Persian Mithras cult and the Indian mysteries of

IV. The Spiraling Dynamics of the Universe

Brahman—one and the same experience was cultivated in the mystery schools: the experience of the "midnight hour."

> [The pupils of the mystery schools] learned that something happens at this time in the Earth's inner being—the overcoming of death by life that is present in death. This was shown them in the conquering light. This they felt and experienced when they saw the light arise and shine in the darkness. They beheld in the stone cave the sprouting life arising in splendor and abundance out of the seemingly dead.[111]

This initiatory threshold was referred to in the esoteric tradition as "seeing the Sun at the midnight hour":

> Whoever is really initiated learns to experience the Sun at the midnight hour, *for in him all matter is obliterated*. The Sun of the spirit alone lives in his inner self and radiates over all the darkness of matter. This is the moment of highest bliss in the evolution of man, when he has the experience that he lives in the eternal light freed from darkness.[112]

The moment will arrive at some time for everyone, Steiner adds, as it did for the prepared Initiate of ancient mysteries, when "the active forces of the world of the spirit around him will be perceived in colored radiance and brilliant light and he will behold the world around him filled with spiritual qualities, with spiritual beings."

Many ancient temples and monuments across the globe were centered on an aperture calculated to receive the first rays of the winter solstice sun deeply into their core chamber. The midnight of the year would thus, year after year, light up a channel between, symbolically, the "cave" of Earth and the solar cosmos, a channel that heralded the initiatory opportunity to cross the threshold from earthly to solar, "enlightened," consciousness.

The recognition that the midnight of the year is not only the pivotal return–rebirth of the Sun in the cycle of the year, but the *heart* of our Sun–Earth–Moon vortex and the source of its yearly pulse shines new light on the unique character of this time of the year. To see in the solstice the hub and heart of our planetary cycle and system is to realize that it also represents a once a year alignment with the *"light hole" of the grand pulse of Life at the vortical heart of all forms*. This alignment of the solstice with the source of our system's life makes it a potential passageway beyond it.

The suspended stillness at the end–beginning of the yearly cycle of our (Sun–Earth–Moon) system harbors the mystery of its origin, the place beyond "place" from which Sun life pulses anew through our world, as it does at the heart of every

---

111 Steiner, *Original Impulses for the Science of the Spirit*, "Signs and Symbols of the Christmas Festival," pp. 177–178.

112 Ibid.

form of life, great and small. It points to the universal hub of "being" at its core, the hub that transcends the space and time of its form cycle. The life pulse of any organism pertains to its particular form, and it also transcends it. At the timeless pause of its "heartbeat," a system experiences itself beyond the time and space of its cyclical experience—it experiences itself as its live essence—its "self." At this form-transcending hub, all forms of Life, knowingly-unknowingly, commune in "beingness."

The hub of life at the heart of all forms illumines the special opportunity for humankind to commune in "being" during this yearly return to the source and heart of its cosmic being. In the suspended pause of the cycle lies the possibility for the planet to experience itself as one timeless "self"—hence the possibility to transcend self-separateness for a few days—and eventually to transcend altogether the separateness of the unenlightened "self."

Says the Chandogya Upanishad, "There is a light that shines beyond all things on Earth, beyond the highest, the very highest heavens. This is the light that shines in your heart."

Indeed, the solstitial stillness of the Sun, as it is about to reverse direction and turn upward in the sky of human experience, is aligned with the transdimensional breath and pulse of Life that passes through the heart of all forms. It is aligned with the passageway of Life—and potentially consciousness—from one smaller vortex-heart system to a greater one. A channel is thus opened and lit up between the Platonic cave of the Sun-Earth-Moon vortex and the radiant Sun-centered planetary-zodiacal vortex in which the Earth is nested and has its being. For the indwelling self-awareness of the system—humankind—the pause of the Sun is the cosmic moment when the experience of the whole as "self" may, if the necessary heart "potential" is present, lift and expand one's earthly consciousness through the timeless hub of the cosmic heart, and initiate it to a vaster consciousness and being.

A gateway between worlds, the midnight of Capricorn heralds the possibility for the dual consciousness of the ordinary self to pass consciously into a "lightened," solar consciousness that no longer alternates between the pairs of earthly opposites. The major initiation on the horizon of humankind hinges indeed on the capacity, based on wholeness of being in the earthly realm, to transcend the threshold of separation between earthly self and cosmic being and enter the continuity of consciousness and seamless dance of being—intimated by Michael and Shiva's graceful whirling.

When the dual consciousness of form life is transcended, the inside–outside, subjective–objective, material–spiritual, conscious–unconscious, you–me, asleep–awake chambers of the soul become a single heart of cosmic-self awareness, and consciousness one spiritual "Day" pervaded with the radiance of solar "being."

IV. The Spiraling Dynamics of the Universe

Having crossed the earthly vortex of separate identity, the "initiated" consciousness knows: I am you, and you are "I." Soul consciousness per se is transcended and "the Sun of the spirit alone lives in the inner self"—and lives in the self *as Be-ing*.

How fascinating to reread in this new context Plato's cryptic comment quoted earlier, considering how "one, two, and three" are the three principles involved in a dual vortex—oneness, duality, harmonious interplay—spirit, matter, consciousness:

> The little matter of distinguishing one, two, and three, is the sort of knowledge which draws the soul from becoming to being.... [It] has the power of effecting the *turning round of a soul* passing from a day which is little better than night to the true day of being, that is, the ascent from [the world] below [to the gods], which we affirm to be true philosophy.

By what collective insight, the question arises, did the incarnation of Jesus Christ become quite literally the turning point of historical time? One can only observe, in the light of what we have been contemplating, that this mysterious insight of history into itself points to the being of the Christ as the heart-vortex of our current evolution—the solar Life that would "resurrect" matter and reorient humankind to its Sun/soul core by awakening it to love as the cosmic essence of its "self." "Love the other as thy self" is the injunction to realize the other as "self" and turn the love core of the self toward the "self" in "others"—which requires that we turn around from protective self-centeredness to emittive world-centeredness. How humankind would know to reverse the course of time in accord with a major turning tide of its collective consciousness is a mystery.

The same insight aligned the birth of Jesus Christ with the yearly rebirth of the Sun. The being who reversed death into life and matter into light was recognized to embody "the same great ideal of all"[113] as Sun Heroes of the ancient mysteries, all of whom epitomized (differently for different phases of human evolution) the human potential to pass consciously through the Capricornian gate of initiation, transcend the divided consciousness of form life, and cross the threshold of self-separateness into the matter illuminating, self-resurrecting solar simplicity of being, one with the center and periphery, spirit and matter of all forms of life on Earth.

The Christmas festival began to celebrate as one the birth of the Christ and the re-ascent of the Sun, thus keeping open with remembrance, lighted with celebration, and alive with love, the cosmic threshold into greater Life it harbored in its midst. So do all solstice time festivals, knowingly or unknowingly, highlight this yearly

---

113 "Signs and Symbols of the Christmas Festival": "In the ancient mysteries, before human beings spoke of a *Christos,* they spoke of a Sun Hero who embodied the same great ideal of all humanity as is connected with the *Christos* of Christianity" (trans. revised).

gate into solar life for the Sun-hero-in-progress in every human being—a threshold of disidentification from the individual form of selfhood and rebirth into a greater sphere (and vortex) of identity.

The 12 "holy nights," which have from ancient times unfurled a sacred trail around the Capricornian gate, add their glow to the mysterious Passageway, reflecting as they do the 12 folds of the world soul and bringing them to bear as day-seeds upon the 12 months of the upcoming Sun cycle—in a way reminiscent of the hovering of the world soul above the Platonic unfolding of creation.

More than individual, the task of the Christ–Sun hero was planetary. More than a teaching, it was a transformative deed. The Earth was changed—quickened across one rung of the Gate. Steiner describes from his clairvoyant sight how, "at the turning point of time," the Earth began to emit light. With the passing of the solar being through the heart of the Earth, a star-like radiance pierced through the planetary darkness (planets do not emit light) and began to light up the Earth with solar Life. Similar to the quickening of the Egoic "heartbeat" when the human "self" awakens to the soul dimension, the quintessential heart of humankind was quickened with the unifying and life giving power of the Sun.

The turning point of time is a turning point in selfhood, heartened by the lemniscating pulse of the golden fifth—the mediating, integrative, and enlivening life of love in the "I." This turning around in progress is where humankind finds itself in this long turning phase of its evolution, "panting" and groaning, like Virgil and Dante—disoriented and reorienting itself, failing then gaining grounds—as it steers its divine Comedy around and across the critical gate.

The pulsating heart, which every kingdom enacts with its own choreography and every entity adorns with unique features, is the flow form of the Logos in cosmos. As an organ or a form of nature, it is the archetypal formula of the Verb that creates: "Love." What all forms of nature and cosmos literally do is "verbalize" the love essence of the universe in whirling "hearts" whose deeds and sufferings become revelations about love. "O Arjuna! The Lord abides in the heart region of all beings, whirling all beings by his power, [as if they were] mounted on a machine." The dual vortex may well be the whirling "machine" of the song of creation—the *Bhagavad Gita*—the Song of the Lord of love who abides in the harmony of the golden proportion central to the spherical vortex. Thus the three are one as the spirit (love), soul (harmony), and body (vortex) of creation.

V. A ONCE AND FUTURE COSMOGONY

Know that through lucid knowledge, one sees in all creatures a single, unchanging existence, undivided within its divisions. —Bhagavad Gita

The universe is fundamentally and primarily alive. —Teilhard de Chardin

The universe is alive in all its manifestations, like a thinking animal. —Nikola Tesla

The view of our universe as a multidimensional vortex of whorls embedded within whorls is intimated in ancient philosophies and underlies a number of contemporary healing practices (from acupuncture to reflexology, from auriculotherapy to iridology). It is also a view emerging on the horizon of scientific findings and theories. With the advent of the new physics, walls of alienation between two radically opposite worldviews, which one could define as matter-centered and consciousness-centered, have begun to dissolve. The matter-centered reasoning of modern science traditionally views consciousness as a chance emergence made possible by increasingly complex arrangements of matter. By contrast, the consciousness-centered framework, which harkens back to ancient traditions, views all forms of cosmos, nature, and beings as manifestations of an intelligent creative energy referred to as spirit, god, or universal being. These two seemingly incompatible perspectives are bound to find their invisible common ground, as the search for a unified science of the universe, the only satisfying science of the intertwined reality we experience, continues to press for emergence in the collective mind. Each is relatively correct in its own right, but only to the extent that it does not exclude the other. Both become quite literally "unsound" and stuck in incompleteness when they exclude what is but the other pole of a comprehensive grasp of reality. The "spiritual" view lacks the precise grounding in physical factuality and experimental analysis provided by science, and the materialistic view lacks the vivifying breath and intelligence of a unitive order apt to make sense of the cosmos and its beings, their structures and behaviors, their evolutionary processes and experiences. What this exploration has uncovered supports the opposite movements that the two perspectives identify with. As much as physical forms are manifestations of spiraling down, densifying divine intelligent energy, self-awareness emerges when sufficiently complex organisms offer appropriate vehicles for the "higher note" of a more integrated consciousness (including self-awareness) to indwell physical forms.

While the Platonic cosmology may, in and of itself, lack the ballast necessary to land on our mechanistic tarmac, multiple developments across the world consciousness hint at the not-so-remote possibility of a coming together of empirical science informed by unitive reasoning and perennial wisdom sharpened by the penetrating eye of science.

Einstein's relativity marked a giant step in this direction by revealing that energy and matter are two sides of one equation, two aspects of one reality. A continuum. Clearly, the "energy" that physicists refer to is not generally associated with intelligent life or being, nor is light associated with consciousness, and a definite gap remains. Nonetheless, the dissolving of the solid borders of matter precipitated a questioning of the classical mechanistic mindset, which led a number of theoretical scientists to search for a new framework that could accommodate the enigmas and hints delivered by the revolutionary findings of relativity and quantum physics.

The objective of this next stage of our exploration is to walk concomitantly the two cosmological tracks—ancient and modern—of our collective knowing, and see how distant or not so distant they really are from each other. How close might we come to a unitive cosmological viewpoint? Guiding us will be resonances between the ancient view of a divine Being manifesting itself into a physical world and the modern view of the emergence of particles and organic forms from a chaotic quantum field. Halfway between the two—between ancient intuitive knowledge and modern materialistic science—the Pythagorean Plato offers a threshold of reason and logic, which points to a unifying axis.

We will see that ancient Hindu Wisdom carries in its mythic grounds the features of our logo-rhythmic universe, and that the same features are also arising, even if not identified as such, from recent scientific findings and perspectives. When one goes back to Eastern sources predating Plato's philosophical framework, then fast forward to recent scientific expansions beyond the Newtonian framework, one notices reverberations between the two fields. These echoes are the subtle field awaiting the leap that our epoch is prompted to take, by the momentum of its own discoveries, into a more wholistic worldview—a worldview in which material and spiritual sciences, sciences of matter-energy and ancient wisdoms of the world consciousness are allowed to interpenetrate each other and, like a grand dual vortex, enter a unitive flow with fresh nodes of joint understanding and mutual growth.

To step back beyond Plato to Hindu wisdom on one hand, and forward beyond Newton to quantum mechanics on the other hand, is to move beyond the

V. A Once and Future Cosmogony

metaphysical–mathematical understanding of the universe and the mechanistic discernment of its physical laws that Plato and Newton offered respectively to the culture of their times, to the once and future unity of self and world deployed in Hindu wisdom, and intimated anew by the quantum paradigm. One of the great impacts of quantum physics has been to challenge the clear line of separation between subject and object, observer and observed, soul and cosmos, gradually drawn by 2,500 years of intellectual development, and given its utmost definition by the laws of Newton and the philosophy of Descartes.

While theoretical physics is brushing up against, or shivering with, a realization that the materialistic split between world and consciousness gets in the way of the unitive theory it is passionately searching for, the educated spirituality of our times is seeking to uncover how the oneness it intuits fits together with the multiplicity it sees, how the mind it identifies with is related to the body in which it reads traces of its psychological patterns. The heart seeks an intelligence of the subtle order that it recognizes and experiences all around. Our science will not gain its wings unless it includes a science of consciousness, nor our spirituality its feet lest it can gain insight into the intelligent constructs and concrete processes of the transcendent source of life, and thus intervene co-creatively in the physics of the world. While Logos and cosmos are in some ways at their point of greatest alienation in history, they are also pulled into novel encounters, blending, and bindings, by the very developments of our science and requirements of our intuitive intelligence.

The endeavor is evidently a complex one and our goal is simply to allow the main articulations of a joint science of world and self to emerge from the gifts of past and present. We shall see that not only are modern sciences and ancient wisdom beginning to show points of resemblance that corroborate the Platonic picture, but that they also vivify it with fresh insights and flesh it out with novel formulations and perspectives.

The attempt to bring ancient teachings and modern science to bear on each other is certainly not new, and not new to scientists themselves. Already in the 1920s, Niels Bohr, Erwin Schrödinger, and Werner Heisenberg, three of the great minds of quantum mechanics, began to ponder a marriage between Western scientific thinking and Eastern philosophy. They wanted, Schrödinger said, "some blood transfusion from the East to the West to save Western science from spiritual anemia." They saw in Eastern thought a possible way out of the numerous paradoxes generated by quantum mechanics. Heisenberg remarked, "it may be easier to adapt oneself to the quantum theoretical concept of reality when one has not gone through the naïve materialistic way of thinking that still prevailed in Europe in the first decades of this century." They read from the ancient Hindu scriptures, the Vedas, and resonated with their teaching that we are more than physical bodies defined by the laws of physics and chemistry, that we are inherently connected to the greater whole, and that this connection is not coming from matter and can only be accounted for by a dimension that transcends matter. Faced with "the lesson of atomic theory"—the influence of the act of observation on observed phenomena—and with "trying to harmonize our position as spectator and actor in the great drama of existence," Bohr advocated "turning to quite other branches of science, such as psychology, or even to the kind of epistemological problems with which already thinkers like Buddha and Lao-Tse were confronted."[1] Both David Bohm and Niels Bohr took interest in Hindu wisdom. Bohm was a student of Krishnamurti and Bohr was influenced by a series of dialogues with Rabindranath Tagore. Einstein reportedly kept a copy of H. P. Blavatsky's *Secret Doctrine* on his desk. After his death, his niece donated the book to the theosophical library in Adyar, India. It was heavily annotated and underlined, and the margins covered with scribbles. Robert Oppenheimer learned Sanskrit in 1933 and read the *Bhagavad Gita* in the original, citing it later as one of the most influential books to shape his philosophy of life. In India, science and religion are not fundamentally opposed, as they often seem to be in the West. Instead, they are seen as parts of the same great search for truth and enlightenment. In the Hindu scientific approach, understanding external reality depends on understanding the Godhead.

In order to gather what might become the features of a unitive cosmology, we will go back and forth between ancient wisdoms and latest sciences, with just enough depth of details to show how they may flesh out from their respective side what in the end points to one and the same understanding of reality.

---

[1] The philosophical writings of Niels Bohr, vol. 2. *Essays 1933-1957 on atomic physics and human knowledge;* quoted in Trinh and Ricard, *The Quantum and the Lotus,* p. 114.

V. A Once and Future Cosmogony

FUNDAMENTAL ONENESS

It is by the One that all beings are beings...for what could exist were it not one? If not a one, a thing is not. No army, no choir, no flock exists except it can be one...It is the same with plant and animal bodies; each of them is a unit.... Health is contingent upon the body being coordinated in unity; beauty, upon the mastery of parts by the One; the soul's virtue, upon unification into one sole coherence.—PLOTINUS[1]

The universe that we are in is an intelligent, self-organizing, learning, participatory, interactive, non-locally interconnected evolutionary system. It's all of those words. So to me, the universe is the body of God, and God is still learning. The evolutionary mind, the consciousness that exists in the universe, is the mind of God.—EDGAR MITCHELL, NASA astronaut[2]

Contemporary science and ancient wisdom agree on two fundamental points about the universe: the single source of its origin and the vibratory, wave-like nature of its ground. Breath field of Brahma or radiatory field of the Big Bang, the ground of the universe is a vibrational matrix emanating from One source.

Georges Lemaitre, the Belgian priest and physicist who proposed the Big Bang theory in 1927, came to the conclusion, by reversing the expanding nebulae backward in time to the precise moment the universe began, that everything must have fit into an infinite point in space comprising an infinite amount of energy. He called this primordial moment the Singularity. "We might expect from these discoveries," Richard Merrick remarks, "that science had at last found its way back home to natural philosophy, ending happily ever after in a marriage of spirituality and science. But strangely nothing happened."[3]

The ancients perceived the universe as a living being. Better said, they *knew* it to be alive. The world they experienced was his vibrant body, and creation whirled out of his primordial breath-sound. "In the Hindu worldview," Alan Watts explains, "the coming and going of all worlds, all beings and all things, is described as the eternal out-breathing and in-breathing of this One Life."[4] The same breath opens the biblical Genesis of the world: "In the beginning...a *wind* from God swept over the face of the

---

[1] Enneads, *The Essential Plotinus*.
[2] See http://ascentmagazine.com/articles.aspx%3FarticleID=195&page=read&subpage=past&issueID=30.html.
[3] Merrick, *Rediscovering Duality*; see http://www.interferencetheory.com/Articles/files/9b204821d5d8e2fae6b9aca120c37885-1.html
[4] Watts, *The Two Hands of God*, p. 68.

waters...then the Lord God formed man from the dust of the ground, and *breathed* into his nostrils the breath of life; and the man became a living being" (Gen. 1:2, 2:7).

In ancient cosmologies, creation is a *sacred sound* composition. In Hindu teachings, it is the "Bhagavad Gita"—the song of God. The vibration of a cosmic Being is what sets in motion the unfolding of cosmos. Tantric teachings describe the ultimate reality as "unfathomable creative vibration"—a description no longer foreign to physics, with the recent advent of string theory. This all-pervasive field of vibration or sacred sound is "the breath of Brahma," the primary reality underlying all forms, subtle and physical.

Consequently, the Indian Vedic tradition views consciousness "not as an emergent property that comes into existence through material structures such as the brain and the nervous system, but as a vast field that constitutes the primary reality of the universe."[5] Brahma, the creator God of the Hindu trinity, is the "Self" of the universe—the Creative Intelligent ground of being and the power of evolution at work in all natural processes. The Anahad Shabd, the soundless sound-breath of Brahma, the Unitive Field that pervades everything, is what the West refers to as the "Word of God" or Logos—the Word that was and is in the beginning, and by whom "all things were made that were made." John's experience of the world's beginning resonates with the Hindu worldview that creation is an echo of Aum, the Logos or the divine Word, "a moment to moment emanation from the Self."[6] Indeed, rather than a chronological indicator, world beginning points to the ongoing passage of Being into time and space formations.

In *The World as Power,* John Woodroffe calls the primordial soundless breath of Life the wavelength of the experience of God. "The causal vibration," he writes, "is the undifferentiated soundless sound, the wavelength of the experience of God. Parashakti is his slumbering energy." "Parashakti" is the "Other" of Platonic cosmology. The "Nature" of the natural philosophers, the Sophia of esoteric Christianity, speak to the same archetypal "resounding" field of nature-cosmos in which the godhead is asleep, as of yet unaware of itself.

Einstein and Tesla echo this ancient view unequivocally: "Everything in life is vibration," said Einstein. "If you wish to understand the Universe," enjoins Tesla, "think of Energy, Frequency, and Vibration."

Also in agreement with ancient wisdom, the contemporary scientific model of creation acknowledges absolute unity at the beginning. Oneness. The Big Bang

---

5   Laszlo, *Science and the Akashic Field*, p. 120.

6   Burger, *Esoteric Anatomy*, p. 121.

underlying field is an information field. Electrons are carriers of (life) information, as we know from electronics. Before becoming carriers of the "electronic" information we want them to hold, they carry informative energy, which is to say that they carry intelligence. A number of other proponents of digital physics view information rather than matter as the fundamental reality.

For David Bohm, quantum physics leads to "a new notion of unbroken wholeness that denies the classical idea of analyzability of the world into separately and independently existing parts.... We say that inseparable quantum interconnectedness of the whole universe is the fundamental reality, and that relatively independently behaving parts are merely particular and contingent forms within the whole."[14] Physicist Fritjof Capra, author of *The Tao of Physics*, echoes Bohm in laying out what has emerged as the invisible subtle ground of physics: "The quantum field is seen as the fundamental physical entity; a continuous medium that is present everywhere in space. Particles are merely local condensations of the field."[15]

In *Dawn of the Akashic Age*, Laszlo and Dennis take Bohm and Capra's now scientific view that we live in an interconnected, intrinsically nonlocal world, to its logical conclusion:

> The meaning of nonlocality is much more than the classical idea that one thing affects another.... True nonlocality is that one thing affects all other things, instantly and enduringly. Every single thing makes a lasting impression on the whole world. New findings in the sciences tell us that we are intrinsically interconnected—in the final analysis, we are one.... Everything that happens in one place can affect what happens in another place, and in some sense happens everywhere.[16]

Even though the realization of this underlying oneness has not brought with it a serious questioning of what is at work in such "wholeness" and what holds the world together as a system, it represents a momentous step.

From the perspective of Hindu wisdom, Capra's "continuous medium present everywhere in space" is a *living* medium, filled with the conscious, intelligent breath of the universe's being—the "breath of the Macranthropos, or Cosmic Person" (Brahma or Logos). It is a vibrant etheric matrix, whose living energy is referred to as *prana*.

In calling the primordial field "the breath of Brahma," Hindu wisdom affirms in it the medium of the universe's Life. The definition of the vibrant wavy ground of the phenomenal world as *medium-of-life* is precisely the understanding yielded by the

---

14 Bohm and Hiley, *On the Intuitive Understanding of Nonlocality as Implied by Quantum Theory;* in *Foundations of Physics,* vol. 5; quoted in Capra, *The Tao of Physics,* p. 138.

15 Capra, *The Tao of Physics,* p. 210.

16 Laszlo, *Dawn of the Akashic Age,* ch. 2.

harmonic model of creation. If the vibratory matrix is the mediating breath of Life, it is because harmonic resonance, and the associated geometry of proportionality, is the way for Oneness to pass into multiplicity—hence *the* way for Life to breathe its vibrancy throughout the body of the universe's being. Plato's mathematical perspective froze the waves to reveal the apportioning of regular, harmonic intervals created by their interferences, and the resultant wholistic fractioning underlying the formations of nature. We see how these ancient sources of wisdom, which were to become condensed and crystallized in Plato's concepts, have the vitality to thaw and enliven his mathematical cosmology by re-insufflating in the fractions ("portions of the world soul") the vibrant breath-waves that generated them and continue to do so out of the breathing fount of the universe's being.

In fact, the notion of waves of energy, common to both the Western scientific and Eastern spiritual models of cosmology, orchestrates with different "instruments" Plato's description of the harmonic ground of the universe. The modern notion of a continuous wave-like matrix of the world certainly echoes the vibrant ground of Brahma's universe, even though the latter includes the breathing consciousness of a Being and the other does not. Halfway between the two, Plato represents the evolutionary emergence of a faculty of rational thinking apt to objectify the world enough to ponder the intelligent mind of its creator, while still experiencing the source of cosmic order as a Being.

It remains that Big Bang and Breath of Brahma, which happen to agree, quite literally, on the "initials" of the world beginning, both point to vibration as the initiating event and foundational ground of our universe. This idea of vibration as the primary cause of everything is present in pre-modern cultures with surprising regularity, as David Tame points out:

> According to the Hindu view of creation, it was sound and not light that appeared first. In Vedic parlance it is called Nada Brahma or the Sound Celestial. This sacred vibration or cosmic Sound, infused with the essence of consciousness, is also referred to as AUM, OM, AMEN, I AM.... the Logos, the lost Word.[17]

Vedic Rishis believed that the Brahmand or universe was a result of "Bindu Vsphot," an atomic explosion that produced infinite waves of sound. In accord with the Vedic seers, physicist and cosmologist Jude Currivan emphasizes in the "Big Bang" birth of the physical universe the surge of a "vast radiating wave of space–time."[18]

---

17 Tame, *The Secret Power of Music*, p. 205.
18 Currivan, *The Wave*, p. 12.

V. A Once and Future Cosmogony

The notion of space–time as the wave-like fabric out of which the universe is fashioned would have been inconceivable before the advent of relativity and quantum field theory at the beginning of the twentieth century. Einstein's insight revolutionized the Newtonian concept of time and space. Since Newton, time and space were regarded as absolute and distinct realities, each existing "without reference to anything external." Einstein's special relativity theory demonstrated that the flow of time slows down for an observer in motion. Time stops when one travels at the speed of light through space. In other words, there is no passing of time at light speed. With Einstein's demonstration came the revelation that space and time are interdependent, relative, and compensatory to each other and that they constitute one continuum. "Space and time are individually relative; space–time is the absolute entity."[19] Like electricity and magnetism, space and time are two sides of one coin. Space–time is a single entity, "apportioned" differently by observers moving at different speed.

The reader may realize, remembering our initial penetration into Plato's harmonic ground, and the accompanying revelation that space–time is a corollary of vibration, that the Einsteinian conception of space–time ties in seamlessly with the notion of the vibrational wave-ground of the universe. As we saw early on, the harmonic field involves a complementary intertwining of wavelengths and frequencies, hence time and space (lengths of waves and cycles of vibration per second). The higher the frequency (time), the shorter the wavelength (space)—a fact of inverse proportion analogous to the space–time relationship discovered by Einstein. In Jude Currivan's words, "wavelengths and frequencies embody the intrinsically energetic and wave-like nature of space–time. The nature of space–time is innately energetic." The same idea is implicit in the Tantric conception that "underlying all the forms of the manifested world are the oscillating wavelengths of primeval sounds in varying combinations."[20] Trinh Xuan Thuan speaks of the Big Bang *creating* time–space, even though he does not connect this birth of time–space with the vibrational character of the event: "In our present state of knowledge, the Big Bang is the theory that best explains the origin of the universe. We think the universe was created about 15 billion years ago when an unimaginably small, dense and hot concentration of energy exploded, *in the process also creating time and space*."[21]

This recent view of space–time nudges the mind to seriously consider the Platonic cosmology of a world created by harmonic partitioning of a primordial oneness, and, beyond Plato, the ancient wisdom of a universe born of sound and soundless

---

19   Greene, *The Fabric of Cosmos*, p. 62.
20   Woodroffe, *The Garland of Letters*, pp. 214-227.
21   Trinh and Ricard, *The Quantum and the Lotus*, p. 23.

breath, a world whose fundamental medium is a rhythmic field of wavelengths and frequencies—a time–space field, *born of* the *"periodicity"* of vibration. The concept of "period" is in itself an expression of the unitary reality of time and space. Space and time are the warp and weft of the vital field of the entity-universe born into existence at the "Big Bang." Space–time is the fabric of its biofield.

Converging with this understanding are the conjectures of superstring theorists, who advance that the fabric of space–time is a field of innumerable strings engaged in a coherent pattern of vibration.[22] It is as if, Brian Greene suggests, individual strings are "shards" of space and time, and only when they appropriately undergo sympathetic vibrations do the conventional notions of space and time emerge. "The hope," he adds, "is that...the theory will describe a universe evolving to a form in which a background of coherent string vibrations emerges, yielding the conventional notions of space and time."[23] This would come very close indeed to establishing the Platonic cosmology as a science of the formation of cosmos.

Taken to their conclusion, the convergences between modern science and ancient Hindu teachings invite us, in faithfulness to science's striving for a unified understanding of reality, to seriously consider that the "continuous quantum medium," the undulating field of world beginnings, is pervaded with the "ordering" breath of a cosmic Life that transcends the world manifestation as its source—a "divine" pneuma that is the Mindful breath-medium of the universe.

POLARIZATION AND TRIANGULATION OF THE ONE: THE ANCIENT VIEW

> *The One becomes Many. The Unity becomes Diversity. The Identical becomes Variety. Yet the Many remains One; the Diversity remains Unity; and the Variety remains Identical.*
> —MAGUS INCOGNITO

> *The Tao gives birth to One. One gives birth to Two. Two gives birth to Three. Three gives birth to all things.*
> —LAO-TSE

As we penetrate further into the three pictures of cosmological unfolding provided by Hindu wisdom, Greek metaphysics, and modern physics, what we see emerge are different formulations of a common pattern of world-unfolding, involving a process of dualization and triangulation. The fundamental One re-"sounds," or "Big

---

22 Greene, *The Elegant Universe,* pp. 377–378.
23 Ibid., p. 379

Bangs," and the polarity thus generated between the One and its vibrational time-space matrix, the "Singularity" and its wave field, engages a threefold dynamics, which Hindu wisdom expresses in terms of cosmic Powers (Shiva–Shakti–breath of Brahma), Chinese wisdom in terms of energies (Yin–Yang–Tao), Platonic thought in rational concepts (Same–Other–Being), and modern science in objective categories (energy-matter-light) or forces (electricity-magnetism-electromagnetic field). Such a convergence of root principles between ancient wisdom, Greek philosophy, and modern science suggests a potential common ground.

These major formulations of universal principles emanating from distant historical periods and cultures correspond to profound shifts in the evolution of human consciousness—from an experience of the world as a living presence, to a metaphysical reasoning about nature and universe, to a techno-scientific relation to material objects and facts. These shifts reflect the gradual withdrawal of human consciousness from its early identification with cosmos and nature as the intellectual faculty and personal sense of egoity began to develop, reaching fruition in the rational-philosophical thinking of Ancient Greece. The European Renaissance marked the next major step, ushering a new chapter of scientific objectification of nature and cosmos, in association with an increased sense of individuality and autonomous will. Essentially, this arc of transformation outlines a shift in the rapport between self and world—from identification, to reflective lucidity, to objectifying separation. The next stage may well be a lucid and applied awareness of all interconnectedness. Components of such a view have begun to light up in the collective consciousness.

Turning first to the past, we can see how the pattern of dualization and threefolding we explored in the mathematical metaphysics of Greek philosophers, appears already, *live,* in ancient Hindu cosmology—alive with the ontological features of deities. We will see later on how a similar pattern can be discerned in the categories of contemporary science.

In Hindu cosmology, the cosmic intelligence, Brahma, is the One source of the fundamental duality of spirit and matter, self and world. Radhakrishnan, a philosopher and former prime minister of India, speaks to this ancient knowledge-experience of the seamlessness of cosmos and consciousness at the heart of the Upanishads, "God does not create the world but becomes it. Creation is expression. It is the self-projection of the Supreme."[24] Mediating this dual, yet unitary field is the dynamics of Brahma's breath.

---

24 Radhakrishnan, *The Principal Upanishads*, p. 82.

Two names, two powers, stand behind the reality of Brahma's breath: Shiva and Shakti. Out-breath and in-breath. We might say that they "personify" these two universal currents of energy. However, it is closer to the consciousness of these earlier times, who experienced the formative beings at play in cosmos and nature—and perhaps closer to the truth—to say that Shiva and Shakti are cosmic beings indwelling these two archetypal currents and giving them their specific qualities and functions. Shiva is the emittive–creative spirit-polarity, and Shakti the receptive-cohering matter-polarity of the universal field, which is the breath of "God." The breath field itself is a divine milieu of loving intercourse between Shiva and Shakti, which gives birth to all forms in cosmos.

This dual universal current is as close to us as our breath, yet it takes an act of divine attention and spiritual imagination to experience its trans-human, universal scope, and feel in it the power of a grand Life at work in the quiet midst of our being. And it takes a further act of inner summoning to ride these waves by raising our vibration to a higher frequency and our consciousness to a universal ground.

The living powers of Shiva and Shakti reappear in Plato's abstract principles of Same and Other, as humankind's developing intellect strives to *think* world creation and conceive metaphysically the polarization of Being into matter and soul—out-sounding and resounding. "The Other being riveted into the Same begets diversity and disagreement," Plutarch elaborates, "and the Same being fermented into the Other produces order."[25] Similar to the primordial couple of Shiva and Shakti, Plato's polarity of Same and Other engages the archetypal interplay of cosmic externalization and soul internalization fundamental to our universe—involuntary out-breath into matter and world, and evolutionary in-breath into consciousness and self-awareness. *Shiva and Shakti reveal in Same and Other the two-directional dynamics of life* at the source of manifestation—one example of how cosmologies can vivify each other. They reveal that spirit and matter, the concepts most frequently associated with the fundamental polarities of the universe, point to relative (hence related) states of being, better conceived as opposite and complementary *directions* of energy: outgoing and incoming, centrifugal and centripetal. Teilhard de Chardin's intuition that matter and spirit are two directions of energy is a luminous insight into this dynamic ground of reality.

Hindu teachings go on to describe a secondary *split* of Shakti into *two fields of force,* bindu and nada, whose interplay holds the world together and generate the Tattvas, the building blocks of the universe. Woodroffe explains:

> A primal shudder splits shakti into two fields of electromagnetic force, nada and bindu. The centrifugal, positive male force (1), bindu, is the ground from which nada operates. Nada, the centripetal, negative, female force (2), unfolds

---

25 Plutarch, *Plutarch on Plato's Procreation of the Soul in Timaeus*, section 24.

the manifest universe. This duality of poles in the substratum of manifested shakti provides the electromagnetic force (3) holding together the molecules of the physical world in a state of vibration."[26]

Bindu and Nada replicate in forms the "divine," unmanifest polarity of Shiva and Shakti. They stand behind the dual phenomenon of extension of spirit-energy in space and cohering-in of matter in time, which constitutes forms. We recognize in "this duality of poles in the substratum of manifested shakti" what we have called the second trinity of form: 3fold spirit, 12fold soul, and 7fold body. We also see the dual vortex emerge from the resonant field. Electricity and magnetism replicate the polarities of (radiating) spirit and (cohering) matter in the subtle energy field that sub-stands the dense physical world. The electromagnetic force is the quasi-physical version of the ontological dynamics of attraction–repulsion at play in the loving intercourse of Shiva and Shakti. Atoms attract atoms and form molecules. Beings attract each other and generate beings. The electromagnetic field points to the vibrant currents of polarized charges and fields that "hold cosmos together." For the sake of clarity, we will postpone a deeper look at the correspondence that Woodroffe brings up between Nada–Bindu and the electromagnetic force, and return later on to the subject of electromagnetism.

The correspondences between ancient Hindu wisdom and Greek metaphysics highlight different perspectives on the same reality—each reflecting the consciousness of a particular cultural time–space. Plato does not focus on the energetics but on the mathematics and geometries of the universe, not on divine powers but on divine intelligence. He does not contemplate the breath of life and waves of love underlying spirit-matter but the harmonic-geometric key to their universe-constructing conjugation—the golden mean. Like the splitting of Shakti into nada and bindu, *golden secting* engages the dynamics of separating-uniting, or pulsing out-drawing in, that generates and sustains the universe. The divine "love"-union of Shiva and Shakti, which produces the universe, becomes in the *Timaeus* the "earthly" ("geo"-metric) "mean" to harmoniously—proportionately—sect-and-bond the polarities of Sameness and Otherness into a soulful world.

From mythical, cosmology has become metaphysical-mathematical, reflecting the collective evolution from "participation mystique" to rational conceptualization and intellectual objectification. Plato's focus is on the rational engineering of the passage of the One into multiple forms through the harmonic conjugations of the fundamental wave–breath of Being. The Hindu picture, on the other hand, highlights behind the golden secting–bondings of Plato's demiurge the love energy of the one breath, the rhythmic interplay of moving away and returning—conceiving

---

26 Woodroffe, *The World as Power*.

forms then withdrawing into self—so that every "particulating" bond or every particle is filled with universal love energy and abreath with the same vibrant Life.

While the harmonic cosmology of Pythagoras–Plato reveals the mathematical facts underlying the vital cosmology of ancient India, this cosmology resurrects the soulful ground of Greek thought and reinvests its concepts with the shimmering life and warmth of divine beings.

The three energetic channels of Ida, Pingala, and Sushumna replicate in the micro-cosmos of our etheric bodies the dynamics of shakti, nada, and bindu—itself the replica in forms of the shiva–shakti breath of Brahma, the subtle breath of the soul of cosmos. They suggest how our small lives partake, directly and centrally, of the vast macrocosmic life and evolution of the universe, which is indeed *our being*— our *universal* being.

Hindu teachings also explore and describe the subjective counterpart of the cosmic powers of Shiva–Shakti–Brahma and shakti–nada–bindu as three fundamental aspects of consciousness—three "gunas": rajas, tamas, and sattva.[27] The gunas designate three primordial qualities of psychic energy whose interplay underlies the entire process of creation and evolution. Sattva (originally "being, existence, entity") is the quality of balance and harmonizing consciousness between the polarities of rajas (energetic activity, dynamic, generative motion) and tamas (lethargic darkness, entropy). The gunas are associated with the three fundamental types of motion— forward, rotary, spiral cyclic—underlying the three operations of manifestation: creation (*rajas*), preservation (*sattva*), and destruction (*tamas*). They correspond to the Trimurti of Shiva, Brahma, Vishnu, the trinity of Father (rajas), Mother-matter (Tamas), Son (Sattva) principles—Plato's Same, Other, and Being. The value of the notion of the gunas is to inform us about the subjective qualities associated with each of the great Three of every threefold, each of the three aspects of the Logos: Rajas speaks to the creative, *sounding/intoning* active spirit energy, Tamas to the formless but in-formable and transformable inertia of matter, Sattva to the form-building energy of soul light, which organizes, sustains, and reproduces wholeness through a creative harmonizing of polarities.

The biblical Genesis offers a similar picture of the cosmogonic breath or pneuma hovering above the waters, separating the polarities of Heaven and Earth, darkness and light, male and female, and assembling dimensions, kingdoms, and species.

---

27 Literally, Guṇa = string, or a single thread or strand of a cord or twine. Abstractly, it might mean "a subdivision, species, kind, quality, property"; also "what binds."

V. A Once and Future Cosmogony

With regard to the "third" that surges from the union of the two, Hindu and Greek cosmology offer two mutually illuminating versions of what becomes the form-generating core of creation.

Rajasic and tamasic currents cross at still points of harmony—"sattvic" points of creative union called bindus. Bindus are the generative points, the seeds of creation at the center of every undulating conjugation of Shiva and Shakti. Bindu is the non-dimensional, oneness-point at the source of any form of manifestation, the conception point also referred to as a transdimensional "singularity"—a definition reminiscent of the term used by Lemaitre in reference to the Big Bang. Bindu is described as a Circle with Void inside—O. The confining Circle is Shakti; the Void is Shiva. "What the Hindus called the bindu or seed," Lawlor remarks, "we call the geometrical point, [which is] the limit between the manifest and non-manifest, the spatial and non spatial."[28] This source-point, which is the replica of the center of the universe in every form, reproduces the archetypal entrance in manifestation of the nondual source of life. It symbolizes circumscribed oneness—the self-contained, self-circumscribed wholeness of being.

The primordial creation point, the form transcending at-onement of Shiva and Shakti, is called Mahabindu, or Para Bindu. As such, it contains what Randolph Stone calls the "primordial mind pattern" of cosmos. In Hindu cosmology, this point is a featureless point of pure energy. Only with the descent of ancient theogonies into the metaphysical–mathematical geometric landscape of Platonic cosmology do the delineated features of a pattern appear within the point of creation. What Plato's harmonic *"geo-metry"* reveals are indeed the *"earthly measurements"* of the Maha bindu. Within the sattvic point of creative intercourse between the world polarities, Plato discloses the features of the golden mean—the principle of creative harmonious rapport, the proportion that sects-in-a-way-that-bonds. Within the Mahabindu stands the tri-unity of the Logos.

Two other names given to the Mahabindu confirm the correspondence between bindu and the vehicle of the Logos. It is called Tribindu—threefold bindu, and Saguna Brahma (Brahma with the Three Gunas). Mahabindu, the offspring of the archetypal intercourse (maithuna) of Shiva-Shakti, is analogous to the fruit of the creative engagement of Same and Other. At once unmanifest union of Same and Other (Shiva-Shakti), and world manifesting "seed-sound," the Logos-bindu is the Word-origin of our dual worlds—of cosmos and soul. Tribindu is defined as *Bindu* (Shiva), *Nada* (Shiva-shakti) and *Bija* (Shakti), which speaks to the threefold seed of the supreme, subtle, and gross levels of spirit, soul, and body. The Platonic

---

28 Lawlor, *Sacred Geometry*, p. 21.

cosmology lights up in Mahabindu the archetypal "mean" of creation, the threefold dynamic harmony that ushers life into physical forms.

While the golden mean of the Greeks reveals harmony and proportion at the core of the ubiquitous bindu point, Hindu cosmology revives in the Platonic dualities of Same and Other the currents of the breath of Being. If the vibrancy of reality is implied in the harmonic nature of the Platonic operation of creation, the live breath animating it is muted and stilled by the growing imperative, in the Hellenistic culture, to discern, elucidate and delineate the *mind pattern* of creation. The Breath recedes behind the Word, the pneuma behind the concept.

The Platonic cosmology of harmony also clarifies the notions of bindu and parabindu through the distinction it provides between the fifth interval of the third harmonic and the other, secondary, intervals. Parabindu is the golden seed-Word at the core of manifestation, while bindus are secondary notes-nodes of harmonic conjugation, which generate the different dimensions of a form. The distinction between whole human being and chakric dimensions, plant and leaf or flower illustrates the distinction between parabindu and bindu.

A remarkable fact further supports this "harmonic" understanding: Not only is parabindu associated with the triune Logos and the third harmonic, but it is also associated with 12foldness. While each chakra is a bindu point, the term *bindu* is especially associated with the twelve-petal chakra at the center of Sahasrara, the crown chakra, which is referred to as "Bindu Chakra." This chakra is also called the "heart-in-the-head," revealing the correspondence between bindu, the seed-pattern of "selfhood," and the chakra that the zodiac represents in the cosmic Being in whom our solar system has its being. This correspondence confirms the understanding we reached earlier, that the reality of "self," soul, logos is associated with a 3fold/12fold structure, and that the zodiac is the prototype of *selfhood*, the "template" of its structure and the temple of its evolution.

While parabindu refers to the nuclear "selfhood" of the universe, bindus are the innumerable nodes of at-oning polarities, which usher life into secondary forms and organs. In the sunflower for instance, seeds unfold at the bindu-points of intersection of centrifugal and centripetal spirals. Every seed is a replica of the archetypal offspring of the maithuna. Every point of emergent life in creation is a "bindu," and parabindu is the archetypal creative point—the 3fold Logoic point.

In human beings, Ida and Pingala, the replica of Shiva and Shakti, cross at every chakra (bindu), generating the *elemental vehicles*: physical elemental (base chakra), watery emotional elemental (sacral), fiery will elemental (solar plexus), airy mental elemental (throat), etheric quintessential selfhood (heart). At the Ajna center, they set the stage for the subtle *light* of the universal mind to dawn, as evolution brings

V. A Once and Future Cosmogony

a gradual transcending of the existential pull of opposites—essentially, personality and soul, then soul and spirit. At the crown chakra, the potential awaits for the great polarities of spirit and body to transcend their dichotomy in a radical "inburst" of fully actualized Being, when the rising of the kundalini Shakti through the chakras delivers a *self-realized being* identified with the Logos-universe. This ultimate at-onement takes place at the parabindu. Not only does it complete the harmonic range of a human being, "crowning" him with a full chord of wholeness, but it produces the Shiva–Shakti point through which he is born into a next octave and order of being. This ultimate oneness no longer generates a *form*, nor a quintessential self-radiance, but it gives birth to a new order of "self."

In the end, Hindu and Platonic philosophies propose a similar view of the birth of the universe through a dynamic triunity. The names Shiva–Brahma–Vishnu, Shiva–Shakti–bindu, Father–Mother–Son, Same–Other–Being, as well as their numerous equivalents across ancient cultures designate the *identity* of the universe—its triune soul, whose spirit essence is love. The Egyptian trinity of Isis, Osiris, and Horus personifies the same principle of creational love. One of the gifts of Greek and Hindu texts is to illumine in the trinities present in one form or another in all world cosmologies, the world creative intercourse of dualization and tri-union, differentiation and integration, outbreath and inbreath, which existentializes the universe.

The Chinese Taoist symbol of yin–yang engages the same fundamental duality of expansion-contraction, light–dark, centrifugal–centripetal motion, into a triune whole, whose mediating, balancing and inclusive third is a breath like wave. Similarly, each *kua* of the *I Ching*, the ancient Chinese book of changes, encodes the binary attributes of yin and yang into a creative triad.

In essence, all of creation emerges from a Being who deploys itself in a triune soul-milieu of love-harmony, which is the world creative germ of all forms, grand and small.

Modern science rediscovers the tri-unity of the world

$E=MC^2$—Albert Einstein

As we turn to the current picture of the birth of the universe, we find that twentieth-century physics has been nudging the modern mind toward a conception of the ground of reality that evokes key features of Hindu wisdom and Pythagorean metaphysics. The puzzling discoveries of quantum physics, which dematerialized solid matter into elusive particles and shimmering waves, even vibrating strings, ushered a new paradigm of quantum field interconnectedness that offers intimations of the resonant ground of the universe the ancients have familiarized us with.

What appears to the modern mind as mythical musings and metaphysical abstractions about the grand energetic breath and harmonic structure of the universe may well be, in the end, the magical bow that draws a deep and full sound out of the latest discoveries of science—from electromagnetism to relativity, quantum mechanics to string theory, from DNA structure to morphic fields and fractals—and allows us to hear the fundamental voice and song of the universe, grounded anew in the beauty of scientific facts and the truth of mathematical equations.

Quantum physics and relativity have had two major impacts on our worldview. On one hand, they have shattered our perception of matter as a solid, compact mass. On the other hand, they have challenged and blurred the absolute separation between subject and object (hence consciousness and matter) that has dominated our relation to the world, particularly since Newton and Descartes.

The realization that matter is not a solid mass, as previously conceived, is one of the most significant revelations of twentieth-century physics. The atom is not a compact grain of indivisible matter with a definite border, but an energy field in which electrons orbit like standing waves around a nucleus. Even more confounding is the fact that, under observation, elementary particles sometimes appear as particles, sometimes behave as wave. At the subatomic level, says Capra, the solid material objects of classical physics dissolve into wave-like patterns of probabilities, which are not probabilities of things, but probabilities of interconnections.[29] "The success of quantum mechanics," Brian Greene elaborates, "forces us to accept that the electron, a constituent of matter that we normally envision as occupying a tiny, point-like region of space, also has a description involving a wave that, to the contrary, is spread out through the entire universe."[30] To conceive of a particle as an incorporeal wave field involving the entire universe is clearly a challenge to the materialistic–mechanistic mindset that has governed science and dominated our thinking for the last three centuries. A new horizon has arisen, and a vaster sphere of vision is inviting expansion of mind.

When we ponder the advances of early twentieth-century physics in the light of the cosmological picture we have been contemplating, what we come to discern, which we might not see without this lens, is the central emergence, in a new form, of the threefold principle at the heart of all wisdom traditions.

---

29 Capra, *The Tao of Physics*, p. 68.
30 Greene, *The Fabric of Cosmos*, p. 90.

V. A Once and Future Cosmogony

The puzzling immaterial and shape shifting features of matter brought to light, ironically, by the cutting edge of *materialistic* science, gradually coalesced into a "face" that was to become the face of the new scientific paradigm: the mathematical trinity $E = MC^2$. Einstein's mass-energy equivalence formula expresses the fact that mass and energy are two aspects of one reality and can change into each other. Energy equals mass multiplied by the squared speed of light. The speed of light is a *time–space* factor. One could say that mass and energy are relative manifestations of a space–time matrix defined by the speed of light.

A major landmark in our cosmological history, Einstein's famous three-term equation of energy–matter–lightspeed loomed in the global sky as a modern version of the Trinity, its three facets lining up with those of the old alchemical triunity of sulfur–salt–mercury, and behind them, the great archetypal trinities of Shiva-Brahma-Vishnu, Shiva-Shakti-bindu, Same-Other-Being, Father-Mother-Son—their mystery X-rayed into a mathematical truth and scientific fact. To align Einstein's equation with the threefold ground of our being and the dimensions of spirit-matter-consciousness they embody may seem rather hurried of a leap. The leap is clearly a major one, since it makes consciousness "part of the equation." Yet, when we read the interpretation of Einstein's equation by astrophysicist Trinh Xuan Thuan, we gauge how close we are to seeing this leap turn into a bridge already behind us: "Energy and matter," he says, "which together embody the universe, are reconciled through the mediation of the speed of light."[31] Undoubtedly, this statement can be read in a purely materialistic way, yet if *the universe as a whole* is indeed embodied in energy and matter, where is the life of our being (and all beings of nature) if not behind its "energy," and where is consciousness, our awareness and passionate search for its order, if not behind its "light"?

The fact is that Einstein's formula heralds the new "transcendence" of a paradigm of tri-unity beyond previously perceived radical dichotomies. The iconic logo of the early-twentieth-century scientific revolution opens the way to conceive the unconceivable: how the universe could unfold from a tiny speck of energy. It holds as one the outside and inside of our reality.

Although it is improbable that Einstein ever looked at his equation in this way, the following anecdote lets us entertain the thought that he might have been pleased with it. During a seminar on physics he gave at the university of Berlin in 1920, a young woman asked Einstein: "Master, what do you look for in your equations?" He replied: "I want to know how God created the universe. I am not interested in this phenomenon or the other, I want to know the thought of God. The rest is only details."[32]

---

31 Trinh and Ricard, *The Quantum and the Lotus*, p. 25.

32 Quoted in Arnould, *Sous le voile du cosmos*.

Realities once regarded as separate and alien to each other: time and space, energy and matter were found to be interdependent. Mass proved to be a slowed down form of energy and energy, a light-sped form of mass. Rather than two separate realities, they are *relative* aspects of a single one. "We have been all wrong," Einstein exclaimed. "What we have called matter is energy, whose vibration has been so lowered as to be perceptible to the senses. There is no matter.

Concomitantly—and, in the light of cosmology, quite logically—physics was being exposed to a deeper ground of reality that challenged the unquestioned separation between matter and self. The discovery that the observer has an influence on the "objective" phenomenon he observes goes hand in hand with the groundbreaking revelation held in Einstein's equation. It unveiled a unitive ground most puzzling to our dual mindset, but intriguing and compelling to our spiritual-scientific quest for wholeness. That such a unitive field would dawn with the emergence of a threefold principle is not a surprise in this cosmological context, which exposes how threefoldness carries the unity that precedes the formation of organisms and systems, and transcends the separation between world and self.

In counterpoint to the metaphysics of old, which contemplated the becoming two, then three, then many, of the One, science was unveiling for the modern mind the entrance to the return path: *from duality to tri-unity and oneness.*

As time and space were proving to be relative to each other and to the motion of an observer, as matter and energy were turning out to be relative ends of one equation—two forms of one and the same reality—quantum physics was revealing that human observation plays a key role in the character of the "matter" observed. Matter can "appear" as particle or wave. Instead of being-what-it-is, matter is an "appearing" phenomenon involving the perceiver. If we try to observe the "particle" in its wave state, it becomes a particle. Even if we keep to the most concrete explanation that the photons projected by the observing eye interfere with the wave-photon probability, the fact remains that phenomena are influenced by the observing consciousness. "The act of measurement is deeply enmeshed in creating the very reality it is measuring."[33] The observer is part of an *interdependent* process that produces an observable phenomenon. John Wheeler summed it up best: "No elementary phenomenon is a phenomenon until it is an observed phenomenon." There is simply no phenomenon outside of its observation. "As we penetrate into matter," Capra elaborates, "nature does not show us any isolated "basic building blocks, but rather

---

33 Greene, *The Fabric of Cosmos*, p. 94.

V. A Once and Future Cosmogony

a complicated web of relations between various parts of the whole. These relations always include the observer in an essential way."[34]

Physicists began to reflect on the artificial nature of the division between self and world. Schrödinger wrote, "Subject and object are only one. The barrier between them cannot be said to have been broken down as a result of recent experience in the physical sciences, for this barrier does not exist."[35]

Referring to Niels Bohr's principle of complementarity, Trinh Xian Thuan notes: "I think that the mind complements matter, just as the 'particle' aspect of matter complements its 'wave' aspect."[36]

Schrödinger, who had a lifelong interest in Vedanta philosophy, also entertained the thought that individual consciousness is part of a unitary consciousness that pervades the universe. What happens at the subatomic level intimates that we are modifying and creative agents in the "outside" world we observe.

Subject and object were emerging as relative ends of a single phenomenon, which engaged and transcended their relative distinctness. Bohr remarked, "The notion of 'object' is subordinate to the 'measurement,' hence to an event."[37] This event engages matter and self, it is a matter-self happening. The classical dichotomy of subject-object recedes behind the relativity of subject and object to "an event" involving them.

Captured in this "event of measurement" and the revelation of the entanglement of self and object is a vaster, universal "Event." This universal "happening" is what theoretical physicist Amit Goswami contemplates in his book *The Self-aware Universe*. It is the coming to self-awareness of the universe. This fundamental happening that, prior to the confrontation of science with the inseparability of self and object, would have been considered pure speculative metaphysics, may now be seriously thought to be the ongoing Event behind the existential duality of self and world. Through its dual manifestation as Logos and cosmos, inside and outside, self and world, the universe is generating self-awareness—in us.

In the course of his work, Goswami quotes mathematician G. Spencer Brown,

> We cannot escape the fact that the world we know is constructed in order to see itself, but in order to do so, evidently it must first cut itself up into at least one state which sees, and at least one other state which is seen.[38]

---

34 Capra, *The Tao of Physics*, p. 68.

35 Schrödinger, *Mind and Matter*.

36 Trinh and Ricard, *The Quantum and the Lotus*, p. 167.

37 Bohr, quoted in ibid, p. 124.

38 Brown, quoted in Goswami, *The Self-aware Universe*.

To Brown's remark, which is nothing less than a modern formulation of the dualization of ancient cosmology, Goswami adds,

> I have said that the brain-mind is a dual quantum system/measuring apparatus. As such, it is unique: It is the place where the self-reference of the entire universe happens. The universe is self-aware through us. In us, the universe cuts itself into two—into subject and object.[39]

The coming to self-awareness of the universe in us is the Event-principle behind the "event of measurement" scientists have been witnessing and trying to come to terms with. Coming to "terms" with the befuddling ambiguity of a reality appearing both as formless wave and formed particle, and defying the separation of self and world, called for a conceptual ground that could embrace these dualities. The notion of "event" provides such a ground, as it captures a dynamic reality that engages the two as one, and transcends the rift generated by their representation as static, separate "things." This transitional reality called "event," in which whole and particle, oneness and duality, measuring self and measured thing come together as one, is a reflection of the archetypal event-advent at play in any "phenomenon" of nature—which, as the Greek root of the word indicates, is fundamentally an "appearing." Appearing is the epiphanic core of every "phaino-menon." This "appearing" dynamics is the archetypal event in the event of measurement, where the "captured" photon appears as particle, and also behaves as a wave. Indeed, "No elementary phenomenon is a phenomenon until it is an observed phenomenon"—until, that is, it "appears" in the eye of an observing mind. The epitome of such "appearing," the "phainomenon" of all phenomena, is the birth of the universe. This birth launches the adventure of self-awareness of the universe—the "appearing to itself" of Being. Similarly, what appears to itself in the event of measurement and the "particularization" of the wave, is the human self.

The movement of the universe toward self-awareness is precisely what animates scientific research. The observation that photons could behave as waves and appear as particles opened a new chapter in this self-awareness. As the divide between particle and wave, subject and object, vanished under their eyes, scientists were exposed to the nondual core of all phenomena—the fact that a phenomenon is only a *relatively* objective reality, which does not exist outside of a mind objectifying it, an "appearing" in the field of a mind. Becoming aware that the subjective mind is an inescapable part of the world as we know it, and that the world is an appearing in the mind, is a significant step toward a more unitive conception of reality and a realization that our strivings to understand the universe serve its coming to self-awareness.

---

39 Goswami, *The self-aware universe*, p. 190.

V. A Once and Future Cosmogony

We see how the new paradigm that science is stepping into begins to reflect the ancient cosmological landscape. The centerpiece of this new paradigm, affirms Goswami, is "the recognition that modern science validates an ancient idea—the idea that consciousness, not matter, is the ground of all being."[40]

The archetypal Appearing–Event is the epiphanic Word of *Phanes,* articulating the primordial breath (wave) into form–words (particles), similar to how the event of measurement "formulates" wave into particle. This event is reflected in language, in the function of the verb, which is to expose, externalize and objectify something about the "subject."

As light to mass and energy, the self-knowing of the universe, its "appearing to itself" in the self-awareness of the human being, is the "constant" beyond the relativity of self and object. For the consciousness trained to expand beyond the duality of the intellectual mind, this conceptual glimpse is an actual experience. To Trinh Xuan Thuan's question, "Is the knowledge of enlightenment a higher level of knowledge than the rational knowledge of scientific thought?" Buddhist monk Mathieu Ricard responds, "Enlightenment isn't only knowledge of apparent reality, but of the essential nature of reality. The false divide between subject and object vanishes, and reason is replaced with direct, clear, enlightened awareness, which mingles with the ultimate nature of phenomena until it is united with it."[41]

Synchronicity is another type of experience that conjures up a blurring of boundaries between inner and outer world, psyche and events. An external event resonates and mirrors an internal occurrence. It startles and enchants us, as does any experience that "miraculously" transcends the order of cause and effect, which is the ordinary logic of the human mind. It quite literally "blows" our mind. What it simply does, however, is reveal its limitation.

The concept of synchronicity emerged in the collective awareness around the same period of the early twentieth century. A famous, seminal example of the phenomenon is the experience Carl Jung recounts he had with one of his patients, a highly educated and intelligent woman, whose intellectual adeptness at rationalization was keeping her disconnected from her emotional and instinctual depths.

> A young woman I was treating had, at a critical moment, a dream in which she was given a golden scarab. While she was telling me this dream, I sat with my back to the closed window. Suddenly I heard a noise behind me, like a gentle tapping. I turned round and saw a flying insect knocking against the window-pane

---

40   Goswami, *The Self-aware Universe,* p. 2
41   Trinh and Ricard, *The Quantum and the Lotus,* p. 230.

from the outside. I opened the window and caught the creature in the air as it flew in. It was the nearest analogy to a golden scarab one finds in our latitudes, a scarabaeid beetle, which, contrary to its usual habits had evidently felt the urge to get into a dark room at this particular moment. I must admit that nothing like it ever happened to me before or since.[42]

Jung handed the scarab to the woman, saying, "here is your scarab." He would later comment how the uncanny experience shocked his client out of her rigid intellectual framework, so that "the treatment could now be continued with satisfactory results."

What happens in such "meaningful coincidences" is that an external happening mirrors or echoes a subjective content in a way that strikes one as deeply significant. It may also be that two unrelated events with similar contents create a coincidence we experience as charged with meaning. Domains that the mind perceives as unrelated and separate, such as inside/outside, thought/event, self/world, appear to be mysteriously connected, without any cause and effect connection. If the coincidence transcends tangible causality, it is because its source transcends the realm of duality in which causality operates. An inkling of transcendent oneness appears and recedes, setting on fire existential coordinates. The separation we take for granted between inside and outside, psyche and world, collapses. For a brief yet timeless moment, one is swept up into a magical sense of intimate rapport with the universe. The outside world seems to conspire in the human quest for meaning. Jung referred to the oneness that collapses existential opposites as the Self. He borrowed the term *Self* from ancient Hindu philosophy to designate the center and totality of the psyche, the *universal* core of consciousness that transcends, not only conflictual psychic contents, but also the duality of subjective and objective worlds, and activates their integration.

The "dual unity" of subjective and objective happening experienced in synchronicity reflects the relative duality of conscious and unconscious in the psyche. Synchronicities reveal that conscious and unconscious, as alien to each other as they may appear to the conscious ego, are *relatively one* from the perspective of the Self. We are only relatively conscious, and we are as well only relatively unconscious. At work in the engagement of the two is evolutionary growth, the gradual expansion of consciousness into the universality of a self-aware Self—which is the gradual self-awareness of the universe in a human self.

In the end, the phenomenon of synchronicity points to the *threefold unity* of conscious-unconscious-Self. This triune principle, which arose from Jung's penetrating observations of the psyche, illumined the sky of psychology with a wholistic

---

42   Jung, *Synchronicity*, p. 22.

V. A Once and Future Cosmogony

"equation" analogous to Einstein's relativity formula in the sky of physics. One did for the science of the psyche what the other did for the science of the world.

Since Jung, scientific experiments have revealed how "simultaneously occurring events in our space–time world can be related meaningfully to a common cause that resides in a nonlocal realm outside space and time." Experimental physicist Alain Aspect was acclaimed for his 1982 demonstration of what is now referred to as "quantum nonlocality." His experiment proved what Schrödinger had recognized as early as the 1930s—that the quantum state of every particle from a pair or group of particles previously correlated cannot be described independently of the state of the other(s), even when the particles are separated by a large distance. Instead, a quantum state must be described for the *system as a whole*. "The outcome of what you do at one place," Brian Greene elaborates, "can be linked with what happens at another place, even if nothing travels between the two locations—even if there is not enough time for anything to complete the journey between the two locations."[43] Einstein derisively, yet evocatively, referred to the phenomenon as "spooky action at a distance." While Schrödinger did not demonstrate it, he defined the notion, coined the term and recognized the importance of the concept: "I would not call [entanglement] one but rather *the* characteristic trait of quantum mechanics," he states, "the one that enforces its entire departure from classical lines of thought." Swiss theoretical chemist Hans Primas takes Schrödinger's comment one step further, "The system which appears to consist of two particles from a traditional viewpoint, is in truth *one* undivided whole."

These observations on nonlocal correlation between elementary particles in physics echo Jung's insights on synchronicity and the correlation of matter and psyche:

> Synchronistic phenomena prove the simultaneous occurrence of meaningful equivalences in heterogeneous, causally unrelated processes; in other words, they prove that a content perceived by an observer can at the same time be represented by an outside event, without any causal connection. From this it follows either that the psyche cannot be localized in time, or that space is relative to the psyche.... Since psyche and matter are contained in one and the same world and moreover are in continuous contact with one another and ultimately rest on irrepresentable, transcendent factors, it is not only possible but fairly probable, even, that *psyche and matter are two different aspects of one and the same thing*.[44]

With Einstein's relativity, Jung's depth psychology, and the nonlocality experiments of quantum physics, the same "event"-revelation has dawned on the horizon of the modern mind. Fundamental dualities have been stirred out of their separateness

---

43 Greene, *The Fabric of Cosmos*, p. 114.
44 *The Portable Jung*, p. 518.

and found to be relative to each other in reference to a third reality that transcends them and integrates their relativity in an all encompassing tri-unity. This trans-dual, threefold oneness brings into view the new reality of a subtle unitive field that precedes the distinction between subject and object, between particle and particle. The step taken by the collective consciousness, synchronistically indeed, in these various domains, is a step from duality to tri-unity, from separation to a threefold dynamic interplay that involves a higher, integrative order of reality. This is the step Einstein spoke to when he made the famous remark, "No problem can be solved from the same level of consciousness that created it." The world of duality is the world available to our perception and concrete mind. Deeper, beckoning the depth psychologist, the quantum scientist, the metaphysician, the meditator, is a matricial oneness from which the two-ness arises.

Einstein would later write in his autobiography,

> Gradually I despaired of the possibility of discovering the true laws by means of constructive efforts based on known facts. The longer and the more desperately I tried, the more I came to the conviction that only the discovery of a universal formal principle could lead us to assured results.... How, then, could such a universal principle be found?[45]

Einstein became convinced that the true laws could not be assembled purely from data "from below" and that only by taking hold of a core universal principle could the coherence of facts reveal itself from this organizing center. While he did not say how, he undoubtedly discovered such a universal principle—"relativity"—relativity of space and time, mass and energy. In bridging the gap between matter and energy, and grasping their interdependence in reference to a fundamental time-space unit (the squared speed of light), Einstein's equation formulated a triadal principle of relation between realities that were regarded until then as incommensurable.

Clearly, the dynamic principle of threefold unity implicit in relativity has not been perceived as significant per se, and has therefore not taken on its full illuminating power. Little if any attention has been given to the threefold dynamics of the equation, because it has not struck any chord of meaning. A sense for its meaningfulness could only come from questioning how the three factors of light, energy, and matter are engaged, and how consciousness, particle, and wave can possibly be related and relative to each other. Rather than the whole reality of such an equivalence, the focus has been on the two main terms for which the third provides an equivalence. As long as the searching eye engages the opposites of energy and matter,

---

45 *Einstein: Autobiographical Notes*, p. 9.

wave and particle exclusively and materially—as opposite "things"—more inclusive laws will continue to elude science.

The passage from a dual to a threefold (one) conception of reality is the essence of the paradigmatic shift at work in contemporary science. The recognition of this essential principle as the very hub of "relativity" depends upon the scientific consciousness hoisting itself out of a one-sided mechanistic and materialistic mindset. The name "relativity" Einstein gave to his "universal principle" highlights its most obvious and radical impact on our consciousness and science: the equivalence of mass and energy. However, the oneness that frames this equivalence, and the triunity it unveils, have remained a blurred reality in the background. The emphasis and attention has been on the momentous opposites of matter and energy being brought to commensurability and compensatory relativity.

The term *relativity* is quite fascinating in that it states the fact of interdependence and interconnectedness, but does not quite engage the dimension of relation. It alludes to relation, as communication hints at the potential for communion. Relativity is, one could say, the shell of relation, and relation the unawakened soul of relativity. Relation is the larger totality implied in relativity. Relation is the threefold dynamic "unit," of which relativity is a manifestation. This dynamic unit becomes alive and meaningful as such only when the third is brought forth as the encompassing "one" that relation indeed is, and when the higher order it represents is pondered and engaged.

Einstein's threefold equation mirrors, from the physical end of human experience, the metaphysical pattern of dualization and threefolding, separation and union, articulated by ancient cosmologies as the fundamental pattern of world beginnings and world foundations. As matter and energy to the space–time unit of C^2, world and self are relative expressions of a triune third, which is the all-encompassing world soul matrix that transcends and precedes their separation.

The Holy Grail of physics—the unified field theory it has been searching for—would be a unitive understanding of all phenomena. The term *unified field theory* was coined by Einstein, who hoped to prove that electromagnetism and gravity were two manifestations of a single field. Later on, when quantum theory (which explains microscopic phenomena) was found to be incompatible with his general theory of relativity (which deals with gravitation and macroscopic phenomena), the notion of a "unified field theory" that would work on all levels and describe all the forces of nature became the new and ever receding horizon of science.

Such a theory can no longer exclude the human component of the "measuring, observing" mind. The confounding interference of the observer with the observed has signaled that in its depths, "matter" is relative to "mind." It has revealed in the fabric of solid matter a vibrant quantum field of waves and particles of energy in which what we regard as objective reality loses its absolute independence from subjectivity. "The mind–self cannot remain outside a genuine 'unified field,' but has to be part of it," says David Bohm. "In quantum mechanics," continues physicist Freeman Dyson; "matter is not an inert substance but an active agent.... It appears that mind, as manifested by the capacity to make choices, is to some extent inherent in every electron."[46] A unified field theory would have to embrace the trans-dual field intimated by quantum physics. It would have to spell out the co-arising of subjective and objective reality, consciousness and world, from a single matrix, as well as the relative evolution of self and world within this one universal consciousness called world being. Subjectivity would have to be uncovered in the outside world as the intelligent patterning of nature, and objective operations acknowledged in the subjective realm of the psyche.

The unitive field beckoning science resonates with the primordial field of the Vedic and Platonic paradigms. Transcending the division between mind and matter, this field is not only the source of physical elements but also of self-awareness. German Indologist Georg Feuerstein refers to the dimension in which objective and subjective realities are absolutely identical as "the transcendental dimension." Transcendental, indeed, to our ordinary representation and perception of reality, the quantum field of the physicist is literally cosmogonical. It takes us beyond and before the solid formations of matter, to what precedes the appearance of elementary particles: the universal wave-field of energy-consciousness. This cosmic field is at once the fundamental (spatially) and primordial (temporally) ground of world beginnings.

Feuerstein proposes a logic of stages of consciousness to account for the different experiences we may have of the relation between "matter" and "self"—from the total independence of mind and matter characteristic of ordinary perception and akin to the physical dimension, to the relative interdependence associated with intuitive mindfulness and akin to the soul dimension, then to the unitive identity experienced in deep meditative states and akin to the "transcendental dimension" of world beginning. "In the transcendental dimension, objective and subjective realities are absolutely identical, in the subtle realm, they are barely distinct, and they manifest as seemingly separate lines of evolution only in the visible material dimension."[47]

The cosmological perspective of life's involution into quasi separate forms answers the question raised by the quantum theory: why are there no grossly detectable interferences between observer and observed when we deal with the visible

---

[46] Laszlo, *Science and the Akashic Field*, p. 110.
[47] Feuerstein, *Tantra*, pp. 61–62.

forms of nature and cosmos? This discrepancy was one of the reasons Einstein objected all his life to the strange properties and assertions of quantum physics. He contended emphatically that the universe is completely independent of human observation. The Moon, he said, does not depend for its existence on our looking at it. Indeed. The cosmological perspective, and its replication in the stages of consciousness outlined by Feuerstein, helps to understand how, while solid forms of matter do exist independently of human observation, at the more primordial, atomic level, mind and matter are only *relatively* independent, and interchangeability is revealed. The stuff studied is also the stuff with which we study: when we observe photons, we precipitate the becoming-photon of a wave by projecting photons ourselves. Light of matter and light of consciousness become indistinguishable. Goswami addressed this question of the discrepancy between the behavior of quantum particles and forms of nature by proposing that if the phenomenal world looks overwhelmingly objective, it is first of all "because classical bodies have huge masses, which means that their quantum waves spread rather slowly."[48]

It is safe to say that our collective mindset and its dualisms have stumbled upon the triune ground of reality, psychic and cosmic. Remarkably, the first scientific inklings of this "trinity" happened very soon after God was proclaimed dead, most famously by Nietszche, his metaphysical trinity having faded away behind the blinding physicality and square facts of scientific materialism.

Inklings about Dualization and Threefolding in Contemporary Cosmology

Based on the equivalence of energy and matter, scientists could now explain how, as the universe cooled, energy began to be converted to matter according to Einstein's formula $E=MC^2$. Trinh Xuan Thuan explains how elementary particles, such as quarks and electrons, "rose out of the primordial quantum vacuum, a vacuum seething with energy, and came together to form atoms, then molecules and finally, hundreds of thousands of years later, the stars."[49]

It is rather remarkable, having in mind the principles of dualization and threefolding, to read a scientific description of the earliest process of particle formation out of the radiation of the Big Bang:

> Within only 100,000th of the 1st second after the Big Bang, which marked the beginning of time itself, *two forms* of quarks, the elemental building blocks of nuclear matter, were clustering together in *groups of three,* to create the protons

---

48 Goswami, *The Self-aware Universe*, p. 144.
49 Trinh and Ricard, *Quantum and the Lotus*, p. 26.

and neutrons which comprise atomic nuclei. These two forms of quarks, which are known as "up" and "down" by physicists, carry electrical charges of +2/3 *and -1/3* respectively. The combination of the two "up" quarks and a single "down" quark forms a positively charged proton, whereas the combination of one "up" quark and two "down" quarks creates a neutral neutron.[50]

What we discover in these primordial steps of world formation is nothing less than the great processes we have become familiar with: dualization (positive and negative charge polarization), golden fractioning of the charge (2/3–1/3), and tripartite wholeness (three quarks form a proton or a neutron).

The threefold principle replicates itself at the next level—the atom. The atomic structure (1) is based on the balancing masses and charges of proton-neutron (2) and electron (3). Much later in the course of evolution, the code of life's biological instructions in the nucleus of any cell, the DNA, will reveal a similar model of two- and threefoldness: a binary structure with triadic substructures of amino acids. Physicists have also come to the conclusion, from probing the structure of matter to scales of about a billionth of a billionth of a meter, that everything in the universe is made up of *three* families of four fundamental particles each (3 x 4 = 12) and their combination.[51]

From under the vastly different descriptions of our cosmological beginnings presented by Plato, Hindu wisdom, and modern science, the same numbers emerge, signaling the same dynamic process: 2 of separation–polarization, 3 of re-union via mediation, 1 of original oneness and wholeness making. The Greek referred to such inner dynamics as the numbers' "demiurgic or fabricative power."

"Over the following 100,000 thousand years," Jude Currivan describes, "shock waves, akin to sound, rippled through the early universe, as it continued to expand and cool, creating eddies of primarily hydrogen gas [that were] seed points for future galaxies and the first generation of stars."[52] These descriptions suggest a process of wave interferences on a macrocosmic scale, leading to vortical formations and bindu like points of phase conjugation: galaxies and star systems. Currivan concludes,

> The contemporary understanding of the origin of the universe is remarkably consistent with the metaphorical and symbolic perceptions of the ancient sages: From its beginning as space–time wave, we now consider that the primeval ocean of energy was pervaded by sound, and subsequently materialized by the manifestation of light.[53]

---

50 Currivan, *The Wave*, p. 13 (italics added).
51 Greene, *The Elegant Universe*, p. 150.
52 Currivan, *The Wave*, pp. 13, 61.
53 Ibid., p. 14.

V. A Once and Future Cosmogony

THE TRIUNE GROUND OF MATTER: FIELD, MASS, ENERGY

> *The three-fold number is present in all things whatsoever...we did not ourselves discover this number, but rather nature teaches it to us.* —OVID

In photons (units of light energy), as well as in quarks, electrons, atoms, and, much later, stars and beings, we are looking at the emergence of *particles* out of a primordial, *wave-like* vibratory field of energy. We are also, and by the same token, witnessing the advent of *forms of energy*—all via the wave phenomenon.

The theory of relativity and the modern Trimurti it revealed—the triunity of energy-mass-light—opened the way for science to conceive how masses (nuclei, atoms, stars) could emerge cosmologically from an energy field. It also highlighted the three-facetedness of this fundamental wavefield–breath of the universe: wave medium (light) – energy emission—particle formation (mass).

To retrace the emergence of quantum physics is, we saw, to retrace a paradigmatic shift from the dual world of mechanistic materialism to a triune world of dynamic processes of mass formation and energy emission. In this triune worldview, the issue is less one of "particle versus wave" as one of *wholistic field science* versus *particle science,* science of energy-mass dynamics versus science of separate "things." From the threefold field perspective, the essential polarity is not particle versus wave, but particle versus energy—the twofold potential of the *wave field*.

The paradigm shift heralded by relativity and quantum physics may be traced back to Faraday and Maxwell's theory of electromagnetism in the 1860s. Maxwell's equations, which followed Faraday's discovery in 1845 that electricity and magnetism were not different forces but complementary phenomena, induced a major shift, at once subtle and huge. They induced a shift in perspective—from the classical focus on "separate" forces to the awareness of a common ground-field of energetic interplay—the electromagnetic field. "Faraday and Maxwell replaced the concept of force by that of a force field and in doing so they were first to go beyond Newtonian physics."[54] A few decades later, a similar shift in perception would take place with regard to space and time, then with matter and energy. The focus moved from space and time as separate realities to the "space–time" field of their relative engagement, later from particle and energy as incommensurable realities to a wavefield productive of both.

Two of Maxwell's equations predicted the wave nature of the electromagnetic field and also defined the behavior of these theoretical waves. They all traveled, he

---

54 Capra, *The Tao of Physics,* p. 59.

found, at a speed that coincides with the speed of light, which led him to infer that light itself is a type of electromagnetic wave. The realization that light is an electromagnetic field traveling through space in the form of waves was the culmination of Maxwell's theory of electrodynamics. It was also the first indication of the existence of the entire electromagnetic spectrum. His equations predicted an infinite number of frequencies of electromagnetic waves, all traveling at the same speed—the speed of light.[55] The fact that the speed of light is the universal speed of electromagnetic waves implies that, in Einstein's equation, the speed of the electromagnetic wavefield as a whole is the fundamental constant in the relative equivalence of mass and energy. Maxwell's historical breakthrough was to identify the electromagnetic *field* and to establish its wave-like character.

A diagram of the electromagnetic field brings to mind the earlier illustrations of dual vortices, from apple to atom to heart (fig. 58, page 266). Indeed, any living form is the materialisation of an etheric dual vortex, the densest aspect of which is electromagnetic, and the electromagnetic field associated with any living form moves in a toroidal manner. The heart emits the body's most powerful and most extensive electromagnetic field. Similarly, shells, as well as fruit, show an interplay of two fields at 90° of each other (fig. 65, page 323).

Einstein's equation established that the (squared) speed of the electromagnetic field is the constant of interchangeability between mass and energy. Speed is a space–time referent—a distance of space traversed in a unit of time. And a wavefield is also, by definition, a field of frequencies and wavelengths—a space–time field—with a unique signature value: its speed. The speed of electromagnetic waves, the space to time ratio that defines the wavefield, remains constant in a medium, irrespective of frequency and wavelength. As frequency increases, the wavelength decreases and vice versa. Gamma radiations, for instance, have a higher frequency than visible light, and radio waves a higher frequency than electrical waves. There are many frequencies and their corresponding wavelengths—but one speed.

---

[55] Electromagnetic radiation is broadly classified into the following categories: gamma radiation; X-ray radiation; ultraviolet radiation; visible radiation; infrared radiation; terahertz radiation; microwave radiation; and radio waves. All of these, from the lowest-frequency radio waves to the highest-frequency gamma rays, are fundamentally the same and referred to as "electromagnetic radiation." They all travel through a vacuum at the same speed (the speed of light). The electromagnetic spectrum extends from below the low frequencies used for modern radio communication to gamma radiation at the short-wavelength–high-frequency end, thereby covering wavelengths from thousands of kilometers down to a fraction of the size of an atom. The limit for long wavelengths is the size of the universe itself, while it is thought that the short wavelength limit is in the vicinity of the Planck length.

V. A Once and Future Cosmogony

 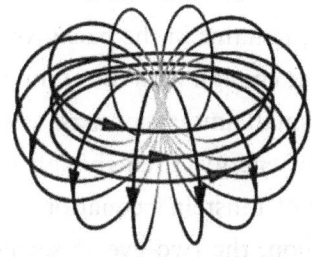

Fig. 65. Similar lines of forces can be observed in a halved strawberry as in a diagram of the electromagnetic field

According to Einstein's equation, the constant of equivalence, C^2, "lightspeed squared," the speed of the electromagnetic field squared, is the factor needed to calculate the energy equivalent of any mass and vice-versa. It serves to convert units of mass into units of energy, no matter what system of units of measurement is used. However, Einstein's equation does more than reveal the relative equivalence of mass and energy and break open a new horizon of practical applications by giving the mathematical key to convert one parameter into the other. What it does is point to the actual unitive ground of mass and energy. In other words, the equation is not only a mathematical equation but a "physics" equation, which exposes the equivalence of energy and mass in term of a common underlying source-field characterized by a specific space-to-time ratio. While the speed of light squared is the quantitative key of convertibility between mass and energy, it is also the particular space-to-time ratio that *qualifies* their common matrix. The wave field moving at the speed of light is the *one* reality underlying the duality of mass and energy and sustaining their relative identity. Beyond their apparent "incommensurability," mass and energy are in fact two facets of manifestation of this field, and they are commensurable in term of its speed squared—a space to time constant. C is the signature of the wave field underlying all mass/energy phenomena. It points to an ultimate space–time signature of the world. We might then wonder if the "squaring" of this space to time ratio (the speed of light) may not represent a constant of space–time two-dimensionality at the foundation of the three-dimensional universe. While beyond conceptualization, this C^2 is not unlike the dyadic reality of "appearing" that is personified in Phanes, and precedes the archetypal triad of the world being. The fact that C = Planck's length divided by Planck's time highlights in C the fundamental constant of the space–time field of the universe, and in C^2 the primordial ground and ultimate solvent of the great dichotomy of mass and energy, spirit and matter.

This qualitative dimension of the equation remains untapped. While the equivalence of energy and mass loomed with the power appropriate to its significance in the center of world culture, the nature of their common ground, which the speed of light points to, remains in the background.

We are familiar enough with the atomic age awareness, associated with Einstein's equation, that a small amount of mass is equivalent to (and can potentially release) a huge amount of energy. Even though the process of atomic fission was arrived at independently of Einstein's equation, and the equation not strictly necessary to develop the weapon, the two events seemed to merge in the collective perception. Indeed, the equation $E = MC^2$ is indispensable to calculate the amount of energy released in a fission reaction. In an earlier 1905 paper, "Does the Inertia of an Object Depend upon Its Energy Content," Einstein focused on the fact that if a body gives off the energy L in the form of radiation, its mass diminishes by L/C^2. In this case, "radiation" means electromagnetic radiation or light, and mass means the ordinary Newtonian mass of a slow-moving object. Conversely, the slight loss of mass experienced in the physical reactions of nuclear fission or nuclear fusion generates a large amount of heat, which can be used to produce electricity.

The key to turning mass into energy is "light speeding" it. The reverse is equally powerful and suggests that a high enough level of energy can manifest as mass. This could explain the "miraculous" occurrences of materialization that run through the life stories of advanced—"en-lightened" indeed—beings. The universal, lightspeed like vibrancy such souls are attuned to would have the "natural" potential to turn energy into mass and manifest objects out of thin air, as certain yogi have been known to do. During a picnic with a few close friends in the Indian countryside, Blavatsky reportedly materialized a teacup after she heard that one was missing in the set brought for the afternoon ritual. The new teacup was identical and as finely designed as the others. She commented that it was a simple operation, once you understood how atoms move and interact.

The same notion makes it conceivable that the Logos of our age, who is referred to indeed as "the light of the world," could manifest bread and fish to feed 5000 people, turn water into wine, resurrect Lazarus, and ultimately "resurrect" his own body, so that, galvanized-dematerialized by solar life, it would enliven the body of the Earth and fill it with Sun radiance. The lighting up of the planet that Rudolf Steiner perceived when he contemplated clairvoyantly the event of crucifixion suggests the energy of "light-sped mass" that ignited in the Earth a spark of the *self*-emittive radiance characteristic of Sun and stars, quickening the solar heart and star becoming of the planet. Indeed, "This is my body" was the radical affirmation of the incarnated Logos. The Logoic "I" of the Earth was awakened. The deed of the Christ was to solarize the matter ("energize the mass") of the planet with the "light of the world."

V. A Once and Future Cosmogony

Closer to day to day experience, Einstein's equation grounds in a mathematical formula the conversion of Sun energy into earthly mass that takes place in the magical "appearing" of plant and tree in nature. What surges in pure mathematical nakedness, stripped of his grand soulful glow and X-rayed to the bones of an equation, is none other than the solar god Phanes, the *"phaino-menon"* of creation—the epiphany of a Logos in the veils of cosmos, the darkening and densifying of the light of being into mass. Mass is energy divided by the speed of light squared. $M = E : C^2$.

It must be emphasized that electromagnetic, etheric and Logoic energy, like the polarities they mediate, of energy mass, solar life–earthly matter, spirit body, speak to different levels of one reality: physical, etheric, mindful–cosmic.

In connecting three realities—energy, mass, and electromagnetic wavefield—which are the quasi-physical counterparts of spirit, matter, and consciousness, Einstein's equation stands out as a cosmological equation—a physical formula of the essential threefoldness of reality.

How intriguing that Trinity was the code name for the first detonation of a nuclear weapon, conducted by the United States Army on July 16, 1945, as part of the Manhattan Project. The code name "Trinity" was assigned by Oppenheimer, the director of the Los Alamos laboratory, who later explained that he felt inspired by the poetry of John Donne and its references to the Christian notion of the Trinity—the threefold nature of God. Oppenheimer would write in 1962:

> Why I chose the name is not clear, but I know what thoughts were in my mind. There is a poem of John Donne ["Hymn to God My God, in my Sicknesse"], written just before his death, which I know and love. From it a quotation:
>
>> As West and East
>> In all flat Maps—and I am one—are one,
>> So death doth touch the Resurrection.
>
> That still does not make a Trinity, but in another, better-known devotional poem, Donne opens,
>
>> Batter my heart, three person'd God.[56]

A subtle, poetic awareness of the oneness of our globe beyond the duality of East and West generated by "all flat maps," and of the oneness of resurrection beyond the dichotomy of life and death, seems to have hovered like a prescient inspiration above the nuclear transcending of the dichotomy of mass and energy. Trinity was

---

56 Quoted in Twiss, *Explorations of Global Ethics*, pp. 307–308.

the perfect code name for the revelation aglow in the explosion lit up by scientists in the collective psyche.

The step from a dual to a threefold scientific paradigm heralded by the equation $E=MC^2$ mirrors in reverse the cosmological step of the second day of Genesis. It heralds a reunion of Earth and Heaven (mass and energy) in light and through light. The two speak to the involutionary (world creative) and evolutionary (consciousness creative) version of the same fundamental 1-2-3 articulation of creation: becoming two and three of the One in the process of manifestation, and becoming one (threefold) of the two in the evolving consciousness. Genesis describes the appearing of light out of the separation of Heaven and Earth; Quantum physics reveals that the radiatory electromagnetic wavefield, which is the dense, quasi-physical counterpart of the Logoic light of world consciousness, is the unitive field of mass and energy.

The movement from Two to a triune One also reverses the Eastern cosmological splitting of Shakti "into 2 fields of electromagnetic force, nada and bindu" evoked earlier. The trio of charge–particle–field reflects and lights up with modern understanding the triunity of bindu–nada–shakti.

> The centrifugal, positive male force (1), bindu, is the ground from which nada operates. Nada, the centripetal, negative, female force (2), unfolds the manifest universe. This duality of poles in the substratum of manifested shakti provides the electromagnetic force (3) holding together the molecules of the physical world in a state of vibration.

The physical universe is a three-in-one reality of wavefield-energy-mass. Energy and mass prove to be polarities whose interchangeability points to the two-directional wavefield—centrifugal and centripetal—from which they proceed.

Science has uncovered, in what was thought to be the compact density of matter, a wave field analogous to the breath field of ancient cosmologies, and evocative of the time–space fabric of world beginnings. Reaching to the minute peripheral "ends" of matter, modern physics has touched the metaphysical "beginnings" of the world.

The electromagnetic wavefield is a space–time fabric by virtue of its wave-like periodical character. It represents also a singular pulse of time and space, given its speed signature. Einstein referred to the "time–space continuum [as] the scene of all events in our physical world." Indeed.

In *The Evolution of Physics,* he articulated quite clearly the fundamental shift in conception and focus we have been examining, and saw its challenge:

V. A Once and Future Cosmogony

A new concept appears in physics, the most important since Newton's time: the field. It needed great scientific imagination to realize that it is not the charges nor the particles, but the field in the space between the charges and the particles, which is essential for the description of physical phenomena.

The field concept proves most successful and leads to the formulation of Maxwell's equations describing the structure of the electromagnetic field and governing the electric as well as the optical phenomena. The theory of relativity arises from the field problems... [it] stresses the importance of the field concept in physics, but we have not yet succeeded in formulating a pure field physics. For the present we must still assume the existence of both: field and matter.[57]

Einstein was not satisfied with the probabilistic nature of the quantum mechanics formulas meant to determine the elusive position and momentum of the particle/wave electron.[58] He knew that quantum mechanics gave accurate results, but he could not accept the idea that the only way to characterize particles would be probability. The universe, he felt, could not operate under purely probabilistic laws. There had to be a more fundamental law. "Quantum mechanics is certainly imposing. But an inner voice tells me that it is not yet the real thing. The theory says a lot, but does not really bring us any closer to the secret of the 'old one.' I, at any rate, am convinced that He is not playing at dice."[59]

Schrödinger himself, who came up with the major equation in this regard, was not entirely comfortable with its implications. The wave function associated with his equation only gives the probability of finding the particle at a certain position. He wrote about the probability interpretation of quantum mechanics, "I don't like it, and I'm sorry I ever had anything to do with it." Later in life, he turned away from the mainstream focus on the wave–particle duality to promote the wave idea alone—interestingly.

Einstein's equation, and quantum physics in general, continue to call attention to an immaterial wavefield that transcends and precedes the divide between energy and mass, and, beyond it, the divide between subjectivity and world. After all, while matter is revealing itself as waves, thought is being registered as brainwave activity; subjectivity (from involuntary emotion to deep meditation) is proven to alter the heart rate and, more profoundly, cause epigenetic changes in gene expression—toward

---

57 Einstein and Infeld, *The Evolution of Physics*, pp. 259–260

58 Quantum-scale particles do not have classically measurable properties. Quantum mechanics developed formulas to evaluate the probability that a particle or particles in a particular state will be found to have a given position and momentum. This probability is derived from the wave function describing an electron's wave packet, which is defined as an "envelope of localized wave action that travels as a unit" and is made of a range of waves with phases and amplitudes constructively interfering over a small region of space.

59 Einstein, *Letter to Max Born*, Dec. 4, 1926.

illness in the case of morbid patterns, and toward greater health when engaged in positive heart–mind activity. The resonance between the notions of wave field or space–time matrix and the ancient harmonic cosmology suggests that the laws of harmony might offer a more satisfying grasp of reality than probability equations, as they speak not to the Newtonian mechanistic laws of motion applicable to masses and forces, but to the Pythagorean laws of waves.

A growing number of independent scientists (Milo Wolff, Geoff Haselhurst, Daniel Winter) are validating the view of some of the founders of the quantum theory (Schrödinger, deBroglie, Dirac), and suggesting that we live in a wave-based universe. Matter is simply the focal point of vibrational vortices in a sea of subtle energy that permeates the entire universe. The condensed center of these vortices is what creates the illusion of a separate particle. Milo Wolff's theory concerning the wave structure of matter proposes that matter is the focal point of a standing wave composed of two interfering waves: one inbound wave moving toward the center, and the outbound wave moving away from the center. Time is caused by wave motion. The waves are spherical waves in the fabric of space and the discrete "particle" is an effect created by their point–center.

Laszlo articulates this paradigmatic shift in a way that reaches beyond purely physical-scientific considerations and invites a new consciousness of "reality" and our relation with it:

> We normally think of the things we experience as real, and the space that embeds them as empty and passive, a mere abstraction. We need to turn this around. It is the space that embeds things that is real, and the things that take place in space that are secondary. They are the manifestations of space.[60]

---

60   Laszlo and Dennis, *Dawn of the Akashic Age*.

V. A Once and Future Cosmogony

THE QUANTUM PARADIGM: EMERGENCE OF A LOGARITHMIC ORDER

> *If nature leads us to mathematical forms of great simplicity and beauty... we cannot help thinking that they are "true," that they reveal a genuine feature of nature.... You must have felt this too: The almost frightening simplicity and wholeness of relationships which nature suddenly spreads out before us and for which none of us was in the least prepared.* —WERNER HEISENBERG[61]

The revolution begun by the introduction of the "field" was by no means finished, Einstein wrote in 1949. Around the turn of the century, a second fundamental crisis set in, the seriousness of which was suddenly recognized owing to Max Planck's investigation into heat radiation.[62]

In 1900, a few decades after Maxwell's discovery, Max Planck made the bold hypothesis that was to give birth to quantum mechanics. He postulated that the energy carried by an electromagnetic wave in an oven is emitted in quantized form. In other words, it comes in discontinuous lumps that can only be one time, two times, three times... an elementary unit. He posited that this minimum unit-lump of energy of a wave is proportional to its frequency: higher frequency (shorter wavelength) implies larger minimum energy; lesser frequency (longer wavelength) implies smaller minimum energy. Consequently, the radiation emitted by an oven is limited not only in terms of the maximum half-wavelengths that fit the size of the oven, but in terms of the highest frequency for which the minimal energy lump still falls within the range of the temperature setting. The "lumping factor" limits the radiation phenomenon to the range allowed by a certain rapport between energy and frequency/wavelength. When the minimum energy quantum of a particular wave is bigger than the expected energy contribution, it can't contribute because it cannot actualize its energy emission potential. Planck found that the minimum whole packet of energy emission for an electromagnetic wave is proportional to the frequency of the wave concerned. He established that the elementary unit of energy of a wave equals its frequency multiplied by a constant, referred to as Planck's constant, $h = 6.62606957 \times 10^{-3}$ Joule-seconds (m^2-kg-s^{-2}).[63]

---

61 Conversation with Einstein, quoted in Thiessen, *Bittersweet Destiny*.

62 Einstein, *Autobiographical Notes*, p. 35.

63 In the early 1900s, when physicists sought to calculate the energy carried by the electromagnetic radiation inside an oven using classical thermodynamics combined with Maxwell's electromagnetic theory, they reached the untenable result of "infinite" energy. The success of Planck's hypothesis was based on the combination of two limitations on the

The incompatibility of Planck's proposal with classical physics was a challenge for physicists, including Planck himself, but experimental measurements confirmed its spectacular validity. Planck first considered quantisation as "a purely formal assumption. Actually, he added, I did not think much about it." Subsequently, he tried to grasp the meaning of energy quanta. "My unavailing attempts to somehow reintegrate the action quantum into classical theory extended over several years and caused me much trouble." Other physicists attempted to set Planck's constant to zero in order to align with classical physics, but Planck knew well that this constant had a precise non-zero value.[64]

Indeed, Planck's constant is essentially—and most significantly in our context—a *proportionality* factor between the frequency of a wave and its smallest packet of energy radiation. It is a three parameter constant of gram-square centimeter-second, which involves the same players as Einstein's equation: mass/energy, space and time. It is a threefold constant of proportional packaging of the three—into wholes of action. Planck's constant revealed, at the core of the phenomenon of quantic radiation, the existence of a "quantum of quanta," an irreducible factor of proportionality between minimum quantity of energy radiation and frequency. Planck's constant is referred to as the *unit of action*. It is a primordial bundling principle of space, time (frequency) and energy/mass, which conditions energy radiation. The higher the frequency of a wave, the higher the size of its minimum quantum of energy radiation, hence the higher the energy required to supply the vibration.

In 1905, following Planck's intuition, Einstein postulated that "light, or more generally all electromagnetic radiation, can be divided into a finite number of energy quanta that are localized points in space, move without dividing, and can be absorbed or generated only as a whole." He showed that Planck's guess actually reflects a fundamental feature of electromagnetic waves: "they are composed of particles—photons—that are little bundles, or quanta, of light." Following Planck's

---

number of waves involved in an oven set to a certain temperature. First, Maxwell's theory, applied to oven radiation, showed that only waves whose half-wavelengths fit exactly the length of the oven could be part of the radiation. Despite this limitation, however, the number of energy producing waves remained infinite, since any minute whole fraction of the oven width was included. It was Planck's genius to conceive the possibility of a similar limitation in terms of energy. He hypothesized that only the wavelengths whose frequencies carried energy within the range of the energy/temperature set for the oven could be involved. He postulated that energy can only come in multiples of a minimal energy carried by a particular wavelength. This smallest unit-base of energy is proportional to the frequency. A higher frequency implies a larger minimal energy, which—and this is the critical point—may exceed the appropriate temperature setup. A whole range of high frequency was thereby eliminated, solving the puzzling question of "infinite" energy. The experimental results proved Planck right.

64 Max Planck entry on Wikipedia.org.

V. A Once and Future Cosmogony

lead, Einstein also proposed that the energy of each photon, which represents the smallest possible "packet" of energy in an electromagnetic wave, is proportional to the frequency of the light wave (f) according to the following equation: $E = hf$[65]— Energy equals Planck constant multiplied by frequency.

In 1924, Louis de Broglie proposed that all matter, not just light, has a wave-like nature, and that particles can exhibit wave characteristics. Generalizing the Planck–Einstein relation, he postulated that Planck's constant represents the proportionality between the momentum and the quantum wavelength of any particle, not just the photon. In 1926, Erwin Schrödinger used this idea to develop a mathematical model of the atom that described the electrons as three-dimensional waveforms rather than point particles.

Planck's revolutionary and disconcerting proposition rings a clear bell in our "logarithmically-minded" cosmological context. To say that electromagnetic energy comes in discontinuous lumps, which can only be multiples of an elementary unit, is to say that energy emission in a contained space is logarithmic. The parallel with the emission of harmonic "lumps" of sound energy, released by a field of harmonically embedded waves, is quite clear. As Capra pointed out, "the electromagnetic spectrum is a logarithmic scale of frequencies."[66]

Just as, in the realm of sound, only waves with wavelengths that are whole number fractions of the fundamental can enter the emission of a musical note, the only waves that radiate heat in an oven are those (and whole fractions of them) whose half wavelengths fit between the walls of the oven (thus generating standing waves), and whose frequencies (times h) result in a minimum unit of energy delivery that fits the range of the temperature setting.

According to Einstein, energy emits "wholes." Energy radiates in wholes or quanta, the size of which is determined by the frequency of the vibratory field "times" Planck's constant. The "wholeness" formula of such wholes, Planck's "constant" is a "co-standing" *trio* of energy/mass, space, and time combination common to all quanta of energy emission, whatever their magnitude. It defines what characterizes the energy emission process itself: the principle of quantic-ness of "action." Planck's constant ("h") is the unit of energy delivery, the quantum of "action."

Arthur Young emphasized that it is not energy, but action that comes in wholes. "It was Planck's epoch making discovery that action comes in wholes. Action is constant, energy is proportional to frequency."[67] Action cannot take on an arbitrary value.

---

65 Greene, *The Elegant Universe*, p. 97.
66 Capra, *The Tao of Physics*, p. 60.
67 Young, *The Reflexive Universe*, p. 19.

Instead, it must be some multiple of this quantum. The quantum of action evokes the life pulse and ontological act of existing at work, we saw, in logarithmic Phi.

✪ The quantum of action combines three dimensions of measurement (length, mass, time) into a whole of relational constancy. This "whole" points to an order that transcends the physical measurements that define it—a non-physical order of *proportionality*. What a constant proportion of dimensions does, which transcends measurements, is to create a keystone of *relational* constancy between these dimensions. Such a constant is an organizing principle for the three-dimensional reality concerned. Planck's constant reveals the existence of an irreducible principle of proportionality (between energy/mass and frequency) at the core of all quanta of energy/mass particles—at the core of matter. It speaks to a "formatting" principle of energy quanta, which transcends physical measurement. H = E/f.

The quantum of action is not a quantitative unit but a unit of pulsation; it is the fundamental bundling formula of space–time–energy/mass that defines the phenomenon of energy/mass emission—the phenomenon of creation. What meets the physicist in the quantitative depths of matter is *not an ultimate quantum of quantity, but a "quantification" principle* and its formula. The quantitative values given to Planck's constant (ex: 1.05...), based as they are on human-created units of measurement such as gram, centimeter, second, are only relative expressions of an invariant principle and qualitative fact—proportionality. Only the proportionality principle captured by Planck's constant is an absolute of our three-dimensional, physical world. ✪

Threefold units such as Einstein's equation and Planck's constant herald a new triune paradigm. What Planck encountered in the space–time-mass/energy constant of electromagnetic pulsation is an indivisible triune whole—an indivisible trinity. This constant is now regarded as the number that rules *techno-logy*. It represents indeed the most objective and *"technical"* expression of the ubiquitous constancy of the *Logos* (literally, "proportion") in everything that exists. Taking what may seem like a leap, though it simply harnesses the noble logic of our context of nested dimensions, one can say that the quantum of action is the quantum of love-in-action, the unit-pulsation of the Logos in time–space (quasi-) physicality. One might even venture to name what Planck's constant captures at the source of every photon the "Logon" of time–space–energy. Whereas quanta are whole wave packets or units, whose energy is proportional to their frequency, the Logon is the principle of proportionality common to all quanta, the archetypal "quantum" of action or manifestation at the heart of form making.

V. A Once and Future Cosmogony

This archetypal quantum of "action" takes us to the beginnings of cosmogenesis. The emission of photons (with no charge nor mass, only frequency/wavelength), followed by the formation of nuclear particles (electrons and protons endowed with mass and charge), then atoms and molecules, take us to the primary, as of yet immaterial stage of world beginnings. Having gone, we thought, further and further away from the living source of the world, we find ourselves shockingly close to the creational activity of a primeval intelligent source encrypting itself into time–space photons of light. In a toroidal fashion, the movement to the peripheral ends of matter has also circled us back to its beginning.

Arthur Young illumines what is indeed the heart of the story:

> The heart of our story, like the beginning of creation, lies in the nature of light. For the physicist, light is unique in that, unlike everything else that exists in actuality, it has no mass (no rest mass). It has no charge and...it has no time.[68]

As the soul light of the world is the Logoic trinity intrinsic to all beings, "physical" light is the essential threefoldness of all *forms* of being. What Planck's quantum of action gives us to apprehend is the pulse of "the light of the world."

Moving deeper into the heart of the story, Young muses:

> When understood, quantum physics can provide the compass that in former times required revealed religion: it can tell us of first cause.... Genesis tells us that first cause is fiat: Let there be light! What is "fiat"? It is action (decision). So when Planck discovers that light is a quantum of action, he is discovering "fiat," the light that was created on the first day.... Quantum physics has discovered in the quantum of action *[the] innate, intrinsic activity* behind [the] evolution of the entities of nature into higher and higher orders of organization—from particles to atoms, to molecules, to cells, organisms and animals.[69]

The quantum of action is the physical formula of the dynamic essence of all forms—their Fiat.

Like master keys into the central chamber of the universe, Planck's and Einstein's formulas invite us beyond the dual worlds of mass/charge, matter/life to the mediating event–advent of "light" that ushers Life into the lightspeed–mass–energy trios of world particles and forms.

Young's luminous insight orchestrates the invitation, implied in a universe of nested dimensions, to extrapolate Planck's discovery to the psychospiritual realm. In a fractal universe, quantum physics would "provide a compass" for the spiritual scientific truths of our being. We may not need it to understand that greater radiance in a person intimates a higher frequency of psychospiritual energy. Halos and auras around sages and

---

68 Young, *The Reflexive Universe*, p. 10.
69 Ibid., pp. 26, 29.

saints point to the subtle radiation emitted by a fully harmonized, *hence* spiritualized being. However, Planck's insights illumine the fact that the wholeness of soul radiance characteristic of accomplished human beings intimates a fundamental proportionality of intelligence, love, and strength—a constancy of rapport among *soul parameters* analogous to Planck's constant in the realm of physics. Such triune integrality of being is indeed the ground necessary for the light of the monad (lit. the *"one"*) to begin to *vibrate* in the human individuality—reversing the cosmological step by which all forms of cosmos emerge out of the primordial trinity born of the One. Reflecting the implications of Planck's oven experiment, the triune constant of spiritual being, which, for most of humankind, is yet to emerge as a constant, is the minimum quantum of spiritual energy radiation required to supply the high frequency vibration of monadic life. The monad cannot vibrate in a form conditioned by lesser frequencies. Whereas the quasi-physical quantum of action is god given, the quantum of "spirit body" has to be accrued from within. As primordial spirit light tended the emergence of the mineral, then the plant kingdom, humankind tends the emergence from within itself of a "kingdom of soul" by tending this triune "fiat" of being. The emission of photons of light at the beginning of world creation has its evolutionary counterpart in the emission of the triune life of love at the beginning of a kingdom of souls. God created light, humankind generates love. What quantum physics suggests is that such emission depends on a minimum quality (a monadic "quantum") of soulful intensity at the heart of one's "actions."

Teilhard de Chardin, in whom this monadic constant had clearly begun to emit its radiance, reflected on the emergence of a higher quantum of action within human activity, a capacity "to unite directly with the divine center through one's action, whatever its form...so that every activity, if I dare say, *"amorizes itself."* Every reality, he suggests, offers a way to grow near the divine through the "line of centers":

> Like multiple nuances blend in nature to give a single white light, the infinite modalities of Action, without being muddled, blend into a unique tonality under the powerful influence of the universal Christ—love as a superior form, universal and synthesizing, of spiritual energy, in which all the other energies of the soul become transformed and sublimated.[70]

Rendered more vivid by the findings of quantum physics, Teilhard's evocative prescience impels us to realize how every bit of human action, when imbued with the discreet quantum of soul unity of action—wisdom, love and strength—is entrained by the heartbeat of the whole, thus charging the human being's Logoic potential, and imbuing heterogeneous activity with a unitive, Logoic "constant."

The Islamist notion of "act of Being" is an apt designation for this supreme constant of humankind, since at this essential level, the quantum of action is

---

70   De Chardin, *Science and Christ*, p. 210.

V. A Once and Future Cosmogony

precisely a quantum of "being" and no longer a quantum of "activity." The act of being is the "only effective reality," says the Iranian philosopher Molla Sadra Shirazi, who contemplates its gradual intensification through the kingdoms as the inner side of the evolutionary progression of "being" through forms. "Nothing is real but the act of being. Man configures progressively his act of being.... God is pure act of being."[71]

Along a parallel line, Teilhard pondered the "gradual increase in psychic temperature" he viewed as characteristic of the evolutionary process.

> The irresistible "Vortex" which spins the whole Stuff of things into ever more comprehensive and more astronomically complicated nuclei...this structural torsion results, under the influence of interiorization, in an increase of consciousness, or a rise in psychic temperature in the core of the corpuscles that are successively produced.[72]

The threefoldness of the quantum field brings us, imperceptibly yet inexorably, to the rim of a universal core of wholeness and One-ness. Not a zero, but a universal unit or "One." The quantum of action, in which Young sees the fiat of the beginning, is indeed the core-formula of space–time–energy/mass proportion that represents the physical expression of the quantum act of being—the precise set of parameters that characterizes the externalization of Being. Young carefully points physics' search for a unified grasp of reality in the direction of this quantum of action.

※

Planck's discovery led him to define a set of universal units—of mass, length and time. "When you get down to the Planck length and Planck time and try to partition space and time more finely," explains Brian Greene, "you find you can't."[73] Every unit captures a primordial "wholeness," which is a primordial proportion of physical parameters, a fundamental rapport between defining dimensions of physics, reminiscent of the ontological wholeness of the triune rapport (of same, other, and Being) that defines the world soul.

The inherent "granularity" of mass/energy, as it has been evocatively described, is counterintuitive to our everyday perceptual assumption that it is possible to "make things a little bit hotter" or "move things a little bit faster." However, it perfectly mirrors the scientific fact that our perception of increase/decrease, in weight for instance, is limited to logarithmic increments or "lumps" of heaviness or lightness—rather than a continuum of "little bit heavier or lighter."

---

71 Shirazi, in Jambet, *Se rendre immortel et traite de la resurrection*, pp. 143, 40, 95, 28.
72 De Chardin, *The Heart of Matter* p. 33.
73 Greene, *The Fabric of Cosmos*, p. 350.

Remarkably, Planck's "bold guess" zeroed in on the logarithmic principle embedded in harmonic wave interferences: the *addition* of discrete bundles of self-*multiplication* of a base-unit of energy. By trying to resolve the question of the quantity of energy radiating in an oven, he discovered how energy delivers itself in quanta or bits of space–time dynamics conditioned by vibratory rate, thus further unveiling the laws of "the field."

Given that Planck was a gifted musician, who not only played the piano, organ and cello, but also composed songs and operas, it would be natural to think that the principles of harmony might have loomed in his mind in association with Maxwell's "waves." Testimonies, however, and Planck's own comments, negate the possibility of any conscious connection between his hypothesis and the laws of harmony. "Planck," Brian Greene remarks, "had no justification for his pivotal introduction of lumpy energy. Beyond the fact that it worked, neither he nor anyone else could give a compelling reason for why it should be true."[74]

The impression remains, like a visual harmonic in the background of Planck's pondering by his oven the *quantum law* of radiating energy, of Pythagoras' silhouette standing by the blacksmith's forge, captivated by the quanta of consonance that were to lead him to the laws of harmony. How fascinating that Planck and Einstein, who both pondered the physics of the field and began to unravel its laws, played music together! Einstein confided,

> The longing to behold...preexisting harmony drove both Planck and myself. Glimpsing such beauty turns on a peculiar form of devotion.... If I were not a physicist, I would probably be a musician. I often think in music. I live my daydreams in music. I see my life in terms of music.[75]

Watching Planck place his finger so aptly, albeit unwittingly, on the pulsating character of energy, and figure out the unit-base of pulsation—the unit-base of space–time–energy—lends a special weight and luminosity to what he wrote in 1944:

> As a man who has devoted his whole life to the most clear headed science, to the study of matter, I can tell you as a result of my research about atoms this much: There is no matter as such. All matter originates and exists only by virtue of a force that brings the particle of an atom to vibration and holds this most minute solar system of the atom together. We must assume behind this force the existence of a conscious and intelligent mind. This mind is the matrix of all matter.[76]

---

74 Greene, *The Elegant Universe*, p. 94.

75 Levenson, *Einstein in Berlin*, 2003. Following is the full quote: "The state of mind which enables a man to do work of this kind is akin to that of the religious worshipper or the lover; the daily effort does not originate from a deliberate intention or program, but straight from the heart."

76 Planck, *Das Wesen der Materie* [The Nature of Matter], 1944 speech in Florence, Italy (Archiv zur Geschichte der Max-Planck-Gesellschaft, Abt. Va, Rep. 11 Planck, no. 1797).

V. A Once and Future Cosmogony

We begin to see how the major discoveries of the twentieth century converge with the ancient cosmological view to acknowledge a waveform order underlying the universe at all scales of manifestation. The ground of the universe is not unlike the breath of a Being, whose Shiva–Shakti like polarity manifests on the physical plane as the electromagnetic wavefield.

In showing how matter has wave-like features, and electromagnetic waves particle or bindu-like eventualities, physics has begun to orchestrate the perennial wisdom with a lucid and sophisticated analysis of physical facts. Energy is not continuous as was assumed, nor is matter made of solid, isolated particles. Light comes in discrete units (photons), so does sound (tones and harmonics), and so does mass-energy (quarks, electrons, atoms). Matter is the "other" of energy, their equivalence mediated by the speed of light.

It is interesting to place the apparently revolutionary character of quantum physics in the light of the ancients' insights into the cosmological secrets of the "oven" of creation. We saw earlier how, in addition to limiting themselves to integers, or natural numbers, the ancients consistently emphasized the smallest range of integers that could express the tonal system of the octave.[77] They gave great value to its simplest wholistic expression—most of all, to the smallest whole-number frequency version of the chromatic or diatonic whole that the scale represents. The intent at work in such number systems is to express in simplest "integers"—that is, integral "units"—the *wholistic* differentiation (chromatism) of a *whole* (octave). Plato himself described the Demiurge as creating "a whole composed out of wholes."[78] What is remarkable is that the numbers that appear again and again in the narratives of world Scriptures, carrying the secrets of cosmological cycles, are these very paragons of wholistic differentiation. This intriguing fact loses its mystifying character when viewed in the light of the noble logic of a harmonic cosmology, as such numbers epitomize the lawful, harmonic periodicity engaged by the finite time and space "whole-and-parts" of creation.

For instance, as McClain uncovered, the smallest array of whole number-frequencies for the diatonic scale produces a scale bound by the numbers that dominate Hindu cosmology.[79]

---

77 See earlier section, "Modern insights into the harmonic structure of Platonic solids and atomic elements."

78 Plato, *Timaeus*, 33a.

79 McClain, *The Myth of Invariance*, pp. 61, 99. McClain points out that Hindu cosmology knows 4 yugas or "ages of the world"—all simple multiples of 432: Kali Yuga = 1 = 432,000 years; Dvapara Yuga = 2 = 864 000; Treta Yuga = 3 = 1,296,000; Krta Yuga = 4 = 1,728,000;

432	486	512	576	648	729	768	864
D	E	F	G	A	B	C	D

McClain reflected on the surprising way in which music and astronomy find themselves unified in the numbers of the Rig Veda. Joseph Campbell referred to the number 432 as the most important mythological number in history. 432 Hz is also increasingly advanced as the legitimate basis of just tuning, rather than the conventional 440. Music lovers have noted that music tuned in A = 432 Hz is "not only more beautiful and harmonious to the ears, but it also induces a more inward experience that is felt inside the body, in spine and heart. Music tuned to A = 440 Hz offers a more outward and mental experience, and is felt at the side of the head and then projected outward."[80] Steiner noted, "Music based on C = 128hz will support humanity on its way toward spiritual freedom. The human inner ear is built on C = 128 Hz."[81] A = 432 Hz corresponds to a tuning of C = 256 Hz, which is one octave higher than 128.

In essence, what whole number scales do, in this harmonic cosmological context, is "quantify" the logic of the world as diverse-and-universe—diversity in wholeness. They articulate its ontological blend of indivisibility and divisibility in terms of partial wholes or harmonic quanta—which are discreet cosmological quanta. Such *wholistically particulate* systems as whole number scales speak to the very quantum logic which has been challenging the modern intellect simply because we are no longer—or not yet—reasoning in a context of "units" of wholeness and oneness. Clearly, the quantum order of harmonic interferences harbors a promising ground of convergence between the once and upcoming understanding of the universe.

QUANTUM ORDER OF MATTER: QUANTUM DYNAMICS OF EVOLUTION

Quantum is the Latin word for "amount." In its modern usage, introduced by Planck in 1900, it designates the smallest possible unit of any physical reality, such as energy or matter. Brian Greene defines quantum as "the smallest physical unit into which something can be partitioned."[82] For example, photons, the quanta of light, are the smallest bundles of electromagnetic radiation. Similarly, waves can

---

Total: Maha Yuga = 10 = 4,320,000. The number 4,320,000 also represents the number of syllables in the Rg Veda. Also, 432 x 60 = 25,920, the precessional number.

80   See http://www.collective-evolution.com/2013/12/21/heres-why-you-should-convert-your-music-to-432hz.

81   Quoted in Maria Renold, *Intervals, Scales, Tones, and the Concert Pitch c = 128 Hz*. She claims conclusive evidence that A=440Hz tuning "disassociates the connection of consciousness to the body and creates antisocial conditions in humanity."

82   Greene, *The Elegant Universe*, p. 420.

only be partitioned harmonically into sub waves with half-wavelengths in whole number fractions of the string or space involved. The quantum distribution of energy in closed systems (whether a string or an atom) points to the phenomenon of constructive wave interference and to the wave nature of "matter."

The size of the energy lumps is typically very small. This is why, Brian Greene explains, "it seems to us, for example, that we can cause the energy of a wave on a violin string—and hence the volume of sound it produces—to change continuously. In reality, though, the energy of the wave passes through discrete steps, a la Planck, but the size of the steps is so small that the discrete jumps from one volume to another appear to be smooth."[83]

The notion of "quanta" introduced a disorienting new concept, which challenged the most basic assumptions of physics and pressed to re-envision the nature of the universe and its laws. The unusual features of the deeper ground of "matter" were puzzling to the classical understanding of the physical world and they eluded its measurements.

On the other hand, the quantic character of "matter" resonates in a most heartening way with the ancient harmonic cosmology. The world of harmony is a quantum world. The condition of discontinuity and "quantized restraints" is the logical manifestation of the logarithmic bundling we have come to recognize as the universal principle of self-replication of the "unit core" of the universe into a multiplicity of self-similar forms.

In 1913, physicist Niels Bohr proposed that the electrons of an atom were confined into clearly defined, quantized orbits. They could jump between these, but could not freely spiral inward or outward in intermediate states. Furthermore, the electrons were found to exist as standing waves around the nucleus—which led to the determination that "the lowest possible energy an electron can take is analogous to the fundamental frequency of a wave on a string. Higher energy states are then similar to harmonics of the fundamental frequency."

Analogies with harmonic laws were beginning to emerge. In his book *The Evolution of Physics*, Einstein ventured a comparison between quantum physics and "a classical example having nothing to do with modern physics."[84] This "classical phenomenon" Einstein was referring to was the phenomenon of standing waves. "It has often happened in physics," he wrote, "that an essential advance was achieved by carrying out a consistent analogy between apparently unrelated phenomena." Having showed how the possible wavelengths of standing waves between two fixed points can only be whole fractions of the distance between the two ends of the oscillating cord, he reasoned:

> The wavelength can only change discontinuously. Here, in this most classical problem, we recognize the familiar features of the quantum theory.... Strange

---

83 Ibid., p. 93.
84 Einstein and Infeld, *The Evolution of Physics*, pp. 286–291.

as this analogy may seem, let us draw further conclusions from it and try to proceed with the comparison, once having chosen it. The atoms of every element are composed of elementary particles, the heavier constituting the nucleus, the lighter the electrons. Such a system of particles behaves like a small acoustical instrument in which standing waves are produced...we have thus discovered some similarity between the oscillating cord and the atom emitting radiation. In the same way that a standing wave produced by a violin player is a "mixture" of several wavelengths pertaining to a defined spectrum, an element emitting radiation admits a limited spectrum of wavelengths.

If Planck did not make the link between his findings and the laws of harmonic standing waves, Einstein did. Remarkably, his brief incursion into harmonic laws ended on the following open-ended note:

Fundamental ideas play the most essential role in forming a physical theory. Books on physics are full of complicated mathematical formulae. But thought and ideas, not formulae, are the beginning of every physical theory.

Einstein did "intone" such a fundamental idea. It hovered, and still does, discreet and unobtrusive like a poetic thought, yet a fully formed insight, waiting for the proper ground to land: "Human beings, vegetables, or cosmic dust, we all dance to a mysterious tune intoned in the distance by an invisible piper."

Quantum physics also shed new light on phenomena for which there had been no prior theoretical framework. It pointed, for instance, to an ordering principle behind what is commonly observed as sudden changes in states and forms of matter. Scientists have coined concepts such as bifurcation points, quantum jumps or leaps, critical thresholds, phase transitions, stages of "punctuated equilibrium," emergent properties, to describe the sudden radical shifts in states or shapes in the continuum of a system—be it a living system engaged in a process of growth (a plant for example), or a substance reacting to outside influences. Water offers the most obvious example of such changes of state as it passes through critical thresholds induced by outside temperature—from ice to liquid to gaseous steam. Between the critical thresholds of 0° and 100° Celsius, not much happens. Biologists have noted than an increase in organizational complexity accompanies the leap of a fundamental "unit" into a next stage or state, be this unit a type of substance or an organism. While "water, when subjected to increasing temperature, goes from shapeless liquid to a highly organized bubbling convection cells... biological systems have a hierarchy of organizational levels, and on each new level, new behavior emerges."[85] These observations have led to a whole new perspective on evolution.

---

85 Trinh and Ricard, *The Quantum and the Lotus*, p. 154.

V. A Once and Future Cosmogony

Many biologists now believe that evolution happened in a similar way, progressing from one bifurcation to the next, becoming increasingly organized in sudden jumps, thereby ascending the scale of complexity from the inanimate to the animate.... Living species remain unchanged for a long time, but then undergo radical changes in a relatively short period. Like the "quantum leaps" of atomic physics, evolution proceeds in "evolutionary leaps."[86]

The quantum perspective suggests that what our senses perceive as distinct, separate aspects of nature or components of cosmos may be phases in an overarching unfolding process. The new quantum ordering of reality supports and potentially illumines what Eastern wisdom posits to be the evolutionary process—from the inorganic to the organic, and, within the organic, from plant to animal and animal to man, and also from one phase of a plant to the next. "In India, John Woodroffe noted, it has always been held that there are no partitions or gulfs between the various forms of existence, and that, for instance, the difference between man and animal is not a difference of kind but of degree."[87]

Teilhard De Chardin's scientific expertise and mystical intuition led him to a similar understanding of the evolutionary process as a successive emergence of increasingly complex forms of organization. His view of evolution, which began to take form around the same period of the early twentieth century, captures the history of the world and its resulting constitution as a series of quantum steps into successive aggregations-formations of increasingly complex spheres: from matter organizing itself into a *cosmosphere* of atoms and electrons, elements and minerals, stars and galaxies, to the emergence of living organisms and the formation of the *biosphere*, to the birth of reflection and thought initiating the formation of a *noosphere*—a mindsphere that has, since Teilhard's time, become a tangible reality, quasi-corporealized into "the worldwide web." This is how he describes the "birth of reflection":

Between the animal world of the Pliocene period and the human world which succeeds it, there is a change of order or of state...it is undeniably a new envelope appearing on the ancient biosphere....Something intervened in the general process of the vitalization of matter: something so subtle that at the very beginning its coming produced no apparent stir, and yet something so violently active, fundamentally, that after several hundreds of thousands of years the face of the Earth has been completely transformed by it. What then, was that something, if not the birth of reflection....Thought suddenly bursts in, to dominate and transform everything on the surface of the Earth....Reflection does more than generate ratiocinative reason: it recasts and transforms animal psychism in its entirety.[88]

---

86 Ibid., pp. 168–169.
87 Woodroffe, *The Garland of Letters*, pp. 292–294.
88 Teilhard de Chardin, *Activation of Energy*, pp. 323–325.

The next step in the universe's growth, Teilhard envisioned as a sphere of fuller, more integrated and authentic humanity, which he named the Christosphere—the "soul" stage of the human kingdom. Teilhard's lifelong passion to bridge science and religion led to his unique grasp of a unitary evolution of Earth and consciousness, nature and humankind, world and self. Behind the apparent duality of inner aspiration and outer evolution reflected in the parted streams of religion and science, he saw the two facets of One being (cosmic Christ and Omega-point) engaged in a threefold dynamic spiral of historical stages and compounding states of involvement–emergence in physical, *bio-logical, psyche-logical,* and ultimately psychospiritual *self-actualization* and self-realization.

The gaps we perceive between the kingdoms of nature—"cosmosphere" of the mineral kingdom, biosphere of the plant and animal kingdom, noosphere of the self-aware, human kingdom—are partitions between quantum states of the universe's being—between major harmonic intervals of the fundamental octave of the Logoic tone. These gaps, which are replicated within each kingdom, are analogous to the areas of chaotic noise, dissonant sound and absence of tone that are exposed between major intervals (notes) when one slides a finger on a string in a continuous way and plucks the segmented string with another finger. In the plant kingdom, the passage from root to leaf to calyx to flower to seed, in which Goethe perceived the unfolding of *one* archetypal organ (the leaf), illustrates this discontinuous quantum harmonic order of unfoldment from one formation to the next. Returning to Einstein's analogy, one could say that the various aspects of the plant express the consonant conjugations of its wave components. The leap of the plant across its consonant spectrum, as it gathers energy from the Sun, is reminiscent of the description of electrons leaping from one orbit of an atom to the next in correlation with a gain or loss of energy, an absorption or emission of electromagnetic radiation. In the human kingdom, itself a quantum of the world being, chakras represent quanta of human being-ness, *whole* bundles of psychospiritual consciousness associated with physiological functions and glands–organs.

When a certain entity—be it a world being, a plant seed or a human being—completes a partial harmonic expression of itself, a critical threshold is reached and radical change looms—as radical as the appearance of a flower on a stem, yet as relatively continuous as well. Bundled in a completed form, having fulfilled a state, the system leaps into a next harmonic key and ground of becoming, a further stage of integrative unfoldment, in which increased complexity of organization allows for more of the "self" of the system to reveal itself. Similarly, it is an established fact that our brains emit distinct ranges of vibratory frequencies, which correspond to graded *states* of consciousness.

V. A Once and Future Cosmogony

As relative as they are real, the partial expressions of "being" at work in the graded steps of the forms and kingdoms of nature, represent distinct "keyings" of matter to the major consonances of the fundamental octave, which is the wholeness of its ideal form.

Spin, whirling holomovement, and double helix

> *The resemblance of the modern views to those of Plato and the Pythagoreans can be carried somewhat further.... In modern quantum theory there can be no doubt that the elementary particles will finally also be mathematical forms but of a much more complicated nature. The Greek philosophers thought of static forms and found them in the regular solids. Modern science, however, has from its beginning in the sixteenth and seventeenth centuries started from the dynamic problem. The constant element in physics since Newton is not a configuration or a geometrical form, but a dynamic law.... The mathematical forms that represent the elementary particles will be solutions of some eternal law of motion for matter. This is a problem which has not yet been solved.*—WERNER HEISENBERG[89]

Another discovery brought yet another convergence between quantum physics and the cosmological view we have been contemplating: the notion of spin. As significant as quantum radiation was the discovery that all particles of matter spin. In 1925, a group of Dutch physicists found that, "somewhat like the Earth, electrons both revolve and rotate. Every electron in the universe spins at one fixed and never changing rate." Spin is an intrinsic property of the electron, much like its mass or its electric charge. Physicists would subsequently demonstrate that all particles of matter have spin equal to that of the electron.[90]

In the context of the logarithmic model emerging with quantum physics—of standing waves involving a limited spectrum of "harmonic" wavelengths—this discovery of universal spinning extends to the realm of subatomic particles the fundamental spinning-spiraling dynamics of nature and cosmos. It intimates that the subatomic world and its minute formations participate of the same spiraling dynamics as plant growth and planetary motion. Most significantly, it supports the view that particles may be the apical points of spinning vortices, whose converging sub waves co-spin to a point of concentrated stillness and quasi-mass.

---

89 *Physics and Philosophy*, "Quantum Theory and the Roots of Atomic Science," pp. 71–72.
90 Greene, *The Elegant Universe*, p. 171.

While modern science was watching solid matter dissolve into a frenzy of spinning and revolving particles, it was also beginning to fathom in the huge spiraling formations of cosmos "imprints of the sound waves that formed the universe." It was starting to regard stars and galaxies as "spiral blast patterns, residual imprints of standing shock waves from the thundering voice of the universe." Sound, standing waves patterns, and spirals spell out the essential components of the universe according to the ancient harmonic cosmology. In Lawlor's words,

> Modern astrophysics views the universe as a vast vibrating field of ionized, pre-gaseous plasma, an image evocative of Nun, the cosmic ocean of Egyptian mythology, or the Prakriti of Hindu cosmology. Within this field, gravitational influences are triggered, which cause a warp and densification into nodal patterns—galactic mass-centers that release compound ripples, causing violent changes in the cosmic plasma referred to as galactic "sonic-booms." These whirling sonic shocks create a spin in the entire galactic cloud, and within the inner regions set up by this spin, stars are born. This clearly restates the ancient image of universal creation through sound waves or the "Word of God."[91]

Whirling sound waves, gravitation, stars. The similarity between the macrocosmic picture and the microscopic formation of particles is clearly striking. Contemporary insights into macrocosmic and atomic worlds show fundamental similarities with the perennial view: Breath or Big Bang of a primordial vibration, wave interferences leading to the formation of standing waves, whose combined holomovement results in the coalescing of atomic particles and later celestial bodies. Particles and stars appear at the spinning end of vortices of embedded standing waves—still eyes of a vortex "a-*massed*" by the whirling energy. This focal point of the vortex-pattern seems to be precisely "the point of maximum intensity" where the probability equation of quantum mechanics suggests that the particle is likely to be found. Based on this equation, the probability of finding a quantum particle at any point is "proportional to the *energy density* of the field at that point."

With the notion that electrons spin and "revolve" around the nucleus, forming standing waves like "orbitals" around it, the abstract logarithmic principle at the core of the quantum phenomenon deploys and reveals its internal gesture and rhythmic-creative power. The grand logo-rhythmic dynamics that runs through the forms of nature and cosmos shows itself at work in the depth beginnings of matter's "particulation." Hinted at in the spin and revolvings of the quantum world is the spiral cyclic holomovement of life. The recognition of this fundamental motion

---

91   Lawlor, *Sacred Geometry*, p. 23.

quickens with life the equations that describe its mathematical features but choke its soul and freeze its breath.

※

Hindu wisdom speaks of the essence of the holomovement of creation as the breath of Brahma and the love interplay of Shiva and Shakti. We saw how this breath rests, "still live," in physical structures, and how it begins to stir and move in plant life and biological growth. The same breath is now intimating its presence, beyond the apparent stillness of matter, in the minute undulating-spiraling (logo-) rhythmics of the electromagnetic and quantum field of particles physics. The interplay of two energy currents is the common ground of ancient cosmologies. In Taoism, Yin and Yang move simultaneously from within to without (involutionary motion) and from without to within (evolutionary pull) in perfect balance. Rather than opposite principles associated with polar qualities such as heat and cold, generativity and receptivity, yang and yin are complementary tendencies—such as cooling down and warming up—in the flux and character of energy.

What the Hindu notion of the breath of Brahma emphasizes, which is also implied in Platonic cosmology, but recedes from view to a somewhat dried out abstraction in modern science, is the fundamental bidirectional dynamics involved in spinning and standing waves. The standing waves into which the vibrational energy of creation weaves itself before turning into particles and stars, result from the constructive, harmonic interferences between the outgoing and incoming breath of the universe replicated in the infinite multiplicity of centrifugal and centripetal currents that make up the energy field.

This awareness is implicitly present in the science of electricity and magnetism, but with a significant shift in focus from bidirectional currents to bipolar "charges," from moving currents to opposite "poles." The wave motion exists only as an abstract, almost virtual reality in the background of "opposite" poles, points, and charges. This subtle shift from a wholistic look at a dynamic phenomenon to an analytical grasp of it as a set of mechanistic factors is a determinant one. It highlights the paradigmatic gap between the ground of modern science and the roots of ancient cosmologies. Concrete factors blot out dynamic processes. Instead of opposite principles associated with polar qualities such as positive and negative charges, heat and cold, electricity and magnetism, what ancient cosmologies emphasize are complementary movements within a (relatively) finite whole. "Space is an entity," they teach. The notion that space with its wave-like properties is an open, but finite universe–entity nested in vaster entities, is the necessary premise to conceive the bidirectional holomovement of "creation" behind its electromagnetic charges and polarized phenomena.

The only contexts, in our time, where the awareness of the dual movement of life energy is regarded as essential are healing modalities and areas of spiritual scientific research. Phenomenologists of nature such as Doczi and Schwenck have shown how dual spiraling currents underlie organic formations throughout nature. Doczi coined the term *dinergy* to designate this fundamental dynamics in the delineation of forms. Dinergy refers to the energetic duality of spiraling currents moving in opposite directions—centrifugal and centripetal, clockwise and counterclockwise—and generating nodal points of harmonic interference, which define the forms of nature.

The primordial manifestation of this dual current is the ontological ground of the universe referred to, in Hindu wisdom, as Parabrahman and Mulaprakriti, Father and Mother, and in Plato's cosmology as same and other, spirit and matter—1 and 2. This primordial pair generates the vibrant Word of our world being—the triune Logos.

The next and secondary level of dinergy is the dual current of the breath of Brahma exhaling itself as cosmos and inhaling itself as self—in a dual operation of corporealization and ensouling, body making and soul becoming. Everywhere in nature, expanding vortices combine with contracting vortices, actualizing in living systems the triune bonding of spirit and matter.

Randolph Stone's "polarity therapy" is founded on this notion of an underlying dual field. Stone spoke of "the polarized field of attraction and repulsion at work in atomic structures as well as magnetic relationships as the underlying reality that determines all physical phenomena including health." He chose the term *polarity* to designate the pulsating movement of energy within all forms: a movement of pushing outward from a neutral source through an expansive, radiant, electric, positive phase; then pulling back to the source in a releasing, magnetic, negative, contracting phase.[92]

If contemporary physics ponders the spinning of particles, the orbiting of standing-wave electrons around the atomic nucleus, and the spiral blasts of galaxies, it does not seem to inquire into the source of these movements: the dynamics of centripetality and centrifugality. These deep motions of a live universe, frozen solid under the Medusan eye of mechanistic science, emit their ghost-like phosphorescent glow in abstract notions such as electromagnetic charge and current, gravitational and antigravitational force, elusive particle or spinning property. Yet the bidirectional momentum of the field of life is the clue to what "turns"

---

92 Burger, *Esoteric Anatomy*, p. 136.

V. A Once and Future Cosmogony

it, quite literally, into particle-mass formation on one hand and energy release on the other hand.

From a larger perspective, centrifugality and centripetality speak to the dynamics of involution and evolution. We saw that in Sanskrit, involution is pravritti, which means spiraling forth—or centrifugally. Evolution is nivritti, which means spiraling back—or centripetally. "Centripetality–centrifugality" speaks to space and spatial deployment. "Evolution" speaks to time. Practically, the dimension of time seems to have largely receded in the background of modern physics. Yet everything is a dynamic process involving time and space, concomitantly and inseparably, as physics is perfectly cognizant of since Einstein's relativity. The time dimension, slowed down to a halt by the analytical intent to isolate particles and measure forces, is pressing its reentry into scientific awareness through such quantum challenges as the impossibility to localize electrons. While they amount to an overall stillness, standing waves are dynamic, vibrant processes, which fit together space with time—harmoniously and rhythmically.

Fig. 66. Vishnu holding his four vorticular attributes

The centrifugal descent of energy-life into matter and the centripetal organization of matter into forms of consciousness are two sides of the cosmological coin—in the same way that particle formation and energy emission are two potentials of the universal wavefield.

The dynamics of energy pulsating centrifugally from a center and returning centripetally to its source sustains the *double spiraling helix*, which defines, we saw, the fundamental pattern of forms. Creation as a whole is a multidimensional dual vortex created by the centrifugal and centripetal energies of Shiva and Shakti.

As the laws of harmony reveal, this fundamental interplay blends the creational impulse toward replication (body) with the Logoic impulse toward integrative consciousness (soul). What governs the operation of manifestation is the imperative to create living forms—self-similar, fractal replicas of the "living one," the triune Logos. The way to create such whole, living forms is to combine octavian replication and quintessential mediation. Dual vortices are the most perfect actualization of this triune logo-rhythmic blending of self-replicating bundling

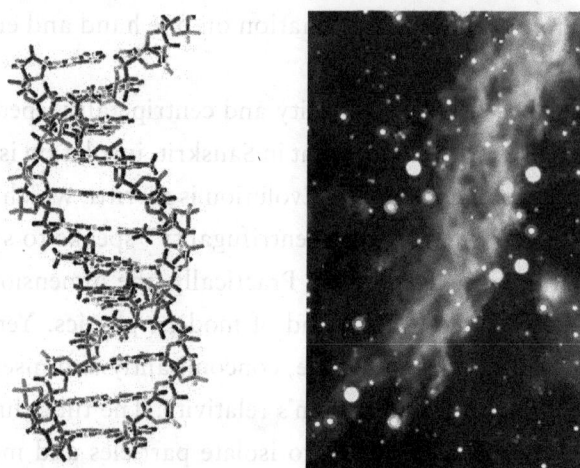

*Fig. 67. Double helix in DNA (left); and in a Milky Way nebula (right)
(wikimedia.org and www.nasa.gov)*

and quintessential unfolding (the mediating fifth)—which makes of the torus the archetypal template of living forms.

The dual vortex form is a central symbol associated with the divinities of ancient India. Most representations of the second person of the Hindu Trimurti—Vishnu—the all-pervading preserver of the cosmic system, feature as his main attributes a conch shell, a chakra (a spinning, disk-like weapon), a lotus and a spindle like mace (fig. 66). These four spinning objects represent different facets of one and the same form-process that whorls the world being into manifestation. The conch alludes to the sound of creation. Its shape epitomizes the dual vortex created by the resounding into forms of the world being. It is indeed the archetypal "instrument" or mean of the creator god.

In the domain of science, the discovery of the DNA and its double helix structure, along with the awareness that the Earth, orbiting around the Sun, traces a helix as the Sun itself moves around the Galactic center, have activated glimpses of the fractal replication of such a fundamental pattern from macrocosmic to microcosmic systems—from whirling galaxies to DNA helices, to spinning atoms. Figure 67 shows a graphic illustration of the double-helical structure of the DNA, next to a photo of an elongated double-helix nebula, near the center of our Milky Way galaxy.

Except for a small number of scientists who recognize its fundamental character, the dual vortex has not been retained, as of yet, as an object of study by mainstream physics or biology. It would take a reorientation of the scientific focus to the formation-concentration of particles or living systems out of the matricial field for this core component of a unitary cosmology to be given a central place. However,

V. A Once and Future Cosmogony

the fact that the description of electrons, atoms, solar system, and galaxies as standing waves is increasingly regarded as plausible indicates that this understanding is surreptitiously gaining ground. Einstein's equation suggests that particle-mass formation and energy radiation are two complementary ends, hence two directions of manifestation of the electromagnetic field—energy radiation speaking to the centrifugal direction, and particle formation to the centripetal direction. Voices from the margins of mainstream science are using the term *in-wave* and *out-wave* to designate the two currents.[93]

The unitary view suggests that what we perceive as particle, plant, body, or star, is a dual vortex of energy. The still center of embedded standing waves forms a relatively separate particle. A small region of space becomes isolated, generating the effect we observe as mass. Standing-wave patterns sustain the vibratory formations that define any stable energy system and *"make possible the birth of matter."*[94] Reminiscent of Rudolf Steiner's revolutionary view of the heart as a vortex created by interfering currents of cosmic-and-earthly energy—not the pumping cause of circulation, but its effect—this perspective on the particle requires a reversal of cause and effect not unlike Copernicus' pivotal shift. The same repositioning could lead one to ask whether gravity is not an effect of space–time, rather than "space–time, and the way it warps and curves, [being] an embodiment of the gravitational field."[95] Might it be that the gravitational field is the expression of the vortical warping into galaxies and "solar systems" of the space–time wavefield of a vibrant universe?

This cosmological view is at the heart of Randolph Stone's scientific philosophy, which exemplifies the theoretical coherence and practical efficacy that result from holding ancient, "consciousness" oriented awareness and contemporary, "matter" oriented views in creative tension, thus allowing the two sides of the human tale to spin into one:

> Sound as supersonic and ultrasonic energy is the creative principle in nature and the cause of all creation. "God spoke" sound, not light, is the first principle of *whirling*, fiery action, [manifest as] wheels of energy in space, creating rivers of whirls, entities and *individualizations* from the All-ness of the One, in its *centrifugal* primal outpouring.[96]

---

93 For instance, Geoff Haslehurst's "Wave Structure of Matter" cosmology or Milo Wolf's theory of the spherical, standing-wave nature of the electron.

94 Winter and Friends, *Sacred Geometry*, p. 92.

95 Greene, *The Elegant Universe*, p. 75.

96 Stone, book 1, p. 22; quoted in Burger, op. cit. pp. 124–125 (italics added).

Stone's focus on the swirling geometries of spiraling waves complements Plato's abstract focus on the harmonic-geometric fractioning of unity by which "the *one*" spirals down into wholistic parts, particles and bodies.

Lawlor suggests how they articulate one and the same picture:

> The transcendent Word is only a vibration (a materialization) of the Divine thought which gives rise to the fractioning of unity which is creation. The Word (the Logos of Christians and Gnostics), whose nature is pure vibration, represents the essential nature of all that exists. Concentric vibrational waves span outward from innumerable centers, and their overlappings (interference patterns) form nodules of trapped energy which become the whirling, fiery bodies of the heavens.[97]

Current physics wrestles with the discontinuity it finds between the "quantum weirdness" of the subatomic world, the classical laws governing perceptible masses and forces, and the gravitational laws that govern the macrocosmic bodies. So far, except for the window of possibilities opened by string theory, quantum physics is regarded as incompatible with classical physics as well as with Einstein's theory of general relativity. Brian Greene asks:

> Somewhere between the tiny world of individual atoms and subatomic particles and the familiar world of people and their equipment, the rules change because the sizes change.... But exactly where is the border? In recent years, physicists have expended much effort in trying to figure out precisely how the atomic and subatomic shed their magical weirdness when they combine to form macrocosmic objects.[98]

But do they? Doczi's works certainly suggest that they don't, showing instead how the same harmonic order of standing waves that explains the subatomic quantum order as well as the structuring of cosmic systems, also governs the composition of living organisms (animal and humans). A logarithmic continuum swirls the minute quantum order, the natural order around us, and the cosmic gravitational order. The seemingly different, "incompatible" laws, which govern the minute and the cosmic, point to different levels and scales of the same formative process. This unitive view suggests that the atomic and subatomic quantum order (or "weirdness") becomes embedded in larger orders of *quantum-like* logarithmic systems. When we identify the logarithmic principle as the key of quantum order, the continuum between the minute quantum world, the "familiar world" of mineral, plant, animal and human forms, and the macrocosmic fields of gravity readily appears. The

---

97   Lawlor, *Sacred Geometry*, p. 22.
98   Greene, *The Fabric of Cosmos*, pp. 202, 199

V. A Once and Future Cosmogony

continuum *is the universal dynamics of harmony, articulated mathematically in the logarithmicity of Phi*. Logarithmic bundling is the common denominator of forms.

Doczi's meticulous measurements of animal and human bodies, tree and leaf structures, insect and bird skeletons, offer a unique access to this continuum. They illustrate how the all-pervasive breath of Brahma organizes living systems and informs their growth with its dinergy, which generates the logarithmic spiraling dynamics that is the great Constant of nature—its triune ground—the trinity reflected in Planck's constant of energy/mass–space–time, and Einstein's equation of energy–mass–light. Every organism proves to be a wholistically fractioned, hence integrated whole of nested standing waves. These bundles of embedded standing waves twirl energy down into etheric-material corpuscles that constitute forms of life—forms of being.

A modern view that comes close, not only to the ancient Hindu notion of omnipresent vibrancy, but to the harmonic cosmology of quantum wave interference and embedded fields of resonance we have been contemplating, is the view proposed by one of the great theoretical physicists of the twentieth century, David Bohm. Bohm explains that his view of reality "can perhaps best be called Undivided Wholeness in Flowing Movement. [It] implies that flow is, in some sense, prior to the 'things' that can be seen to form and dissolve in this flow."[99] He refers to this Undivided Wholeness in continuous flux as the "holomovement" (movement of the whole) from which all things arise and into which all things eventually dissolve. "In this view, there is no ultimate set of separately existent entities, out of which all is supposed to be constituted. Rather, unbroken and undivided movement is taken as a primary notion."[100] Everything is to be explained in terms of forms derived from this holomovement and in-formed by it—"subtotalities" derived from a deeper order of unbroken wholeness. He compares reality to a flowing stream, in which vortical proto-patterns and standing waves arise and vanish:

> On this stream, one may see an ever-changing pattern of vortices, ripples, waves, splashes, etc., which evidently have no independent existence as such. Rather, they are abstracted from the flowing movement, arising and vanishing in the total process of the flow. Such transitory subsistence as may be possessed by these abstracted forms implies only a relative independence or autonomy of behavior, rather than absolutely independent existence as ultimate substances."[101]

---

99 Bohm, *Wholeness and the Implicate Order*, p. 11.
100 Bohm, "On the Intuitive Understanding of Nonduality," p. 77.
101 Bohm, *Wholeness and the Implicate Order*, p. 48.

He points to Einstein as the first inspiration for this significant challenge to the mechanistic view:

> Einstein saw deep contradictions in the notion of an independently existing particle, and proposed that what we normally think of as a particle is actually a temporary localized pulse emerging from a larger field, very much as a vortex temporarily forms from the dynamic flowing of a stream.[102]

Bohm views the universe as a "series of interpenetrating and intermingling elements in different degrees of unfoldment all present together." His central intuition is that there is a pattern of wholeness enfolded within all being, a universal pattern he calls the "implicate order," according to which "everything is enfolded into everything." This implicate domain, he says, "could equally well be called Idealism, Spirit, Consciousness. The separation of the two—matter and spirit—is an abstraction. The ground is always one."[103]

Implied in these short statements is a radical reversal of perspective from the mainstream scientific outlook on reality. Instead of going from parts to whole, Bohm approaches reality from what he views as its underlying wholeness to its emerging parts. In contrast to the classical mechanistic approach that focuses on measurements and equations in order to capture elementary particles and their relations, with the hope that an overarching coherence will eventually emerge, Bohm, like Plato, starts from the fundamental ground of wholeness and watches it flow, enfold and implicate itself, as a whole, into quasi-separate (sected-and-bonded) forms that externalize, unfold, and "explicate" it. His view of forms as quasi-separate "explicates"–unfoldings of the "implicate whole" takes the focus away from the forms per se to their formation, and reorients the scientific inquiry from the dead end conundrum of particle versus wave to the process that "turns" the whole wavefield into particles.

The dominant orientation of modern science to seek ultimate insights into world beginnings by isolating and measuring particles runs contrary to the Big Bang perspective and seems to ignore its implications. The Big Bang implies that all things, when traced back in time, merge into *one* field of background radiation, then *one* point of explosion-birth. The resistance to consider that reality has two poles—fragmentation and oneness—hence two directions of approach, which it behooves science to integrate into a third, such as David Bohm proposes, gets in the way of the splendid possibilities that loom beyond the wall that mechanical physics, a glorious accomplishment in its own right, finds itself up against as it meets the limits of its

---

102 From Nichol (ed.), *The Essential David Bohm*.
103 Quoted in Talbot, *Holographic Universe*, p. 271.

V. A Once and Future Cosmogony

ring of application, while seeing on the horizon the signs of a more fundamental order of reality, which its framework cannot accommodate. Indeed, suggests Bohm,

> We have reversed the usual classical notion that the independent "elementary parts" of the world are the fundamental reality, and that the various systems are merely particular contingent forms and arrangements of these parts. Rather, we say that inseparable quantum interconnectedness of the whole universe is the fundamental reality, and that relatively independent behaving parts are merely particular and contingent forms within this whole.[104]

In this view, "things, such as particles, objects, and indeed subjects" exist as "semi-autonomous quasi-local features" of an underlying activity. "The electron itself can never be separated from the whole of space, which is its ground."[105]

Bohm's "holistic" view is rooted in the intuitive feeling for wholeness he experienced in early childhood. In an interview with physicist and friend David Peat, he recounted,

> When I was a boy, a certain prayer we said every day in Hebrew contained the words to love God with all your heart, all your soul, and all your mind. My understanding of these words, that is, this notion of wholeness—not necessarily directed toward God but as a way of living—had a tremendous impact on me. I also felt a sense of nature being whole very early. I felt internally related to trees, mountains, and stars in a way I wasn't to all the chaos of the cities. When I first studied quantum mechanics, I felt again that sense of internal relationship—that it was describing something that I was experiencing directly rather than just thinking about.[106]

When Bohm states that everything is to be explained in terms of "forms derived from this holomovement," he intimates that everything is, literally, in-formed by it. He goes even further to propose, "the whole notion of active information suggests a rudimentary mind-like behaviour of matter."[107] He points out that thought is intricately connected with physical reactions, as we know from everyday experience, and suggests that mental and physical are two "poles" of a unified whole, and that, at each level, "information is the bridge or link between the two sides.... Even the electron is informed with a certain level of mind."[108] He once remarked that he frequently had the impression that the sea of electrons was in some sense alive.

It is only logical that Bohm would embrace the fractal framework to which harmonic cosmology leads as well. Bohm worked with Pribram on the theory that the

---

104 Bohm, "On the Intuitive Understanding of Nonlocality."
105 See http://www.fdavidpeat.com/interviews/bohm.htm.
106 Ibid.
107 Bohm, "A New Theory of the Relationship of Mind and Matter."
108 Hiley and David, eds, *Quantum Implications: Essays in Honour of David Bohm*, p. 443.

brain operates in a manner similar to a hologram. The fractalization process, which has been regarded as the "holopattern" of creation—the pattern of unfolding of an [implicate] Being into world dimensions—mirrors the ancient conception, confirmed by the experience of accomplished Tantrikas, that "the material world, not only is a 'fraction' of what is, but constitutes the lowest vibratory level of cosmic existence."[109]

The "new order" of fluid movement that Bohm formulated between implicate wholeness and explicate parts is rich in insights that could move physics beyond the unsatisfactory interpretations of quantum physics and dissolve the traditional border of mutual exclusion between continuous and separate, particle and wave, matter and consciousness.

String Theory and the Logarithmic Model

> *If string theory is right, the universe is akin to a cosmic symphony.* —Brian Greene

The arrival of string theory on the scientific stage in the 1970s has brought mainstream contemporary physics one step closer to the harmonic model of the universe. A number of theoretical physicists (among them Stephen Hawking, Edward Witten, Juan Maldacena, and Leonard Susskind) believe that string theory may hold the correct fundamental description of nature. "String theory is definitely revealing the deepest understanding of the universe which we have ever had," says Curum Vafa, a leading string theorist.[110] Though widely considered to be mathematically consistent, it is still regarded as speculative. Many hope that it will fully describe our universe and become the theory of everything.

According to string theory, the elementary ingredients of the universe are not point particles, but tiny, one-dimensional filaments or loops vibrating to and fro—"tiny strings whose resonant patterns of vibration are the microscopic origin of particle masses and force charges."[111] These infinitely small vibrating strings of energy measure 10^{33} centimeters, which is none other than Planck length.

Trinh Xuan Thuan explains,

> Just as the strings of a violin vibrate, these strings vibrate and generate tones and harmonics that are detected by our measuring instruments as protons, neutrons, electrons, and so on.... The strings are all basically the same. All that varies is how they vibrate. For example, a proton is simply a trio of vibrating strings, each

---

109 Feurstein, *Tantra: Path of Ecstasy*, pp. 61–62, 75.
110 Greene, *The Elegant Universe*, p. 213.
111 Ibid., p. 206.

of which corresponds to a quark. Just as musicians charm us by playing a piece by Brahms, the combined vibrations of these three strings produce the music of a proton. When our measuring apparatus captures it, the music comes out as a mass, a positive electric charge, and a spin.[112]

What physicists find particularly compelling about string theory is that it offers a unifying framework. First, it proposes that all particles arise from one basic ingredient—oscillating strings—executing different vibratory patterns. Second, it appears to resolve the conflict between general relativity and quantum mechanics. Brian Greene explains that the problem surges "when the central tenet of the former—that space and time constitute a smoothly curving geometrical structure—confronts the essential feature of the latter—that everything in the universe, including the fabric of space–time, undergoes quantum fluctuations that become increasingly turbulent when probed on smaller and smaller distance scales.... The spatially extended nature of a string is the crucial new element allowing for a single harmonious framework incorporating both theories."[113]

From the perspective of harmony, the "smoothly curving geometrical structure" of macrocosmic systems is not in contradiction with the ultramicroscopic waveforms that make up the quantum field and condition energy radiation in an oven. "Quantum fluctuation" (involving minute wavelengths and corresponding high frequencies) is quantum order, even if at such a minute scale, the perception is one of "turbulences."

Greene's presentation of string theory resonates fully with the harmonic understanding derived from the Pythagorean–Platonic perspective:

> Each string can undergo an infinite variety of vibrational patterns known as resonances. These are the wave patterns that the string can support by virtue of *their evenly spaced peaks and troughs exactly fitting along its spatial extent* between the string's two fixed endpoints.

We recognize the principle of harmonic wave interference at the basis of standing wave formations. Indeed, Greene remarks, "With the discovery of superstring theory, musical metaphors take on a startling reality."[114]

According to string theory, the "stuff" of matter and forces is the same. Each particle is a single string, and all strings are absolutely identical. What appear to be different elementary particles are actually different "notes" on a fundamental string. "Just as the different vibrational patterns of a violin string give rise to different musical notes, the different vibrational patterns of a fundamental string give

---

112 Trinh and Ricard, *The Quantum and the Lotus*, p. 106.
113 Greene, *The Fabric of Cosmos*, p. 18; *The Elegant Universe*, pp. 136, 152, 211.
114 Greene, *The Elegant Universe*, pp. 143, 135.

rise to different masses and force charges."[115] Calculations even show that "the masses of the string vibrations follow a series analogous to musical harmonics: they are all multiples of a fundamental mass, the Planck mass, much as overtones are all multiples of a fundamental frequency or tone."[116] If string theory is right, "the universe is akin to a cosmic symphony."[117] Logarithmicity, which we have found to be the central cosmological principle, surges anew in this contemporary context. "The energy embodied in a string vibrational pattern is a whole number multiple of a minimal energy denomination."[118] Brian Greene does not hide the fact that this leads to gargantuan masses, which, he says, does not rule out string theory but is a challenge it must overcome.

The notion of *standing waves* illumines the continuum from quantum weirdness to organic systems to general relativity, intimating how all levels of the universe participate indeed of one symphony. Missing from the insights of string theory, yet implicit in the very notion of minute standing waves or vibrating "strings" (though not in that of "loop" or filament) is an awareness of the bidirectionality of currents involved in "strings"—an awareness necessary to understand the dual potential of energy emission and mass accretion intrinsic to a standing wave. Although implicit in Einstein's equation, this principle of bidirectionality may remain unrecognized until the mathematical equation is related to the geometry of waves as its phenomenological underpinning. Similarly, even though the notion of logarithmic embedding is implicit in quantum mechanics, its significance cannot be realized as long as it remains a purely abstract, mathematical concept, as removed from its expression in the spiral-cyclic dynamics of vortices as vortices are from quantum physics. Vortical motion, as Bohm intuited, is the link between whole and part, and it can be regarded indeed as the vital informative principle frozen in the banal mathematics of logarithmicity.

Werner Heisenberg observed with prescient lucidity, "The conception of objective reality has evaporated into the transparent clarity of mathematics that represents no longer the behavior of particles but rather our knowledge of this behavior."[119]

The gap between the contemporary scientific view and the Platonic cosmology of harmonic fractioning clearly grows smaller with the advent of string theory. Particles become strings, which evoke three-dimensional eddies. The distance between

---

115 Ibid., pp. 143, 145.
116 Greene, *The Fabric of Cosmos*, p. 357.
117 Greene, *The Elegant Universe*, p. 145.
118 Ibid., p. 149.
119 Cited in Popper, *Quantum Theory and the Schism in Physics*, p. 85.

V. A Once and Future Cosmogony

the two worldviews remains the difference between an effort to identify physical particles and quantify their behavior and an effort to understand their formative processes and dynamic geometries. It is the difference between a physical scientific approach, which takes its start from a complex array of facts, observations and data, and a spiritual scientific approach, which starts from the whole and asks how the one whole becomes many parts. As great scientists before him, Einstein trusted theory and vision as much as observation, and even gave precedence to the former. In a famous statement worth quoting again in this new context, he affirmed, "On principle, it is quite wrong to try founding a theory on observable magnitudes alone. In reality the very opposite happens. It is the theory which decides what one can observe."

It is intriguing to note how Einstein's equation of energy–matter–light resonates with one of the ancient Tibetan stanzas of Dzyan cited at the beginning of H. P. Blavatsky's *Secret Doctrine*, which he kept on his bedside table. The following extract is from a section entitled *Cosmogenesis*:

> Father–Mother spin a Web, whose upper end is fastened to Spirit—the Light of the One Darkness—and the lower one to its shadowy end, Matter; and this Web is the Universe, spun out of the Two Substances made in One, which is Svabhâvat. [120]

The Sanskrit term Svabhâvat designates the spirit or essence of substance. It evokes the alchemical notion of the "lumen naturae," the light of nature, the ethereal fabric of the world soul, which is the quintessence of the elements (matter). The primordial web "spun" and swirled by the Father-Mother polarities of spirit and matter, which Plato referred to as "Same and Other," and the laws of harmony translate into fundamental and octave, is the vibrant "sound" matrix of space–time–consciousness—from which all things emerge. By engaging qualitatively the three realities of wavefield, mass and energy (web-spinning, spirit, matter), Einstein's equation invites into modern science the etheric holomoving web that precedes and transcends physical forms and forces. Tesla, whom Einstein held in high regard, was quite declarative, "All attempts to explain the workings of the universe without recognizing the existence of the ether and the indispensable function it plays in the phenomena is futile and destined to oblivion."[121]

Einstein stays with this formative process when he suggests turning around the classical perception of particles to see in them *localized vortices temporarily forming from the dynamic flow of a larger field*. It is with such vortical "particles" that energy and mass, whose equivalence his equation reveals, first emerge as two facets of "particularization" of a field of wave conjugation, in which duality is implied as

---

120   Blavatsky, *The Secret Doctrine*, vol 1, p. 29.

121   When asked how it felt to be the smartest man alive, Einstein replied, "I don't know; you'll have to ask Nikola Tesla."

polarization. Polarity interplay (space and time, electricity and magnetism) precedes the energy/mass "duos" of particle physics—replicating the Shiva–Shakti dynamics of emission-containment that gives birth to all the organic/lives, human/beings, and cosmic/entities of the universe.

✪ The web from which "particles of energy" surge is a dense, quasi-material expression of the solar etheric field from which "living organisms" emerge. This etheric, vital field of nature is in turn a relatively dense expression of the quintessential "mental" field of the world soul from which self-aware entities arise. These three orders of field-matrices, which underlie the realms of elementary, organic and individual forms of being, are analogues of each other. They replicate at denser levels of corporealization the primordial milieu of creation, the web of polar interplay and creational harmony that begets the world soul "compound," the milieu of the great (ana)logue, the Logos—the matrix from which the multidimensional whorls of harmonically spun particles and forms of cosmos unfurl, and into which they eventually turn. ✪

One of the oldest mantric prayers, a verse from the Rigveda referred to as the Gayatri, directs the inner gaze and sets the inner breath to the radiant flow of the life giving, life sustaining spiritual Sun alive in this dual motion of worlds—which is His breath:

> O Thou who givest sustenance to the universe,
> from whom all things proceed, to whom all things return,
> unveil for us the face of the spiritual Sun, hidden by a disk of golden light...

THE FOURTH COSMOLOGICAL STEP:
THE DEPLOYMENT OF THE TRIAD INTO A 3–7–12FOLD FORM

> *There is a certain natural number belonging to every animal. For things of the same species would not be distinguished by organs after the same manner, nor would they arrive at puberty and old age about the same time, or generate, nor would the foetus be nourished or increase, according to regular periods, unless they were detained by the same measure of nature.* —IAMBLICHUS[122]

The fourth step of cosmogenesis is implied in the third, as Cosmos is in Logos, and as the fourth interval is implied in the emergence of the fifth within the octave. The fourth is the "other side" of the fifth. The note G is at once a fifth from the

---

122 Iamblichus, *Life of Pythagoras*, p. 226.

V. A Once and Future Cosmogony

fundamental C below and a fourth from its octave above. From the cosmogonical perspective, the fourth is but the counterpart of the "descent" of the fundamental into its third harmonic. The "logoic" fifth, the offspring of the octave like intercourse of the world parents—Same and Other—already implies the cosmic quaternary (the fourth). Light implies darkness. Light declares shadow. "Otherness" declares separativeness.

Carl Jung's statement placed earlier as a revelatory caption under Plato's "X" scheme of corporealization may be at this point meaningfully brought up in its totality: "All things which are brought forth from the Pleroma by differentiation are pairs of opposite; therefore God always has with him the Devil."

Two sons are born of the Biblical parents, Adam and Eve: Abel and Cain. The first is a contemplative, lighted figure; the other will be a "builder of cities." Abel carries the spiritual grace of the triad; Cain bears the seal of the quaternary, the conflictedness and arduous labors of the fourfold world. Is it not then a cosmogonic necessity for the fifth and its Logoic light to be eclipsed and "killed" by the fourth in order to enter the *elemental* stage of a physical world?

The cosmology of the *Timaeus* describes the fourth step of cosmogenesis as a deployment of the nuclear triune principle (father-mother-son, spirit-matter-soul) into a twelvefold–sevenfold cosmos composed of one quintessential and four dense elements. This quintessential element of life, the twelvefold ether body of the soul, the deployment of the triune fifth, is the intermediary between the formless Triad and the elemental Quaternary.

The step from the third to the fourth harmonic, which yields the fourth interval (4/3), is the harmonic equivalent of the corporealization of the divine triad into the quaternary of elements. It takes the triune world of spirit into the fourfold world of matter. Remarkably, the musical interval of the fourth (4/3) holds in its mathematical reality the 7/12fold features of the fourfold cosmos to be born of the rapport it represents, as the 4/3 ratio, between three and four—the conjugation of triad and quaternary via addition and multiplication: $3+4=7$; $3 \times 4=12$.

The two numbers that characterize the fourfold world—7 and 12—dominate ancient cosmologies. They are sealed in the collective stories of world beginnings and pertain to the mathematics of world foundations. We saw that they stand at the beginning and end of the Biblical story: involutionary movement of Genesis centered on the seven days of creation and the seven pillars of wisdom ("Wisdom has built her house, She has hewn out her seven pillars" [Prov. 9:1]), and evolutionary apotheosis of the Book of Revelation, centered on the 12fold heavenly Jerusalem, the Woman crowned with the twelve stars and the tree of life bearing twelve kinds of fruit.

The Hindu, Zoroastrian, and Japanese creation myths all describe a seven-stage process. We will also see that the two numbers appear in the cosmology of the Dogons of Mali. In the Rig Veda, The Maruts, the sons of Heaven and Earth, also referred to as the children of Shiva, "are born in every manvantara (round) seven times seven." The Ramayana recounts how, with his thunderbolt, Indra divides the embryo in their mother Diti's womb into seven portions, and then divides every such portion into seven pieces again, which become the Maruts. Blavatsky remarks that Diti—Aditi or Akasa in her higher form—is the Egyptian seven-fold heaven. Her embryo is the primordial body in the universal womb. She adds, "The septenary doctrine or division of the constitution of man is a very ancient one, and was not invented by us."[123]

The principle of organic endings and beginnings associated with the 7/12fold octave stands behind the Biblical story of the destruction of Jericho. The Lord, the Bible recounts, handed the city to Joshua (Josh. 6). What he did was to give him the key to "unbind" its wall—the power of the 7fold octave. He prescribed first that the warriors circle the city silently once every day for six days. On the seventh day, they should march around the city seven times, followed by the seven priests blowing the seven trumpets of ram's horns, ahead of the Ark of the Covenant. Then, at the sound of the trumpet, all the people should utter a great shout, and the wall of the city would fall down. Indeed, "The people raised a great shout and the wall fell down." Divine binding and unbinding is tied with a 7fold octavian completion. Centuries before the time when the "son of God," charged with the love principle, would hand Peter the secting-binding "key" (love) to the kingdom of soul, the "father" God gives Joshua the unbinding key of creation and evolution, the Octavian self-resounding Word of the beginning (shout, Om, Big Bang), which is the secret of 7fold completion behind the major initiatory leaps of evolution.

The following verse, extracted from an ancient Indian text, speaks to the same sevenfold grand world opening:

> Space and time are one. Space and time are nameless, for they are the incognizable THAT which can be sensed only through its seven rays—which are the Seven Creations, the Seven Worlds, the Seven Laws.... The seven secret emanations, the seven sounds, and seven rays [are] the spiritual and *sidereal models of the seven thousand times seven copies of them* in later aeons.[124]

---

123 Blavatsky, *Secret Doctrine* vol. 2, pp. 613, 640.
124 Ibid., pp. 612–613.

V. A Once and Future Cosmogony

The descent of the twelvefold Logos into the quaternary deploys a sevenfold cosmos, and the evolution of cosmos into self-awareness involves a twelvefold patterning. The sevenfold speaks predominantly to a process of differentiation of the whole into levels or dimensions, and the twelvefold to a process of union into a whole "self." This polarization of 7 and 12 along one of these two functions happens only when the two processes are engaged together in the 7/12fold constitution of the archetypal form, as reflected and illumined by the musical scale. Only when the two become relative to each other in the operation of scale and cosmos formation does the function of differentiation becomes associated with seven, and the function of union with twelve. On their own, each process of sevenfolding and twelvefolding involves, we saw, a division-and-bonding operation, which engages the two polarities.

Although the two patterns of differentiation and union are clearly intertwined from the beginning—as the golden dynamics of the first three harmonics illumines—they reveal themselves sequentially: one after the other. The order of appearance is reversed depending on whether we trace the cosmological process from Plato's involutionary perspective or from the modern perspective of evolution. The reason Plato describes first the unfolding of the triune One into a twelvefold is that, from the perspective of the world being, the world soul precedes and informs all cosmic formations, whereas from the evolutionary perspective of the human observer, the sevenfold manifests itself first, being the organizing principle of matter, and the twelvefold principle of soul appears only with the emergence of living systems and "entities." Living organisms attest to a gradual integration of autonomous "selfhood," as reflected in their emerging properties of self-growth, self-regulation, and self-reproduction. Twelvefoldness appears first with the plant kingdom—as the quintessential, dodecahedral element of life embodied in the chlorophyll.

To contemplate the evolution of the world from both perspectives at once is to watch the twelvefold soul make its descent and eventually appear in the growing complexity of the forms of matter and kingdoms of nature. It is to watch the Logos wrap itself ever more deeply and fully in forms of matter shaped by his golden mean. Invisible and mute in the harmonic structure of atoms and atomic elements, physicalized-materialized in the perfect solidity of the mineral, the soul of the world continues to penetrate deeper into matter, softening solidity into green, "solarized," vegetal substance, then animating it into sensitive flesh, and finally breaking through the lightened flesh of the human eye to reveal itself to itself and look at its world. Physicalization, vegetalization, animalization, humanization is the journey of emergence of the soul through increasingly sophisticated forms of matter.

The unifying view of a consciousness informed universe finds little echoes in mainstream science. The notion of a 7fold /12fold differentiation of the primordial triad, which the spiritual scientific approach reveals as the next step in the organization of matter-life into forms, is absent as such from modern science, though facts of common scientific knowledge and ordinary perception hint at this subtle order—from the 7 rows of the periodic table of elements to the 7fold color spectrum of light and the 7 planets of the solar system, from the 12fold chlorophyll and DNA to the ubiquitous presence of the golden mean in natural forms. It is also an admitted fact of particle physics that, preceding and underlying the 7fold periodic table of the elements, 12 particles have been identified as the fundamental constituents of matter. Significantly, all twelve are described by a single equation—a very evocative mirroring in physical matter of the twelve-in-one archetypal form of the world soul described by Plato.

Singular gazes and voices have discerned and pondered the 7fold patterning of forms and systems. Babbitt's view of the atom involved 3 major spirals and 7 spirillæ. Leadbeater and Besant's clairvoyant research would confirm this picture two decades later. This first appearance in the universe of a 7fold patterned "particle" may justify after all the name "atom" given to it by the Greeks—"that which cannot be divided"—beyond the modern discovery of smaller particles (protons, neutrons, quarks). Smaller particles do not appear, at least to current knowledge, to form indivisible, "whole" complexes.

"Every atom has seven planes of being," states Blavatsky. "The atom is a concrete manifestation of the universal energy. Atoms are vibrations. Absolute intelligence thrills through every atom. The atom belongs wholly to the domain of metaphysics." Alice Bailey added in the 1940s, "Every atom, we are told, contains within itself three major spirals and seven lesser of which ten are in process of vitalization but have not yet attained full activity. Only four are functioning at this stage, and the fifth is in process of development."[125] This fifth spiral alludes to the unfolding of the solar/soul mind in humanity.

The sevenfold leitmotiv appears next in the systemic organization of the chemical elements, brought to light by the periodic table. Mendeleev's classification of the atomic elements by increasing atomic weight revealed a 7fold "periodic" repetition of properties. The sequential arrangement of the elements into 7fold rows produces vertical columns ("groups") of elements sharing recurring ("periodic") physical and chemical properties. "If all the elements are arranged in the order of their atomic weights, a periodic repetition of properties is obtained. This is expressed by the law of periodicity."[126]

---

[125] Bailey, *Treatise on Cosmic Fire*, p. 246.
[126] Mendeleev, *Principles of Chemistry*, vol. 2, p 17.

V. A Once and Future Cosmogony

Elements of the same period have the same number of electron shells; with each group across a period, the elements have one more proton and electron and become less metallic. There are eighteen main groups, classified into seven kinds, which range from metal to noble gases—in other words, from solid to gaseous substance: Metalloids, Alkali Metals, Alkaline earth metals, transition metals, rare earth elements, non-metals, noble gases. The implication, Schwaller de Lubicz points out, is that there are at most seven electronic stratas around the atomic proton, evocative of the sevenfold planetary system, the musical scale, the colors, etc.[127]

The logarithmic character of the atomic system further emerged into view with Moseley's law of 1913. British physicist Henri Moseley established that the frequencies of certain characteristic X-rays emitted by chemical elements are proportional to the square of a number close to the element's atomic number. In other words, the charge of an atomic element is proportional to its atomic number. Up to that time, the notion of atomic number referred to the place of an element in the periodic table according to its mass. From Moseley on, it also referred to the exact number of protons found in the nucleus of an atom of that element, in other words, its charge number. Moseley showed that Mendeleev's ordering according to atomic weights actually corresponds to the increase in atomic charge and number of protons. Substituting atomic numbers for atomic weights produced an integer-based sequence of the elements.

"We have a proof," he explains, "that there is in the atom a fundamental quantity, which increases by regular steps as one passes from one element to the next. This quantity can only be the charge on the central positive nucleus, of the existence of which we already have definite proof."[128]

Two years later, Moseley died in World War I at the age of twenty-seven—"the most costly single death of the war to mankind generally," Isaac Asimov would later say, thinking of what Moseley might have accomplished.

The significance for us of Moseley's demonstration is that it shows how the system of chemical elements displays a sevenfold order of arithmetic increase (in number/weight) as well as proportional increase (in charge) of a base unit, which is to say that it is ordered logarithmically. The two-dimensional spiral cycling arrangement proposed in 1900 by Erdmann in reference to chemical properties and atomic weight, reproduced in a previous section, already suggested this logarithmic deployment of the elements.

---

127 Schwaller de Lubicz, *Du symbole et de la symbolique*, p. 33.
128 Moseley, *Philosophical Magazine*, vol. 26, 1913, p. 1030.

We saw earlier that a similar sevenfold classification has emerged with regard to crystal systems. Modern researchers have also pointed out the remarkable fact that the arrangement in a scale of the traditional seven metals (lead, tin, iron, gold, copper, mercury, and silver) according to their key physical properties of lustre, resonance, malleability, and conductivity results in the same scale as an ordering of the corresponding planets in terms of their speed of movement.[129] These scales show an increase in inner mobility from lead to silver, which parallels the increasing angular speeds of the planets. Rudolf Hauschka, who first described this, concluded: "We see then that planetary movement is metamorphosed into the properties of earthly metals." Hauschka and Wilhelm Pelikan viewed the seven metals as the expression of the seven planetary characters: "The seven fundamental metals represent something like the seven notes of a scale. As there exists a great variety of intermediate tones within the scale, so one can recognize intermediate tones between the metals."[130] Biochemist Dr. Frank McGillion elaborates, "The orbital motion of the planet correlates in sequence with its corresponding metal's conductivity.... The slower a planet moves, the less able its corresponding metal is to conduct electricity."

The next level of organization of matter combines atoms into molecules. About this molecular stage, Leadbeater observed, "The simplest unions of atoms, *never, apparently, consisting of more than seven,* form the first molecular state of physical matter."

In his book *The Reflexive Universe*, Arthur Young introduces a sevenfold division of the molecular kingdom, which he obtained from the chemist Dr. Charles Price.[131] In Young's view, the molecular kingdom is itself at the center and apex of a remarkably integrative sevenfold model of evolution, which charts along a V-shaped arc the descent of spirit into matter (from Light to Particle to Atomic to Molecular stage) and the corresponding ascent of matter into spirit (up from Molecular to Vegetable, Animal, and Dominion–human stage). Each of the seven stages thus delineated comprises seven sub-stages whose characteristics echo the corresponding major stage or kingdom. The seventh stage represents the culminating expression of a kingdom, which Young calls the "dominion" or *principle* of the series. Of particular interest to us is to find that "crowning" the 7fold molecular kingdom in Young's model is the *12fold* DNA molecule. This molecule corresponds, in the molecular kingdom, to the causal body in the human kingdom—at once its core and culmination.

---

129   See tabulation by N. Kollerstrom (alchemywebsite.com/kollerstrom_sevenfold.html).

130   Ibid. Hauschka, *The Nature of Substance*, p. 162.

131   Young, *The Reflexive Universe*, p. 67.

V. A Once and Future Cosmogony

The connection of mainstream science with a 7foldness of systems and forms is as of yet limited to the periodic table, and there is little chance that this world structuring principle will be verified through purely empirical means—unless clairvoyant perception comes to be regarded as an empirical approach in its own right. If the 7foldness of the chemical elements revealed itself to the intuitive intelligence of a few dedicated scientists, the 7foldness of atom, plant, and human being lies beyond concrete observation and quantitative measurement. So, we must at this point leave mainstream science behind in order to let the numerous insights gained into a universality of nature's forms and formative process inform our understanding.

FROM THREEFOLD FIELD TO 7/12FOLD FORM: FROM RESONANCE TO CONSONANCE

> *Everything going on in nature is permeated by a hidden music, the earthly projection of the "music of the spheres." Every plant, every animal actually incorporates a tone of the music of the spheres.* —RUDOLF STEINER[132]

> *To me the world of perfect forms is primary—its existence being almost a logical necessity—and both the other two worlds are its shadows.* —ROGER PENROSE[133]

Key to the momentous step of form building and its 7/12fold character is a distinction that emerged at the end of the second phase of this journey—the distinction between resonance and consonance. We saw how these two levels of harmony speak to the difference between simple harmonious sound and musical interval.

A harmonious tone (in contrast to a "noise") is a sound that combines a fundamental wavelength with subsidiary "harmonic" waves entrained by *resonance*. What characterizes resonant sub-waves is that their half-wavelengths are whole number fractions of the fundamental. A standing wave always includes "resonant" sub-waves. Resonance is the expression of constructive interferences among waveforms. Waves that are not in phase will not do, because they cannot nest constructively within each other and swirl into whole vortical "forms." There is an infinite number of harmonic standing waves entrained by the fundamental vibration of a string and undulating within it. This "harmonic series" is the sequence of smaller and smaller whole number fractions of the unity:

---

132 Steiner, *Balance in Teaching*, p. 18.
133 Quoted in Ward, *Pascal's Fire*.

1, 1/2, 1/3, 1/4, 1/5, 1/6, 1/7, 1/8, 1/9, etc.

Consonance involves a next grade of harmonic complexity, the secret of which is held in the Pythagorean "musical proportion": 6:8:9:12. While there is a potentially infinite number of harmonics engaged in a harmonious sound, as innumerable as there are fractions of "1," there is a limited number of "consonant" ratios apt to create musical notes and intervals. These special ratios are the major intervals of the musical proportion, also referred to as super particulate ratios: 1/2, 2/3, 3/4. These ratios and their octave equivalents are the only absolute consonances of the Pythagorean system. In his book *Through Music to the Self*, Peter Hamel mentions the fact that the Byzantines regarded octave, fifth, and fourth as spiritual sounds, pneumata, while they called somata, bodily sounds, the intervals of the third and fourth octaves of the harmonic series (i.e., the seconds and thirds).[134] This insight into a lost knowledge echoes the ontological logic of primary numbers we have been exploring. It also charges with new depths of significance the awareness we have gained of the cosmological ranking order of the first harmonics and of the levels of Being at work in the corresponding musical intervals.

Whereas the phenomenon of resonance is based on the harmonious blend of *waves* whose half-wavelengths are whole fractions of the fundamental, consonance involves a combination of *standing waves* whose frequencies/wavelengths are in harmonic proportion with each other (octave, fifth, and the resultant fourth). "Consonance [in Greek, *sumphonia*]," says Ghyka, "is a concord of intervals."[135] Consonance is *relational* harmony—"sym"phony—such as implied in the intervals of the musical scale. The notion of a fifth, for instance, or the note G, is in reference to a fundamental and its octave. It expresses a consonance between the three intervals of fundamental, octave, and fifth. The same applies, even if to a lesser degree of harmony, to the second, third, fourth, sixth, and seventh intervals, which generate the notes of the diatonic scale. Tone implies resonance. Note implies interval. Musical interval implies consonance.

Pythagoras emphasized how concord is associated with small numbers. "Complex fractions," he says, "produce discord." Jamie Jeans elaborates,

> It is a general law that two tones sound well together when the ratios of their frequencies can be expressed by the use of small numbers, and the smaller the number, the better is the consonance. The farther we go from small numbers, the farther we go into the realms of discord.... Like Pythagoras, Chinese philosophers of the time of Confucius regarded small numbers 1, 2, 3, 4, as the source of all perfection.[136]

---

134 Hamel, *Through Music to the Self*, p. 121.

135 Ghyka, *Le Nombre d'Or*, p. 32.

136 Jeans, *Science and Music*, p. 154.

V. A Once and Future Cosmogony

The octave fraction (1/2) represents a whole "unit" of vibration—a whole enclosure of life. The fifth (2/3) is the golden formula of integrative relation between whole and part. These simplest ratios preside to the constitution of living forms—shells, vegetables, and bodies. Jointly, they preside to the finite wholeness of a form (the octave), and its harmonious weaving of parts into a whole, visible in the logarithmic spiraling of shells or the Fibonacci coupling of spirals shaping vegetables, flowers, and trees.

Standing waves are the primary "components" of the space–time matrix of the universe, which is the resounding field of the primordial vibration. From this space–time matrix of the breath of Brahma arise all the systems of cosmos and forms of nature.

Emerging from the fundamental wavefield of resonance, the archetypal Form is a consonant or "symphonic" composition of standing waves.

✪ Consonance is the key to the form-building stage of cosmology. A form is a sustained consonance—a symphony. Standing waves per se do not involve consonance, only co-vibrancy or "sympathetic" resonance. Dual vortices or toruses do. *The transition from energy field to forms of matter is a transition from a resonant wave field to a consonant composition of standing waves.* The torus is defined as "two periodic motions interacting with each other in such a way as to become locked together into a repeating rhythm." The torus is to the simple vortex as consonance is to resonance—a next level of complexity, which sustains an enclosure of life, an inwardness of being—an autonomous system apt to sustain and reproduce itself. According to Jude Currivan, it is speculated that certain of the vibratory patterns of quantum strings may also take such a form.[137] ✪

What happens when elementary particles, then atoms and chemical elements emerge from the vibrant, quantum field of the pre-form, pre-mass stage of cosmos, followed, eons later, by organic living forms, is that harmonic standing waves enter "consonant" interferences. Atoms and chemical elements appear to be the first complex combinations of elementary particles—which are the "particle effect" of standing waves. If the atom is not the ultimate particle the term *a-tom* would indicate, it is nonetheless the primordial form of organization of spirit-matter, the primordial sevenfold waveform "unit." Leadbeater and Besant's findings revealed that the shapes of the chemical elements replicate the four basic Platonic solids, and the Platonic solids involve, we saw, the major ratios of musical consonance (1/2, 3/2). The chemical elements are physical manifestations of the musical proportion. Observations (with the electron microscope) of isometric virus shells have

---

[137] Currivan, *The Wave*, p. 92.

consistently revealed icosahedral designs.[138] Viewed from the modern, bottom up perspective of organized matter, the Platonic elements combine to constitute the building blocks of the world's sevenfold forms out of the "nuclear" triunity of energy, mass and the space–time constant C^2 ($E = MC^2$). Platonic solids display the basic units of a symphonic world architecture, generated by the Brahma–Logoic sound. They display the ladder of consonant (partial) "wholes" that sustain the multiple forms of life, which are part of the seven subdivisions of every kingdom.

In the same way that musical notes arise as tonal wholes from the ground of vibrational possibilities, colors step out of the world-light in 7fold rainbows; chemical elements emerge out of the energetic sea of world beginnings in 7 groups, and organic forms unfurl their 7fold compositions (plants and human beings) out of the time–space breath of the universe. The quantum field, as it precedes the formation of chemical elements and minerals, is reminiscent of the wordless yet active preverbal babble of infants, whose echolalia precede the emergence of the 7 vowels, then consonants, with which they begin to express themselves in words and speech. These 7fold "units" epitomize the archetypal template of the world form swirling into "itself" the primordial wavefield. The mythological emergence of Venus, the beautiful human form, from the seawaves, speaks to this archetypal advent—the emergence of the quintessential "Form" of the Logoic universe. Like the atom, the human form is 3fold (head, heart-lungs, limbs) and 7fold (chakras).

When quantum physics meets harmonic cosmology, we see more clearly how, while the waves engaged in energy emission in Planck's oven were selected by resonance, and limited by oven size and temperature, the waves entering the formation of elements and forms are further selected according to harmonic consonance.

The passage from field to particles, atomic elements, then organic forms, which is the process of world formation and evolution, is not a process that science ponders. Compartmentalized into separate "bundles," which rightly mirror the quantum character of the universe's unfolding—such as particle physics, astrophysics, chemistry, biology, physiology, psychology, cosmology—science loses the chance to catch glimpses of structural replicas and parallel profiles across realms. The steps of the spiral of cosmic unfolding vanish into gaps analogous to the absence of musical sound between harmonics. Yet, a single, coherent process underlies nature's discontinuous formations.

---

138 In Amy Edmonson's book, *A Fuller Explanation* (p. 239), she relates how Dr. Aaron Klug, who first observed the geodesic structuring of viruses, told Buckminster Fuller of his discovery in 1962.

Although science is now largely open to the notion of electrons as standing waves, for it to entertain the idea that atoms and organic forms might be *consonant* compositions of standing waves seems like a remote possibility.

If the discovery that the electron behaves both as wave and particle has invited a shift in focus from particles to the vibrational ground that precedes their formation, scientific research is still primarily oriented to isolate and define the smallest components of matter and their features (from atoms to genes to Higg bosons), rather than their formative processes. And yet, it would seem logical, in the wake of the Big Bang theory, for mainstream research to wish to investigate the process of mass-energy "particulation" out of the primordial wavefield, thus repolarizing the scientific compass from particle to field. The materialistic assumption is so strong, and the tangibility of sense perception so blinding, that the process of currents of energy whirling into particles of matter remains blurred from sight and relevance by the value placed on quantitative distinctions and measurements—of time, temperature, rate of expansion.

Only a complete revolution in perspective may eventually make sense of the universal leitmotiv of sevenfoldness at play in every kingdom of nature and system in cosmos. Such a revolution would imply a paradigm shift from matter-centric to "logo-centric." The notion of "information" field proposed by Laszlo may be a more acceptable conceptualization of the "Logoity" shaping all living forms—one devoid of metaphysical overtones, yet one that carries the same notion of a "wording" of the world's beginnings. The question remains, of course, what intelligence is thus "in-forming" the universe?

Ironically, the sevenfoldness that physics does not inquire into is most visibly displayed at the human scale (from music to colors to solar system), halfway between the minute and cosmic realms on which physics has been predominantly focusing (particle physics and astrophysics). Physics does not regard the intermediary realm of organic structures and living systems as relevant to its investigation of particles, and if biology studies the chemistry and physiology of such systems, it rarely inquires into their formative laws and structures as *wholes of matter and units of life*. Particles of matter and organisms of nature are not only "small parts," but *whole forms*. In a fractal world, small particles are analogous to larger particles, so that insights at one scale inevitably shed light on another. Failing to ponder the chain of analogous forms across astronomical, organic, and atomic units foregoes the possibility for structural and dynamic similarities to emerge. Yet the road to a unified understanding of the universe passes through such analogies.

The great absent of mainstream physics, the notion of organic wholes and of a whole-making principle at work in nature and cosmos, might well be the elusive nucleus of a unified theory of physics. The success of a physics of particles depends on a physics of their formation. Understandably, the quasi-subjectivity of the universe implied in the notion of a whole-making principle is the great hurdle for a science based exclusively on the objective, the empirical and the quantifiable. Yet how far is it really from the notion of a singular energy source and space–time "universal" point of origin tacitly acknowledged by the Big Bang theory? That the exclusive focus on isolating and measuring separate particles ignores the golden principle of secting-*bonding* at the center of the creational process might not impress materialistic physicists, yet it is the root of what makes the goal of a "unified theory" simply unattainable by mechanistic means: "Divide" may be key to understanding, but it is only half of the key. The other half is "Unite." The existence of *whole* self-organized systems—from atoms to galaxies—depends on this dynamics of bonding.

A unifying approach to reality would seek to engage the realms of form, color, tone, plant geometry and growth, biology, psycho-physiology (chakras), as relevant to a comprehensive science of the physical, and as necessary to gain insight into the fact that forms—from elementary particles to organisms to cosmic systems—are "informed" by a whole-making principle that points to a source as mysterious yet as reasonable as the highly ordered point-power singularity of the Big Bang.

Erwin Laszlo embraces in a comment what remains a blind spot of modern physics:

> The level of coherence in the organism suggests that in some respects it is a macroscopic quantum system.... The quasi-instant connections that occur throughout the organism suggest that distant molecules and molecular assemblies resonate at the same or compatible frequencies.[139]

However, the reality of "shape" transcends the basic laws of resonance of quantum physics—a fact noted by Heisenberg:

> The interest of research workers has frequently been focused on the phenomenon of regularly shaped crystals suddenly forming from a liquid, e.g. a supersaturated salt solution.... Even when the state of the liquid is completely known before crystallization, the shape of the crystal is not determined by the laws of quantum mechanics.[140]

The convergence of pioneering scientific works (Doczi, Hans Jenny, Agassiz, and Stoney, to name a few) and ancient sources of knowledge (Hindu Scriptures, Pythagoras, Plato) around the notion of a harmonic universe suggests that it may only be a matter of time before science, carried by its own currents, runs into the profound rationality of this view and takes itself to task to trace it in the organization of living systems and to understand its own discoveries in this light.

---

139 Laszlo, *Science and the Akashic Field*, p. 146.
140 Heisenberg, Nobel lecture, Dec. 11, 1933.

V. A Once and Future Cosmogony

The geometries of crop circles are offering discreet hints. Astronomy professor Gerald Hawkins first noted that the patterns of crop circles displayed "exact numerical relationships" similar to those found in the musical scale (i.e., diatonic ratios). A number of patterns had concentric rings like a target. Comparing the areas delimited by the rings, he found ratios such as 3/2, 5/4, 9/8. Something fundamental to our universe is clearly being communicated through these vibration induced harmonic geometries, some of which evoke the forms obtained through cymatics.

In 1952, the composer Paul Hindemith, who based his opera *Die Harmonie der Welt* on the life of Kepler, wrote:

> The science of music deals with the proportions objects assume in their quantitative and spatial, but also in their biological and spiritual, relations. Kepler's three basic laws of planetary motion... could perhaps not have been discovered without a serious backing in music theory. *It may well be that the last word concerning the interdependence of music and the exact sciences has not yet been spoken.*[141]

What Amit Goswami refers to as "idealist science" may be more factual than the term *ideal* suggests. It may be harmonic.

> Idealist science not only heals the schism of the mind–body relationship but also answers some questions that have puzzled idealist philosophers for ages—question like, how does the one consciousness becomes many? And how does the world of subjects and objects arise from one undivided being? The answers to such questions are found within such concepts as tangled hierarchy and self-reference—a system's capacity to see itself separate from the world.[142]

According to the Vedas, self-reference is precisely the principle of self-awareness, the essence of consciousness—which, etymologically, means "co-knowing." It is the knowing that arises from the reflecting-reversing of the universe on itself—what Plato called its "conversing" with itself. "And he made the universe a circle moving in a circle, one and solitary, yet by reason of its excellence able to converse with itself."

Goswami continues, "The time and context for a strong anthropic principle has come—the idea that observers are necessary to bring the universe into being... We are the center of the universe because we are its meaning."[143] Indeed, we are the universe experiencing itself in the reflexive chamber of consciousness, which continues to evolve, under its transcendent triune model, from the form-organizing bonding of harmonic wavefields, to the soul-organizing conjugation of body and spirit in the human self.

---

141 Quoted in Ruff, *A Call to Assembly* (italics added).
142 Goswami, *The Self-Aware Universe*, p. 147.
143 Ibid., p. 140.

The Mystery of Sevenfoldness:
Law of the Octave and Principle of Self-Replication

Reflecting the self-existent Lord like a mirror,
each becomes in turn a world. —Blavatsky[144]

If consonance is the selective, "symphonic" level of harmony that informs and sustains cosmic systems and living organisms, it is because it creates the conditions necessary for a whole unit of life to involve itself in a wholistic form of matter—a "perfect solid." While resonance is the fundamental ground of the world being, consonance is the harmonic conditioning of its forms. While resonance weaves the space–time field of the world soul, consonance shapes its corporealization into forms.

From the point of view of physics, the first necessity in the externalization of the primordial vibration into a vibrant matrix, out of which forms and systems will emerge, is that only half-wavelengths that represent whole number fractions of the fundamental be part of the wavefield or breath of Brahma. The next imperative, which concerns the shaping of forms of energy/mass (atoms), then living organisms (plants), in which the world soul may abide *as a whole*, is that consonant standing waves combine into "symphonic" constructs.

✪ The basic unit of *resonance* is the self-reverberating space of the octave. Common to all musical systems, the octave has been referred to as the "basic miracle of music." The "eighth" note of the musical scale repeats the fundamental on a higher or lower level. We tend to hear fundamental and octave as essentially the same, only higher or lower. The octave symbolizes the primordial act of Being, the projecting of the One Life into what Plato called "Same" and "Other." It is the model of self-sameness, the symbol of an "other that is same." By the same token, the octave designates the *whole* musical space or bundle bound by this sameness. The return of the same at a new level defines an "octavian" space of sameness, hence wholeness. The octave is the archetype of *resonant* wholeness—the replication of the "fundamental," the duplication of the original "one."

The basic unit of *consonance*, on the other hand, is the musical interval, which speaks to a triune rapport of harmony and intermediation between prime, octave, and "consonant interval." The fifth is the archetypal interval of consonance and consonant wholeness at the basis of living forms. Seed point, sphere and organic form are the expression of prime, octave, and fifth. ✪

While resonance weaves the *octavian* space–time field of the world soul, consonance shapes its corporealization into *golden sected* forms.

---

144 Blavatsky, *Secret Doctrine*, vol. 1, p. 30.

V. A Once and Future Cosmogony

The British chemist John Alexander Reina Newlands pointed to the harmonic underpinnings of the sevenfold leitmotiv in the structures of systems and forms of nature when, four years before Mendeleev announced his periodic table, he wrote in *Chemical News*,

> ...The numbers of similar elements differ by seven or multiples of seven. Members stand to each other in the same relation as the extremities of one or more octaves of music. Thus in the nitrogen group, phosphorus is the seventh element after nitrogen and arsenic is the fourteenth elements after phosphorus as is antimony after arsenic. This peculiar relationship I propose to call *"The Law of Octaves."*

The prevalence of sevenfoldness in the structures of nature and cosmos leads one to wonder what necessity bundles the world into "octave"-like forms and systems. What makes the octave the universal model of forms of nature and systems of life?

Two archetypal phenomena are actually engaged in the octave, and they address two cosmological questions. One is the phenomenon of finite wholeness, which speaks to the question we examined earlier: What limits the spiraling out of organic growth? What bundles up whole forms of life? The second is the phenomenon of sevenfoldness, which characterizes organic forms and cosmic systems, and raises the question: Why seven?

What defines the octave interval is the repetition of "the same" note at a higher or lower level of pitch. The octave is a principle of harmonic *identity*, fit for the coming to be of any form of life as a new, living, indivisible "entity."[145] The octave epitomizes the primordial interval of wholeness. A "pocket" of vibratory space, it represents the space–time ground of any entity, be it a universe or an atom—the fundamental space–time "identity" of a unit of life. The notion of "wholeness and sameness" intrinsic to the octave mirrors the archetypal wholeness of living forms.

The octave is literally a "duplicate" of the fundamental—"double" its frequency and half its wavelength. Such is the mathematical literality of the notion of *"self-duplication"* or self-reproduction. The externalization of the self-core of the universe rests on the principle of self-sameness carried out by "self-duplication." The first cell of the human embryo divides itself into two "sames." The reproduction of the primordial "Same" in an *"other same"* opens the *octavian* space of universe and forms. The *wholeness* of any living system lies in this principle of self-replication, which underlies the concepts of hologrammatic and fractal self-similarity. However partial a manifestation of the world being, every form of life is a form of the universal "self" and the outcome of a self-replication. The octave symbolizes the unit of "existence."

---

145 Fideler, *Jesus-Christ, Sun of God*, p. 90.

The "octave" is the "eighth" note of a seven-interval whole. The name "octave" given to the fundamental resonance of the unison intimates that the wholeness it defines encompasses seven major intervals. The term *octave* captures into one the two fundamental principles of forms: wholeness and sevenfoldness. It designates both a note and a 7fold span of notes. The octave-as-note epitomizes the duplication of "wholeness." The octave-as-span represents the interval and sub-intervals that model the organization of forms, organisms, and bodies. Every form of nature is a whole unit of life born of self-duplication, and it is a 7fold "unit." The octave interval is called "diapason"—from the Greek "across all"—which emphasizes its encompassing the entire range or scope of musical intervals. Every form reproduces the seven dimensionality of the world being symbolized in the diatonic scale. Every unit of being unfolds within an invisible mandorla-octave. When we speak of living forms, we speak of *wholes*—hence of *octaves*—hence of *sevenfolds*.

Musicologist Hans Cousto orchestrates this view. "The octave as a unit of measurement can be applied to astronomic periods as well as vibrational microbiological phenomena. Whatever the field, the uniting factor is always the octave.[146]

Reflecting on the experience of the octave as a simultaneous experience of sameness and difference, Lawlor places his finger on the source of the profoundly satisfying, harmonizing, relaxing experience that self-sameness and fractality have been found to induce in human beings—the blend they represent and elicit, of "precise discernment" (otherness) and "harmonious integration" (same).[147] This blend reconnects, re-attunes, and recharges human beings with the golden principle at the heart of creation. To participate at once of the part *and* of the whole is to enter the great chamber of consonance of the universe.

The self-externalization epitomized in the octave is the foundation of the cosmogonical process. *The standing wave constitutive of the octave is the unit-base of resonance* between center and periphery, the fundamental "interval" of the living breath that underlies and sustains all phenomena in cosmos, and the unit of polarity conjugation that "makes possible the birth of matter." *The sevenfold spherical octave or torus is the unit of consonance,* the diatonic unison that informs every system in the universe and makes it a unit of existence. The template of the universe is a sevenfold symphonia or consonance, and a sevenfold standing wave structure swirls in the filigree of every organism. Like spirillæ in the atom, some consonances might lie dormant until awakened by evolution.

---

146 Cousto, *The Cosmic Octave*, p. 25.
147 Lawlor, *Sacred Geometry*, p. 82.

V. A Once and Future Cosmogony

World traditions offer analogous outlines—at once different and identical—of the cosmogonic process. In the beginning was *vibrancy*. Being (1) resounded and *resonated* with itself (2), and space–time–soul was conceived, the logoic trinity of *world–self-awareness* (3)—a triune Oneness, which deploys itself in sevenfold forms.

As cited earlier, the ancient Stanzas of what Blavatsky refers to as the "Parent Doctrine" describe: "When the one becomes two—the 'three-fold' appears. The three are (linked into) one; and it is our thread, O Lanoo, the heart of the human plant, called *Saptaparna*."[148] In Sanskrit, *saptaparna* means "seven leaves." Theosophy represents Saptaparna (sevenfold human) as a triangle (or triad) above a square (or quaternary). Remarkably, Goethe also saw in the plant a seven-leaved entity.

We saw how the octave archetype can be traced in every kingdom of nature, from crystal to plant to human form, from snowflake to solar system. E. J. Pace remarks how the snowflake provides us with "the best design for illustrating the law of the octave, since each flake is the outgrowth in six directions of an original central nucleus. This central nucleus, always present, binds the sixfold ramification into a unity, thus giving us the number seven."[149]

One could say that a plant is a unison defined and "bound" by seed. Its sevenfold form unrolls within the virtual unison of the seed-fundamental and its seed-octave at the end of the cycle. Similarly, the human being is a seven-chakra unison, a finite octave-like system sustained by consonant standing waves, and harboring an evolving self. Ultimately and primordially, the universe is a unison. Like every lesser form, it constitutes a self-organized whole of space–time self-awareness, a spherical vortex expanding out of its center and returning to itself, exhaling a cosmos that is drawn back to its source. Conscious breathing gives us a direct experience of this fundamental gesture of creation.

In the midst of every unison—*as its midst*—dwells an I-dentity, the "I" of the entity sounding itself out as unison, the self of the "self-sameness." From the interplay between self and not-self arises self-awareness: the unfolding substance of I-dentity, which is the triunity of self, world, consciousness—fundamental, octave, golden mean.

The *I Ching*, the ancient Chinese "book of changes," states the following: "By the constant return of the seventh day, we may discern the mind of heaven." Indeed, the mind of the universe reveals itself in the sevenfold bundling that characterizes its manifestation in forms and cycles. The return of the seventh day is the return of the self to its self—the dominant of the scale, the dominus of the world, the *"dimanche"* (or Sunday) of the week.

---

148 Blavatsky, *Secret Doctrine*, vol. 1, pp. xliv, 231.
149 Pace, *Moody Bible Institute Monthly*, vol. 22, p. 962.

In ancient Greece, notes Rudhyar, "the seven Fundamentals were called arches (Beginnings, principles). Each of these had a function to perform in an organized life field. Each arche, however, could act as the origin or initial tone of a mode that dealt with the unfoldment of the arche's special function in the sevenfold scale of fundamentals (the grama)." Rudhyar rightly cautions against associating the seven cosmic forces with any particular overtone, emphasizing that they refer to seven aspects of the original power of the One, which manifest as seven fundamentals at the foundation of most musical cultures... and "centralizing areas of tone potency in human beings."[150]

The names given to the seven notes of the diatonic scale a thousand years ago by Guido d'Arezzo suggest a still vivid awareness of the correspondence between the sevenfold scale of consonant sounds and the cosmological ladder of Being. Beginning with the higher *DOH*, taken from *Dominus*, which means "Lord" (the absolute), the scale descends through *SI*, from *Sider*, the stars; to *LA*, from *Lactea*, the Milky Way galaxy; then to *SOL*, the Latin name for the Sun; *FA* or *Fateas*, the Fates, the ancient name given to the planets; *MI*, from *Microcosmia*, the Earth; *RE*, from *Regina Coeli*, the Queen of the Heavens, the Moon; and finally the lower *DOH*-*Dominus*, Lord—to be revealed in man.

❋

The question remains, what makes self-replication a sevenfold phenomenon? What makes the cosmos of a Logos a sevenfold "octave"? Different perspectives have been proposed on this archetypal structure of the universe and its forms. All converge to reveal that the key to the mystery of sevenfoldness lies in harmony.

1. One approach to "why seven?" is that of Kepler, who splendidly fathomed in the sevenfold the "pre-designed cosmological clausura" (musical enclosure) of the heavenly motions:

> The heavenly motions are nothing but a continuous song for several voices (perceived by the intelligence, not by the ear); a music which through discordant tensions, through syncopes and cadenzas, as it were (as men employ them in imitation of those natural discords) progresses toward certain predesigned, quasi-six-voiced clausuras, and thereby sets landmarks in the immeasurable flow of time. (*Harmonice mundi*, 1619)

Kepler's major insight into the law of the octave was based on experience. He found that there are six and only six possible ways to divide the space of the octave (half a string) in a completely harmonic way. He called them harmonic divisions or harmonic means: $5/6, 4/5, 3/4, 2/3, 5/8, 3/5$ (fig. 68). With the octave half point of

---

150 Rudhyar, *The Magic of Tone and the Art of Music*, pp. 68, 79.

the string, 1/2, these harmonic divisions amount to seven. These are not mathematically obtained means, nor do they follow the harmonic series, except for the initial overtones. They are harmonic divisions identified by hearing. The simple experiment of lightly moving a finger up a string, barely touching it, reveals that certain regions give out a sound, while others don't. Over time, cultures throughout the world would have found these natural stopping places and singular points within the octave. Just about every musical culture discovered the fifth, which meant they also knew the fourth.

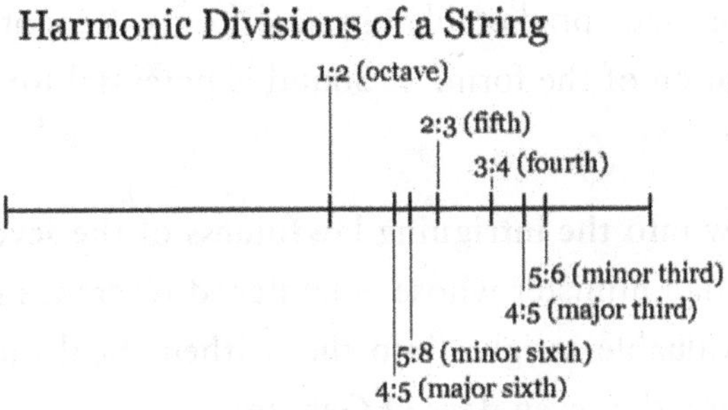

Harmonic Divisions of a String

1:2 (octave)
2:3 (fifth)
3:4 (fourth)
5:6 (minor third)
4:5 (major third)
5:8 (minor sixth)
4:5 (major sixth)

Fig. 68. Kepler's ratios for the intervals he found to be consonant with the whole and with each other. (www.keplersdiscovery.com)

These six singular points identified by Kepler are points of consonance within the octave. Each has the significant property that it produces a *threefold* harmony: "the larger part is consonant with the whole, the smaller with the whole, and the two parts with each other, so that if all three tones created by this single division are sounded together, the result is completely consonant with itself."[151]

A vibrational matrix, space is not uniform. Analogous to the standing waveforms of electrons around the atomic nucleus, points of potential consonance along a string or within a finite space represent singular zones of wave conjugation apt to generate secondary tones. In the sevenfold systems of nature, these *bindus* points of "union" are wombs of partial, secondary replication of a fundamental wholeness into sub-formations. Such places of concord (proportional rapport) between part and whole are where planets orbit, where leaves grow and flowers appear, where chakras unfold the seven dimensions and organ systems of being.

The imperative of wholeness folds in and organizes any wavefield along internal bondings that culminate into forms of consonance. Musicologist Joscelyn Godwin captures this process both factually and soulfully:

> One could regard a tone not in the usual way, as a point radiating harmonics, but as a musical space created in chaos by the formative forces of harmony:

---

151 See http://science.larouchepac.com/kepler/harmony/book3/3.

forces that become more and more clearly focused as they proceed from complex to simple ratios, culminating in the perfect fourths, fifths, and octaves around the tone itself. In an analogous way, Steiner had his students imagine flowers not as the result of interior, vegetative forces pushing the petals outward, but rather as molded by exterior, cosmic forces pushing inward. And here, too, the unimaginable complexity of forces in the environing cosmos resolves itself at the flower-head into the simple geometry of petals.[152]

As Chladni's experiments show, matter aggregates around zones of consonance, confirming what Tantric wisdom taught long ago, that "sounds as physical vibrations are able to produce predictable physical forms. Repetition of the exact note results in a duplication of the form.... Sound is potential form, and form is sound made manifest."[153]

2. Another window into the intriguing lawfulness of the sevenfold is proposed by E. J. Pace, a Christian minister whose enlightened reverence for the world and the Word led him to valuable insights into the mathematical rationality of the seven notes of the scale and the seven days of Genesis:

> The arrangement is neither arbitrary nor due to mere coincidence. On the contrary they are based on laws to which the ear of man is so nicely adjusted as to result in pain if these ratios are disturbed. The ratios in sufficient number to illustrate are here given: do is to do (an octave above) as 1 is to 2. Do is to sol as 2 is to 3. Do is to fa as 3 is to 4. Do is to mi as 4 is to 5, while do is to re as 4 is to 4 1/2. We need go no further, for the remaining notes are similarly related.[154]

Pace also highlighted in the seven notes and the seven colors triadal symmetries around a center reminiscent of the Platonic Lambda, to which we will soon return.

Dane Rudhyar complements Pace's insights into the five fundamental intervals at the basis of the octavian structure:

> The structure of the entire tonality scale rests upon five intervals or types of relationships: the octave, which defines the wholeness of the whole; the fifth, which is the organic factor of centrifugal expansion inherent in all living wholes; the fourth, which seeks to reintegrate the centrifugal elements within the organic whole; the whole tone which is the building block of the organism; and the semitone which refers to the circulation of sonic energy, the fluidity of life as well as of psychic feelings (the aspirations, longing, suffering and traumas of the individualized consciousness).[155]

---

152 Godwin, *Harmonies of Heaven and Earth*, p. 189.
153 Woodroffe, *The Garland of letters*, pp. 224–227.
154 Pace, "The Law of the Octave," in *Moody Bible Institute Monthly*, no. 22, Sept. 1921.
155 Rudhyar, *The Magic of Tone and the Art of Music*, p. 95.

V. A Once and Future Cosmogony

3. A third perspective sheds light on the relation between five and seven, and illumines in sevenfoldness the house-like template of the universe: fourfold base (material elements) and threefold roof (spiritual quintessence).

This perspective is based on the correspondence between elements, major musical ratios and Platonic solids. Ancient wisdom, echoed by Tantric teachings and modern researchers such as Randolph Stone, views the elements as the major fields of consonance that underlie the formations of nature and entrain all of creation through sympathetic vibration. We saw earlier that Hindu wisdom describes these five elemental fields in terms of motion: radiatory (ether), cohering (earth), upward (fire), downward (water), and transverse (air). This fivefold may be viewed as the deployment, in the world of polarized forms, of the essential threefold motion—forward, centrifugal, centripetal. Mirrored in the cosmic scaffolding of nested Platonic solids intimated in the *Timaeus* and "cosmographed" by Kepler, Ether, Air, Fire, Water, Earth represent the major quantum phases of harmonic conjugation between the materializing and spiritualizing currents of the universal breath. They constitute the archetypal fields and dimensions of the world being. "All solidity, regardless of the phenomenon, resonates with the earth harmonic, all liquidity with Water, all heat with Fire, all movement with Air, and all space, the element of the universal breath, with Ether."[156] Jill Purce offers a similar insight, "The soul as the nucleus emanates five fields of force. All creation is attuned to the harmony of these elemental keynotes of vibration."[157] The four elements are moments in a cycle of wholeness, phases in a single process, relationships within a whole.

Ether is the quintessential, wholistic element, in which the plenitude of the soul can abide. The four elements differentiate and ground this quintessential ether of space into the four corners of matter. The world soul, which lives and breathes in the ether, is a threefold Logos, so we can see how the "five" imply seven: four tangible elements plus a triune quintessential one. The first three of the sevenfold are implied in the triune fifth—hence the notion of the fivefold (versus sevenfold) stepping of spirit down to matter. The lesser four are also implied in the quintessence. Being the triune quintessence of the four elements, ether may be viewed as implicitly sevenfold. Using the correspondences proposed by Plato, this sevenfoldness may be laid out as follows: Point (Father-source), sphere (mother-matter), dodecahedron (Logoic ether), tetrahedron (fire), octahedron (air), icosahedron (water), cube (earth).

In Tantric teachings, ether is said to "sustain the space in which the other forces operate." The "Space" in which the other forces operate is the vibrant quintessential ground of the physical world, the Shiva-Shakti-breath of Brahma. At once the essence

---

156 Burger, *Esoteric Anatomy*, p. 149.
157 Purce, *The Mystic Spiral;* in Burger, op. cit. p. 11.

of physical reality and the "element" of the Logos, ether is literally the "milieu" where metaphysics (the triune Logos) and quantum physics ($E=MC^2$) could begin to converse with each other, the dynamic field where mind and body are engaged in constant co-creation and co-regeneration. Ether is the *mindful* element of the physical world as well as the grounding, *soul-filled* element of the universal Life. It is the intelligent biofield that shapes particles, organizes forms of matter, and sustains all living systems by means of vibratory interplay and harmonic consonance. It is the milieu of the universal psyche, as its 5fold/12fold dodecahedric signature indicates. With ether and its dense substratum, the electromagnetic field, we step out of physicality and mass into the quintessence of manifestation, which is indeed a psychic reality, being the wholistic carrier of the harmonic intelligence at the source of all living forms. "The biofield, the causative energetic template for the organism, is mediated by coherent electromagnetic fields," states Jude Currivan. "A deeper understanding of electromagnetism and the nature of biofields appears to be the most likely key to the interface between consciousness and the physical world."[158]

Le Mee's research on the Platonic solids brings a concurring insight on the relation of the 5fold to the 7fold archetypal field of the world being. We remember his demonstration that the Platonic forms, which represent the only five possibilities of fractioning Euclidian space equally, are defined by the ratios of the first three integers. These ratios (1/1, 1/2, 2/1, 2/3, 3/2), which correspond to the three intervals of unison, octave, and fifth, are associated with the internal angles that produce, respectively, the octahedron [1:1], the icosahedron [2:1], the cube and tetrahedron [1:3] and finally the dodecahedron [2:3]. Le Mee's next thought is what interests us here: "Starting with the geometric point as origin (spirit), the Platonic solids constitute five steps toward the complete sphere (matter) which, becoming the seventh step, links them as in a musical scale in an octave progression."

The point and the sphere correspond to the first and second harmonic (fundamental and octave boundary of a musical whole). Similarly, the Pythagoreans regarded the tetrahedron, octahedron, cube, and icosahedron as the shapes of, respectively, the elements of fire, air, earth, and water.

Le Mee's study shows that the octave-like ladder of Platonic solids, which informs natural and cosmic forms, from the solar system (Kepler) to the atomic elements, from the constitution of plants (Goethe) to the chakric human energy system (yoga), issues from the first three numbers. He relates his finding to Socrates' cryptic statement in Plato's *Republic*, "All this based on "the little matter of distinguishing one, two, and three."[159]

---

158 Currivan, *The Wave*, pp. 215, 186.
159 Plato, *The Republic*, vii 522.

V. A Once and Future Cosmogony

Unlike the sharp and smooth steeples or straight towers of European cathedrals, the ogival towers of the Angkor Wat temples, erected around the same period, narrow up in graduated steps to the top—lotus bud or pine cone like—or even, one might say, pineal gland like. When I asked our Cambodian guide if he had any thought about this unique architectural feature, he replied without a pause: "Because this is the way to heaven." Indeed.

4. Fabre d'Olivet, an eighteenth-century French poet and composer who combined a love of music with a deep interest in Pythagoras, offers an illuminating fourth perspective. His insights into the foundations of the musical scale echo major themes of the present work. He emphasizes first the unanimously recognized influence, among the peoples of the world, of the numbers 7 and 12, produced by the simple addition or multiplication of the numbers 3 and 4—the numbers associated with the trinity of spirit and the quaternity of matter:

> The number 12, formed from the ternary and the quaternary, is the symbol of the universe and the measure of tone... a scientific and sacred dogma accepted in all ages and among all nations from the north of Europe to the most eastern parts of Asia. The Egyptians, the Chaldeans and the Greeks attributed the government of nature to 12 principal gods.... The Scandinavians named these 12 rulers the Ases.... The ancient temples all bore the same number and the same division. The Peruvian architects had ideas in this regard no different from those of the Egyptians, the Persians, the Romans, and even the Hebrews.[160]

As for the number 7, Fabre explains that the seven-holed flute placed in the hands of Pan, the god of the universe, figured the distribution of the soul of the world among the seven spheres. The rationale he offers for the diatonic septenary reaches right into the golden essence of harmony. He traces it to the two identical sets of tones produced by the descending fourths and ascending fifths unfolding from two notes, which he refers to as "the fundamental principles"—B and F:

BEADGCF

FCGDAEB

"It is this identity," he says, "that constitutes the musical septenary."

While all seven diatonic notes figure in this septenary, they appear in a different order. Instead of the succession of the scale, they follow the series of fifths and fourths. To realize the significance of this new order, we need only to note that the fifth and the fourth are the dual manifestation of the third harmonic within the span of the octave. The golden third, the basis of the harmonious division of the whole

---

160 D'Olivet, quoted in Godwin, *The Harmony of the Spheres*, pp. 345–346.

(octave), divides the octave into a fifth and a fourth. The fourth complements the fifth. Fabre also points out that the note on which the two strings meet is D, which he thus regards as the archetype of unison.

The meaningfulness of this different ordering highlighted by Fabre is that it reveals in the septenary the direct descendant of the third (as fifth-and-fourth)—into the "whole form" of an octave. This dual descendance honors the two sides of the third harmonic, which produces a fifth in reference to the lower octave, and a fourth in reference to the higher end of the octave.

The 12fold chromatic scale is also, we saw, the deployment of the third harmonic within the octave milieu. However, what distinguishes the 7fold from the 12fold is the unique extract of identical ascending fifths and descending fourths that it (the 7fold) represents. This integrated division of the whole by fifths and fourths holds as one the duality of above and below, soul and body, involution and evolution, ternary and quaternary that fifth and fourth symbolize. Directly derived from the fundamental triad of harmony, it expresses the two sides of this trinity: the solar fifth and its earthly, corporeal shadow, the fourth. This perfect conjunction of the downward steps of creation and upward steps of evolution makes of 7 the most integral and concise expression of creation, a splendid essence of the fundamental harmonic steps of descent/ascent between spirit and matter.

This dual septenary is also the bidirectional series of sharps and flats associated with the keys of music composition. Sharps or flats are the modifications necessary to reproduce the intervals of the diatonic scale for scales based on different fundamentals. As every musician knows, the order of the sharps is F, C, G, D, A, E, B, and the order of the flats is B, E, A, D, G, C, F. Thus B and F are the true extremes or "fundamental principles" indeed of the dual septenary produced by the coinciding notes of the descending series of fourths and ascending series of fifths—within the octave.

Fabre's dual septenary of fifths and fourths is the regal profile of the sevenfold, the essence of what the diatonic scale lines up into a sequence of seven notes. The secret of its noble birth from the third—as both fifth and fourth—reveals in the sevenfold the steps of deployment of the trinity into the centrifugal and centripetal currents of an octave like form—whole and alive. This precious fact of harmonic law opens a window of insight into the dual advent of body making and soul arising—suggesting how the world soul enfolds itself in forms of matter that are also emerging forms of consciousness.

V. A Once and Future Cosmogony

5. An equally luminous and essential perspective on the sevenfold is the mathematical approach offered by Plato. The unfolding of the first trinity into a septenary and a dodecad, which is how Plato articulates mathematically the deployment of cosmos, highlights another access to the harmonic ground of the seven and their mathematical inevitability.

Plato's Lambda, at once a mathematical and geometric formula, captures succinctly and powerfully the world's formative step we are attempting to grasp from the physical end of reality, as we did earlier from the metaphysical end: the unfoldment of the trinity into a sevenfold cosmos generating septenary forms and systems. In the *Timaeus*, Plato describes the step that follows the creation of the soul compound (the primary mixture of *Same, Other,* and *Being*) as an apportioning of the triune "one" (the world soul) according to the simple "powers" of 2 and 3. The demiurge, he says, makes *seven portions* based on two geometrical series generated by the simple powers of 1, 2, and 3 raised to their own powers ($1, 2, 2^2, 2^3$, and $1, 3, 3^2, 3^3$): 1, 2, 4, 8 and 1, 3, 9, 27.

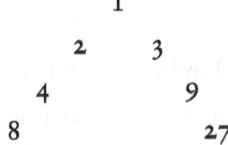

Arranged in two branches, the seven portions correspond to the series of the powers of 1, 2, and 3:

> 1 (the 1st);
> 2 (the 2nd), 4 (the 4th), 8 (the 6th);
> 3 (the 3rd), 9 (the 5th), 27 (the 7th).

The first branch contains the "double intervals," i.e., the powers of 2; the second one the "triple intervals," i.e., the powers of 3.

The Lambda conveys in number and graphic form the cosmological steps from oneness to two-ness (the two progressions), three-ness (two triads forming a triangle: 2, 4, 8 and 3, 9, 27) and sevenfoldness. We see quite literally the one (the threefold one) unfold into seven portions (seven numbers to the lambda) by way of "apportioning according to the *simple powers of one, two, and three*—a quintessential rationale for the cosmological necessity of seven.

The threefold principle (Logos) of the universe is what makes self-replication a sevenfold phenomenon. At work in the three, as well as in the seven, is the harmonizing of the power of two and the power of three—power to divide and power to integrate. The sevenfold is the universal quantum of 3fold Logoity—the quantum of unitive diversity.

The etymology of the word *music* analyzed by Fabre d'Olivet infuses with the vitality of meaning asleep in the depth of words his and Plato's insights into the harmonic secrets of cosmology:

> The word *music* came to us from the Greek *mousike*, via the Latin *musika*. It is composed, in Greek, of the word *mousa*, muse, which comes from the Egyptian *mas ou mous*, [and] signifies *"generatio,"* the production or external unfoldment of a principle—in other words, the enactment or manifestation in form of what was in potential.... So the Greek word *mousa* (muse) originally applied to any unfoldment of a principle, any sphere of activity in which spirit transited from potential to action and sensible form. From this etymology of the word *music*, one easily understands why everything which serves to externalize thought, render the intellectual sensible, translate potential into action, and clothe it in an appropriate form, was regarded to pertain to music.[161]

Stated simply, *music means genesis*. Genesis is an essentially musical phenomenon.

The Twelvefold Inwardness of the Sevenfold Form

The other beautiful facet of the lambda, which makes it a most pithy insight into cosmology, is the mathematical portal it offers into the twelvefold core *of the sevenfold*. It shows in mathematical terms how the 12fold emerges from within the 7fold as its complementary reality—as inside to outside. The supreme meaning of the 12fold radiates discreetly through the factuality of numbers when we see it emerge from a "filling in" of the 7fold scaffolding by mediating "means." Filling in with mediation reveals the "inside" of the sevenfold, which is the reality of the twelvefold: the soul "milieu."

Although this insight into the inwardness-bearing character of the 12fold in the midst of an external 7fold frame was introduced earlier on in the book to facilitate clarity for the reader, I did not have this clarity at that early stage of the journey. This fuller grasp of the Lambda only emerged when an unanticipated return to the *Timaeus* brought the complementarity of "seven" and "twelve" to the light of day. At the risk of creating a feeling of repetition, it seemed important at this point to recollect and behold this central blossom of 7 and 12 held in Plato's Lambda as a gem like emblem of the essential contemplation of this book.

Here, again, is the way Plato describes this next creational step: "He [the demiurge] went on to fill up the intervals in the series of the powers of 2 and the intervals in the series of powers of 3" by inserting two means (harmonic and arithmetic *means*) between the terms of the double geometric progression (by 2 and 3). Once

---

161 D'Olivet, *La musique expliquée comme science et comme art*, p. 47 (author's translation).

V. A Once and Future Cosmogony

reduced to their simplest denominator, these means amount to the three intervals/ratios of 3:2, 4:3, and 9:8. We recognize in these ratios of 3:2 (arithmetic mean), 4:3 (Harmonic mean), 9:8 (the relation of the harmonic to the arithmetic mean) the fundamental musical intervals of the fifth, the fourth, and the tone. Plato continues: "He went on to fill up the 4:3 intervals with 9:8 intervals [the tones]. This still left over in each case a fraction, which is represented by the terms of the numerical ratio 256:243 [the semi-tone]. And thus the mixture, from which He had been cutting these portions off, was now all spent."[162]

In the end, the "inside" of the sevenfold, the mediating means between the 7 portions of the world construct is a 12fold "filling" of semi-tone intervals. Like the 7, the 12 unfold out of the 1 at the apex—the three-in-"one" at the source of manifestation. This triune one, the third of the primordial trinity, is initially "apportioned" into the 12fold world soul compound, which the Demiurge then splits, says Plato, into the X of 7fold planetary system and 12fold zodiac—outside physical construct and inner quintessential structure of selfhood—body and soul. Similarly, the third harmonic blossoms into a 12fold circle of fifths within a 7-octave cycle. The triangle of 3, 7, 12 is the second trinity, the *trinity of manifestation* abiding in the forms of nature and cosmos.

If we bring the Hindu perspective on the elements in resonance with Plato's description of the splitting of the world soul compound into two circles crossing each other and revolving in opposite directions, fresh insights vitalize the understanding we reached earlier:

One circle is the *radiating*, life-giving 12fold zodiacal circle of ether—the zodiacal quintessence of our physical cosmos, the animating "circle of life," which "fills space with the 'Hairs of Shiva,' nonobstructive motion radiating lines of force in all directions."

The other is the substantiating, *cohering* 7fold planetary circle that involves (in-volutes) the world soul in densifying spheres of elemental substance and gives it physical, earthly form, "through a motion (Earth, *Prithivi*) which produces cohesion and obstruction, the opposite of the nonobstructive Ether."

In the human being, the polarity of causal body (the 12fold soul's body) and chakras (the 7fold etheric–physical body) replicates the cross of 12fold zodiac and 7 planetary spheres. Plato said it plainly, "the soul of the world is crucified on the body of the world." The great Platonic cross separates and binds the 12fold soul and the 7fold body of the world.

---

162 Plato, *Timaeus*, 36b

From the perspective of world beginnings, the emergence of the twelvefold soul precedes that of the sevenfold form. From the perspective of evolution however, the twelvefold eventually arises (as chlorophyll, then DNA, then causal body) from within the sevenfold.

Twelvefolding is at once the first operation of creation and the goal of human evolution: the blossoming of the world soul in humankind's consciousness. Twelvefoldness carries the evolutionary, "soul-ward" direction of cosmology, while sevenfolding carries the "world-ward" involutionary stream of cosmos making.

The 12fold–5fold does not appear in the formations of nature until matter reaches a stage of integration of the four elements sufficient to host their "one" quintessence, the *ether of life*, and so reveal in *living* organisms the soulful milieu that "fills" the 7fold forms of cosmos and nature with a 12fold organum of autonomy and "selfhood." This 12fold organum signals the emergent capacity of matter to organize itself in resonance with the whole, live breath of the world soul.

The plant kingdom marks the apparition of physical forms endowed with life, cycles of growth, and self-reproduction. By contrast, particles and chemical atoms, crystals and minerals have no "entity" principle and no autonomous self-regenerative biofield—no processes of growth and reproduction. A quantum leap from the mineral kingdom, organic systems represent the next stage of complexity and coherence in the evolution of matter.

In terms of the underlying laws of harmony, while atomic elements are shaped by the secondary harmonic intervals on which the first four Platonic solids are built, the plant kingdom integrates the "quintessential" interval—hence the ubiquitous presence of the golden mean in the plant kingdom. With trees and plant, and, much later, animal and human forms, the fifth element, the twelvefold ether, appears in singular "forms of life" as the *essential* element it is, the element of *life*. Life passes into forms, wrapped in vortical biofields that are at once "field" and "bio"—forms of substance and forms of life—countless in quantity yet one in essence. The biofield ushers life into forms of substance by way of the golden mean—the "earthly measure" (*geo-metry*) of the Logos. While the plant kingdom is sevenfold in its archetypal shape, it is also imbued with the golden, twelvefold quintessence of the world. In the mineral, the world soul becomes aware of itself as structuring power. In the plant, it becomes aware of itself as life-giving energy.

While the plant kingdom represents the evolutionary synthesis of a *life-bearing* twelvefold molecular structure, the human kingdom carries the emergence of the *love-bearing "molecule,"* the twelvefold blossoming of Logoic "I"-ness in the human psyche. In the same way that the plant kingdom achieves the synthesis of

V. A Once and Future Cosmogony

chemical nutrients out of sunlight energy, the human kingdom is engaged in "synthesizing" soul forces. Plants achieve photosynthesis, humans—*philosynthesis*.[163]

If the fifth interval (third harmonic or golden mean) does not enter the composition of elementary particles and chemical atoms, it is nonetheless implicitly present from the beginning as the keystone of the universe—the threefold principle of harmony-love. However, it takes eons for forms of matter to evolve to the degree of integrated complexity necessary to house and express it. Similarly, the human being is potentially present from the beginning as the bearer-to-come of the Logos. The Logos is implied in cosmos, unhoused and unrevealed in form until beings appear that are endowed with a sufficiently integrated fourfold constitution to vehiculate and express the fifth principle of creative love-intelligence.

The fifth, the "dominant" of the scale, expresses in sound the "dominus" of creation, the triune Logos who intones the principle of selfhood. Although golden mean and *penta-dodeca* structures, the signatures of the third harmonic, appear only at the organic level of evolution, the third harmonic is implicitly involved in the sevenfolding of any system in nature.

To avoid confusion, it is important to note that the etheric is the *fifth element of the physical* plane, which is itself the seventh and densest level of cosmos, and that it is only with the advent in form of the *fifth principle of the world being*, the principle of mind, that the "dominant" of the grand octave of cosmos, the human soul, appears—dominant in that, like the third harmonic, it is at once physical and divine, humus and Being. It is the true golden mean.

The fifth of individual mindedness and soul (the twelvefold causal body) defines humanity as the golden mean between particulate physicality and universal being. A new instrument of expression was born with the human kingdom: language and speech. Speech emits the fundamental Breath, which language shapes into a mediating space of communication–communion among humans. Spoken language is meant to articulate the mediating reality of the fifth—Logoity. It harbors the potential to voice the "Word" of life. Human beings emulate the harmonic cosmological act of creation when they emit vowels and shape "consonants"—by "*partly* obstructing the breath." Similar to the way musical instruments are set up, via frets or keys, to

---

163 Robert Monroe recounts the revelation he had of this central task of humankind—to produce love. He relates how, during one of his out of body experiences, he was given a glimpse of the human kingdom's contribution to the universe—one that translated in his mind into a picture of cattle raised to produce milk. For two weeks following the experience, he found himself in a deep depression. Although love in its many forms—love between beings, love for the universe—is a creative, joyful, life-giving participatory experience, it may nonetheless feel sobering for the intellectual self to realize that it is the "product" our kingdom is meant to contribute to the whole.

emit notes, the human throat trains itself to create the proper intervals and shapes for standing waves to format the air element into "units of mindedness"—vowels, words, phrases—that convey the fifth dimension of Being. For long, these formations happen only "up in the air," yet Steiner indicates that there will come a time when "man will speak man." A central feature of human evolving is the journey from the multiplicity of speech to the simplicity of Word. The word is divine; speech in its many diversifications is human. Wrong speech separates. Right speech contributes "to that great harmonizing chord or unifying word which it is the function of mankind ultimately to utter."[164]

In his 1970 book *I Seem To Be a Verb*, Buckminster Fuller wrote, "I live on Earth at present, and I don't know what I am. I know that I am not a category. I am not a thing—a noun. I seem to be a verb, an evolutionary process—an integral function of the universe." The verb, the middle third of any sentence of subject–object–verb, is indeed its animating "logos."

If the plant kingdom heralds the appearance of golden mean geometries apt to vehiculate for the first time the etheric wholeness of life, the human kingdom marks the revelation of the *inner, Logoic reality* of the golden mean in the self-aware, self-causing human being. Teilhard de Chardin refers to this momentous event as the birth of reflection, Alice Bailey as the event of individualization. Since then, the Logoic quintessence has been unfolding in stages: After a long build up of the intelligence aspect of "selfhood," the second, love aspect—the heart of selfhood—seeded into Earth evolution by the incarnation of the "Logos," continues its two-thousand-year gestation and approaches a threshold of collective birth in the human race as a whole. Clearly, a large number of human beings have already crossed this threshold as individuals.

---

164 Bailey, *A Treatise on White Magic*, p. 143.

VI. FACETS OF A UNITIVE SCIENCE

> *What we call truth lies in the rational harmony*
> *between the subjective and objective aspects of reality,*
> *both of which belong to the super-personal man.*
> —Rabindranath Tagore[1]

Allowing the historical extremes of our collective understanding of the world to echo each other has hollowed out the hub of a potential unitary view centered on the forms of nature, their formation and their evolution. To envision these three facets more clearly, we will take a brief recapitulating look at the first two and explore the third.

Toward a Harmonic Science of the Forms and Systems of Nature

> *I recommend that we consider the comparison of the periodic table of elements with our harmonic system of partial-tone coordinates in the same way that all comparisons should be considered: as a parallel, incomplete but touching upon the innermost core, between a well-known phenomenon (periodic table of the elements) and an as yet unknown phenomenon (partial-tone coordinates).*—Hans Kayser[2]

What emerges from bringing together the complementary perspectives of modern research and ancient wisdom—matter's evolution into forms and spirit's involution into forms—is the notion that the *universal unit of form* is an octave-like structure, a sevenfold dual vortex, whose inner core, triune and twelvefold, is the vessel of an indwelling "self principle." This archetypal "inside" of nature's forms is, in essence, the harmony of power–intelligence–love, which in-forms with its unitive

---

1 From a conversation between Rabindranath Tagore and Albert Einstein in the afternoon of July 14, 1930, at the latter's residence in Kaputh. Tagore was responding to Einstein's statement: "The problem begins whether Truth is independent of our consciousness."

2 See http://www.sacredscience.com/archive/HarmonicsScience.htm.

logarithmic dynamics—of centrifugal and centripetal spiraling—the biofield of the fourfold physical realm.

The world-creative operation of self-*replication-and-mediation* defines seven dimensions, seven steps of involution of spirit in forms of matter imbued with a 12fold soul potential, through which the world consciousness makes its way from the cosmic stage of "world soul" to the systemic stage of etheric autonomy (plant form), then, much later, to the individualized stage of causal selfhood (human form).

The seven-dimensionality of the universe replicates itself in sub-septenaries, which are seven types of consonance among dimensional polarities: matter-energy for the mineral, Earth–Sun for the plant, self-world for the human realm. Whereas the plant kingdom represents the creative vitality of the world-being between Sun and Earth, the human kingdom represents the self-aware milieu of mindedness between spirit and matter.

Each kingdom is, in turn, sevenfold, and replicates each of the seven primary dimensions at a lesser or greater level of complexity: 7fold periodic table of the chemical elements, 7fold composition of the plant, 7fold chakric differentiation of the human being. Each "fold" resonates with one of the seven primary dimensions and expresses one type of self-world creative consonance, one key proportion of matter-spirit—from the base consciousness of self-preservation to the crown consciousness of self-world identification. Whereas the flower, for instance, is the expression of the fifth or third, "soulful" stage, of a plant, the sacral center is the sixth (from above) or second (from below) dimension of our being—its etheric, generative, and regenerative dimension.

Every level of a sevenfold, every qualitative "quantum" leap in the forms and systems of nature shows an increasing order of complexity. The increasing number of harmonics present in the successive octaves of the harmonic series suggests a corresponding increase in complexity. "The human form being the highest development of the external universe," Babbitt advances, "it should have the highest manifestations of harmonic features."[3]

These considerations lay outside the field of mainstream science. However, stretched by the questions and possibilities raised by the new framework of a quantum or string-like ground of reality, the scope of inquiry of contemporary science is widening, and the exploration of formative processes engaging matter and consciousness (even if consciousness is clothed as "information") is no longer such a distant proposition. In their musings about matter, theoretical physicists of the quantum era came close to the view of the ancient Greeks. Einstein and Schroedinger, Weyl and Clifford agreed that the secret of matter would be found in the structure

---

[3] Babbitt, *The Principles of Light and Color*, p. 19.

VI. Facets of a Unitive Science

of space, not in point-like bits of matter. "What we observe as material bodies and forces are nothing but shapes and variations in the structure of space. The complexity of physics and cosmology is just a special geometry."[4] To investigate such geometries and their source seems central to a next vital leap in science. As long as the notion of a science of forms is not gaining momentum, the dream of a theory of everything may remain a dream, and the revelatory power of such apparently marginal sciences as cymatics remain cut off from the possibility of further theorization and practical elucidations.

The discovery of the electromagnetic field provided the first scientific inkling of an invisible wavefield underlying natural phenomena. The subsequent revelation of the quantum field reality of matter, followed decades later by the demonstration of non-local correlation in physics and morphogenetic fields in biology dealt a final blow to the hope that a mechanistic analysis of parts and particles interactions could master the mysteries of matter and organisms. A collective step had been set in motion toward a recognition that particles, as well as cells and organs, are parts of an undivided whole and respond to its laws.

The increasing presence in contemporary culture of various forms of energy healing (acupuncture, radionics, reiki, homeopathy) implies a tacit acknowledgment of a formative field at the source of organic forms—an energy field known for thousands of years under the names *chi, ki, prana, ether,* and *akasha.* A bridge between physical and psychospiritual realms, the etheric field steps down into physical organisms the psychic wholeness we are and the universal beingness of which we participate—of which we are "parts." Randolph Stone portrays the step-down of soul energy into the physical fields of the body as caduceus-like—"a dual force of positive and negative life breaths conveyed to all the tissues of the body." The interweaving of Ida and Pingala in the human system is indeed Caduceus-like, and so are the spiraling currents of Babbitt's atom, the strands of DNA in the cells, or the torus-like streams of seeds rippling out from the center of a sunflower.

From minerals to plants, from animals to humans to higher entities, the entire world participates of one consciousness, which is what a worldwide chorus of ancient traditions and modern teachings affirm. "All is consciousness—at various levels of its own manifestation. This universe is a gradation of planes of consciousness," wrote Sri Aurobindo. Modern scientific minds have joined in: "the stuff of the universe is mind-stuff...the source and condition of physical reality," noted Sir Arthur Eddington.[5] What differentiates what we commonly call the kingdoms is the degree of consciousness they express—from the deep-sleep-like consciousness of the mineral, to

---

4 See http://www.blazelabs.com/f-p-element.asp.

5 Quoted by Laszlo, *Science and the Akashic Field*, p. 118.

the dream-like consciousness of the plant kingdom, to the sentiency of the animal, to the self-aware wakefulness of the human being, to the enlightened, and eventually universal consciousness of advanced individualities, to cosmic consciousness.

Toward a Science of Nature's Formative Process: A Logarithmic Bundling of Cosmic Quanta

> *This is the kind of thing toward which we should strive in any kind of a real science: not to define phenomena in terms of abstract concepts, but rather to define phenomena in terms of phenomena.... What we need is not empirical data, but rather new ways of synthesizing that material.* —Rudolf Steiner[6]

World creation is a symphonic composition in the making. The externalization of Being into a self-consonant whole implies the internalizing of a soul milieu in which awareness unfolds. The complementary secting-and-bonding deployment of the fundamental trinity of being into 7/12fold living systems is the simplest mathematical, harmonic, and psychospiritual combination of the powers that make up this triune one: the powers of One, Two, and Three: fundamental, octave, and third. The 7/12fold archetypal form of the universe is the expression of the power of the triune One to de-multiply itself harmoniously and unitively. To "demultiply" speaks to the duplicating power of 2; "its self" speaks to the whole making power of 1; "harmoniously and unitively" speaks to the mediating, triangulating power of 3. The power of One is the power to be one self, the power to BE (power of identity and will). The power of Two is the power to divide, differentiate, discern (intelligence). The power of Three is the power to unite (love). Their harmonic combination is the logos-rhythmic dynamics of creation, a dynamics that defines the fundamental unit of the universe—*the universal quantum of spirit, body, and soul: 3, 7, 12*.

The fundamental "Reason" underlying the sevenfold structure of the world body and twelvefold structure of the world soul is the logarithmic bundling of the three Logoic principles—being (the power of one), manifestation (the power of two), and self-bonding or consciousness (the power of three)—oneness, differentiation, and integration; being, form, and consciousness; spirit, body, and soul. The bundling of the Word into flesh.

In the end, world-creation is as extraordinarily simple and elegant as the launching of the combined powers of 1-2-3 (self, resonance, consonance) into

---

[6] Steiner, *Interdisciplinary Astronomy*, pp. 243–244.

VI. Facets of a Unitive Science

nested hierarchies of space time bundles of consonance, which are the universe's quanta of being.

Systems of cosmos and forms of nature are partial, yet wholistic, or consonant, replicas of the universal whole. What bundles the great oneness into finite 7/12fold sets (and subsets) of dimensions, kingdoms, systems, is the combination of the principle of self-replication, which generates self-similar wholes, with the principle of integrative union, which generates the consonant milieu of "con-sciousness" intrinsic to any whole. The coming together of the octavian cycle of repetition of the same and the twelvefold cycle of quintessential mediation in the diatonic/chromatic scale is the harmonic template of the living forms of nature and cosmos. All forms are bound by a 7fold self-same replication and filled with a 12fold milieu of consonant bonding. The 7fold and 12fold differentiations of the 3fold one are intertwined and mutually binding, as are the outside structure and inner core of the forms they define. They speak to the intertwining of body and consciousness (2 and 3), cosmos and psyche, nature and soul, world and self. The process of self-replication is limited to a 7foldness by the cycle of mediation (the 12 fifths) and the process of mediation is bundled into a 12fold by the 7fold cycle of octavian replication.

Different worldviews tend to emphasize one power over another. While Plato, in the *Timaeus*, emphasizes differentiation as he focuses on ratios and intervals, Hindu wisdom emphasizes union and bindu points of generative consonance. Goethe's approach to color distinguishes itself from Newton's in an analogous way. Newton used the prism to *differentiate* light into a sevenfold spectrum of colors. Goethe produced the same color spectrum by bringing *together* darkness and light. His conclusion was that colors resulted from the interplay-union of darkness and light, rather than the division of light.

The periodic and quantum-like character of everything that exists stands out as a fundamental phenomenon. The fabric of all forms, as well as their shape, is a rhythmic intertwining of space and time. Time is predominantly related to 12 (hours, months), space to 7 (structures of solar, human, atomic elements system), which points to their correspondence with the inside and outside of the universe (12fold soul and 7fold cosmos). Time may be regarded as the milieu of the universe's consciousness, and space as its body—intertwined and "relative" indeed to each other, as body and soul.

This perspective is a relative point of view, given that time and space are so inherently intertwined. Rudolf Steiner, for instance, proposes what appears to be an opposite view, when he associates 7 with time and 12 with space. "Seven is a clue for everything that happens in time. On the other hand the number twelve is a clue for all things that co-exist in space."⁷ However, the contradiction may be more apparent than real. While 12 undoubtedly holds the space for the soul's unfolding in a zodiac-like causal body, such unfolding implies time, and "whatever has to do with time," says Steiner, "is filled with inner life."⁸ And while 7 governs the great cycles of time, these cycles reflect the evolution of consciousness through the corresponding chakric and planetary (spatial) spheres.

Steiner brings attention to the inner power of numbers in general, and, quite specifically, to the key numbers, 1, 3, 7, 12:

> Just as we place images, Imaginations, before the soul, so on still higher levels the inner power of numbers is placed before human beings. Human beings have to learn to experience the inner proportions of numbers as spiritual music. Of particular importance is the proportion 1:3:7:12....⁹
>
> When Pythagoreans wanted to express the four members of the human being, they expressed the harmony in the ratio 1:3:7:12. That signifies the sound wherein the four numbers harmonize in the same way as do the four parts of the human being.¹⁰

I was unaware of these indications Steiner gave until I reached the very end of this work. The confirmation his statement offers of the connection of time and space with the two numbers, and even more significantly, his highlighting of the key proportion of 1:3:7:12, was undoubtedly heartening. However the contradictions seemed blatant and hard to reconcile. At least, initially.

In addition to his reverse association of time and space with seven and twelve, Steiner's interpretation of the four numbers appears to lead to a different correspondence than the one reached along this journey—of (1)-3-12-7 with the triune spirit(3)-soul(12)-body(7) constitution of forms—in particular, the human form.

> If you understand the proportion of these numbers as a musical relationship, in the sense that one oscillates three times in a given period, another seven, and still another one twelve, then you will find expressed in these numbers the relationship in spiritual music of the "I," astral body, etheric body and physical body.... Everything in the "I," astral body, etheric body, and physical body resounds in tones. One tone resounds in the "I," three tones in the astral body, seven tones in the etheric body and twelve tones in the physical body. Altogether this results in harmony or disharmony.¹¹

---

7   Steiner, *East in the Light of the West*, ch 9.
8   Steiner, *Interdisciplinary Astronomy*, p. 52.
9   Steiner, quoted in Huseman, *The Harmony of the Human Body*, p. 127.
10 Steiner, *Reading the Pictures of the Apocalypse*, lect. 3.
11 Ibid.

VI. Facets of a Unitive Science

As Steiner further develops these correspondences, however, it becomes apparent that the picture is not really dissimilar: "What comes forth from the Earth, Sun, and Moon," he says, "sounds together in our astral body." Indeed, Earth, Sun, and Moon are the replica, in our earthly context, of the primordial Trinity of Father-Sun, Mother-Moon, and Son-Earth—the triadal oneness of 1:3. This trinity is reflected in the threefoldness of the astral "self": will, thinking, and feeling.

> But what comes forth from the planets sounds in our etheric body. There is a sevenfold influence from the planets on the etheric body, as there is from the seven musical intervals: the unison interval, major second, major third, perfect fourth, perfect fifth, major sixth, major seventh—Saturn, Sun, Moon, Mars, Mercury, Jupiter, Venus.[12]

This correspondence between the septenary and the etheric constitution is in accord with our picture. As for twelve, which he associates with the physical body, Steiner explains:

> There are twelve influences from the signs of the zodiac that resound into our physical body. The seer experiences twelve fundamental tones on the devachanic plane. They influence our physical body.[13]

A complementary statement, however, points to an ultimate ground of concord with the view we have arrived at: in addition to the physical body being the echo of the zodiac, he also says that the Ego or causal self is the "perception of the echo of the zodiac."[14] Indeed, the evolution of the human soul happens through the physicality of a body that reflects its cosmic, zodiacal template and "incarnates" its individual, causal replica.

Another key notion for a science of formative processes based on a unitive view of the universe is one that Steiner and Teilhard de Chardin point to in different contexts but similar terms. They arrive at a parallel sighting, not only of the fundamental bidirectionality of formative currents, but of the unique dynamics of each current. Stemming from the same wholistic view of a oneness of flow behind the archetypal gesture of the universe and its forms, the radical insight they propose enriches the simple grasp we gained of the principles of centrifugality and centripetality with a finer understanding of a secondary polarity within the first—that of radial (centrifugal) and spherical (centripetal) movement, says Steiner; of radial and tangential energy, says Teilhard. Schwenck's contemplation of the dynamics of water led him

---

12   Ibid.
13   Ibid.
14   Steiner, "The Alphabet," p. 9.

to a similar observation of the fundamental creative tension between the "spherical tendencies of water and the gravitational forces"—which is the tension of creation between the gesture of the whole and the gesture of the part.

In other words, in addition to the simple directional distinction we made between the two currents, they illumine in each a distinctive form of motion: radial for the centrifugal activity, and spherical for the centripetal current. These two types of motion are clearly evocative of the dual vortex and they highlight in each of its "halves"—hence in the two cosmic directions of their flow—a different gesture: radial and spherical. A dynamics of issuing forth and wrapping around; a dynamics of filling in and hollowing out. This polarity is quite evident in the electromagnetic field of all living forms—and its graphic illustration in the field produced by electrically charged objects (fig. 65, page 323)—where the radial flow of electricity combines with the spherical flow of magnetism into the toroidal flow of the field.

Teilhard views the universe as a gravitational flux of "love-energy," galaxy-like, whose spiral motion has two interdependent components—radial and tangential—which he refers to as energy of "without" and energy of "within." While tangential energy, he explains, "links an element with all others of the same order," radial energy draws an element toward "ever greater complexity and centricity—in other words, forward."[15]

Focusing on the human experience and constitution, Steiner points to the reflection of this polarity in the contrast between the radial character of the will and the spherical character of ideation. He shows how this contrast expresses itself in the polarity between the earthward-outward raying out of long, tubular bones and the cosmic spherical inflow associated with cranial bones—between the "shooting out of itself" of our limbs and will, and the "influence that makes itself felt in the entire circumference of our life of consciousness."[16] Steiner also points to the interplay between inward directed sensory perception leading to ideation and outward directed metabolism—between processes of cognition and processes of reproduction.[17] He takes the polarity even further into the realm of science to emphasize the interdependence of astronomy and embryology for instance, or mathematics and the reality of the heart. In general, he stresses the necessity, if one is to gain a genuine understanding of reality and develop a science of formative processes, to ponder the radial-spherical interplay at work in the polarities of the world being.

---

15 De Chardin, *The Phenomenon of Man*, p. 64.

16 Steiner, *Interdisciplinary Astronomy*, p. 154.

17 Ibid., pp. 13, 55, 61.

VI. Facets of a Unitive Science

Toward a Science of the Evolution of Consciousness as a Process of Harmonization: From Yoga to Alchemy and Jung's Depth Psychology: from Four back to Three, Two, and One

> *The unfolding of man's spiritual nature is as much an exact science as astronomy, medicine or jurisprudence.* —Manly Palmer Hall[18]
>
> *The essence of the journey becomes the achievement of harmony, finding not so much the path of least resistance, but a constant move along the path of greatest benefit, indicated by the degree of harmonic resonance with all that is connected with each action in each moment.* —Laszlo and Dennis[19]

Bringing ancient and modern science to bear on each other yields a unitive view of the formative process of the universe from primordial oneness to twofold polarization (wavefield of time–space), to threefold triunity, to the fourfold elemental corporeality of 7/12fold forms. Combined with the higher three, the elemental fourfold produces the sevenfold forms.

Among the fundamentals of the universe encompassed in Pythagoras' famous Tetractys, a most significant one is this overarching arc of involution of the primordial One into a twofold, threefold, and fourfold, and the complementary evolutionary ascent of consciousness from elementary fourfoldness to universal oneness.

The term *Tetractys* designates the equilateral triangle formed by the sequence of the first ten numbers aligned in four rows (fig. 69, page 398). With this supremely simple symbol, at once mathematical and metaphysical, Pythagoras planted in the West a seed of the world mysteries he inherited from the East, a seed charged with the principles of world beginnings. It was referred to as the "fountain and root of everlasting nature.... Not only are all symphonies found to exist within it, but it appears to contain the nature of all things."[20]

The first four numbers symbolize the unfolding constitution of the cosmos:

1. Oneness—Monad and uroboros
2. Dyad – Lux and Tenebrae (Light and Darkness, Spirit and Matter)
3. Harmony (Tri-unity of soul)
4. Kosmos (Tetrad of the four elements; Quaternary of the body)

---

18   Hall, *The Secret Teachings of All Ages*, p. 120.
19   Laszlo and Dennis, *Dawn of the Akashic Age*, p. 184.
20   Theo of Smyrna, in Iamblichus, *Life of Pythagoras*, p. 235.

Other representations of the Tetraktys highlight these four dimensions geometrically: Circle of oneness, two-ness of center and circumference, triunity of the triangle, and cube-like volume of the 7fold form (6 hexagonal points + central point). We will come back to this central sevenfold, which will prove to be a geometric representation of the universal form: the dual vortex.

The Tetractys also represents the four dimensions of space:

1. Point (preceding dimensionality);
2. Line (defined by two points);
3. Plane (defined by a triangle of three points);
4. Volume (a tetrahedron defined by four points).

Ancient philosophers, Pythagoras first among them, revered the Tetractys (sometimes called the "Mystic Tetrad") as the symbol of all there is. It was the object of invocative prayers:

> Bless us, divine number, thou who generated gods and men! O holy, holy Tetractys, thou that containest the root and source of the eternally flowing creation! For the divine number begins with the profound, pure unity until it comes to the holy four; then it begets the mother of all, the all-comprising, all-bounding, the first-born, the never-swerving, the never-tiring holy ten, the *keyholder of all*.[21]

The four rows add up to 10, which points to a higher order of oneness: the decad. Reminiscent of Arthur Young's sevenfold vortex, the Tetractys contains all that has been undergone and integrated along the descent/ascent through the dimensions: 10 is a fully conscious, self-aware "one."

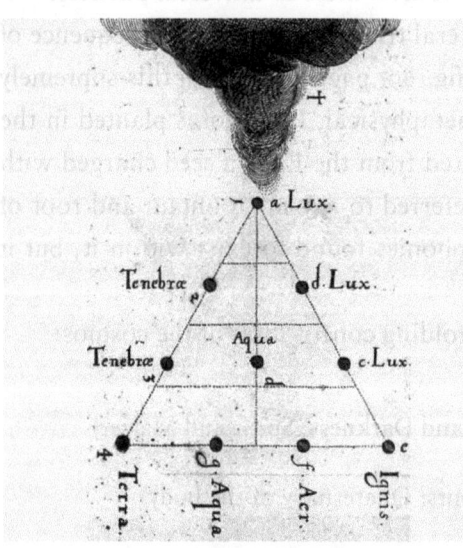

Fig. 69. The Tetractys

The Tetractys also contains the essentials of the Pythagorean musical system as *musica universalis*: the rows can be read as the ratios of 4:3 (perfect fourth), 3:2 (perfect fifth), 2:1 (octave), which form the basic intervals of the Pythagorean scales.

It even outlines the fractioning process of world creation from vibration to form and the corresponding grades of etheric fields. This fractioning, the Tetractys suggests, is *vibratory* (2-ness of the polarized field, the "light ether") before becoming *tonal*

---

21 Dantzig, *Number, the Language of Science*, p. 42.

VI. Facets of a Unitive Science

(3fold phenomenon of a "unit" of tone, and "tone ether"), tonal before becoming elemental ("life ether") (4). In other words, tones and tone intervals (3) underlie the shaping of physical forms (4), and the vibrational field (2) of the polarized Breath underlies the consonant emissions of tones or tone-zones. The Tetractys captures in one impressively simple figure the descent from Logoic unity to dual wavefield to threefold tones to fourfold forms.

The cycling of humankind through expanding spheres of consciousness to the universal "circle of life" completes the cosmological arc. Ancient philosophies, as well as more recent psychological paradigms, such as Jung's depth psychology of individuation, light up the inner side of the universe's spiral cycling back to oneness—which is in fact a cycling forward to *conscious* oneness and self-aware universality.

Alice Bailey's *Treatise on Cosmic Fire* ends with this solemn finale:

The morning stars sang in their courses.
The great Paean of creation echoeth yet, and arouseth the vibration.
There comes cessation of the song when perfection is achieved.
When all are blended into one full chord, the work is done.
Dissonance in space soundeth yet. Discord ariseth in many systems. When all
  is resolved into harmony, when all is blended into symphony, the grand
  chorale will reverberate to the uttermost bounds of the known universe.
Then will occur that which is beyond the comprehension of the highest
  Chohan—the marriage song of the Heavenly Man.[22]

The evolution toward increasingly complex forms of life takes a significant leap, we saw, with the emergence of the human kingdom. From quantum particles to elementary mineral forms, from living organisms (plant kingdom) to independently moving entities (the animal kingdom), the penetration of the world soul into matter and the adaptation of matter to life reach their next quantum stage with a new form of being, one that vehiculates the fifth principle of mind and Egoity: the human form. The human being is the first externalization in an autonomous form of the quintessence of the universe: selfhood—the potential noosphere.

At the human stage of world unfolding, the 12fold harmonic core of the universe, charged with 3fold logoity, begins to replicate itself into the Egoic forms of human monads incarnated in 7fold bodies. For the first time in Earth evolution, the 12fold soul begins to interact with the 7fold body *from within*. In the human being, the world soul is engaged in a process of conscious integration with the four elements. An increasingly self-aware interplay with emotions, bodily instincts and outside world begins to "fill" the 12fold template of selfhood with individuated soul forces

---

22 Bailey, *Treatise on Cosmic Fire*, p. 1283.

and qualities. Self-awareness, the fifth essence, seeks to harmonize and unify into *one*—indivisible—"individuality" the four elemental vehicles: physical earth, vital water, connecting air, and personality fire. The universe becomes individualized in the human being. While it is present in one degree or another in all forms, the archetypal template of a 7fold structure with a 12fold "self" milieu can only actualize itself in full with the advent of the human kingdom.

The emergence of humankind marks the coming to self-awareness of the universe as a creative unit of self-directed thinking, feeling, and will—and not only of form-weaving intelligence, chemical sentiency, and elemental powers as manifest in the first three kingdoms. A self-aware 7fold-12fold form of being has appeared, in whom the Logos-cosmos finds its fullest replica and promised land of self-awareness.

Indeed, if we look imaginatively at the human being from above, we see the flower-like display of the 12fold heart-in-the-head chakra, and, next in incarnational density, the 12fold heart chakra. Looking from the side, we see the 7fold *vertical spiral* of the chakric spine. Similarly, if we imagine peeking into our universe from above, looking into the solar system through the opening of the zodiac, we see the 12fold flower-like vortex of the heart-in-the-head chakra of a grand cosmic Being and, next, the heart chakra of the solar system. Looking at the solar system from the periphery, we see the planetary spheres line up as the 7 major chakras of the Sun-entity we refer to as the "solar system." The heart and spine of the world being are mirrored in the heart and spine of the human being.

At this point, the evolution of material life on Earth becomes centered in the journey of humankind—from intelligent to loving to self-directed universal Egoity. The lesser kingdoms surround the emerged "dominant" of creation and find themselves, as current global issues expose most poignantly, bound to its slow progress beyond the separative stage of self-protective, self-serving, exploitative intelligence that threatens their existence.

The journey of humankind from personality structuring to soul development to universal self-awareness is laid out in the philosophy of yoga, which can be described as a physico-psychospiritual science of development based on the subtle constitution of the human organism, its etheric currents and chakric centers, and the harmonizing character of the work of evolution. Like spiritual teachings in general, the purpose of yoga is to enlighten the mind, ground the spirit, and liberate the consciousness, by providing the map and training to facilitate the journey of becoming fully human.

As the term indicates, *yoga* aims at "union"—the union of body and soul—ultimately the ecstatic union of the great polarities of spirit and matter, Shiva and Shakti,

VI. Facets of a Unitive Science

which is consummated when the essence of earthly elemental forces, the kundalini shakti, rises and eventually unites with the spirit self at the crown chakra. "The Yogi unites within himself his own Principles, female and male, which are the 'Heart of the Lord' or Shakti and Her Lord Consciousness or Shiva. It is their union which is the mystic coition (Maithuna) of the Tantras."[23] Ultimately, the human being stands as feminine, negative polarity to the great generative "masculine," and the maithuna fully unites self with Self. Yoga lays out a path of physical, philosophical, meditative, and devotional disciplines to this golden conjugation of the poles of Being.

The yogic path shows how the evolutionary process reverses the involutionary operation of world creation in a movement from 7/12foldness (human nature) to 3foldness (body, spirit, kundalini; or Ida, Pingala, Sushumna) to oneness of union. The yogic way to spiritual union involves a gradual harmonization of physical and spiritual being, which has to be worked out at every level of the body–psyche. The chakric levels of consciousness are the major harmonics of the fundamental vibration of a human "I." In Steiner's words, "seven planets resound into our etheric body."[24] The first five chakras resonate with the five elements—though not necessarily in a sequential way. The fifth element, for instance, may be associated with the 12-petal heart chakra, rather than with the throat. The 6th (or 2nd) chakra, the ajna center or "third eye," resonates with the unitive mind of the Logos, the Light of consciousness, and the first or crown chakra replicates the oneness of Being.

> In the course of spiritual practice the chakras are increasingly harmonized until they vibrate in unison. It is then that the subtle sound OM, which is reverberating throughout the cosmos, can be heard in the state of ecstasy. This coincides with the balanced functioning of the body-mind.[25]

What was perceived as separate (body and mind) is realized as one and resounds as one. Body–mind, the microcosmic replica of the cosmos–Logos, Shakti-Shiva polarity—a faithful mirror of "the entire architecture of the universe"—is experienced as one in a perfect harmonization of Heaven and Earth within. "It is through the microcosm," Feurstein comments, "that we find the doorway to outer cosmos." And it is through the microcosm, one might add, that the cosmos finds a doorway to the central chamber of self-awareness.

The experience of "unison" described by Feuerstein is a state of resonance with the fundamental vibration of the solar systemic being. Identification with any of the partial, chakric dimensions of what is involved in "being human" is

---

23 Woodroffe, *Shakti and Shakta*, ch. 29.
24 Steiner, *Reading the Pictures of the Apocalypse*, lect. 3.
25 Feurstein, *Tantra*, p. 150.

transcended. Attunement to the primordial "sound," the matrix of universal time–space–consciousness before and beyond forms, heralds the dawn of the fully vibrant, "unison" body–soul, which Feuerstein refers to as the "primordial body, coextensive with the universe itself."[26]

The ancient Chinese notion that the human goal is to embody the cosmic sound echoes the view from India. David Tame explains:

> He who succeeded in harmonizing the discords within his mind, emotions and body could become a more perfect embodiment of Cosmic Sound, an incarnation of the Word.... Even as the perfectly tuned pitch pipes could be the standard for the tuning of all other instruments, therefore bringing earthly music into conformity with universal harmony, so too could the perfectly "tuned" or self-realized man become the standard for all other men to follow.[27]

In both cultures, spirituality is a matter of vibration and harmony.

The ecstatic state of "unison" announces the eventual *stage* of identification with the world being alluded to in the mantra "Tat tvam Asi" (You are that). Attunement to the fundamental sound announces the completion of the path of spiritualization of the body and corporealization of the monad (as micro-logos) exemplified by the great representatives of humankind. Monadic oneness pervades what has become a trans-dual body–mind reality. The diamond body of the Buddha and the resurrection body of the Christ epitomize the "meta-form" of a "body clothed with Sun": the Logos made micro-cosmos, and the cosmos brought to self-awareness in a human being.

While yoga offers an especially clear and scientific mapping of this "harmonic" path, every spiritual discipline means to facilitate what is simply the natural course of human evolution: the complete interpenetration of Logos and cosmos in the singular body-soul-spirit of a human being. The Christian path, and its central sacrament of Eucharist, speaks to the same holy grail of the human labor, brought to a next step of evolutionary unfolding: transmutation of cellular matter into the vibrant body of the Logoic "word," attunement of energies to solar harmony, heartening of self with cosmic love. Teilhard de Chardin formulated his experience of this process as "amorization of the universe and universalization of self."

As the expanding self gains in universality, finding its intelligence, feeling life, and motivation identified with larger spheres of "selfhood"—from family to community, to kingdoms of nature, to planet, to universe—these non-personal "worlds" are experienced as one's being and body. Ultimately, the words from the Tantric Hindu scriptures, "we are the world, the world is our true body" become experience.

---

26   Ibid, p. 143.
27   Tame, *The Secret Power of Music*, p. 49, 57.

Shiva-Samhita, a seventeenth-century Hatha Yoga manual composed under the influence of Tantra, says:

> Within this body exist Mount Meru, the seven continents... in it also dwell the seers, the sage, all the stars and planets.... In it whirl the Sun and Moon.... Likewise, it contains ether, air, fire, water and earth.... He who knows all this is a yogin.[28]

The uplifting influence of universalized individualities upon the global consciousness attests to the vibrant power of the human apotheosis. So does the resonance of the kingdoms of nature to those who experience them as *their* very nature. Examples abound of the rapport such "realized" human beings entertained with the natural world. A number of well-known stories, such as the peaceful encounter between Francis and the wolf, or Sergius and the bear, attest to wild animals growing tame and gentle in the presence of saints. The mineral, plant, and animal kingdom famously stirred, moved, and danced, says Ovid, when Orpheus played his seven-stringed lyre; even Hades, the Lord of the underworld, whom Orpheus implored to let him see Eurydice, could not resist the power of his harmonies, as they gently bent his rules, dissolved his gates and whirled out a vortex between worlds. Similarly, the tempestuous elements re-harmonized at Jesus' command. At his "word," the bread became his body, the wine his blood—the natural, "logical miracle" of the descent of a Sun Logos in the Earth: "This is my body." Waters mounted at the death of Al Hallaj, then subsided when covered with his coat.

The primordial harmony of the Logos finds its fulfillment—the Alpha, its Omega—in the zodiac like rose-lotus of a universalized self, who knows itself as the Self—coextensive with the solar system and in complete resonance with zodiacal plenitude. Rudolf Steiner recommended the following ancient verse as a morning meditation: "More radiant than the Sun, purer than snow is the Self. I am that Self, that self am I."

In Sanskrit, the omnipresent cosmic sound experienced in deep meditation by the advanced yogi is called *anahata,* which is also the name of the heart chakra—the twelvefold center of selfhood—"the anchoring point within man of the Word of God."[29] By contrast, audible sound is referred to as *ahata*. The Word of the Logoic heart of the universe can only resound in full in the fully open lotus of the soul. The heavenly Jerusalem of Christian cosmology hints at this unfolded cosmos–logos of universal selfhood.

---

28 Quoted in Feuerstein, *Tantra*, p. 61.
29 Tame, *The Secret Power of Music,* pp. 171, 185.

The path of individuation, says Jung, begins in anahata with the awakening of the (12fold) heart center.[30] It ends in identification with the 12fold heart center in the head. "It might be stated," echoes Alice Bailey, "that the awakening of the heart center leads a man to consciousness of the source of the heart center within the head. This in turn leads a man to the twelve-petal lotus, the Egoic center on the higher levels of the mental plane."[31] Zodiacal universality of soul is the full deployment of what spiritual teachings refer to as "the birth of the Christ in the cave of the heart"—the birth of soul consciousness within the separate "me"-awareness—a nascent realization of the other as "self." From the perspective of harmonic cosmology, this awakening of the heart center and Christ principle within points to the activation of the third harmonic/fifth interval of individuation in the octave of consciousness. A critical proportion (2/3) of self-centeredness and other awareness, personality and universality is reached, intoning in the person the quintessential note of humankind—empathic consonance.

The Greek myth of Hercules, the only mortal in the Hellenic world to become an immortal at death, illustrates the path to zodiacal universality. At the completion of his famous twelve labors, the great hero "put on the divine" and was delivered to a next octave of being. His teacher greeted him with the following words, "Welcome O son of God who is also a son of man. The jewel of immortality is yours. By these labors have you overcome the human and put on the divine."[32]

Remarkably, the 12fold character of human completion is inscribed in the whole edifice of the chakras—the 7fold human octave. The petals of the first six chakras, Alice Bailey points out, add up to 12 x12, the "squaring" of the soul sphere—the incarnation of the Logos in cosmos—evocative of the heavenly Jerusalem of the Book of Revelation:

> Symbolically, if the sum total of the forty-eight petals of the five centers is added to the ninety-six petals of the center between the eyebrows, the number one hundred and forty-four appears. This number signifies the completed work of the twelve creative Hierarchies, twelve times twelve, and thus the bringing together of the subjective soul and the objective body in perfect union and at-one-ment. This is the consummation."[33]

The 12fold character of the complete "squaring" of the soul in a fully embodied being is also inscribed in the fact that the first square in the Fibonacci series, which is in essence the Phi series of the soul of the world, is 144—the square of twelve. The

---

30 Jung, *The Psychology of Kundalini Yoga*, p. 39.
31 Bailey, *The Light of the Soul*, p. 295.
32 Bailey, *Labors of Hercules*, p. 197.
33 Bailey, *A Treatise on White Magic*, p. 199.

number 144 is also the twelfth term of the series: 1, 1, 2, 3, 5, 8, 13, 21, 34, 55, 89, 144. Literally, the "squaring" of the soul is "figured" in the square of twelve.

Adding the last, the thousand-petal lotus of the head chakra, Alice Bailey continues, gives "the number of the saved in the Book of Revelation, the one hundred and forty-four thousand who can stand before God...before the very Presence Itself." This suggests how there still awaits, unachieved in the completion of the individual journey, the achievement of one's "kind"—the enlightenment and universalization of humankind as a whole. Only the human race *as a whole* can meet the full Presence of "God." Individual achievement is forever incomplete without the achievement of the "last pilgrim."

> *Welcome, O life! I go to encounter for the millionth time the reality of experience, and to forge in the smithy of my soul the uncreated conscience of my race.* —JAMES JOYCE

Jung's depth psychology brought the "science of the psyche" very close to the completeness of the ancient Yogic view. It also enriched the understanding of this philosophy with penetrating insights in "psychological" complexities that were not relevant until centuries of intellect and ego development opened the appropriate milieu of awareness in the collective psyche. The psychological "inside" of evolution, which yoga harnessed *energetically* but not psychologically, is thus a relatively recent field of inquiry and science. As recent as quantum physics, it points to the other, inner side of the same unitive understanding.

The process of individuation, the central theme of Jungian psychology, refers to the psychospiritual path of human evolution, the centripetal current of the cosmological process. This is how Jung defines individuation:

> Individuation means becoming a "single, homogeneous being," and insofar as "individuality" embraces our innermost, last and incomparable uniqueness, it also implies becoming one's own self. We could therefore translate "individuation" as "coming to selfhood" or "self-realization."[34] An innate urge of life is to produce an individual as complete as possible. For instance, a bird with all its feathers and with the size that belongs to that particular species.... So the urge to realization naturally pushes man to be himself.[35]

Jung's great intuition was that the human self unfolds to full potential by individuating the *Self*—by which term he referred to the center of the totality of the psyche. The "total psyche" is the universal psyche. "Individuation does not shut

---

34 Jung, *Two Essays on Analytical Psychology*, p. 171.
35 Jung, *The Psychology of Kundalini Yoga*, p. 4.

one out from the world," he points out, "but gathers the world to oneself."[36] The Self is implicitly the cosmic self in whom we have our being, the universal psychic wholeness that engages our personal consciousness by way of circumstances and life events, challenging it, as dreams often allude to symbolically, to expand its limited framework to integrate non-ego, "foreign" elements—from unconscious forces to unlived potentials to collective developments—and gradually widen, deepen, and universalize its mind, heart, and will. "The self is something exceedingly impersonal, exceedingly objective," affirms Jung.[37] The "unconscious" is coextensive with the universe, and individuation is a process of universalization of selfhood and personalization of the universe. Psycho-logy—literally, the logic of the psyche—is in essence the logic of an ego-cosmos dialogue, whose finality is an at-onement of self and cosmos. In yoga, as in any other spiritual scientific philosophy, the ultimate goal of individuation is the wholeness of the universalized individuality.

Even more than yoga, Jung became deeply interested and engrossed in a Western Tradition that carried through the centuries the same mysteries of the world's constitution and evolution: alchemy. Founded upon the primordial oneness of spirit and matter and aimed at their ultimate re-union, alchemy explored and conducted the mysteries of their intercourse as a unified "metaphysical science" of matter-psyche. A unique model in the history of science, its opus was focused at once upon external operations of material transmutations and concomitant psychospiritual processes within the alchemist. This dual task was reflected in the motto, "Ora and labora—Pray and work." The great endeavor, the transmutation of raw material, or *"prima materia,"* into gold, rested on a parallel dedication of the alchemist to inner transformation and sanctification.

Briefly summarized, the alchemical process is marked by a series of transitional operations through distinct states of matter-psyche—from the dark stage of nigredo, to the white stage of albedo, to the red stage of rubedo, with a number of smaller intermediary steps. Beginning with the dissolution of the raw material of experience into its separate, "warring," elements, the "great opus" proceeds through operations of trituration and purification, then balancing and harmonization, which eventually bring about a state of integration-union-coniunctio. The process culminates with the formation of an ultimate, stable component, the philosopher's stone, and the production of a life-giving "quintessence" referred to as elixir vitae. This final union signals the consummation of the alchemical wedding of self and Self in the soul.

The gold of the philosopher's stone suggests integrality of being—the unified condition necessary for the nonduality of life and being to take hold of the

---

36   Jung, *The Structure, and Dynamics of the Psyche.* CW 8, p. 226.
37   Jung, *Psychology of Kundalini Yoga,* lect. 2, p. 40.

VI. Facets of a Unitive Science

fourfold elemental vehicle and abide in it. The harmonic understanding of cosmology explains how the integration-"union" of the four elements in the "stone" seals a transformational step that is more than chemical–material. The integrative stone is at once matter and *life*—a "living stone." As the fifth interval unites fundamental and octave into a third harmonic, the quintessence of the four elements fuses matter and Life into a live stone. Alchemy affirms the trans-physical ground of this union of the elements. Harmonic cosmology illumines how this union, which galvanizes the fully unfolded quintessence with monadic (Logoic) life, is a union of matter and spirit—an *enlivening* of substance. The corresponding step in the soul of the alchemist is the philosophical wedding of self and Self in the soul. The long journey of repeated psychospiritual dissolutions, purifications and harmonizations of one's "nature" eventually results in the state of golden body-soul-spirit consonance that allows the Self, the Logos, to resound in and galvanize the whole unified being (his bride) with monadic vibrancy. "*Homo verus,*" the veritable human being, is essentially "*homo-geneous.*"

The philosopher's stone, the quintessential "gold" of human nature, is at once distilled from nature and structured by spirit. Significantly, in our context, the atomic composition of gold, as seen clairvoyantly by Annie Besant, shows a number of twelve-based structures.[38] In accord with this twelvefold wholistic structuring, alchemy regarded gold as a perfect balance of the alchemical *trinity* of Sulfur, Mercurius, and Salt—the three principles, according to Paracelsus, from which all things are born. Twelve and three line up here again, sealing the archetypal ascent of raw matter from chaotic multiplicity to the twelvefold wholeness in which the primordial creative (tri-)unity of being dwells.

Though they vary in order, the alchemical operations leading to the formation of the twelvefold stone often amount to *seven*. The following sequence is one example of such a sevenfold categorization and ordering: 1. Calcination 2. Dissolution 3. Separation 4. Conjunction 5. Fermentation 6. Distillation 7. Coagulation. The sequence exposes the great rhythm of *solve et coagula,* dissolve and coagulate ("sect and bond"), the gestalt of the whole alchemical process and its motto. Reflected in it is the primordial *two*-ness, the centrifugal (dissolving) and centripetal (coagulating) motion of the cosmic breath, involving itself into multiple forms and evolving, through them, to conscious unity.

According to this unique psycho-physical science, which the father of modern science himself, Newton, studied assiduously, if in secret, and which Jung would use as model for the process of individuation, human evolution culminates in the formation of the stone, a subtle jewel-like radiant structure—which Eastern philosophy

---

38   Fernando, *Alchemy*, p. 128.

refers to as "the jewel in the lotus." The stone or jewel replicates the twelvefold heart of cosmos, the simplest structure suitable to vehiculate the threefold unity of "be-ing." Like the world soul of the beginning, the stone is a soul-"compound" and a quintessential body, a "unit" of matter-life, an intermediary reality between unmanifest being and corporality—a golden mean indeed.

Gold is to the mineral kingdom what the third harmonic is to harmony, and Phi to geometry—a golden mean. The inner, subjective essence of the golden mean is love, which is indeed the secret of alchemical union. Al-Jaldakî, one of the greatest of medieval Islamic alchemists, writes: "The order of the world is kept by love, and love has for its cause the mutual recognition of Lover and Beloved, the homogeneity of their nature, their convenance to each other."[39] It is the soul force of love in the alchemist–philosopher, which "effects the nuptial union between the Son of the Sun and the Daughter of the Moon [spirit and body], [so] she begets from him his semblable, a being as beautiful as he." The merging of Self and self in a human being begets the solar human, who is a radiant, star-like stone, a Sun-filled earthly being.

What love generates, which intelligence or will alone cannot do, is union and unity. The alchemical work is fundamentally a unitary work, which engages matter and soul, world and self *as one*. It calls for the heartful mind, the quintessence of the universe, to bear on the incubation of a "jewel" both outside in the athanor and inside in the soul.

Lawlor expresses the same idea in the language of geometry when he says that the golden section is the regenerative seed, which lifts (spirals up) "the mortal realms of duality and confusion back toward the image of God."[40]

The inner reality of the philosopher's stone is the stage of universalized identity that yoga and ancient spiritual teachings refer to as the "unison" goal of their practices. The self finds unison with its grand universal octave. As living stone and elixir transcend the duality of minerality and life, matter and spirit, the perception of separation between outside and inside vanishes in the realization of universal identity. Reflecting the stone's reputed power to transmute any matter into gold, the universalized consciousness sheds its illuminative and transmutative influence, vivifying those it contacts and uplifting the human race as a whole. So great is the power of a Buddha's vestures, the Tibetans believe, that to be in physical incarnation at the same time as a Buddha is an incomparable blessing, thought to greatly hasten one's spiritual evolution.[41]

---

39   Corbin, *Alchimie comme art hieratique*, p. 84.
40   Lawlor, *Sacred Geometry*, p. 37.
41   Tansley, *Subtle Body*, p. 49.

VI. Facets of a Unitive Science

> *Perhaps one of the ways leading to the wisdom of maturity is the pathway of proportions, of shared limitations, which we need to find beneath the weeds and underbrush that have overgrown it. Proportions are shared limitations. As relationships they teach us the mana of sharing. As limitations they open doors toward the limitless. This is the power of limits.* —György Doczi[42]

Jung saw in the details of the alchemical process an extraordinary source of illuminating symbols and operative insights for psychology. It shaped his understanding of the journey of individuation as an opus of dissolving, harmonizing, and integrating. If alchemy shed light on archetypal "operations" of psychospiritual growth, existential experiences ought to be regarded as opportunities to engage consciously in such operations, analysis being meant to elucidate and facilitate this generic process.

The essential component of an alchemy inspired psychological work is a willingness to develop a conscious relationship and integrative attitude toward the many protagonists of one's existence, which are also the transformative agents of the psyche—from external events and circumstances to partnerships, from intrapsychic components (shadow, animus/anima) to dream characters. The fundamental ingredient of gold making is the recognition that life is our partner, and that wrapped in every seemingly external circumstance or happening of "fate" is our unconscious "other." Every person and life development brings foreign ingredients, hence harmonizing challenges, in the cauldron of becoming. Ultimately, every partnership is a partnership with the Self, and behind every occurrence of love or pain, conflict or failure, is the transpersonal call of the Self to work out a new step of integration. In sprinkling otherness in the same, and throwing unconscious material in the flask of ego awareness, "Life" (the Self) challenges past formations and loosens old identifications to make room for a more encompassing ground of being. "Others" arrive on the stage to universalize us, invite and challenge us to more wholeness of being. The recognition that life in all its manifestations is our most faithful partner on the journey to wholeness activates the conversation between self and Self, which awaits at the core of existential storms and dream images. Giving up resistance to the drowning of crown or the toppling of cherished constructs allows for an augmented ground of selfhood to collect itself. "Until you make the unconscious conscious," Jung writes, "it will direct your life and you will call it fate."

Such a perspective on life developments as invitations to make room for foreign qualities and unconscious (lit. unknown) potencies opens the way to view all

---

42 Doczi, *The Power of Limits*, p. 139.

circumstances from the golden angle of their "rounding out" logic and integrative potential. There ensues a willingness to tune one's instrument to a vaster note of being. Held in the light of this fundamental labor to harmonize the self with the whole, every disconcerting experience can release its ambiguous goods and reveal an intricate stairwell toward greater wholeness. Every life happening gazes upon us with an enigmatic, alchemical glint, beckoning us to engage in yet another harmonizing interplay and participate in the music-making that is the breath of creation and the inspiration of evolution.

The sudden dissolving of a previous life structure is often what brings a person to in-depth psychospiritual work. The breakdown of an existential frame urges a sorting out of the raw material of experience, a labor of "separation" and dissection, which reveals warring elements and exposes them to the alchemical-analytical fire. Psyche and circumstances, the two sides of experience, which, when all goes well, are kept relatively at bay from each other by the ego, have lost their clear distinctness, and the great psychic partners of conscious and unconscious are engaged in a chaotic process of submergence-emergence, breakdown and partial assimilations. While the ego may try to cling to drifting pieces of old "forms," the unconscious present fresh, if raw, ingredients, and alien but powerful forces. Clothed in dream characters or existential "stuff," new contents come to the light of day, ready, if given proper incubating space, to bring their energetic reality to foster the emergence of a next psychospiritual structure, charged with greater fullness of being.

"Prayerful" attentiveness (*Ora et labora*) to the contents presented by life provides the transmutative warmth of the athanor. Engaging the consciousness with foreign, "un-conscious" components is the overall nature of the operation. Unconditional acceptance and attention allows the Self to catalyze unique amalgamations of the "two" and elicit new integrative steps. Dreams will often give intimations about such increments of integration. Recurrent themes pointing to divisive and destructive conflictedness within may well disappear from the dream life. New contents emerge, the inner landscape changes, and so do external circumstances. A critical, yet often inconspicuous quality of lucid surrender within has laid a new threshold of integration around which the psyche reorganizes itself.

One of the first intrapsychic unconscious partners Jung enjoins to discern, relate to and walk with consciously is what he calls the shadow, a composite of hidden facets of the personality cast out of awareness and swept under the rug of the persona to avoid the conflicts it generates with the self-image. The work of retrieving and owning these psychic traits commonly denied, suppressed and projected, widens and enriches the ground of personal identity.

VI. Facets of a Unitive Science

The second major focus of individuation involves the contrasexual inner partners—animus/anima—masculine unconscious component in a woman and feminine unconscious component in a man. Cast in dream characters or real life figures, these significant "others" carry the numinous projection of an unconscious "Other"ness key to greater wholeness of being. They provoke intense interplay and dialogue between conscious and unconscious. The initial fusing "mystique" of falling in love is inevitably followed by experiences of distance and alienation.

> Hatred is the thing that divides, the force which discriminates. It is so when two people fall in love; they are at first almost identical. There is a great deal of "participation mystique," so they need hatred in order to separate themselves. After a while the whole thing turns into a wild hatred; they get resistances against one another in order to force each other off—otherwise they remain in a common unconsciousness, which they simply cannot stand...In the case of an exaggerated transference in analysis, after a while there are corresponding resistances. This too is a certain hatred.[43]

Donned in the ambivalent garb of love and hate projected on them, these special "others" act as golden means between the personal self and the center that transcends all dualities and catalyses their encounter—the Self. In perfect golden secting manner, they are charged with the potential to trigger both ecstatic union and painful rifts and splits. They are the great facilitators of consciousness—they embody the "with" intrinsic to con-sciousness (etymologically, "con-sciousness" means "with-knowing"). Eventually, the profound experiences of fusion and division elicited open access to a next grade of "proportionality" of character, understanding of love, and harmony of being.

Jung describes the individuation process as an endeavor to "make what fate intends to do with us entirely our own intention."[44] The apparent lack of rhyme or reason associated with fate points to the transpersonal wisdom and timing of the Self. What "happens" has its reason in this central, organizing principle, whose "Logoic" essence of *love-harmony* means to bring split off parts and unconscious forces in *relation-reunion* with the conscious self, stretching the ego to reconfigure itself in this light and conjuring up a next level of integrality of being.

---

43 Jung, *The Psychology of Kundalini Yoga*, p. 5.
44 Bergengruen; quoted in Jung, *The Structure and Dynamics of the Psyche*, p. 377.

*One's spirit must widen until it
encompasses the world.* —Goethe

In Jung's view, the anima/animus components of the psyche operate as the royal door to the totality of the psyche. The work of integration they elicit begins to pave the way toward the ultimate stage of universalized consciousness the alchemists called *Unus mundus* (Latin for "One World"). Indeed, the call of the animus/anima is felt as a call to oneness with the universe.

As the golden mean engages polarities into a threefold dynamic union, turning the two into one (a triune one), the experiences associated with the contrasexual partners activate the interplay of conscious–unconscious that leads to integrative expansions of self and eventual at-one-ment with the Self. "When male and female combine," says the Tao Te Ching, "all things achieve harmony."

The integration of the anima/us evokes also the union of Shiva and Shakti along the yogic path, as Feuerstein points out:

> Psychologically speaking, the unitive relationship of Shiva and Shakti can be understood as a symbol for intrapsychic unity or, in Jung's terms, the integration of animus and anima. All the polarities and dualities—notably male and female—that we can possibly encounter in the world are precontained in the Shiva-Shakti dimension.[45]

The same is conveyed in the depth teachings of Christianity,

> Jesus said to them, "When you make the two one, and when you make the inside like the outside, and the outside like the inside, and the above like the below, and when you make the male and the female one and the same, so that the male not be male nor the female female...then you will enter the kingdom." (Gospel of Thomas)

In the language of the old alchemists, *Unus mundus* designates the one reality from which everything emerges and to which everything returns (but with full consciousness—as a *decad*, a 10), the one source invoked in the Gayatri of ancient India, "O Thou who givest sustenance to the universe, from whom all things proceed, to whom all things return..."

In collaboration with Wolfgang Pauli, a theoretical physicist and pioneer of quantum physics, Jung explored the possibility that his concept of synchronicity might be related to the *Unus mundus*. Pauli, who reached out to Jung for treatment at a time of existential crisis, did some analytical work with a student of his. He then began to meet regularly with Jung. While not devoid of therapeutic components, their conversations rapidly evolved into a collaboration on theories related to

---

45 Feuerstein, *Tantra*, p. 83.

VI. Facets of a Unitive Science

their common interests, one of which was synchronicity. As mentioned earlier, synchronicity is the intriguing phenomenon of meaningful coincidences, in which an outside event uncannily mirrors a psychological content. An objective happening converges with a subjective preoccupation in a spark of meaningful, yet acausal connection. Implied in synchronicity, Jung saw, is a level of reality that precedes and transcends the distinction between circumstances and self, world and psyche. Distinct in our psychological experience, these two poles are one at the deepest level of reality. He referred to this transcendental unitary reality as the unified psychophysical reality (*die psychophysische Einheitswirklichkeit*)—Unus mundus.

> That even the psychic world, which is so extraordinarily different from the physical world does not have its roots outside the one cosmos is evident from the undeniable fact that causal connections exist between the psyche and the body which point to their underlying unitary nature.... The common background of microphysics and depth-psychology is as much physical as psychic and therefore neither, but rather a third thing, a neutral nature which can at most be grasped in hints since in essence it is transcendental.[46]

Separating and bonding the opposites of persona and shadow ("good and bad"), ego and animus/anima, allowing mind and heart to hold them together with the contradictions, impossibilities, and inner conflicts they bring, is the royal way for a third reality to emerge, one that transcends the seemingly unsolvable dichotomy and establishes a new and more inclusive ground of consciousness. The long process, carried out through many lives, of periodic dissolution of structures (existential constructs and beliefs, ultimately body and personality) and integration of vaster grounds of being is the overarching work of evolution of consciousness, which eventually brings about the union (wedding) of self and Self. The *"Unus mundus"* stage of identification with the world marks the end of the opus of individuation and the beginning of an order of reality not fathomable for that which "fathoms"—the mind. The mind as we know it is relinquished in the process, and so are time and space.

Every spiritual tradition is oriented to such an "omega point" of unison with cosmos. The nature of this ultimate stage of selfhood suggests that the royal way of growth is a tireless practice of "unisons"—a quintessential practice of unconditional acceptance and deep energetic union with lesser aspects, reactions, dimensions of one's psychic being. Relating to instinctual forces as waves of energy, opening an authentic energetic space for a fear or anger vibration to experience itself, free of thought and judgment, but within the loving intelligence of mindfulness, will have the effect of harmonizing one's being and unifying the psyche on a new, more integrative basis.

---

46 Jung, *Mysterium conjunctionis*, par. 768–769.

Describing the Rosicrucian path, Rudolf Steiner emphasizes the same striving to unison:

> It is in concrete reality that one must plunge into every being and phenomenon and lovingly accept it as part of oneself. It is a concrete and intimate knowledge, far removed from merely indulging in phrases like: "being one with the World Soul."[47]

The legendary experiences of "self-realized" beings, and the mystical poetry born of their language-transcendent experiences, speak to supreme at-onement. Mansur al-Hallaj's after-death elemental epiphanies, Jesus "miraculously" calming the tempest (air), commanding trees to dry out (earth), transmuting water into wine, or taming the fire of wild animal natures, suggest the vastly expanded consciousness of beings no longer identified with individual Egoity, but with the universal spirit at work in the elements—in a way analogous to the experience of the body cosmic that pervades the verses of the Shiva-Samhita quoted earlier: "Within this body exist Mount Meru, the seven continents...; in it also dwell the seers, the sage, all the stars and planets....In it whirl the Sun and Moon.... Likewise, it contains ether, air, fire, water, and earth.... He who knows all this is a yogin."

Similarly, Chinese thought, in pithy statements such as "the term is the germ," intimates how the end fuses with the beginning. The individuality reaches plenitude in becoming cosmic, and Taoist writings are replete with teachings on how to reunite with the cosmic all. Returned to the Tao, the Daoist "immortal" flies above the clouds...beyond the limits of time and space...so "one" with the universe that he no longer knows "whether the wind was riding on me or I on the wind."[48]

Steiner speaks of the last, seventh stage of the Rosicrucian path as a "blissful experience of the life and being of all things," a stage of godliness that is a "blessed repose within all things."

Yoga philosophy blends with Jung's alchemical psychology in Aurobindo's evocation of the "informing" and homogenizing Will at work in all things. Aurobindo's "Will" points to the same reality as Jung's *"Self"*—a term Jung actually borrowed from Hindu sources. "This is surely," says Aurobindo, "that Will in things which moves, great and deliberate, unhasting, unresting, through whatever cycles, toward a greater and greater *informing* of its own figures with its own infinite reality."[49]

As life points to Life, and light to Light, so does the self-organizing core of all living things point to the self-organizing Will of the universe.

---

47 Steiner, *Supersensible Knowledge*, p. 166.
48 Lieh-tzu, book 2; quoted in Bergeron, *La Chine et Teilhard*, p. 166.
49 Sri Aurobindo, *Rebirth and Karma*, p. 44.

VI. Facets of a Unitive Science

Alchemical texts highlight the major stages of psychospiritual evolution with a sovereign simplicity that mirrors the Tetractys, offering yet another illustration of its extraordinary synthesis of the *seven*-stepped cosmological process:

One, Two, Three, Four, Three, Two, One.

Jung was most inspired by the works of Gerhard Dorn, a Belgian physician and alchemist who studied Paracelsus and rescued many of his manuscripts. Paracelsus was one of the first to affirm that some diseases are rooted in psychological conditions—a direct application of the matter-psyche continuum of alchemy. Jung borrowed from Dorn the notion that the alchemical process involves three levels of separation, each followed by a *"coniunctio"* of opposites:

Unio mentalis (union of soul and spirit)
Caelum (re-union of body and spirit-soul)
Unus mundus (*coniunctio* of self and universe)

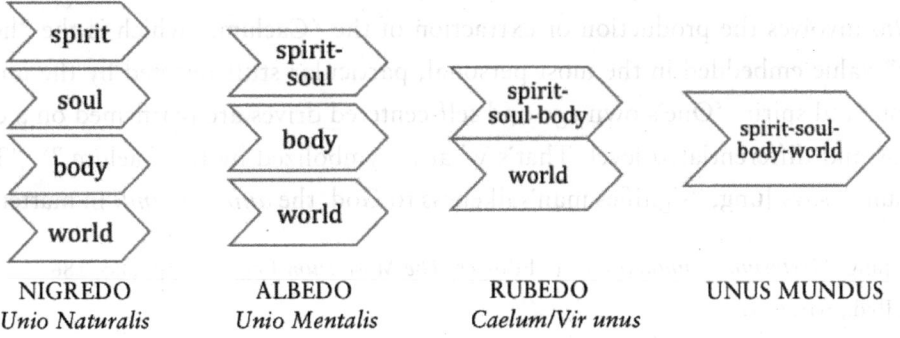

Fig. 70. "Hermetic Caduceus and the Seven Steps" in Le triomphe hermetique *by Limojon de Saint-Didier (1689)*

This process reads as an archetypal model of humankind's individual as well as collective evolution. The diagram below turns the vertical Tetractys on its side to suggest how it frames the alchemical opus delineated by Dorn. From left to right, the successive union stages of the process illustrate the evolutionary progression from Four to One—from the fourfold coexistence of *Unio naturalis*, to the threefold *Unio mentalis*, to the twofold *Caelum* stage of the stone (or *Vir unus*), to the oneness of *Unus mundus*.

NIGREDO
Unio Naturalis

ALBEDO
Unio Mentalis

RUBEDO
Caelum/Vir unus

UNUS MUNDUS

Preceding the first *coniunctio* is a state of unconsciousness, a confused chaos of inextricable interweavings of soul and body, which Edinger refers to as the basic, "untilled state of the soil of the human psyche." At this initial stage, soul and body form a dark unity—*"unio naturalis."* The soul is engulfed in the fourfold elemental nature. The task of the alchemist or analysand is to disentangle and "separate" the confused mass of affects, instincts, thoughts, aspirations, to "dissolve projections, objectify affects and confront instincts with consciousness, in order to free the soul, and to establish a rational, spiritual–psychic counterposition over against the bodily appetites and turbulence of the emotions...[a position] immune to the influence of the body."[50] This first stage of the work was referred to as the black opus, the *Nigredo*. Jung regarded it as a major phase in any psychotherapeutic process.

Along with the alchemists and the Pythagoreans, Jung understood the "golden" truth that "Only separated things can unite."[51] The outcome of the separation work is a first stage of union or *"coniunctio,"* which the alchemists called *"unio mentalis."* It is a union of soul with spirit, which requires a certain separation of the soul from the body. It involves a mortificatio—dying—of the instinctual, "bodily" psyche, at the same time as it brings about a *sublimatio* of the soul. "The whole process of achieving the *unio mentalis* is a spiritualizing, abstracting, generalizing kind of process which depreciates everything that...derives from the desires of the individual ego."[52] It leads at first to a "dissociation of the personality and a violation of the merely natural man."... and is symbolically referred to as *Albedo*, the whitening stage. On a collective level, one recognizes in this process of divorcing mind and soul from the body's instinctuality a central theme of the cultural history of the Judeo–Christian West in the course of the last two thousand years. This second union corresponds to an upward move from level four to level three of the Tetractys: from body–soul–spirit–world to *soul-spirit*-body-world.

The next step is for the *Unio mentalis* created in the first stage to be reunited with the body. "At this level the ego has achieved the acceptance of the opposites and is able to endure the paradox of the psyche's two-sidedness."[53] The next *coniunctio* involves the production or extraction of the "Caelum," which is the "heavenly" value embedded in the most personal, particular stuff rejected by the union of soul and spirit. "One's own ego and self-centered drives are reaffirmed on a conscious and differentiated level. That's what is symbolized by the Caelum."[54] "The Caelum," says Jung, "signifies man's likeness to God, the *anima mundi* in matter."[55]

---

50 Jung, *Mysterium conjunctionis,* in Edinger, *The Mysterium Lectures,* pp. 280, 286.
51 Ibid., par. 671.
52 Edinger, *The Mysterium Lectures,* p. 288.
53 Ibid., p. 281
54 Ibid., p. 288
55 Jung, *Mysterium conjunctionis,* parag. 770

VI. Facets of a Unitive Science

This phase corresponds to the alchemical *Rubedo*, also called reddening. The consciousness achieved by the unio mentalis is brought into "full-blooded reality, so that one lives it out fully in everyday life."[56] The Caelum was regarded as the balsam and elixir of life...a "living stone," a simple and universal union of all opposites, a "lapis" of inner unity.[57] The "actualization of the paradoxical wholeness that has been made conscious" represents, says Jung, "the crux of individuation"—the achievement of the philosopher's stone. It marks the advent of the *"Vir unus"*—the fully realized, whole human being—and it is the unison of being aimed at in yoga.

"It's not easy to get the *unio mentalis* to embrace the body," Edinger comments; "it has previously devoted all its efforts to separating from the body.... What had previously been seen as a bundle of greeds, lust, power-striving and unconscious negatives of all kinds must now be invited back in."[58] The evolutionary step of reengaging instinctual energies, connecting with the energetic ground of one's being and the elemental consciousness of the body has been emerging as a new dimension of psychospiritual work and wholistic healing in the last few decades. One can see in such conscious corporealizing of the psyche and spiritualizing of the body a major cultural move toward wholeness, unity, and indivisibility of being. With the completion of this stage, body, soul, and spirit are reunited, and evolution moves from level three to level two in the Tetractys: from soul–spirit–body–world to *soul-spirit-body*–world.

The alchemical psychologist who facilitates the reunion of soul-spirit with body and the individual pledged to this work, the "Pythagorean quantum scientist" who gathers matter and consciousness into one science, the ecologist restoring humankind's partnership with nature, all are engaged in the same "opus" of Union—one from within, the other from without—an opus that might be regarded one day as the most significant cultural advent of this time of history.

The final step, up from Two to One along the Tetractys, is the union of the soul-spirit-body individual—the *"vir unus,"* the "whole human"—with the world soul. Jung develops, "If Dorn sees the third and highest degree of conjunction in a union or relationship of the adept, who has produced the Caelum, with the *Unus mundus*, this would consist, psychologically, in a synthesis of the conscious with the unconscious."[59] This third stage points to a unitive experience that defies the dual activity of "description," an experience in which the separation between self and world (*mundus*) is transcended. This last stage marks a step from complete harmony

---

56 Edinger, *The Mysterium Lectures*, p. 296.
57 Jung, *Mysterium conjunctionis,* par. 762–763.
58 Edinger, *The Mysterium Lectures*, pp. 290–291.
59 Jung, *Mysterium conjunctionis,* par. 770.

of being to identification with the universe—evocatively reflected in the transition, up the ladder of the Tetractys, from 3/2 (harmony) to 1/2 (octave unison).

In summary, says Dorn, "The first union makes the mental disciple of wisdom, not the wise man. The second union, of the mind with the body, shows forth the wise man, hoping for and expecting that blessed third union with the first unity [the *Unus mundus*]."[60] In modern terms, we could say that Dorn's three unions correspond to the stages of personality integration, individuation, and universalization.

A number of passages in Teilhard de Chardin's writings point to the *"Vir unus"* experience—the "whole, unified man." In the following excerpt, he describes the powerful unison of being he experienced when attuned to the ultimate objective of the totality of his nature.

> Thus, because the ultimate objective, the totality to which my nature is attuned has been made manifest to me, the powers of my being begin spontaneously to vibrate in accord with a single note of incredible richness wherein I can distinguish the most discordant tendencies effortlessly resolved...[61]

Irrupting in the midst of the *"Vir unus"* consciousness, early tastes of the *"Unus mundus"* state come as sweeping tides:

> No brutal shock, no, nor gentle caress can compare with the vehemence and possessive force of the contact between ourselves as individuals and the universe, when suddenly, beneath the ordinariness of our most familiar experiences, we realize, with religious horror, that *what is emerging in us is the great cosmos*.[62]

The alchemical framework of psychospiritual evolution mirrors, in reverse, the involutionary process of creation. While creation "involves" the (threefold) One into the fourfoldness of physical-elemental manifestation, the "self" evolves by harmonizing and at-oning its fourfold reality (spirit-soul-body-world). The central axiom of alchemy, the axiom of Maria, essentializes in a succinct formula this dual movement of descent from One to Four and ascent from Four to One, which is the dual process of involution and evolution: "One, and it is two; and two, and it is three; and three, and it is four; and four, and it is three; and three, and it is two; and two, and it is One."[63]

This evidently represents, Jung remarks, the quartering of the One and the synthesis of the Four in One. Both are held as one in the simple figure of the Tetractys.

---

60   *Philosophica meditativa*, quoted in Jung, *Mysterium conjunctionis*, par. 663.
61   De Chardin, *Hymn of the Universe*.
62   De Chardin, *Writings in Time of War*, p. 27 (italics added).
63   Jung, *Psychology and Alchemy*, par 210, footnote 86, p. 162.

VI. Facets of a Unitive Science

Perfectly complementing the unfolding and refolding of the world being, the Tetractys also displays, we saw, the four major intervals of music (unison, 1/1, octave 1/2, fifth 2/3, fourth 3/4), which are the *inner keys* of the cosmological process—the fundamental fractions involved in the differentiation–creation process and the composition of all things.

The Tetractys emerges as the luminous symbol of a unitary science of matter and consciousness, cosmos and psyche—a cosmo–psychology. One begins to understand the great reverence in which the Greeks, who historically and culturally stood between the ancient subjective world of sentient intuition and the modern objective world of intellect and science, held the Tetractys.

Jung practiced yoga for many years. He also studied and eventually lectured on the philosophy of yoga, from the angle of its usefulness in the psychotherapeutic process. When he researched alchemical symbolism, embracing it as a template of the individuation process, his interest was primarily psychological. He was not essentially concerned or intrigued by the metaphysical ground of these teachings and the cosmological science articulated in the seven chakras or three gunas of yoga, the seven metals and three principles (mercury, salt, sulfur) of alchemy, or its seven or twelve operations. Such was not his task. The focus of his inquiry was on psychological processes. He related to metaphysics and religion as productions of the psyche—which they are, although not more than psyche is a trans-physical production. By not tracing the alchemical and ancient yogic keys he held in his hand to their cosmological ground, however, he stopped short of giving his psychology its full scope. He deprived it of what the cosmological perspective provides: an objective framework for the inner realities of self and individuation.

The fact is that if, as the *Unus mundus* stage intimates, self-knowing eventually expands into universal self-knowing, if self and world (Shiva and Shakti) are *ultimately* one in the fully unfolded consciousness, they are one in the *primordial* reality that precedes and transcends the duality of inside/outside, world/self, in which ordinary consciousness operates: the ultimate oneness points to the oneness of universal beginnings. To reach the *Unus mundus* experience of "self" as world-self, and of universe as self-aware universality, is to transcend the duality of self and world and "realize" the primordial oneness of cosmos–Logos from which the adventure of consciousness began. Indeed, the alchemists regarded the *Unus mundus* as "the potential world of the first day of creation, when there was as yet 'no second.'"[64] With the *Unus mundus*, psychology meets cosmology. The at-onement of subjective

---

64 Jung, *Mysterium conjunctionis*, par. 766.

and objective worlds is the ultimate horizon of a wholistic psychology of integrative dialogue between conscious and unconscious, ego and universe.

If one inquires about the completion of the process of individuation in the writings of Jungian authors, one finds general references to death as the natural termination—though not the ultimate "end"—of the process. To grasp the process of individuation as a whole requires the perspective of its completion, which both alchemy and yoga point to as the universalization of "self"-awareness. For such a complete perspective to be reached, however, the full space–time reality of human experience must be taken into account—namely the entire chakric constitution and the notion of successive lives.

With regard to the latter, Jung implicitly acknowledged what common sense suggests, which is that the course of individuation far transcends the possibilities of a single life time—lest such a life happens to represent for a particular individuality the ultimate step of the process. As for the spatial framework of individuation, Jung limited its chakric topography to what he regarded as relevant for psychology—the first five chakras—leaving out the last two chakras as "superfluous speculations with no practical value."

Jung's lectures on kundalini yoga (1932) make clear how he circumscribed *part of* the human constitution as valid for psychology to focus upon, based on the general level of humankind's experience. "It does not help to speculate about ajna and sahasrara and God knows what; you can reflect about those things, but you are not there if you have not had the experience."[65] While totally consistent with his psychotherapeutic orientation, as well as his professed empirical and pragmatic approach, the limit Jung thus set on the exploration of the path of chakric identification and integration assigned its limit to the philosophical framework of his psychology. It did not diminish the extraordinary value and impact of his depth psychology, but it held it back from taking its full stature as a spiritual scientific psycho-cosmology. Both yoga and alchemy are rooted in an in depth understanding of how inside and outside, psyche and matter, are intertwined. Jung's psychology remained psychological.

Whereas Jung's insights and personal experiences led him very close to the cosmological science embedded in the ancient teachings that lighted his way, he stopped short of considering their metaphysical context as relevant to the quest for wholeness that he regarded as the core of psychology. By setting aside the ultimate oneness of psyche and cosmos that the culmination of the two disciplines (yoga and alchemy) points to, he sealed a dead end for generations of followers.[66] Nonetheless, he did

---

65 Jung, *The Psychology of Kundalini Yoga*, p. 47.
66 The publication of Jung's *Red Book* in 2009 has begun to dissipate this artificial dead

VI. Facets of a Unitive Science

acknowledge that, however "really beyond our grasp" this *Unus mundus* stage may be, in the end, psyche merges with cosmos: "Suppose somebody reaches the ajna center, the state of complete consciousness, not only self-consciousness. That would be an exceedingly extended consciousness, which includes everything... every tree, every stone, every breath of air, every rat's tail—all that is yourself."[67] The significance of this acknowledgment of an ultimate unitary experience that transcends the duality of self and world is that it blows out the purely subjective frame of psychology to expose a wholistic objective-subjective, "cosmo-psychic" ground, a primordial "unitary reality" beyond the distinction of inner world and outer world—a spirit ground. Regrettably, but perhaps wisely, if his psychology was to thrive in the context of the twentieth century, he maintained the separation between psychology and cosmology and did not seize the opportunity to usher psychology into the fully rounded science that alchemy and yoga could have led him to. The goal of science is understanding, the goal of psychology is healing. However steeped he may have been in vaster grounds of understanding, his core motive was that of the physician of the soul—to heal—to make whole. Yet, as Jung was well aware, the split between "the two"—in this case, cosmology and psychology—and the removal of "one of the two" will forever prevent the third, the unified field of the *"coniunctio oppositorum,"* to appear.

What a cosmological psychology does is extend the scope of traditional psychology to include the awareness of soul-spiritual developments taking place in a psyche understood to mature across many lives. It involves an understanding of the developmental steps and issues associated with the seven chakras, a path of insights partially paved by Jung. It includes the recognition that love, the power to unite, is

---

end. The revelation of the initiatory experiences Jung underwent in 1913/14, and later consigned in the texts and pictures of his *Red Book,* has blown away the ceiling that, perhaps wisely for the viability of his work at the time, he placed on a "Jungian" depth psychospiritual understanding and approach of reality. Of his own avowal, these visionary experiences, imbued as they are with Gnostic themes and ancient mystery symbols, were the real matrix of his psychology. "The years... when I pursued the inner images were the most important time of my life. Everything else is to be derived from this.... My entire life consisted in elaborating what had burst forth from the unconscious and flooded me like an enigmatic stream and threatened to break me.... Everything later was merely the outer classification, the scientific elaboration, and the integration into life. But the numinous beginning, which contained everything, was then.... When I began to understand alchemy I realized that it represented the historical link with Gnosticism, and that continuity therefore existed between past and present. Grounded in the natural philosophy of the Middle Ages, alchemy formed the bridge on the one hand into the past, to Gnosticism, and on the other into the future, to the modern psychology of the unconscious. The possibility of a comparison with alchemy, and the uninterrupted intellectual chain back to Gnosticism, gave substance to my psychology" (Jung, *Memories, Dreams and Reflections*).

67 Ibid., p. 59.

the fundamental, "scientific," catalyst of individuation-universalization, the golden "mean" to the undivided *Unus mundus*. It opens a framework of understanding and a therapeutic space for the "vehemence" of events and tides coming from the universal sphere of spirit—the sphere that challenges the mainstream psychological notion that the loss (albeit temporary) of a clear separation of self and world, of within and without, signals mental illness. Psychotic-like experiences might be experiences that point to greater wholeness. When faced with psychotic-like symptoms, a psychology inclusive of the universal dimension of spirit, such as the works of Stanislav and Christina Grof foster, will carefully assess whether the psychological breakdown is signaling a pathological "emergency" or a "spiritual emergence." Cosmological psychology also engages the science that addresses jointly the inner and cosmic side of selfhood—astrology.

When psychology opens its doors to the teachings of yoga and the hermetic roots of alchemy, a more complete and coherent framework emerges, consistent with the facts of synchronicity, and pregnant with insights into the whole cycle of self-and-cosmos. Reinserted in its natural continuum with spiritual teachings, psychology expands its scope to include the field of soul–spirit relation. It recognizes that the individuating human being is inevitably a spiritual seeker, while it also provides a ground of sound psychological navigation for the spiritual seeker who, knowingly or unknowingly, is engaged in a process and labor of individuation. "Tantric metaphysics," Feuerstein points out, "is also a meta-psychology with far-reaching practical implications."[68]

The potential for a fully rounded cosmo–psychological science runs parallels to that of a unified field theory in physics. As the grand universal spiral would have it, circling deeper into the nature of matter has brought the scientific mind to circle nearer the spirit of things, and discover through the most analytical, dissecting study of matter glimmers of a primordial, "elementary," trans-dual dimension, in which observer and observed are no longer unquestionably separate, and mind and world become relative distinctions associated with the dual character of perception. Since the "subject"—the human mind—is already implicated in the "quantum" scientific field, it would seem inevitable that the next installment in the search for the grail of physics—the unified field—would involve the integration of the fact of consciousness, hence a serious consideration of the ancient unitive grounds mused upon by the quantum fathers as the next beckoning horizon of science. If quantum physics reveals a ground of reality that transcends mechanistic reasoning, pointing

---

68   Feuerstein, *Tantra: Path of Ecstasy*, p. 83.

VI. Facets of a Unitive Science

to a paradigm for which ancient wisdoms appear to offer clues, it also promises to grant such wisdoms their utmost splendor by showing the reverberation of their grand principles in the minute formative dynamics of matter.

The scientific revolution of the early twentieth century ushered by quantum physics and Einstein's relativity opened the door to new insights into world grounds and world beginnings. A world of separate materiality was replaced with one of interconnected energy fields. This revolution came on the heels of the firm establishment, by the end of the nineteenth century, of a materialistic and mechanistic cosmos, which went hand in hand with philosophical pronouncements about the death of God. It is remarkable to note that, just when the collective psyche was declaring itself free and clear from "religious illusions," free from the transcendent God of traditional theologies, transcendence surged anew and from the least expected place—matter itself—in the form of trans-physical, trans-dual quantum phenomena that put to question Newtonian physics and Euclidian common sense, rekindling the mysterious and patient smile of divinity in the minute depths of matter. Shifting ground from heavenly heights to the depths of matter, the nondual activity of the "holy spirit" was reappearing upon the waters—of matter's waves—displaying the enigmatic formative process of world beginnings.

What seemed to lead further into materialistic alienation turned out to be an unsuspected access to the transphysical—literally, metaphysical—ground of reality. The etheric vibrancy of being began to shimmer through the thinning walls of matter. The death of a transcendent God may have meant the end of a long chapter of religious separation—paradoxically—of the human from the divine. It ended up heralding a new chapter of relation to the transcendent—one inspired by the expanding psycho-physics of world-self (depth psychology) and "minded" particles (quantum physics) spiraling onward in divine hands and human wills.

Astrology: A Cosmo-Psychological Science

> *In the same way that the outer world, or "external sky," owes its order and persistence to the seven wandering stars or planets and the twelve castles or zodiacal signs, the persistence and order of the spiritual world or the "inner sky" rest on the seven prophets (Adam, Noah, Abraham, Moses, David, Jesus, Mohammed) and the twelve Friends of God, and their higher octave—the "seven Angels ecstatic with love" of the 8th heaven and the twelve angels of the 9th heaven.* —Henri Corbin[69]

The unitive psycho-cosmology in germ in Jung's psychology finds practical grounds in astrology. Alchemists studied and practiced astrology. So did Jung. And so did pre-Newton scientists. Corbin points out how the philosophers-alchemists of Islam regarded the rhythm of seven and twelve as a fundamental law of Being. They pondered these central cosmological numbers in terms of a dual "balance of seven and twelve," at once astronomical and angelological—a balance of world and self, body and soul:

> If angelology and astronomy unveil identical rhythms of seven and twelve, it is because both proceed from the One mystery of theophany. The rhythm of seven and twelve expresses a fundamental law of Being—the very "balance" of Being.[70]

Balance and proportion were, in their view, the essence of music, which, like their Greek predecessors, they defined as the science of proportions.

From tacit assumption rather than a reasoned claim to the validity of astrological wheels to offer a lens into the mysteries of becoming, the astrologer explores as one reality the inner course of a person's life experiences and the cosmic score of concomitant planetary configurations. His/her task is to help each shed light on the essence of the other. Like Plato, s/he implicitly regards the soul as within and without, core and circumsphere: "And *in the midst* thereof He set Soul, which he diffused throughout the body, making it also to be the *exterior* envelope of it."

From the perspective of astrology, personal dispositions are the expression of an objective "circumstance" of planetary relationships. The birth sky offers a signature of the psyche for a lifetime. Circumstances are "transitional" factors (transits) and forces that elicit self-unfolding. Charts of planetary movements guide the journey of individuation by displaying the unique score of attunements and re-tuning brought to bear on the natal planets by the solar systemic "psyche." Individuation

---

69 Corbin, *Temple et contemplation*, pp. 80–81.
70 Ibid.

VI. Facets of a Unitive Science

is displayed quite literally as a "universalization" process. This most subjective of sciences, psychology, acquires a framework that blends theories and insights with factual configurations of planetary motions and zodiacal alignments.

With the emergence of the human kingdom, which represents the involution of the world-soul principle *as a whole* in an autonomous form of nature, the spiral cycling course of electron-waves around the atoms and of plant growth around the Sun-Earth axis "turns into" the evolutionary coursing of the personality around the soul and of the soul around its spirit-monad. When the Logos reaches self-awareness in the human psyche, evolution becomes centered in the human journey. From mineral to biological, from vegetal to zoological, the journey becomes *psycho-logical*. The process of psychological growth parallels the spiraling pattern of organic growth. Astrology charts how we grow by learning "from all angles" of experience.

This latest stage of our exploration brings us back to our starting place, the soul. Our pursuit of a unitive science is running into what led to it, which included an inquiry into astrology as a science of the soul, inclusive of world and self. The adventure that was set in motion by this question has uncovered a whole rational ground in support of the notion that astrology might uniquely qualify as a science of the soul. The initial proposition was based on the wealth of experiences associated with astrology's contribution to self-understanding, in combination with the intriguingly similar structure of the essential (albeit "esoteric") ground of psychology—the 12fold Egoic body of the soul—and the ground of astrology, the 12fold zodiac. Suggestive of a kinship between the human core and its cosmic matrix, this structural similarity also hinted at a potential ground of "depth rationality" for the astrological practice of reading information about the innermost life of the self in its outermost cosmic background.

Like the alchemical snake biting its tail, the course of this journey has taken us onward and around, back to our initial focus—the connection between world and soul. The gradual uncovering of a fundamental 7fold/12fold template of forms has drawn astrology back on the stage, setting in a new light its 7/12fold frame of planets and signs, which engage world and self in one cosmo-psychological discipline. Astrology may well prove to be a central science of this human stage of evolution—a quintessential approach to the "fifth" interval and third harmonic of the universe, the golden blend of world and self that is the core of humankind.

Coming upon astrology again and anew at the end of this expedition, we note how clear and majestic it now stands, its framework jutting out of a fundamental ground of reality instead of hovering from beyond the blurred skies of old and obscure superstitions. Tacitly displayed in this framework are its letters of nobility, which are the numbers of the cosmological deployment of Being into cosmos, and the spiraling-in of self to Egoic fullness. We see how strikingly reflective its charts are of the unitary world-self archetype of forms and systems.

Uncovering the quintessential tracks of the universe's unfolding has exposed the deeply logical—logoic—ground of astrology as a cosmic science of the soul and its existential growth. The strength of wholistic coherence accrued along the way affirms the cosmo–ontological underpinnings (at once cosmic and soulful) of a 7/12fold spiritual science of the soul milieu, its fundamental structure and its transitory modulations. "Aster" (-star) logic is the outside, objective, cosmic partner of psychology, psychology being the internal, subjective side of the science of self. Astrology offers the unitary "physics–metaphysics" appropriate for the human stage of the universe—a *physics of the psyche, and a metaphysics of experience.*

In the light of Plato's Pythagorean cosmology, the operative ground of astrology (12fold wheel and 7/10 planets) emerges as a symbolic design of the matrix of world and self—a self coextensive to cosmos, and a cosmos that is a living self. "The unity of the great Heaven," wrote Paracelsus, "is split into our diversities by the various moments at which we are born. The starry vault imprints itself on the inner heaven of a man."

Astrology regards the horoscopic map of the natal sky as a blueprint of the psyche. The unique configuration of the planets around the vortical matrix of the zodiac at the time of birth is viewed to capture the intention of the incarnating soul and the psycho-energetic signatures of its vehicles. Between lives, the soul consciousness is identified with its grand cosmic analogue, the zodiacal field. Steiner describes how, between death and a new birth, the human being circles the zodiac and looks out from every viewpoint to gather from each section the forces he/she will bring to bear upon his/her next incarnation to best meet his/her evolutionary needs. At the time of birth, this inner heaven, he says, embeds itself in the structure of the brain.

The horoscope harbors a reality that transcends the separation between external world and inner experience. Astrology translates the celestial matrix of zodiac and planets in a symbolic language, which speaks to the experience of an evolving psychological self. This symbolic milieu bonds the subjective and objective ends of the human experience, providing an objective approach to self-understanding, and a subjective, meaningful interpretation of external circumstances of life—a psycho-"logic" at once universal and unique, cosmic and personal, objective and subjective. Astrology abides at the confluence of the two sides of human experience: the outside world that we eventually come to realize we are, and the subjective space of the psyche through which the universe comes to know itself.

VI. Facets of a Unitive Science

The symbolic meanings associated with each planet show logical archetypal correspondences with their position and relation to the sevenfold, be it via chakras, numbers, diatonic ratios, or triadal pairs: Saturn's ring-pass-not and sovereignty resonates with the crown chakra; Jupiter's inclusive wisdom with the Ajna center; Mars's urge to self-expression with the throat center; Sun's sense of identity with the heart center; Venus' bonding impulse with the sacral center; Mercury's networking, entangling and disentangling with the solar plexus; Moon's emotional needs and expression with the base chakra. Again, such correspondences are far from absolute.

Another archetypal logic for planetary meanings is that of a replication, at three different levels, of the primordial triad of father-mother-son, yang–yin–tao, centrifugal–centripetal–breath wrapped in the sevenfold octave of any form of life. With Mercury, the com-"*unicator*," at the center, secting and linking each pair of polarities, we see complementary dimensions of the psyche arise behind the planetary pairs.

- Saturn (social structure)/Jupiter (individual philosophy of life)—collective, societal self
- Sun (conscious self)/Moon (subconscious needs)—individual self
- Mars (separating urge to act, express)/Venus (unifying urge to bond)—personal, instinctual self

We may note how each of the three pairs speaks to one level of the contracting–expanding, centrifugal–centripetal, yin–yang dualities of human experience, and how the three as a whole point to the body–soul–spirit—personal-individual-societal—triunity of the human predicament.

A third perspective on planetary functions and meanings stems from the triad-and-quaternary view of the 7fold. As Blavatsky points out, "The union between the number three (the symbol of the divine triad with all and every people, Christian as well as pagans) and four (the symbol of the cosmic forces or elements), [gives] the number seven [which] points out symbolically to the union of the Deity with the universe."[71] From the Earth's point of view, the triad of Sun (generative identity), Moon (receptive "nature") and Mercury (secting–bonding activity) actualizes itself through the four functions of Mars–fire, Venus–air, Jupiter–water, and Saturn–Earth.

Rather than planets and stars exerting some kind of physical influence[72] on our characters and destinies, what emerges from an exploration of harmonic cosmology

---

71 Blavatsky, *Secret Doctrine,* vol. 2, p. 412.
72 Although the planets do not exert a physical influence per se, they emit a very low electromagnetic frequency, in keeping with the harmonic model of understanding. Astrologer Rick Levine remarks in an interview, "Models based on quantum physics will eventually embrace the as-of-yet unknown mechanisms of cosmic influence. Planetary

is a view similar to that of the last astronomer-astrologer, Kepler. Kepler regarded the soul as bearing the idea of the zodiac, and astrology as a "science of harmonics":

> Inasmuch as the soul bears within itself the idea of the zodiac, or rather of its center, it also feels which planet stands at which time under which degree of the zodiac, and measures the angles of the rays that meet on the Earth; but inasmuch as it receives from the irradiation of the Divine essence the geometrical figures of the circle and the archetypal harmonics... it also recognises the measurements of the angles and judges some as congruent or harmonious, others as incongruent.... Ten years ago I rejected the division into 12 equal parts, the houses, the dominations [i.e., rulerships], the triplicities etc, all of that, keeping only the aspects and *transferring astrology to the science of harmonics*.[73]

In essence, astrology maps the harmony-making potential of a life (natal chart) and of any period of life (transits). It reads existential experiences as uniquely creative encounters of self with cosmos—self with Self. The celestial configuration at the moment of birth represents the fundamental harmonic signature of a life, imprinted as a "psychic chord" on the brain of the entity entering the world and exchanging its first breath with it.

The most organic way to grasp the premises of astrology is to align oneself imaginatively with the grand stairwell, which, according to ancient wisdom, every soul takes on its way in and out of incarnation. Hermetic and theosophical traditions harkening back to Greece, Egypt, and beyond, describe how, at some point between death and rebirth, the soul leaves the twelvefold zodiacal-causal field of Light to spiral down the planetary spheres and successively don the psychic vestures-functions—seven in all—with which each sphere equips the soul for its upcoming existence in a human body–psyche:

Saturnian function to define and structure one's existential frame and line of contribution.
Jupiterian function to tend growth and becoming, inspire existence with meaning, joy, hope, trust and vision
Martial drive to act and self-express
Solar function to kindle one's sense of existential identity and give it its tonality
Venusian function to bond, unite, harmonize, beautify
Mercurial function to link, bridge, communicate
Moon function to foster and nurture the subconscious ground of the psyche

---

cycles will come to be understood as very, very low frequency electromagnetic vibrations, measured in cycles per year or cycles per century or cycles per millennium, rather than in cycles per second."

73 Kepler, *Harmonices mundi*, book 4.

VI. Facets of a Unitive Science

Each planetary function-vehicle is qualified by the zodiacal energy it is aligned with (the sign "it is in") as well as the harmonic relations it forms (or not) with other planets.

When the soul leaves the dying body to journey back to the zodiacal field and beyond, these subtle envelopes of the soul are returned, transmuted by the experiences of a lifetime, to the corresponding planetary spheres of the macrocosmic body. The incarnating descent/ascent of the soul sheds light on the planetary placements of the horoscope as vibrational tonalities, which inform the person's character and destiny.[74]

A repository of the ancient knowledge about the incarnating descent and excarnating ascent of the soul through the spheres, astrology is also a unique portal to a future cosmo–psychological science, rooted in the unitive ground of world and consciousness.

Twelvefold zodiacal matrix and sevenfold spiral-cycling motion speak to the two dimensions of astrology: energetic composition and evolutionary steps of the psyche.[75] It is both a science of individual dispositions (based on the natal chart), and a science of individual growth (based on current planetary motions in relation to natal planets—"transits and progressions"). On one hand, it focuses on the singular harmonic chord imprinted on the newborn's crowning head as the seed of his/her psychic structure; on the other hand, it studies the constantly evolving relationship between this psychic matrix and the cosmic cycles, as a source of insights into current themes of psychospiritual growth (transits and progressions).

What modern astrologers do is interpret in terms of experiential processes and soul intentions the planetary signatures and configurations of a person's cosmo-psychic matrix and their interplay with current planetary movements. Listening to the person's experiences, the astrologer hears the energy dynamics at work within his/her inner sky or between current and natal skies, and offers the perspective provided by the symbolic analysis of these planetary dynamics. By reframing the personal interpretation of events, often reductive or one-dimensional, provided by the ego's habitual beliefs and patterns, these cosmic perspectives open a space of transpersonal understanding and meaningful collaboration between the person and his/her destiny (the Self). By tapping into the musical score underlying

---

74   For insights into the voyage through the spheres after death, see Steiner, *Karmic Relationships*, vol. 5, lect. 5, 6, and 7.

75   The planetary sevenfold refers to the seven traditional planets and the archetypal functions associated with them. The additional three—Uranus, Neptune, Pluto—may be regarded as the planetary outposts of the transcendent trinity, operating as the transpersonal powers of the universe in the psyche: Uranus as universal intelligence, the higher octave of Mercury; Neptune as universal love, the higher octave of Venus; Pluto as universal will, the higher octave of Mars.

experience, astrology helps to attune one's natural response to circumstances to the heart of their matter. It grants us to participate more knowingly in the orchestra of life, modulate one's interpretation of the chords, and breathe willingness into challenging "intervals" of experience. It gives voice to new possibilities and also points to temporary constraints calling for harmonization between personal dispositions and more universal intentions within. Grasping the consonances–dissonances of the time, the mind can suspend its reactive attempts to figure things out, and listen instead to a deeper reason ticking its quiet heartbeat within the shaky walls of the moment.

The unitive (self–cosmos) ground of astrology provides a unique access to the exchange between the personal and the universal within. To become aware of the harmonizing challenges and creative opportunities intimated by the evolutionary "tuning" issuing from our cosmic depths is to enter a more conscious relation with metaconscious intents veiled in what one tends to experience as fate or fortuity.

Plato comments on the kinship between the two sides of experience that astrology brings together as the single object of its science: revolutions of Reason in the sky, evolution toward concord in the soul. Seeing the revolutions of Reason in the sky assists us, Plato says, in developing–evolving reason within. He urges us to avail ourselves of the contemplation of cosmic revolutions as an inspiration toward greater harmony. He compares it to hearing music, the harmonies of which entrain the inner muse, the soul, to restore "concord" and grace in one's being:

> God devised and bestowed upon us vision to the end that we might behold the revolutions of Reason in the Heaven [zodiac] and use them for the revolvings of the reasoning that is within us [soul], these being akin to those.... Concerning sound also and hearing, once more we make the same declaration...music, too, insofar as it uses audible sound, was bestowed for the sake of harmony. And harmony, which has motions akin to the revolutions of the Soul within us, was given by the Muses...as an auxiliary to the inner revolution of the Soul, when it has lost its harmony, to assist in restoring it to order and concord with itself. And because of the unmodulated condition, deficient in grace, which exists in most of us, rhythm also was bestowed upon us to be our helper by the same deities and for the same ends.[76]

Rudolf Steiner echoes Plato's musings when he states that, as we descend from pre-earthly existence to a new birth, we seek to establish harmony between the stars and our earthly life.[77] Also reminiscent of Plato is his comment that we, human beings, who, from a cosmic point of view, are wandering around aimlessly in our

---

76   Plato, *The Timaeus*, 47d–47e
77   Steiner, *Anthroposophical Leading Thoughts*, p. 82.

earthly conditions, can "rise above our disoriented earthly thinking and receive strength in thinking and feeling from the calm movement and regularity of the cosmos.... And in the cosmos we see, as if crystallized, the word in its calmness, and the word in its movement."[78]

Watching the revolutions of Reason in the skies helps fathom a certain order and "reason" in difficult, confusing, even traumatic experiences. Pluto conjunct Neptune, Pluto square Mars, or Saturn squaring Venus are equally noble ratios of soul making as Neptune trining the Sun. Being aware of the cosmic "circum-stances" that bestow transformative opportunities on the consciousness helps to grow in concord—that is, in harmony of the heart—with the current "intervals" of becoming. Astrology is not only a science of self-understanding, it is a science of soul making. Ceaselessly "Reasoned" and tuned to vaster vision, love, or strength across the span of many lives, the human self eventually blossoms into a uniquely universal expression of the world soul. The great extremes of cosmos and self, monadic and "elemental" being, multiplicity of matters and unicity of Life are made one ("indivisible") in the accomplished individuality.

Ultimately, fullness of soul means to know oneself as the world soul and know the world soul as one's (universal) self—to know, as St. Catherine of Genoa said, that "my me is God, nor do I know my Selfhood save in him." The Logos of the beginning becomes self-aware in a "son of God." The images of Mithras in the mandorla, or the Christ surrounded by the twelve apostles symbolize this completed work of individuation, which blends into one sound of zodiacal selfhood—the 12-petal Egoic lotus—the twelve zodiacal "intervals" of soul consciousness. Jung regarded the Christ as the archetype of selfhood. Indeed. If the demiurge made twelve portions of the triune world soul, it behooves the human being to gradually integrate the twelve notes of the soul into a triadal oneness of intelligence, love, and power in which the Logoic vibration of the monad can resound.

The zodiacal fullness of soul that crowns the human journey heralds the dissolution of the causal body and the end of identification with "individuality" as we understand it. The full blossoming of the Egoic vehicle is soon followed by its dispersion–destruction. "The beautiful robe no longer serves," the last words of Jesus Christ on the cross as translated by Blavatsky, highlight this major initiatory landmark of relinquishing the marvelous, radiant Egoic lotus.[79] "He then left...the Identity which had hitherto been His and identified Himself with that which was

---

78 Steiner's address for the first performance of the "Twelve Moods" on Aug. 29, 1915; in the newsletter *Deepening Anthroposophy*, Mar. 4, 2015.

79 Blavatsky, *Secret Doctrine*, vol. 2, p. 613.

then revealed."[80] The monad is no longer bound to reincarnate, in other words to reenter the zodiacal vortex it has transcended from within—though it may choose to do so to serve the Earth.

In the *Timaeus*, Plato pursues a similar course of thoughts. He speaks of the soul's goal to attune to cosmic revolutions and become "cosmos-like" *(eu kekosmêmenon)*:

> Each one of us should follow [the intellections and revolutions of the Universe], rectifying the revolutions within our head, which were distorted at our birth, by learning the harmonies and revolutions of the Universe, [so the soul may] achieve likeness with its original nature and attain finally to that goal of life which is set before men by the gods as the most good both for the present and for the time to come.[81]

This original nature (*"archaian* phusis") is the *arche*-typal nature of the soul. The goal of life, says Plato, is to become "like" this original nature, which means to experience the world soul as "self." This goal does not suggest returning *back* to an original "condition," but actualizing an archetypal potential by "fleshing it out" with soul forces distilled from experience. It means growing into a fully embodied, universally minded, individuality. The goal is to bring to fruition the template of "same-selfhood" we are seeded with, by "filling" it with wisdom gained from experience, embodied strength, and self-generated grace and love. Following the cosmic score of harmonies at play behind personal matters helps sustain, Plato suggests, the daily work of attuning elemental forces and personal patterns to the chords of the world soul.

The Tao Te Ching settles the confusion that might arise between the movement of rectifying return, implied in Plato's words, and that of evolutionary unfolding:

> *Going on means going far,*
> *Going far means returning.*

The right way back to the origin is by going onward, away from the beginning. The world-deploying outbreath of Being turns into the centripetal in-breath, which draws the extraverted movement of differentiation back to introverting integration—in the torus like uni-verting dynamics that "bundles" the universe. When the evolutionary arc brings consciousness "back" to the source, it is not as a dissolving fusion with the origin, but as a conscious expansion in self-aware Oneness. The 10 of the completed Tetractys is not the 1 of the beginning. It symbolizes instead the self-realization of the One in a human likeness realizing itself as one with the One: "I said, 'Ye be gods'" (Psalm 82:6; John, 10:34).

---

80   Bailey, *Esoteric Astrology*, p. 315.
81   Plato, *The Timaeus*, 90c-d.

VI. Facets of a Unitive Science

In speaking of the "goal of life that is set before men by the gods," also translated as "that most excellent life offered to humankind by the gods," Plato echoes old sources of wisdom, reintroduced in modern esoteric teachings, which present the zodiac as twelve great formative powers, twelve Creative Lives or groups of beings who provide the matrix of the universe and "set before men" the ideal of the human soul.

According to such sources, the term *zoe-diac* (from the Greek *zoe*, life, and *zoion*, living being and, by extension, animal) designates a circle of "Lives or Beings" who inform with their singular energies collective "Ages" of evolution as well as facets of individual and world consciousness.

From his clairvoyant investigations into earlier times on Earth, Steiner refers to twelve "World Initiators": "The twelve Signs of the Zodiac were not then externally visible; but instead, twelve great Forms, twelve Beings were present who let their words ring forth from the depths of the darkness."[82] He describes how the ancient Persians pictured "twelve Archangel-Beings working from the twelve directions of the Zodiac into the human head in twelve rays... twelve cosmic radiations... which then densified in the human head into twelve main cerebral nerves."[83]

Haydar Amoli, a fourteenth-century Iranian spiritual thinker, speaks of the twelve zodiacal signs as "the dwellings or forms of manifestation of twelve angelic entities from whom the twelve imams receive the knowledge they communicate to human beings, in the same way that the twelve signs transmit the influx and energies they receive from the twelve angelic entities."[84]

Blavatsky brings a similar perspective. "Esoteric philosophy teaches that the first human stock was projected by higher and semi-divine Beings out of their own essences." She refers to "the twelve great gods, Jayas, created by Brahma to assist him in the work of creation in the very beginning of the Kalpa."[85] These twelve divine "Creators" went through their own human evolution at a much earlier time and offered themselves as world-generating seed-beings.

Alice Bailey lists twelve "Creative Hierarchies"—each associated with a zodiacal constellation—whose complementary qualities and purposes account for the full range of world manifestation.[86] They are, she says, "the true forms of all that persists," from elemental lives (12th) to human lives (9th) to "Divine Flames" (6th) to the "Intelligent Substance" of the first hierarchy. These twelve realms of the world

---

82 Steiner, *Man in the Light of Occultism, Theosophy and Philosophy*, p. 191.
83 Steiner, *Occult History*, p. 89.
84 Quoted in Corbin, *Temple et Contemplation*, pp. 80-81.
85 Blavatsky, *Secret Doctrine*, vol. 2, pp. 87, 90.
86 Bailey, *Esoteric Astrology*, pp. 34-35, 48.

being correspond to the twelve chromatic intervals and tones of the world soul: twelve harmonic fields sustaining twelve forms of consciousness.

Such insights into the inner reality of the great wheel-vortex that forms the space–time womb of our universe reveal the milieu of lofty Lives that constitutes its mental field and tone its kingdoms with different intervals of consciousness. The zodiacal signs are the signatures of these twelve archetypal soul tonalities, which inform the consciousness of all that exists.

Ironically, the zodiac is often regarded as a self-mythologizing projection of humankind upon the surrounding heavens. The ancient notion of a cosmic group entity, projecting its formative energies into the world we are, suggests rather a dual, reciprocal projection engaged between mirroring structures–entities (human soul and zodiac)—formative projection of higher entities' harmonic fields on one hand, imaginative projection of humankind's psychic patterns on the other hand. Twelve great Lives, whose circle models the paradigm of soul plenitude, inform and inspire the individualities-in-the-making we are with their unique harmonic "signatures" and "assignments."

Alice Bailey's statement that "the zodiac is a heart-in-the-head center" suggests that the zodiac is no more (and no less) a projection of the human mind than the human mind–soul is a projection of the zodiac. The projection speaks in one case to the creative self-replication, and in the other case the evolutionary unfoldment, of a universal ground of self-resounding called selfhood, which informs at once the center (soul) and periphery (zodiac) of our being. Recognizing the two currents of the "projection" offsets the reductive meaning of "illusory reality" commonly associated with the term *projection*; and it reveals what is a nondual ground of world and self-becoming.

Unsurprisingly, there were twelve Olympian gods hovering above Ancient Greece, and twelve knights around the round table of king Arthur. The Welsh name Arthur is derived from artos, "bear," and Arthur is associated, in the Tradition, with the constellation of the Great Bear, of which Alice Bailey says that it represents the (7fold) head center of the (12fold) zodiacal heart-in-the-head. The Great Bear is indeed the "head" of a cosmic round table of Twelve. The grand archetype of selfhood landed as well in the "people of God," the group of twelve tribes issued from the sons of Jacob. Jacob was, quite appropriately, the *third* of the great founding patriarchs, and the one to receive the name of the whole-in-formation—"Israel"—at the time his twelfth child was about to be born. In replicating the cosmological process (from one to three to twelve-in-one), the gradual formation of the Judeo–Christian "people of God" emblematizes the organic course of actualization in humankind and by humankind of its 12fold core self. The "I AM-ness" heard by Moses in the burning bush was announcing its incarnating descent in humankind, anticipating how the prototypical "identity" of the human being, its soulful self, would one day be born on Earth as a Logos surrounded by twelve.

This prototypical matrix of the world-self is at work in us, its microcosmic replicas, harmonizing the Chaotic and the Logoic bundled up into our multidimensional being. "The goal of life set before men by the gods" is a curriculum of harmonization-integration of instinctual forces and disjointed psychic patterns with the unifying rhythms and harmonies of the 12fold oneness of being seeded in our core, so that a *whole* of personalized universality may gradually "compound" itself (to echo Plato's depiction of the creation of the world soul) or, echoing the alchemists, "homo-genize" itself. The unfolded Egoic lotus represents this ultimate harmonic "Unit" or dual vortex that fully integrates the duality of matter and consciousness, stone and wisdom (a "philosopher stone"), self and universe.

This curriculum of the soul is remarkably supported and illumined by the symbolic description and navigating insights provided by astrology, whose view of the human course as an on going attunement between personal notes and cosmic tones, personal dispositions and transpersonal purposes, internalized chords and universal music making, points to its true potential as a science of the soul's journey to *Unus mundus*.

As harmonious sound strikes to audibility the twelvefold "organization" of time–space, the zodiac calls to visibility the twelvefold inwardness of cosmos called world soul or world-self. Such a correspondence between zodiacal circle and octavian (chromatic) wholeness suggests a view of the zodiac as a tonal and toning ground. The notion of the *tonal* reality of the zodiac is not new. A number of "tone-zodiacs" have been proposed over the centuries, pairing each sign/constellation with one of the semi-tones of a chromatic scale.

One of the oldest known tone-zodiacs, from Ptolemy's *Harmonics*, established correspondences between the zodiacal circle and the Greek two-octave system.[87]

Rudolf Steiner, who regarded the signs as the "tone-signatures" of the twelve hierarchies of beings under whose care humankind develops, associated their sequence with the succession of notes along the circle of fifths, starting with Aries (C major/A minor), continuing with Taurus (G major/E minor), and ending with Pisces (F major/D minor).

Other models have rearranged this simple chromatic order to support a specific perspective: musical, astrological, or philosophical. As we saw in the case of the

---

87 Godwin, *Harmonies of Heaven and Earth*, p. 140, 142.

sevenfold, rather than invalidating the correspondence, the variations reflect the Protean expression of archetypal patterns. "This very diversity," Godwin remarks, "is characteristic of harmonic thought."

The lineage of tone–zodiac shows the persistent intuition of a correspondence between the harmonic differentiation of tonal space, the cosmological differentiation of the world soul, the astrological differentiation of types of energy, and the psychospiritual differentiation of forms of consciousness—in other words, the correspondence between circle of fifths and circle of "signs."

When we look more closely at each domain—harmony, cosmology, astrology, psychology—we see how in every case the correspondence speaks to the respective formative process. They all share the archetypal foundations articulated by the Platonic–Pythagorean numbers of cosmological unfolding: 1, 2, 3, 4 (7, 12). Every twelvefold can be traced back to a threefold, a twofold, and a One, while being also the essence of a fourth—the core of a form.

Musically, the 7/12fold Pythagorean scale originates, we saw, from the interplay of the "numbers" 1, 2, and 3 through the prime ($c = 1$), the octave ($c' = 2$) and the fifth ($g = 3/2$). The fifth produces the fourth, the archetype of the "form," which will deploy itself as a bundle of seven octaves and twelve fifths.

Astrologically, the twelve signs of the zodiac are traditionally defined as the combination of one of the three modes of divine motion—forward, rotary and spiral cyclic—with one of the four elements of matter. The three motions are referred to as cardinal (forward creative Father aspect of divinity), mutable (adaptable intelligence of the matter–Mother aspect) and fixed (concentrating–evolving motion of the Son–soul aspect). The zodiacal constellations are the twelve harmonic "offsprings" of the union of the two—same (as 3fold oneness) and other (as 4fold differentiation), spirit and matter.

Cosmologically, the "three" are what Plato refers to as "Same, Other and Being," but also Father-source, Mother-recipient, and Offspring, as well as Being, Place, and Becoming—"three distinct things that were existing, he says, even before the Heaven came into existence."[88] He calls Becoming (the third) "the Nurse of Becoming," and describes how it is "liquefied and ignified as well as receives the forms of earth and of air"—in other words, how it deploys itself into the elemental quaternary of the fourfold physical cosmos.[89] The fourfold cosmos displays 7fold systems (atomic, elemental, biological, and psychological) with a core 12fold structure. Plato's demiurge sets the 7fold and 12fold "across" each other, suggesting the

---

88 Plato, *Timaeus*, 50d. "It is proper to liken the Recipient to the Mother, the Source to the Father, and what is engendered between these two to the Offspring."

89 Ibid., 52d-e.

VI. Facets of a Unitive Science

four-armed cross of matter. Le Mee's adquadratum method for generating the Platonic solids demonstrated what the *Timaeus* simply affirms—that the four (solids) issue forth from the three. "In the adquadratum method for generating the Platonic forms, only 1, 2, and 3 are used."[90] The fifth, the dodecahedron, is at once the quintessence of the fourfold and the 12fold archetypal expression of the third. 12 itself, the cipher, is a blend of 1+2 into a "third."

Psychologically, 12(-fold causal body) and 7 (chakras) constitute the fundamental template of individuality. They frame the work of harmonization–union of the 4 elemental vehicles with the 3fold spiritual triad of intelligence, love and will, which the Eastern tradition refers to as Manas, Buddhi, Atma: universal mind, feeling, and will. The monadic core of the triad is one with the Self.

Plato's harmonic process of creation illumines the soulful ground of cosmos and the cosmological ground of the psyche, showing the primordial identity and ultimate inseparability of psyche and cosmos. The logic of the psyche, psychology, rests on the same harmonic ground as the logic of cosmos. While the cosmos is soulful and full of growth, the psyche is cosmos like in structure, with 7 levels of consciousness and a 12fold core.

Plato's cosmogony also illumines the rational grounds of astrology's claim to a cosmos-based science of the psyche. It reveals how its 12fold /7fold framework is born of the primordial secting-bonding dynamics that underlies the becoming cosmos of the world soul, and the becoming soul of cosmos. The 12fold zodiacal matrix, whose primordial harmonies weave the space–time-consciousness fabric of the universe, pulsates with the "heartbeat" of the twelve—the heartbeat of the universe—the logarithmic triunity, which nests the oneness of life in all of nature formations.

The 12/7fold structure on which astrology bases its *cosmic* science of self and *self-growth* is the common ground of cosmos and soul—their *one* matrix—the archetypal pattern of integrative differentiation of inside and outside, self and world, which governs, we came to recognize, the shaping of all forms of life. The "rational," harmonic nature of the common ground of cosmos and psyche laid bare by Plato unveils the rational ground of astrology as a cosmo–psychology.

※

Indeed, the basic components of astrology are best understood in harmonic terms. The natal chart, which maps the zodiacal alignments of the planets and their interrelations at the moment of birth, displays the fundamental tones and chords of a psyche and its matrix of growth for a lifetime. The planetary spheres, which are, we

---

90 Le Mee, *Ad quadratum Construction and Study of the Regular Polyhedra*, p. 144.

saw, vibratory fields "playing out" the major harmonic ratios of the solar systemic octave, correspond to psychic functions—physical-instinctual preservation (Moon), affective bonding (Venus), perceptive and discriminative activity (Mercury), sense of creative identity (Sun), drive to self-express (Mars), quest for meaning and fulfillment (Jupiter), ego-structuring (Saturn), etc. They are qualified or "toned" by their zodiacal placement as well as their relation with other planetary functions. These relations are referred to as "aspects."

The three trans-saturnian planets—Uranus, Neptune, Pluto—represent a triad of transpersonal psychic functions, which may be viewed as the higher octaves of the personal planets Mercury, Venus, and Mars: Uranus is the higher octave of Mercury (universal intelligence), Neptune the higher octave of Venus (universal love), Pluto the higher octave of Mars (universal will). Their discoveries—Uranus in 1781, Neptune in 1846, Pluto in 1930—coincided with the respective entrance into the collective societal consciousness of these three principles—which correspond to the transcendental attributes of divinity: the True, the Beautiful, and the Good. The first coincided with the culmination of the "enlightenment" and the eruption of the American and French revolution, the second with the upsurge of communitarian ideologies and aspirations, the third with the unleashing of nuclear power and the worldwide destructive rise of Nazism and fascism, which would elicit the formation of global instances of planetary goodwill.

Aspects are one of the most informative dimensions of astrology, psychologically. They refer to specific angular relationships between planets, hence psychic functions, in a chart. "Bonding angles" (120°, 60°) are based on the unitive power of 3 and suggest harmonious, mutually supportive and enhancing relations (gifts). Differentiating angles, based on the secting power of 2 (180°, 90°, 45°) suggest conflictedness and tensions, as well as unique music making potentials. Godwin points out how impressed Kepler was, and Ptolemy before him, "by the fact that the three most powerful aspects (trine, opposition, and square) are expressed by the same ratio as the three perfect consonances of music."[91] The trine (120°: 2/3 of the circle) corresponds to the fifth, the opposition (180°: 1/2 of the circle) to the octave, the square (90°: 3/4 of the circle) to the fourth, the sextile (60°) to the tone, the semisextile (30°) and quincunx (150°) to the twelfth interval of a semitone. Bonding and differentiating aspects suggest the complementary experiences of ease/tension, collaboration/conflict, pleasure/displeasure. Zodiacal placements speak to fundamental consonances and colors of the psyche. Aspects (interplanetary intervals) speak to its "deeds and sufferings"—for a lifetime.

---

91   Godwin, *Harmonies of Heaven and Earth*, p. 136.

Transits are *temporary* relationships (of the same harmonic nature as aspects) between natal placements and planetary positions of the moment. They indicate transient "intervals" between ego and Self, which support or challenge personality dispositions. When current planetary alignments create consonances or dissonances with a natal planet, they challenge the "transited" psychic function to let the cosmic ground of being re-attune, re-calibrate, and sometimes radically transform its expression. Transits unroll the individual score of growth and soul making. They hint at alchemical processes through which the soul may spin the golden threads of its "beautiful robe" by allowing one or another self-seeking and self-driven planetary "nature" in the psyche to absorb a touch of transpersonal quality. However powerfully cosmic tones may stir the habitual notes of the psyche, the outcome of the transit depends on one's willingness to listen to the score and interpret it skillfully. The work of harmonization–integration of self and universe, which defines the path of human development, inevitably involves the refinement of coarse personal components and the extraction of more universal sounds of being—wide-minded thinking versus self-seeking calculatedness, generous love versus expectant attachment, conscious willingness versus willful control.

For example, a transit of Saturn to Mars brings the Saturn principle of structure, balance, measure, deliberation, and rigor to play upon Mars' urge to self-assert and take action, and its inclination to competitiveness, impulsiveness, and anger. If the harmonic mode of the transit is an opposition or a square, the experience may initially be one of frustration and obstruction in terms of freedom of movement (on the physical plane), desires and drives, or capacity to act. It may be experienced as hindrance to a project or curtailment of an impulse or initiative. The injunction and opportunity presented by Saturn, however, is to reassess, readjust, and redefine instinctual ways, consciously slow down and reconsider one's approach, question appropriate action, bring a new maturity of deliberative poise to reactions, self-expression, and decisions. As a result, a new gradient of patience—a deep version of the will—may be integrated.

> *It seems quite plausible that much of astrological theory may rest on a basis of figurate rationality rather than upon empirical or special omen lore. In this sense astrology... developed on a very rational basis, with a figurative theory and the associated symbolism at its center.* —DEREK DE SOLLA PRICE

The framework of astrology reflects the 12fold–7fold archetype of all forms, the externalization of Being into a living "body." The archetypal "bundle" defined by the coming together of the 12fold circle of the third harmonic with the 7fold ladder of the second harmonic stands in the filigree of macrocosmic systems (zodiac and planets) as well as microcosmic beings (Egoic lotus and chakras) and illumines the profoundly rational grounds of astrology as a science of the soul, uniquely apt to embrace and access as one the internal and external side of experience and its dual character—self-directed and "fated," conscious and unconscious, personal, and universal.

What astrology offers is a rational lens (based literally on harmonic "ratios") into the trans-dual ground of the human predicament, and the soul making at work in the ceaseless weaving of self and Self. A mirror of soul patterns and cycles, this lens is archetypal, and it opens the way to an *archetypal* analysis of past, present and future developmental themes, rather than a prediction of events.

On one hand, astrology is a psychological cosmology, whose science of cycles articulates in experiential terms the *(meta)*physics of fractioning–bundling of the *world-soul* into consonant planetary spheres, which have their correspondence in human vehicles, psychic functions, and symbolic "meanings."

On the other hand, it is a cosmological psychology, an evolutionary psychology of attunement, harmonization, and reunion, within the unfolding soul, between personality vehicles and Self—the "Lord of the spheres," the Logos of our solar system—between personal tunes and cosmic music. When the soul reaches full blossoming at the end of the path of individuation, the "inside is like the outside, the above like the below," psyche is cosmos, and cosmos psyche—a unitary reality of universal self-awareness, at which point experience transcends external transiting and inner transited.

Built upon the structural interface of the two facets of being (soul and cosmos), astrology has its "laboratory"—literally, its alchemical space of *Ora et labora* (pray and work)—in the two-faceted wheel of world-and-self–becoming that spins out time–space formations and spins in soul-self substance. Its gaze is on the golden wheel where self-becoming and world-creating intents conjugate their opposite

VI. Facets of a Unitive Science

streams—like a dual vortex. Its work is to bring this interplay to consciousness, activate collaboration between the two, and foster their essential oneness.

In the end, the notion of the "music of the spheres" is clearly more than a poetic proposition. It is a spiritual–scientific insight into the harmonic ground of cosmos and the harmony making underpinnings of natural phenomena—from the productions of nature to the rhythms of cosmos. It reveals in cosmos an essentially "poietic" (i.e., creative) orchestra and in every living system, from flower to bird to human being, a complex form of dynamic consonance. Everything comes to existence by virtue of consonance.

The "metaphysical physics" of resonance and consonance, which reveals itself when ancient knowledge and contemporary physics are brought together in the context of a spirit–matter universe, illumines the gradual unfurling of the world being into increasingly complex forms—from the harmonic organization of matter into atoms and molecules to its vitalization into seeds and plant systems, from its self-directed autonomy in animal bodies to its alignment to independent selfhood in Egoic beings, all of which speaks to the evolution of consciousness from the "chemical" separating-bonding proclivities of minerality to the self-organizing, self-reproducing capacity of the plant world, to the desiring/fearing faculty of animality, to the self-aware intelligence, loving faculty and creative "mentality" of humanity.

VII. UNIVERSALITY OF THE HARMONIC MODEL OF CREATION

*We may regard the phenomenal world, or
nature, and music as two different expressions
of the same thing.* —ARTHUR SCHOPENHAUER

FROM THE DOGONS TO INDIAN MUSICOLOGY TO THE SUMERIAN PANTHEON

Our journey into the formative grounds of the universe has brought up significant resonances between cultures, from ancient Greece to Taoist China, from ancient India to medieval Persia. The cosmogony of the Dogon people of Mali offers yet another illustration of the universality of the Pythagorean-Platonic model of creation. A brief incursion in this cosmogony, in association with recent insights into the foundations of Indian music and the ancient Sumerian pantheon, will show that harmonic cosmology is not simply a Greek theory and Western worldview, with the limited relevance to a particular time and culture this would imply, nor is the pattern of 7/12 specific to Western music and "sensitivity." Platonic cosmology and Pythagorean laws of harmony represent rather a cultural version, relevant to the dawn of the intellectual stage of humankind, of a universal and ageless wisdom of cosmological facts.

French anthropologists Marcel Griaule and Germaine Dieterlen, who lived with the Dogon people for twenty-five years (between 1931 and 1956) and were eventually initiated into the tribe, were astonished by the advanced astronomical knowledge they seem to possess, particularly with regard to the star Sirius. Griaule wrote in the preface to his analysis, "The problem of knowing how, with no instruments at their disposal, men could know the movements and certain characteristics of virtually invisible stars has not been settled, nor even posed." In 1976, in a controversial book entitled *The Sirius Mystery*, Robert Temple argued that the precise knowledge the Dogon people had of cosmological facts only known to modern astronomy must be traced back to the Egyptians, and even further to extraterrestrial sources, who transmitted astronomical knowledge to the Egyptians. The Dogon tribes are generally thought to have originated in ancient Egypt, and they claim to have

VII. Universality of the Harmonic Model of Creation

received their astronomical knowledge from amphibious beings, the Nommos, sent to Earth from Sirius for the benefit of humankind.

For the Dogon people, the world begins as a cosmic egg that contains "all the germs or signs of the world." In the beginning was Amma, the intelligent consciousness behind all of creation, called "He who rests upon nothing." Before creating the universe, Amma carved the signs of the universe on the walls of his own placenta, which was divided into four sectors, each containing eight figures, each of which produced eight more. The oval thus contains 8×8×4 or 256 signs, to which are added eight (two per half-axis) and two for the center. So the total of the "signs of Amma" is 266.

Within his egg, Amma began spinning around, inducing the formation of the po seed. "Quickened by an internal vibration, the po seed burst the envelope sheath, and emerged to reach the outermost confines of the universe."[1]

Reproduced in figure 71 is a drawing of the Dogon Egg of the world and prefiguration of the human being within it. As in the biblical Genesis, Griaule and Dieterlen comment, "In the beginning there was Logos or power in the form of Amma, the Creator God. Then there was a vast egg called by the Dogon *aduna tal*, the 'egg of the world,' which contained the germs of objective reality."

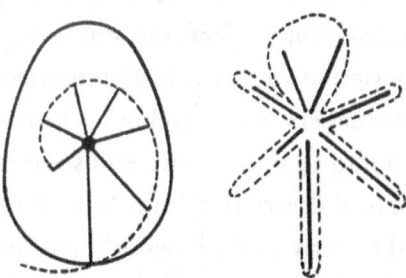

Fig. 71. Marcel Griaule's drawing of the Dogon Egg of the world and the "seed of Man"

The Dogons' cosmogonic process unfolds in a rhythmical pattern reminiscent indeed of Genesis, and along a curve evocative of the logarithmic spiral. The movement of the expanding contents of the seed, explains Chukwima Azuonye, a scholar of African studies, describes a spiral path represented in Dogon drawings by a zigzag line:

> This unwinding spiral movement took place in *seven* vibrations, each longer than the one before it [fig. 71]. At the seventh vibration the envelope broke, releasing creation into outer space just as life hatches from an egg.... The extension of the seventh vibration into a new order caused it to become the first pulse in a new and enlarging series, a continual progression expressive of the infinite extension of the universe.[2]

The parallel between this pictorial description of the 7fold unfurling of the One seed into manifestation and the cosmogonic narrative embedded in the *Timaeus* and in world theosophies in general is impressive. It offers a remarkably universalizing confirmation of the octavian template of world dimensions and forms of life.

---

1   Griaule, *Conversations with Ogotemmeli*.

2   Azuonye, *Dogon*, pp. 25–26.

A few sentences from the ancient stanzas of Dzyan, for instance, expose strikingly resonant themes:

> The mother swells, expanding from within without, like the bud of the lotus.... The vibration sweeps along, touching with its swift wing the whole universe and the germ that dwelleth in darkness.... Light drops one solitary ray into the mother-deep. The ray shoots through the virgin egg, causes the eternal egg to thrill and drop the non-eternal germ, which condenses into the world egg. Then the three fall into the four. The radiant essence becomes seven inside, seven outside...the first seven breaths of the dragon of wisdom...the radiant child of the two, OEAOHOO.... The sparks of the seven are called spheres, triangles, cubes, lines, and modelers.[3]

The name *OEAOHOO* evokes the seven vowels, while the "modelers" suggest the Platonic solids.

However, while similar in underpinnings, the cosmogonies are quite different in presentation. In fact, the similarity between the Dogonic picture and the Platonic description could easily go unnoticed, were it not for the arresting parallel between the numbers central to both schemes—not only 7, but the more intricate differentiation into 12 and 266 we will soon return to. It seems that what the *Timaeus*' mathematical description leads us to deduct somewhat painstakingly is laid out candidly and directly in the Dogons' pictorial narrative. On the other hand, the cosmogony of the Dogons might remain an intriguing enigma—were it not for the *Timaeus*' harmonic keys, which help unravel it. The two cosmogonies complement and illumine each other, further confirming that logarithmic spiraling—the dynamics of the golden mean, and its extension into the 7/12 cycle—is the core of the unfolding universe, the key to its quantum order, and the formula of its form-making process and organic formations.

The Dogons' whole existence reflects their grand cosmogonic picture. "Villages replicate the primordial egg, the gate symbolizing the envelope broken by the seventh vibration, and the main buildings and shrine elements standing for the human prefigured in the egg. Cultivated land is laid out according to the spiralate paradigm."[4]

John Adkins Richardson further elaborates how "the zigzag pattern symbolizes the perpetual helicoidal movement [which results from]...the alternation of paired opposites in reality: right versus left, light versus dark, male versus female, and so on." We recognize the familiar notion of a primordial dualization of Being and polarization of Life into centrifugal and centrifugal currents—Shiva/shakti, yin/yang, electricity/magnetism—whose conjugation weaves the universal breath of creation and the etheric field of forms.

---

3   Blavatsky, *Secret Doctrine*, vol. 1, p. 28.

4   Richardson, *Speculations on Dogon Iconography*, pp. 52–57.

VII. Universality of the Harmonic Model of Creation

This pictorial representation of alternating opposites illustrates a key concept of the Dogons' cosmogony, which is also a central feature of their figurative sculptures: twin-ness. According to the Dogons, the first beings and the first two generations of human ancestors are pairs of twins, and the "ideal couple" is a twin brother and sister pair. The connection that the Dogonic culture brings to light between the theme of the twins and the helicoidal pattern sheds light on the cosmological underpinnings of the cross-cultural mythological themes and rituals involving twin forces or sister–brother pairs—from Zoroastrian Iran (Ahriman and Ahura Mazda) to Pharaonic Egypt, where Seth is the dark brother of Osiris, Isis is Osiris' wife and sister, and Pharaohs married their sister.

MYSTERIOUS NUMBERS: THE NUMBERS OF THE MYSTERY OF CREATION

The number of signs on the walls of Amma's placenta (256–266) is very close to the number associated with the harmonic "apportioning" or fractioning of the Platonic world soul. The ratio of 256:243 corresponds to the interval of the semitone. The 256:255 harmonic corresponds to the last note of the seventh octave,[5] which almost coincides (but by a comma) with the twelfth fifth (259.48) and "bundles up" the joint cycle of fifths and octaves into the 7/12fold template of living forms.

In Plato's account, the 256th harmonic represents the smallest portion of the harmonically differentiated oneness constitutive of the world soul—the oneness that speaks metaphysically to the triune Logos, is displayed musically in the fifth and geometrically in the golden mean. As such, 12fold fractioning defines the ideal of externalized "wholeness," the archetypal Form of the threefold Logos—"oneness" in manifestation—the archetype of cosmos.

In the Dogonic cosmogony, 256 is associated with the eighth power of two ("The oval thus contains 8 x 8 x 4 or 256"—that is, indeed 2^8) $+ 2 + 2^3$, which points to a (7fold) octavian *duplication* like process within the world egg. The Dogons' description of Amma carving 256–266 signs on the walls of the matrix of the world—his own placenta—is remarkably reminiscent of the Platonic description of the demiurge "apportioning" the world soul compound in such a way that 256:243 was the smallest portion—1/12 of the soul compound. The 12fold process of proportional integration, which, as the laws of harmony and musicology revealed, accompanies and determines the cosmological 7fold, is not more explicit in the Dogonic account than it was in the Platonic. However, it is reasonable to guess it might be implied in the limited number of duplications (seven) involved

---

[5] The series considered begins with the second octave (as it contains the first fifth) and ends with the eighth octave and twelfth fifth.

in the primordial extent of the Dogonic universe. We will see very soon that this is the case. Richardson does point out how 12, along with 7 and 10, is one of the predominant numbers in Dogonic culture and art. That the number 12 would be hidden or implicit in both cosmogonies, yet readily emerge, albeit in different ways, under one's inquiry, certainly reflects the inner reality of consciousness and soul it harbors. One may even wonder, given the initiatory secrecy in which the dodecahedron was reputedly held, if the central presence of "twelve" in the creation process was kept intentionally veiled.

In searching for the potential presence of 12 in the Dogonic account of creation, a fortuitous play with numbers was to prove significant. It revealed that the division of 256-266 by twelve gives approximately 22, which is one of the first key numbers in the Dogonic cosmogony. The Dogons refer to 22 yala to designate images, written signs or codes that stand for 22 kinds of life, which encompass everything that exists. "As the first vibrations of the egg occurred, the yala were flung into space. Each fell on the category of life to which it corresponded and brought that category to life." Chukwima Azuonye suggests that the 22 yala may be compared to 22 of the 23 pairs of human chromosomes, which carry the codes of life—the genes. The 23rd pair determines a person's gender.[6] The Dogonic cosmogony also counts as 22 the total number of the first three generations of people.

The two cosmogonies intimate the same fundamental process: 12fold differentiation of the "placenta"-matrix of the world and 7fold deployment of world dimensions. Both also point to vibration as the primordial act and fact of creation, either explicitly or via numbers associated with the phenomena of vibrational interferences and consonance. The morphology of the name Am-mA reflects the mirroring of "Same and Other," the centrifugal (am)/centripetal (ma) dynamics of the Shiva/Shakti breath that makes up the primordial field of "self-resonance"—a perfect name for the primordial Ipseity (its-self-ness) of Being.

The carving of signs on the placenta of Amma mirrors the zoo-graphing of the Platonic sky by the demiurge. Like the Platonic zoographs, the signs of Amma differentiate the vibrant matrix (of Brahma, Logos, Amma) that precedes the emergence of world dimensions and formations. The cosmic egg of the Dogons also shows remarkable similarities with the image of Mithras emerging from a broken egg, surrounded by the zodiacal "signs" on the wall of his "placenta."

The egg shape of the One pregnant with the seed or germ of the world evokes the golden spiral—which is indeed the dynamic path (a 7fold spiral) described by the expanding contents of the seed within the Dogon egg before it breaks open. The Greek word for germ is *speira,* and the archetypal world germ is a spiral—a golden

---

6   Azuonye, *Dogon,* 1996, pp. 25-26.

VII. Universality of the Harmonic Model of Creation

spiral. The world egg is charged with the marvelous germ of all germinations: the spiral—the golden mean—the magical key to the unfurling of oneness into multiple forms. At the core of the 7/12fold archetypal form is the 2/3 ratio, the third harmonic. As the next section will confirm, the 7/266–22 design within the egg is an equivalent of the 7/12fold form. We will see how 22 and 266 are *in fact* related to 12, confirming that the 256–266 signs speak to the wholistic (harmonic) differentiation of One. The gap between 256 (2^8) and 266 evokes the comma between the 7 octaves and 12fold division of the one octave.

It is interesting to note that in computer science, the byte, which is a unit of storage defined as eight bits, can store 256 values, counting from zero to 255. The number 256 often appears in computer applications such as the typical number of different values in each color channel of a digital color image (256 values for red, 256 values for green, and 256 values for blue used for 24-bit color). The presence at the center of technology of the same intelligence of differentiated wholeness that is found in cosmology is certainly intriguing. However, the squaring process that results in the 256-fold unit of stored information (2^8, the octavian squaring of 2) does not show any concomitant presence of number 12, whereas the harmonic 256-fold associated with the cosmological wholeness of world beginnings—12 semi-tones—as well as the Apocalyptic City of human completion, a city "lying foursquare," fundamentally do. "The city lies foursquare, its length the same as its width. And he measured the city with his rod, 12,000 stadia. Its length and width and height are equal.... The city wall was broad and high, with twelve gates guarded by twelve angels.... The twelve gates were twelve pearls, each gate made of a single pearl" (Rev. 21:16, 21:12, 21:21).

The City is a squaring of twelve. By contrast, 256, the squaring of 16, represents the squaring of the square of squares (4), the squaring of fourfold matter—the foundation of a technological City of matter, indeed. It may "lie foursquare," based as it is on 256, but it does not show 12 gates. Division has done its remarkable technological work, but the bonding-unitive dimension of soul consciousness lies outside its field, an absence masked by awesome, though mindless, mechanical, hyper-*linking* activity. 12 involves the multiplication of 2^2 by 3. The sixth-century mystic Dyonysius the Aeropagite pointed out that it is only with the number 3 that the possibility of harmony, as distinct from proliferation and reproduction, begins.[7]

Nonetheless, similar to the Dogonic picture, the 7foldness implied in the 256-foldness of the byte raises the question of what might limit the duplication process to 7 times. In what way might the logic of techno-logy involve, unbeknownst to it, the 3fold–12fold note of the Logos? It seems, intriguingly, that the power of

---

7   Godwin, *Harmonies of Heaven and Earth*, p. 165.

3 is implied in creating a boundary to the duplicating power of 2—the squaring of division—on which computer technology is based.

The 7fold zigzag in the Dogons' world egg brings to mind an equally arresting parallel with Babbitt and Leadbeater's atom. Both describe the atom as composed of seven *spirillæ*. The atom-like Dogonic egg of world beginning evokes the image and term that the "father" of the Big Bang theory, Georges Lemaitre, used to refer to what would be later called the Big Bang. He used the term *primitive atom*. A primordial convergence is suggested between cosmic egg, primitive atom, and Big Bang.

The carving of the signs on the walls of the placenta of the world being point to the harmonic differentiation of the primordial breath into a (world soul) field charged with the power to in-form the twenty-two categories of life. The emergence of this primordial morphogenetic field "assigned" with configurative potential evokes the embryological differentiation of stem cells along "bio-lines" (zoo-graphs) keyed to one particular organ formation—one particular harmonic.

The French philosopher Gilles Deleuze was intrigued by the parallel he saw between the world egg of the Dogons, charged with differentiating potentialities-"intensities," and the peculiar description by the actor and writer Antonin Artaud of his psychotic-like experience of a "body without organs." Artaud's expression is a singularly apt designation for the archetypal world matrix (egg) that precedes, intones and in-shapes the organs of the universe. Indeed, as Deleuze remarks, "The body-without-organs is the egg.... The egg is the milieu of pure intensity.... Zero intensity as principle of production.... It always designates this intensive reality, which is not undifferentiated, but is where things and organs are distinguished solely by gradients, migrations, zones of proximity. The egg is the BwO."[8] Evocative of the proto-world of ancient wisdoms, Artaud's "BwO" also carries intimations of the ultimate world of universal consciousness. His tormented experience of this primordial-ultimate matrix suggests a premature, unwholesome break into transpersonal oneness, possibly drug or psychosis induced. Without the appropriate maturation of consciousness and ripening of "self" into this de-particularized state of awareness, such inkling of the eventual transcending of partial vehicles and partial consciousness is likely to give off the alienating vibes and screeching overtones, as of an overstretched string, one can hear in Artaud's recorded reading of his radio play, *To Have Done with the Judgment of God* (1947), the play where the expression "body without organs" originated: "When you will have made him a body without organs, then you will have delivered him from all his automatic reactions, and restored him to his true freedom."[9] However keen Artaud's insight might be, the

---

8 Deleuze, *A Thousand Plateaus*, p. 164.

9 Excerpt from Antonin Artaud's radio address, Nov. 28, 1947.

VII. Universality of the Harmonic Model of Creation

only sustainable and lasting way to "make" this body without organs is through the existential labor of harmonizing the components of human nature. Only through the balancing processes of daily life can the partials of the psyche eventually become integrated into a whole sound of Being and deliver the universalized body-self, which, like the original matrix, but in an individually conscious way, transcends particles and parts, because the essential simplicity of consciousness achieved is no longer invested in parts, folds, and organs, but is radically attuned to the fundamental "wavelength of God," which encompasses all harmonics.

When the ego structure fails to contain and negotiate the interplay between personal and universal contents (conscious and unconscious), it breaks into self-inflated fragments of transpersonal truth and drifting dysfunctional bits of personality. The psychotic or drug altered psyche experiences in dissociated and disturbing ways what authentic expansions of consciousness and lasting transcending of ego constructs eventually lead to experience as a blissful at-onement with cosmos and divinity. "I am God," the claim for which the great saint Al Hallaj was tortured and killed, spoke to such a state of being.

The *Unus mundus* of the alchemists, as well as the theosis of Christian saints, refer to the body-without-organ experience of universalized selfhood. In contrast to a dissociated "cosmicity" voicing itself through brittle ego parts, the cosmicity of a master of wisdom is rooted in a culmination of selfhood, an actualized plenitude of self-world identity that renders obsolete the partial vestures and "organs" of selfhood, and leads to relinquish the vehicle that subsumes them all, the causal body of "individuality." This radical expansion into universal self-awareness dissolves the beautiful Egoic form, now an obsolete impediment to a next chapter of "identity," unfathomable to the mind because it transcends it. To step out of the organs of selfhood is to step out of the completed cycle of individual human selfhood.

Indian musicology and the Sumerian Pantheon: A radical light on cosmology

The Dogons' description of the birth of the world shows clear parallels with the Platonic process of creation: 7fold vibratory spiral–cyclic unfolding of the world germ from a matrix etched with 266 signs (the Dogons) or toned with 256 harmonic fractions (the *Timaeus*).

The number 22 however, central in the Dogonic picture (22 yala), is not a number present in the Platonic cosmology. And 12 is not directly part of the Dogonic narrative. Yet, the fortuitous realization that the division of 256/266 by twelve gives approximately 22 points to a significant mathematical relation between 266, 22, and 12—hence the possibility of uncovering other grounds of reverberation and resemblance between the two cosmogonies.

Further inquiry into this relation between 12 and 22 proved not only corroborative but revelatory, thanks in large part to the musicological research of an Indian engineer, Sreeni Nambirajan, which opened another source of penetration into the cosmological principles we have been contemplating, and contributed to demonstrate their universality.[10]

The excursion to the ancient Indian musical system prompted by the number 22 will turn out to be an incursion into the common root of Dogon and Greek cosmogonies. It will reveal their profound unity—their shared harmonic ground. This unveiling of a common ground between Western and Indian harmonic systems will confirm the universality of a cosmology based on harmony. It will also powerfully orchestrate the essence of the Platonic cosmogony held in the golden mean, by taking us to an entirely fresh affirmation of the harmonic character of divinity.

Classical Indian music is based on 22 tones called *shrutis*. The source and rationale for this unique division of the octave has puzzled musicologists worldwide. According to medieval Indian musicologists, ancient Indian music (Ghandarva music) was based on 22 microintervals, or shrutis. It had its source in the Vedic Tradition and was regarded as the eternal music of Nature, a pure expression of its life and laws. Ghandarva music was believed to be a direct revelation of the "cosmic sound of truth" heard by ancient Rishis who translated what they heard into something understandable by humans.[11] Indian classical music is based on this tradition.

Tantric anatomy, on the other hand, taught that Sushumna is connected laterally to Ida and Pingala at the chest (Anahata chakra), throat (Vishuddhi chakra) and forehead (Ajna chakra) with the help of 22 horizontal nadis per chakra. The human body was conceived as a "harp," whose chakric sets of 22 nadis, when stirred by the "vital breath" or the "gandharva" music of nature, would resonate and open up the gateway into the Sushmna nadi, allowing pranic "energy" to flow upward to the "sahasrara" or crown chakra.

Etymologically, the term *shruti* means "what is heard." It designates the smallest interval of pitch that the human ear can detect—a "human" unit of pitch. This definition suggests that 22 musical intervals were perceptible, within an octave, to those who established the foundations of Indian music and that the 22-tone Indian system was based on this direct perception.

Indian musicologists theorized that of these 22 tones, only seven were *actually* audible (or still discernable when the bases of classical Indian music began to be

---

10   See http://www.22sruti.com.

11   Jho, *History and Sources of Law in Ancient India*, p. 59.

VII. Universality of the Harmonic Model of Creation

established) and that these seven tones must have been the foundation of the "Sadja grama" scheme, the earliest known format of Indian music. The "Sadja grama" scheme was inherited directly from the Sama Veda, the Veda of Holy songs chanted, from time immemorial, within a framework of "seven tones." Indian musicology quantifies these seven tones in reference to 22 equal gradations or shrutis within the octave. Rather than a spot frequency as in Western music, each "note" represents a *band* within the 22-shrutis spectrum of the octave. These "bands" are dissimilar in size, each defined as "2," "3" or "4" shrutis. Each tone is associated with a certain number of shrutis, which defines its "seat" in the 22fold octave. In Sanskrit, tones are called swara. The meaning of swara is closer to the notion of interval between successive tones than to "tone" as understood in Western music. The following tabulation shows the range of shrutis associated with each swara of the "Sadja-grama."

Sadja =	S (Sa) =	0 shruti
Rishabha =	R (RI) =	3 shrutis
Gandhara =	G (GAo =	5 shrutis
Madhyama =	M (MA) =	9 shrutis
Panchama =	P (PA) =	13 shrutis
Dhaivata =	D (DHA) =	16 shrutis
Nishada =	N (NI) =	18 shrutis
Sadja (upper) =		22 shrutis

Nambirajan emphasizes how, like every tone in the Indian system, these seven tones of Sadja-grama were not arrived at through the mathematics of harmony, but on the basis of *perceived* intervals, or shrutis. "While the West always handled music only as mathematical fractions, not knowing how to organize them as the 'tones' of an octave, Indians handled music only as the 'tones' of a perceptual octave, not knowing what musical fractions are all about." There was no mathematical notion associated with these seven tones. Nor was there any theoretical knowledge of what constituted one basic shruti gradient (i.e., *Pramana Shruti*). Nambirajan would bring to light the mathematical order at work in this perceptual tone system.

In the course of his research into the structure of Indian music and its intriguing 22-fold foundation, Nambirajan discovered the remarkable fact that 22 *simple fractions, and 22 only, can be formulated mathematically between the Natural Numbers "1" and "12"* (the numbers regarded as "sacred" by the ancients) and that these "fractions" lie between the values "1" and "1/2," which correspond to the ends of a musical octave. In other words, the octave (1/2) contains a maximum of 22 simple mathematical fractions: 3/2–4/3–5/3–5/4–6/5–7/4–7/5–7/6–8/5–8/7–9/5–9/7–9/8–10/7–10/9–11/6–11/7–11/8–11/9–11/10–12/7–12/11.

While this discovery lights up for us an essential mathematical link between the key numbers 7, 12, and 22, which further supports the parallel between Platonic and Dogonic cosmogonies, it struck Nambirajan as a possible clue to a law that would "explain" the 22-shruti system, as well as offer a connection between the 22-tone Indian system and the 12-tone Pythagorean system.

To see where this mathematical connection might lead musically, he proceeded to convert the 22 fractions into tonal values by plotting them against an octave calibrated in 22 equal segments. Inserted in a "0" to "22" octave, the 22 fractions acquire the tonal values between "Sadja" (tonic) and "Sadja" (upper octave) shown on the table below (fig. 72).

What Nambirajan discovered is that the seemingly "arbitrary" assignment of numerical values to the seven tones of Sadja grama—still unexplained by musicologists—makes mathematical sense when the seven tones are brought in parallel with the 22 simple fractions and their tone values. The table shows that the quantified, "shruti" values of the traditional Sadja grama Indian scale are the "rounded off" values of seven of the 22 fractions. In other words, the "Sadja-grama" lines up fairly exactly with the scheme of the 22 fractions, revealing the accuracy of Nambirajan's intuition and the mathematical underpinnings of the 22-shruti system.

Swara	Fraction	Sruti value	Traditional Sadja-grama values	Swara	Fraction	Sruti value	Traditional Sadja-grama values
S	1/1	0.00	0.00	M2	11/8	10.11	
R1	12/11	**2.76**	**3.00**	M3	7/5	10.68	
R2	11/10	3.03		M4	10/7	11.32	
R3	10/9	3.34		P	3/2	**12.87**	**13.00**
R4	9/8	3.74		D1	11/7	14.35	
R5	8/7	4.24		D2	8/5	14.92	
R6	7/6	**4.89**	**5.00**	D3	5/3	**16.21**	**16.00**
G1	6/5	5.79		N1	12/7	17.11	
G2	11/9	6.37		N2	7/4	**17.76**	**18.00**
G3	5/4	7.08		N3	9/5	18.66	
G4	9/7	7.98		N4	11/6	19.24	
M1	4/3	**9.13**	**9.00**	S'	2/1	**22.00**	**22.00**

Fig. 72. Tabulation by Sreeni Nambirajan revealing the mathematical underpinnings of the 22 shrutis and 7 swaras of Indian music

Nambirajan's table (fig. 72) shows the mathematical alignment between the notes (*swaras*) of the "Sadja-grama," the 22 fractions between 1 and 1/2 (octave) and the "unexplained" traditional shruti-value associated with each of the seven notes.

Nambirajan also observed how the "tonal" value of two other ancient tone schemes, which constitute, with the "Sadja-grama," the edifice of Indian music, namely the "Madhyama-grama" and "Murchana," were the rounded-off values of the remaining 15 fractions. The "Madhyama-grama" group was composed of the

VII. Universality of the Harmonic Model of Creation

rounded-off values of five more members of the family of 22 fractions, and the remaining members were "seeded" in the 14 Murchanas.

※

What was to grant their full significance to these facts is the correlation Nambirajan uncovered between the basis of ancient Indian music and the mathematical metaphysics of the Sumerians. The earliest civilization known to historians, the Sumerians lived in ancient Mesopotamia, the territory lying between the Tigris and Euphrates rivers, south of modern Baghdâd, Iraq, from 5300 to 1940 BC. They had all the signatures of an advanced civilization: a mathematical system, knowledge of geometry and astronomy, cuneiform writing, law code, pharmacopeia, calendar, even a farmer's almanac. According to the Greek scholar Philo, they were also known to "seek worldwide harmony and unison through the musical tones." Music and musicians occupied a central place in their culture, and their musicology, mathematics and theology formed one fused corpus. Laurence Gardner, a historian on Sumer, remarks, "To this day, everyone concerned is baffled by the sudden, extraordinary emergence of the Sumerians, seemingly from nowhere. But there is no doubt that, upon their advent in southern Mesopotamia, they were already highly advanced, to a level far beyond that recorded or sustained in any place from where logically they could have emanated. Nowhere on Earth was there a culture like that of the Sumerians." In the course of the last century, the decipherment of ancient languages has shed light on how Sumerian cosmology, theology, ethics, and education "permeated to a greater or lesser extent the thoughts and writings of all the peoples of the ancient Near East.... And the Hebrews of Palestine, the land where the books of the Bible were composed, redacted, and edited were no exception."[12]

Speculations abound about the origin of the Sumerians, as they do not appear to be related to the earlier inhabitants of the region. One hypothesis is that they inherited their civilization from beings "from elsewhere who came to Earth around that time"—a mystery reminiscent of the enigma surrounding the origin of the Dogons and the source of their knowledge.

Like many ancient people, Sumerians regarded the first twelve numbers as "sacred," a fact for which there seems to be no known rational basis. They worshipped 12 gods and goddesses. Each deity was the expression of one of the first twelve numbers multiplied by five. The number-gods of the Sumerians were: 60, 55, 50, 45, 40, 35, 30, 25, 20, 15, 10, and 5.

Each god was regarded as a fraction of "An" (60), the supreme God. They wrote the number 60 as a large "I" and used it as their "reference" number for everything.

---

12 Kramer, *The Sumerians*, p. 291.

"An" was 60/60 = 1/1. Enlil (50) was 50/60 = 5/6. Ea/Enki (40) was 40/60 = 2/3, etc. as in the following list.

The 12 Sumerian deities, six male gods and six female goddesses, were thus the essences of the following *fractions of (the) One*:

60	An /Anu (God)	1/1
55	Antu (Goddess)	11/12
50	Enlil (God)	5/6
45	Ninlil (Goddess)	3/4
40	Ea/Enki (God)	2/3
35	Ninki (Goddess)	7/12
30	Nanna/Sin (God)	1/2
25	Ningal (Goddess)	5/12
20	Utu/Samash (God)	1/3
15	Inanna/Ishtar (Goddess)	1/4
10	Ishkur/Adad (God)	1/6
05	Ninhursag (Goddess)	1/12

Each god/ess was a simple, "whole fraction" of the (12fold) One, a wholistic "portion" of divine being. From the insights into 12foldness gained from the Pythagorean harmonic cosmology, one could say that the Sumerian gods were the twelve tonalities of the One. From the ratio of 60/60 (unison) to 30/60 (the tritone) through the 40/60 ratio of the fifth and the 45/60 ratio of the fourth and so on, the gods expressed twelve forms of self-awareness of the world being.

The "figure" of the highest God Anu—60, the multiplication of 12 x 5—and his fractions cannot fail to evoke the harmonic deployment of the Logos, the inner reality of the third harmonic, into the world soul "compound" of *twelve fifths*, followed by its "corporealization" into the *twelve-pentagoned* dodecahedron. 12 times five is the formula of a dodecahedron. The *Timaeus* described the corporealization-physicalization of the world soul as the formation of five solids, the dodecahedron being the quintessential one—the 12 x 5 archetypal form of the Logos. One can see, with the eye of imagination, how "multiplying" the 12 folds (12 gods–fractions) of An by five in an ideal, harmonious way would deploy the One into a 12faceted polyhedron, a quasi-spherical "body" of 12 pentagons—a corpus of 12 *five-pointed* "stars." The form and vehicle of the Logos, the temple of "God," could only be the fairest of the beautiful solids. Here again, with the Sumerians, the supreme "God" is "one" manifesting through a 12fold subtle body.

The term *Anu,* we saw, is the Sanskrit word for "atom" and the designation Leadbeater and Besant used for their "ultimate atom." It is also, Blavatsky points

VII. Universality of the Harmonic Model of Creation

out, one of the names of Brahma—"the most atomic of the atomic, the immutable and imperishable."[13] On the other hand, Lemaître called "primeval atom," or "Cosmic Egg," the initial point or singularity of the emerging universe, later to be known as the Big Bang. We see how the meta-physics of science (for the atom is not a physically perceptible object) and the cosmic wisdom of religions converge in most specific ways on the distant horizons—in time and space—of the human mind. Lemaître was himself a catholic priest and physicist.

The *dodeca*-hedric "figure" of the world soul is at the same time the ethereal *quinte-e*ssence of the physical cosmos, the pentagrammatic integration of its four elements. It is at once 12fold and 5fold. The pentagram or five-pointed star features the quintessential reality of "selfhood"—as mediation between elemental multiplicity and oneness of spirit. Each pentagrammatic facet of the dodecahedron is 1/12 of the world soul, and every god of the Sumerian Pantheon is one constellation of the pleroma. What this contemplation of the 12fold pantheon of "gods-stars" further intimates is that, while the gods are indeed fractions of An, thus partial aspects of deity, they represent, for the One, stellar outposts of rapport with the elemental world—for they are fivefold entities, distillates of the fourfold, and akin to it. Their affiliation with a certain fraction-harmonic of the Logoic sound makes them unique intermediaries between Logos and cosmos, with special aptitudes (associated with their "partial" character) to usher Logoic consciousness into elemental dimensions and foster their evolution. In summary, the multiplication by five points to an operation of subtle corporealization of Being—into a 12fold corpus of Egoic "gods" who could also be referred to as world masters—12 members or limbs for the One to deploy itself in a world being and a world body. The absolute key to this primordial "templating" of the world soul, which is also, by the same token, a pantheon of gods-masters (the 12 world powers evoked earlier), is the life-transmitting principle of the golden mean, which the pentagram precisely embodies. "God" is the grand world soul of the universe. Gods are 12 facets, petals, or virtues of the life-giving love-might of the One being.

The Sumerian pantheon offers a very significant theo-cosmo-logical confirmation of the harmonic nature of the "Divine" principles of world creation. In fact, with the Sumerian civilization, we may be reaching a primary source of the Pythagorean teachings. Sumer preceded Babylonia, which is where Pythagoras was reportedly sent as prisoner after the invasion of Egypt by the Persians, and initiated in their sciences and mysteries. The Syrian philosopher Iamblichus wrote that Pythagoras "reached the acme of perfection in arithmetic and music and the other mathematical sciences taught by the Babylonians." Known originally as Sumer, Babylonia was the

---

13   Blavatsky, *Secret Doctrine*, vol. 2, p. 542.

ancient country of Mesopotamia. The Babylonians inherited much of their knowledge, culture, and traditions from the Sumerians.

If, for us, the Sumerian pantheon confirms the harmonic underpinnings of creation, for Nambirajan, the Sumerian "theology" opened the door to vital musicological "clues." Their pantheon appealed to him as the "symbolization of an ancient musicology." He was intrigued by the notion of metaphorical copulation of male and female numerical arrays, on which Sumerian cosmology and mythology is based—not unlike the copulation of numbers we encountered in the Platonic Lambda. He felt inspired to experiment with extending the list of ratios-"rapport" of every god/dess-to-Anu to all the possible ratios-relations *between* gods and goddesses. Deleting redundancies as well as the upper and lower octave limits (1/1, 1/2) since both reference the whole rather than a fraction of it, he obtained 22 simple fractions—indeed, the only simple fractions possible between 1 and 12 (1/1, ½, 2/3, 3/4,... 10/11, 7/12, 11/12).

It is via this deeply organic and Platonic process of bonding or wedding the twelve among themselves that Nambirajan discovered the 22 fractions, and that, struck by its correspondence with the Indian musical system, he proceeded to confirm, as retraced above, that these 22 fractions are the mathematical underpinnings of the 22-tone system of Indian music. By converting the mathematical values of the Sumerians' divine fractions and bonds into the tonal, perception-based values of Indian music (as shown above), he brought to light the harmonic ground common to Vedic musicology and Sumerian metaphysics. While correspondences could have been noted between, for instance, the Sumerian god Ea/Enki (40, 2/3), the Panchama (swara) of the Sadja Grama, and the fifth of the Pythagorean scale, they remained isolated convergences until the discovery of a mathematical matrix common to Sumerian gods and Indian musicology could reveal that they shared the same foundation, allowing for the reality of this fundamental scheme to be contemplated.

By revealing, on one hand, the metaphysical mathematics underlying the tonal sophistication of the 22 shrutis of Indian music, and, on the other hand, the science of harmony in which the mathematical theology of the Sumerians is grounded, the correspondence uncovered between Indian Sama Vedic tones (and their shruti-subtones) and Sumerian Gods (and their sub-bonding) exposed two versions—one perceptual-tonal, the other mathematical-theological—of the fundamental harmonic matrix that is at once the geometric template of cosmos and the toning milieu of formative powers that Plato called the world soul.

VII. Universality of the Harmonic Model of Creation

The convergence between the wholistic mathematical fractioning of Oneness (octave interval of 1/2, 22 mathematical fractions, 12 numbers) and its perceptual-experiential counterpart (the differentiation of 22 shrutis) is a powerful intimation that the science of harmony is the divine, objective-*and*-subjective, science of the manifold oneness that is the universe.

One can only marvel, at the end of this journey into the harmonic underpinnings of cosmos, at the awesome clear-mindedness of a culture who recognized and worshipped as the divine powers of creation the simple fractions of a 12fold One. Gods are intermediaries, and the fractions are indeed intermediating "ones"—partial, yet whole "units" of the One, by means of which the world being pervades in a unitive way the multiplicity of creation, and creation finds access to its oneness of being.

Behind the 22 simple fractions stand 22 vibrant interplays between 12 "divine" aspects of the One being—22 intervals that found the "music"-making of creation. Based on wholistic bonds between the fundamental 12 "fractions-gods," 22-foldness represents a sub-level of harmonic differentiation of "one," a finer texture of harmonic weave and tonal intercourse.

The 22 bonds among the 12 god-fractions, and the corresponding 22 shruti intervals of Indian music illumine the universal cosmological significance of the 22 yalas of the Dogons. Beyond unique presentations, which, in the end, finely orchestrate a common core, ancient cosmological models (Sumerian, Indian, Dogonic, Greek) show remarkable convergences, all of which were based on the same "natural" ground of harmonic laws. In the filigree of Indian and Western musical systems stand the same divine mathematics, and these mathematics illumine the principles that stand behind the Sumerian Pantheon as well as the Platonic world soul. The same three-layered ground of fractioning–bonding of Oneness (7-, 12-, and 22-based) involved, implicitly or explicitly, in the foundations of Western and ancient Indian music, is the vibrant basis of the formation of the Platonic and Dogonic cosmos. In Dogonic terms, it is what allows the oneness of Life to "enliven" the 22 "categories of life." Such en-livening is indeed the unique and marvelous property of the golden fractioning at the core of the primordial 12fold.

Nambirajan's uncovering, in the filigree of the Sumerian Pantheon, of a mathematical ground common to different musicological systems, strengthens the view that the secrets of cosmogony reside in the laws of harmony. The display of a pantheon, whose gods are fundamental ratios of musical harmony, is a stunning ratification of the harmonic cosmology we have been exploring. The divine architects (gods) of world creation are simply and beautifully major harmonic "fractions" of the One, which means that they are exemplars of highly perfected soul tones, hence also toning Masters, whose sublimely vibrant expression of one of the divine combinations of Intelligence, Love and Power, hums around and into us from the zodiacal peripheries, shaping and tuning, harmonizing and entraining all evolving forms of beings.

Ernest G. McClain, the musicological scholar whose research has shed remarkable light on the mathematics of harmony informing ancient scriptures, exposed the harmonic underpinnings of the Sumerian pantheon:

> For reasons that have been vigorously argued but remain unclear, Sumerians developed a base-60 number system. *Waiting to be recognized within it are the main patterns of harmonical theory* that appear later in India, Babylon, and Greece... Notice that only the most important fractions of 60 are deified (1/6, 1/5, 1/4, 1/3, 1/2, 2/3, and 5/6) and that their pantheon encodes the primary ratios of music.[14]

The greatest god of all, Anu-An = 60 = 1, McClain develops, is father of all, but remote from the affairs of man. He is the fundamental One. The trio of highest gods (40, 50, 60) defines the basic musical triad of 4:5:6 (do, mi, sol, rising; and mi, do, la, falling). The ratio 4:5 defines a major third, and the ratio 5:6 defines a minor third, taken either upward or downward within the matrix of the musical octave.

It is fascinating to find from McClain's brief portrayal of the gods that Enki = 40 (or 40/60 = 2/3) was the Divine Patron of Music. "God of the sweet waters," he writes, "and perhaps the busiest deity in Sumer, he organizes the earth, including the musical scale." These notations speak clearly to the third harmonic as the principle-deity of world building and the "organizer" of the musical scale. The name "sweet waters" suggests the most harmonious blend of wave currents, which the third indeed epitomizes. And what could be more exact for the *main*, golden *mean of creation,* than to refer to it as the busiest deity? One last fact leaves no uncertainty as to the identity of the god: He is, McClain notes, the "first born son" of the great Anu-An = 60. A striking confirmation of the connection between the harmonic third and the metaphysical Son, this sonship at the center of the Sumerian pantheon highlights the golden trinitarian structure that runs through the metaphysical systems of humankind's history, and illuminates its cosmological character and harmonic underpinnings.

From Sumer to Christianity, from India to Egypt, the metaphysical notion of the Son-god redeemer carries indeed and *in fact* the vibrant life-giving Presence, worldwide and world deep, who abides in the proportional dynamics of living forms, the diastolic and systolic heartbeat of animated beings, the separating-bonding rhythms or arrhythmias of the human self, and the individuating-universalizing unfolding of Egoity. With the unfolding of solar/soul selfhood, the Love Mean that is "in the beginning" of everything begins to ray out from the human realm.

---

14   McClain, *Musical Theory and Ancient Cosmology*, pp. 371–391 (italics added).

VII. Universality of the Harmonic Model of Creation

Ultimately, the sonship born of the great (octave) polarities speaks to the long gestation and birth of Logoic consciousness in the soul of humankind. Sonship "redeems" the relative separation of the Two by bonding spirit and matter into the creative heart of reality. The objective face of the world, separation is the diastole of the heart of existence, the pulling apart that exists as long as the universe ex-ists, making space for systolic beauty to emerge, the beauty of what love unites and generates—the inner face of the world.

Even before McClain and Nambirajan's insights into the harmonic ground of the Sumerian pantheon and the universality of musicological schemes came about, a team of scholars from the University of California at Berkeley had discovered that the Assyrio-Babylonian civilization had the same heptatonic-diatonic scale as Western music. In March 1974, following years of research, Professor Anne Kilmer and her colleagues announced that they had deciphered the world's oldest song. They were able to read and actually play the musical notes written on a cuneiform tablet from around 1800 BC, found at Ugarit on the Mediterranean coast (now Syria). They produced a record of the song, called *Sounds from Silence*. Anne Kilmer demonstrated that the music was composed in harmonies of thirds, and that it was also written using a diatonic scale. The song is in the equivalent of the diatonic major (do, re, mi) scale. The evidence that the 7-note diatonic scale existed 3,400 years ago shatters the traditional view held by most musicologists that harmony was then virtually nonexistent and the scale only as old as the Ancient Greeks. "We always knew," the Berkeley team explained, "that there was music in the earlier Assyrio-Babylonian civilization, but until this deciphering we did not know that it had the same heptatonic–diatonic scale that is characteristic of contemporary western music, and of Greek music of the first millennium BC.... This has revolutionized the whole concept of the origin of Western music."

✪ The Sumerian background of the Pythagorean musical system illumines the harmonic underpinnings of the Greek pantheon of 12 gods and, in general, the *harmonic* basis of metaphysical principles. To emerge from this journey to the source of world metaphysics with the realization that Gods are the world creative principles-and-princes of harmony, and that *harmonic fractions are their vibratory vehicles,* is to gaze anew at all of nature, and divine in the astonishingly unitive diversity of forms the sweeping breath and modulating overtones of one majestic Being "in whom we have our being." ✪

The gods worshipped by the Sumerians are the inner realities of the "means" of world creation. Twelve musical intervals point to their countenance, as the figures born of cymatic experiments point to the formative powers of specific tones. Expressed in the twelve simple fractions of One are the tonal qualities and designing powers of intermediary beings of a universal order, "divine" to us, yet second to the Logos, who make up the pleroma of formative originators of cosmos. Deployed as a 12fold pantheon around the Logos, An, they are the toning world powers who inform dimensions of consciousness and categories of forms with the whole fractions–harmonies of the fundamental One, thus allowing the oneness of Life to enliven and ensoul the circle of all beings.

What the modern mind would regard at first as an intriguing divinization of fractions speaks rather to an awareness that "divinity," whose primordial reality is hidden in the mystery of creation, resides in the wholeness of harmonic fractions. Whole "musical intervals" carry the oneness of life, and the Sumerian god-fractions are great Lives.

Major world cultures orchestrate this notion of the 12 "folds" of the creator. Vishnu, the second person of the Hindu Trimurti, defined as "one who enters everywhere" and "the All-Pervading One," has 12 different expressions called Adityas. Each of these expressions of the "Sun-god," says the *Bhagavata Purana*, shines for one month of the year. The spiraling shape of the conch, one of the main attributes of Vishnu, and its secret tone world suggest the harmony inducing function of the god.

The ancient books of the Egyptians depict the Sun God Ra riding through the night in a boat led by 12 Helpers—"twelve star-gods who conduct the Sun at night."

Gods of the months and the hours, the 12 world powers also speak to the 12 creative hierarchies who, says Alice Bailey, stand behind the dimensions of the world, each associated with a specific cosmic function and a zodiacal sign.[15] Rudolf Steiner refers similarly to the inner reality of the constellations as hierarchies of spiritual beings and to a lofty circle of 12 Bodhisattvas. Both teachers specify that only 7 of the Bodhisattvas/ Hierarchies are incarnated and active and 5 are latent and out of incarnation.

The laws of harmony explain why the numbers 7, 12, 22, which form the quasi-universal ground of musicology, appear consistently in ancient cosmologies and define the structure of our universe—7 planets, 12 zodiacal constellations; 7 levels of being, 12 hierarchies—as well as the human constitution—7 chakras, 12 fold Egoic body, 22 pairs of chromosomes. The additional, 22fold fractioning of "one" explicit in the Dogons' cosmogony, and implicit in Sumerian cosmology and Indian music (the first two mathematico-theologically, the third tonally-perceptually) further differentiates Plato's harmonic cosmogony and reveals its universality. In addition to corroborating Plato's description of the primordial process of harmonic differentiation-integration that structures the universe, it brings to light a finer level of differentiation-conjugation, one that goes beyond our current perceptual faculties

---

15 Bailey, *Esoteric Astrology*, pp. 34–35

VII. Universality of the Harmonic Model of Creation

and may well be the key to the structure of such ancient "languaging" systems as the 22 arcanas of the Tarot, the 22 letters of the Hebrew alphabet, the 22 pathways linking the Sephirot (spheres of life experience) of the Kabbalistic Tree of Life. Embedded in the 7/12fold archetypal form, the 22 chromosomes carry the in-formation code between soul (12) and body (7). The 22 letters serve a parallel function between essence and form, and, in language, between self and world.

The Hebrew alphabet itself highlights another significant character of 22: it is simply the sum of the three key cosmological numbers: 3, 7, 12. In remarkable resonance with the present context, the Hebrew alphabet has 3 "Mother" letters: Aleph, Mem, Shin; 7 "Double" letters: Beth, Gimel, Daleth, Kaph, Pe, Resh, Tau; and 12 "Simple" letters: He, Vau, Zain, Cheth, Teth, Yod, Lamed, Nun, Samekh, Ayin, Tzaddi, Qoph."[16] It has been suggested that written language may have its roots in the cymatic geometry of harmonics formed by the spoken word.

One of the earliest and most famous Kabbalist texts, allegedly written by the biblical patriarch Abraham, the *Sefer Yetzirah* (*The Book of Creation*, or *Book of Formation*) outlines as follows the creation of the Universe and all it contains:

> Two forces come from one and form twenty-two basic properties: Three root forces, seven doubles and twelve simple ones. They are all filled with the same light....
>
> II/ 2. He [the living God] hath formed, weighed, transmuted, composed, and created with these twenty-two letters every living being, and everything that is destined to come into being....
>
> III/1. The three mother letters A, M, SH are the foundations of the whole; and resemble a Balance, the good in one scale, the evil in the other, and the oscillating tongue of the Balance between them....
>
> IV/3. The seven double letters He formed, designed, created, and combined into the Stars of the Universe, the days of the week, the orifices of perception in man; and from them he made seven heavens, and seven planets, all from nothingness, and, moreover, he has preferred and blessed the sacred Heptad....
>
> V/2. The twelve letters, he designed, formed, combined, weighed, and changed, and created with them the twelve divisions of the heavens (namely, the zodiacal constellations), the twelve months of the year, and the twelve important organs of the frame of man.[17]

A brief look into the potential parallel between the 22 shrutis and the 22 major arcanas of the Tarot shows that "Temperance," the 14th card of the Tarot, is the arcana that corresponds to the 3/2 interval and 14th shruti (14=2/3 of 21 +1). Depicted on

---

16   See http://www.world-mysteries.com/science-mysteries/life/dna-secrets.

17   *Sefer Yetzirah*; see http://www.sacred-texts.com/jud/yetzirah.htm.

the card is an androgynous being standing one foot on land, one in water, and holding two chalices, one in each hand. A twofold current (which some tarot decks represent as red and blue), flows back and forth, up and down between her hands. She is an iconic portrayal of the operation of harmonic blending (constructive "wave interference") of the sending-and-receiving currents associated with *the two poles of the One*—same and other, fundamental and octave. A human expression of the third harmonic musically, and the triune Logos metaphysically, she portrays the evolutionary counterpart of world unfolding—the folding of duality into unity, the gathering of opposite forces into a standing wave of transdual oneness. She herself stands as a dual vortex, her wing-like aura wrapped, torus-like, around her physical body. The star on her forehead suggests the fully opened transdual, "third" eye, whose traditional representation involves precisely a dual flow between left and right—a stylized symbol of the faculty to bring dualities into creative intercourse. Illuminating what her hands are doing, the radiance of the 3rd eye from "beyond the two" intimates that she embodies literally *the golden mean* between human and divine—the creative harmony between nature's flow from below and divine flow from above. She epitomizes the human-and-divine "one," who, having made herself whole, is adept at wielding the cosmological magic of making the two one. Psychologically, she embodies what Jung called the transcendent function of sym-bolization (which indeed, literally means, to throw together). Temperance was a central virtue in the Pythagorean School, being the golden thread to the ultimate goal of harmonious, well-tempered, rounded-out unity of being.

Fig. 73. Arcana XIV
(Rider-Waite tarot deck)

In the end, the 7-12-22fold Sumerian-Indian-Dogonic fractioning of the One shows the remarkable universality across time and space, history and cultures, of the harmonic model of the world's organization and logarithmic unfolding, partly described and partly implied in the *Timaeus*. The same wisdom-science, conceptual and perceptual, of the soulful symphonic coherence of the cosmos is clearly revealed.

The differentiation of oneness along the numbers 1, 2, 3, 7, 12, 22 represents a successive bundling—first, of oneness into triunity, then of triunity into a 3-7-12fold

VII. Universality of the Harmonic Model of Creation

archetypal whole—a third oneness that is the archetype of world forms. The second triunity of 3, 7, 12 is the externalized replica of the first. A new brightness of evidence lights up this "divine" replication when one considers that the number that essentializes the seven and takes them to the next "octave" is 8, and the essentializing number of the twelvefold is 13—the 13th at the center of the twelve. Viewed as 1, 8, 13, the trinity of the universal form reveals itself as the incorporation into a threefold unit of a ratio of the Fibonacci series, the series that articulates in terms of whole numbers the grand creational curve of life. The new sequence of 1, 1, 2, 3, 5, 8, 13, 21 exposes the divine genius of Phi at the heart of Creation; 3, 7, 12 is indeed the second "round" of the golden trinity, the triune Mean incarnate, the Logos made world. This second trinity represents the unfolding of Phi into a three-dimensional form, which is the vehicle of the Logos. This vehicle of the soul of life is the formula of all living systems.

Figure 74 shows how the Fibonacci sequence of 2, 3, 5, 8, 13 (between sets of black to black, then white to black keys) is embedded in the musical keyboard.

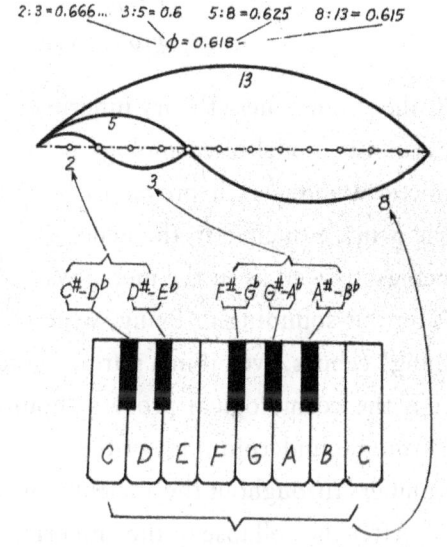

The secret of the sequence of cosmogonical numbers is the secret of Phi—the essential *Temperance* of creation that separates in a way that unites, the grace of the logarithmic curve that keeps turning inward while moving outward, the essential intelligence of proportionality that generates the architecture of unity and diversity, heterogeneity and homogeneity displayed in the depths and heights of the universe, and the essential intent to harmony that engages the intercourse of union and division, sameness and difference at the magical and challenging core of love—from the love that moves the stars to the love that moves human beings.

Fig. 74. *Fibonacci proportions in the musical keyboard (Doczi,* The Power of Limits)

In the end, the reason for the cosmological centrality of 12 and 7 lies in the fundamental *relation* that the two numbers foster, envelop, and articulate—the relation between two intentions of the world being: the intention to separate and the intention to unite—the power of 2 and the power of 3. A major reflection to emerge from this entire exploration, one already steering our collective mindset, is the importance of focusing on relations between components, rather than on isolated parts and particles. Rapport and ratios may in the end reveal more than particles, proportions more than ratios, processes more than proportions, and phenomena more than objects. The secrets of life lie in the relation between things—and "things" are, in and of themselves, realities of interplay. It behooves the intellect to gently disengage from its natural fascination for the object it can distinguish, count, and measure, and begin to imagine instead the formative processes that result in natural phenomena.

THE ANCIENT SIGNIFICANCE OF NUMBERS

As the musician harmonizes his lyre through mathematical numbers, so nature through her own natural numbers, orderly arranges and modulates her productions. —IAMBLICHUS[18]

Chemistry is simply numbers. If you are an atom with one proton, you are hydrogen; two, helium. —CARL SAGAN[19]

Rather than a mere theory inherited from ancient Greece, the Platonic–Pythagorean harmonic model of cosmology proves quite universal. Rather than a purely mythopoetic worldview, it proves to be a spiritual–scientific one. Harmony emerges as the prime principle in the genesis, structure, and evolution of the universe and its beings, as well as in the metaphysical–religious structures of the world. If creation is an out-sounding of Being, which ancient traditions affirm and the modern "Big Bang" echoes, even if in a purely one-dimensional, mechanistic way, it is no surprise that the cosmological process should be illumined by the laws of harmony. The harmonic underpinnings of the world-construct explain in turn the significance of numbers throughout the ancient world—from Sumer to India, from Egypt to Greece.

After the collapse of the Sumerian civilization around 2000 BC, the Babylonians carried on the musical–mathematical–cosmological knowledge received from the Sumerians, and Pythagoras became acquainted with their fount of wisdom during his initiatory travels in Egypt and Babylon. After his return to Greece around 518 BC, he founded his own mystery school in Crotona.

While mythologies offered psycho-dramatic displays of the vital forces of the universe personified in gods and goddesses, the mystery schools of Egypt and Greece instructed their pupils in the realities of these divine powers and their operations. They educated their minds in the cosmic principles and impressed upon their hearts their moral implications. Tested disciples were eventually initiated into an actual

---

18 Iamblichus, *The Life of Pythagoras*, p. 226.
19 Sagan, *Cosmos*, p. 223.

VII. Universality of the Harmonic Model of Creation

experience of macrocosmic realities. Many of the great minds of antiquity were initiates of these secret mystery schools. Plato was one of them. "He traveled widely," says Manly Palmer Hall, "and was initiated by the Egyptians into the profundities of Hermetic philosophy."[20] The school of Crotona was eventually burnt and Pythagoras killed by disgruntled wanted-to-be members. The school ceased to exist and the teachings went underground. They would continue to influence the development of Western culture, from arts to sciences.

In the centuries that followed, the mystery schools gradually degenerated. The teachings, disciplines, and rituals lost authenticity and caliber, a course explained in part by the growing attention given to earthly affairs, material life, and physical phenomena, and a loss of contact with spiritual realities. The final blow came in AD 529, when Emperor Justinian closed the last philosophical school in Athens, the Academy founded by Plato. With the closing of the mystery schools, "spiritual night descended over the Occident." Nonetheless, the ancient wisdom-sciences continued to live quietly in various "esoteric" streams (Kabbalah, Rosicrucianism, Alchemy, Hermeticism, Masonry to name the most prominent) and be carried forward by influential individualities.

These different streams taught the significance of numbers and their involvement in the sacred geometry of nature. However, the *source* of such significance seems to have ebbed away. Harmony and the indivisibility of Being at work in the laws of harmonic fractioning disappeared from view. Yet it is precisely in the articulation of the two facets of numbers—quantitative enumeration of the many and qualitative diversifying of the One—that the physics of matter and the mathematics of consciousness can potentially meet, cognize, and recognize each other as two sides of one reality. Blending their gifts, they might bring forth a science of forms and particles based on formative processes.

Pythagoras was the last great mathematician and teacher of the divinity of numbers. While he is still highly respected for his mathematics—most of all, his famous theorem and his laws of harmony—the regard he had for numbers as divinities tends to be dismissed by the modern mind as part of the pantheistic vagaries and mystical arithmetics of the ancient past. Yet, if we espouse the logic of harmony, we come to sense the profound significance of the Greeks' science of numbers and the wholistic knowledge it contains about the universe.

Pythagoras taught that everything is arranged and defined by number. Number is divine and divinities are associated with numbers. "Pythagoras worshipped the Gods not bound to a human form, but to divine numbers…," Iamblichus writes. "Divine numbers, both according to Orpheus and Pythagoras, are the gods

---

20   Hall, *The Secret Teachings of All Ages*, p. 20.

themselves."[21] "The Pythagoreans," says Proclus, "celebrated number as the father of gods and men, and the Tetractys as the fountain of ever flowing nature."[22] According to Iamblichus, Pythagoras learned that the essence of the gods is Number from "the Orphic writers," the literature associated with Orphism and the mythical poet Orpheus, the founder of the Greek mysteries and patron of music.[23] The message is again that the divinity of numbers is inexorably tied to the facts of harmony. Numbers are divine because they inform the harmonic unfurling of creation and the consonant structures of life. They are the wholistic means by which the One source of Life distributes itself into a vibrant, harmonious world creation.

Mythology tells that Orpheus was the son of Apollo and Calliope, the "beautiful-voiced" muse who presided over eloquence and epic poetry (in ancient Greek, *Kalliopē* means "beautiful-voiced"). Ovid refers to Calliope as the "Chief of all Muses," so ecstatic was the harmony of her voice. She taught her son to make verses for singing, while Apollo, the Sun God of music, gave him a golden lyre and taught him to play. Orpheus traveled to Egypt where he deepened his knowledge about the gods and was initiated into macrocosmic realities, as intimated in his famous "descent to the Underworld." Along with the account of this extraordinary experience, he brought back to Greece the inspiration and practices for the "Orphic" mysteries he founded.

Orpheus' legendary music and singing charmed birds, fish, and wild beasts. It could coax the trees and divert the course of rivers, move stones and stir rocks into dance. It softened the heart of Hades, who relented and allowed Eurydice to follow Orpheus back to Earth. Iamblichus' account of the life of Pythagoras offers similar hints of the great teacher's dominion over beasts and birds by "the power of his voice" and the "influence of his touch." He reportedly subdued an eagle and tamed a savage bear by "stroking it gently with his hand."

These capacities indicate a stage of consciousness of the same order as the one displayed in the "miraculous" events associated with a Francis of Assisi or an Al Hallaj—a mastery of the elements (mineral, vegetal, animal) stemming from a "knowing" of nature as *one's* greater nature, brought about by initiatory expansions of consciousness and identification with the living source of nature. The initiate "subdues" the elements because he knows them as dimensions of his universalized "self" and experiences them as his "own" macrocosmic nature. As a Tibetan master can heat up his body at will, the advanced initiate experiences the body of the world as his macrocosmic body, pervaded with the universal selfhood he is now identified

---

21   Iamblichus, *The Life of Pythagoras*, p. 81, footnote.

22   Proclus, *Commentary on the Timaeus of Plato*.

23   Fideler, *Jesus-Christ, Sun of God*, pp. 74, 384.

VII. Universality of the Harmonic Model of Creation

with. The transcending of personal, then individual, egoity opens the realization of identity with the essential nature of divinity—harmony and love. "Dominion" over creation heralds identification with the "dominus," the living god of love abiding in the dominant, the triune consonance of the fifth at the core of nature.

Orpheus and Pythagoras' communion with stones, animals, and gods intimates a supreme integration of instinctual and spiritual nature—a perfect proportionality of being. After his return to Greece, Pythagoras came to be regarded as a god rather than a man, says Palmer Hall, and so was Orpheus. He was referred to as "son of God." These great human beings were unique mediators between the body of nature and its invisible Intelligence. One was the greatest harmonist, the other was the greatest teacher of harmony.

Such initiates epitomize the full vibrancy of the soul identified with the heart of the planet and resonating with the heart of the Sun. The fully intoned 12fold scale of egoity provides the integrality of selfhood needed for the "fundamental" of Logoic sonship to begin to resound—and to resonate with the whole at the heart of all beings. Sons of harmony in their nature, they are sons of love in their consciousness. Fully attuned to the *essence* of nature and cosmos, they are in harmony with all its beings and denizens.

The ancient knowers–teachers of nature regarded the laws of harmony and the "figures" of the gods as the royal way into the essence of things. The central tenet of Pythagoras' teachings—that "all is number"—was not the isolated proposition of an initiate returning to Greece with the secrets of harmony. The notion of the divinity of numbers was expressed in an elaborate knowledge about the number-reality of divinities, which was part of a science called Gematria. Widely known from Babylonia to Greece, practiced by Jewish Kabbalists and later by Christian gnostics, Gematria consists in translating names and words into number values based on a code of association between letters and numbers. David Fideler, who explored in details "the remarkable canon of Greek Gematria," concluded that the names of the major divinities and mythological figures of the Greeks were consciously codified in relation to the natural ratios of geometry to equal specific numerical values. These names were unanimously regarded as having been revealed by some god or godlike being to humanity. They also related to each other through the primary ratios of geometry.[24] In other words, the names came *after* the numbers, which symbolized the harmonic fractioning of space–time at work in nature, vesseling living

---

24   Ibid., pp. 73, 75.

gods–principles. Indeed, says one of Pythagoras' most famous "akusmata" (epigrams), "Number is the wisest of things, and the name of things the next wisest."

For example, the name of the solar divinity Mithras (Μειθρασ) is equivalent to 365, the number of the solar year. The name Jesus (Ιησουσ) is equivalent to the number 888, which symbolized for the Christian Gnostics the perfection and harmony of the spiritual realm. Fideler points out that .888 is the ratio of the whole tone, the mediating bond (logos) between the two tetrachords of the octave.[25]

What may appear to us as abstract inconsequential play-equivalences between words and numbers lights up with meaning when placed in the context of the cosmological laws of harmony. Language being derived from the one vibrational source, names would be exquisitely positioned to vehiculate cosmological facts. One can imagine the enlightened man of old "hearing" the vibrations of the force-entities of nature and bringing their "consonances" into words-names informed directly by the harmonics of the "Logos."

The word *Gematria*, Fideler points out, is based on the Greek word *geometria*, or geometry.[26] It is a science of the geometry-creating potential of the numbers behind letters. It registers and studies the "poietic," world-architecturing power of numbers as vestments of the gods and instruments of creation. The essentialization of names into geometrizing potentials reveals the formative forces of the gods behind their names. As Platonic cosmology reveals and cymatic experiments demonstrate, numbers and their fractions are geometers and architects: 2 divides; 3 triangulates; 4 creates volume and solids, cubes, and crosses; and 5 stellates and quintessentializes. The name *Gematria* given to the science of numbers-words suggests an awareness of the relation between numbers–divinities and the geometries of nature, an awareness to be traced to the sciences of harmony that informed the Sumerian, Indian, as well as Chinese civilizations. The harmonic mysteries of cosmology led to cultivate numbers as the geometrizing forces of the universe, charged with the tones and toning powers of the divine kosmocratores. The first 12 numbers-fractions of One would thus be the "figures" of the tonal sub-units that constitute the world-creative pantheon of the "fundamental" god-trinity.

Pythagoras' wife Theano felt the need to rectify what must have been a misinterpretation of her husband's teaching. He did not say, she explained, that all things are generated from number, "rather, *in accordance* with number." Numbers are only the harmonic signature of the "generators." Similarly, Plutarch specified that Plato does not assert number to be the substance of the soul, only that soul is ordered and proportioned by number.

---

25 Ibid., pp. 28–29.
26 Ibid., p. 27.

VII. Universality of the Harmonic Model of Creation

The following arc orders the notions of number, tone, geometry, forms of nature, according to the phase each reality represents in the cosmological process. God-vibration, tone, form pertain to the involutionary creational process leading to form. Form, geometry, number-god pertain to the evolutionary process that brings humankind to cognize geometric patterns and, beyond them, harmonic numbers. Geometry and numbers are the reflections in the human mind of the creational operations of the gods.

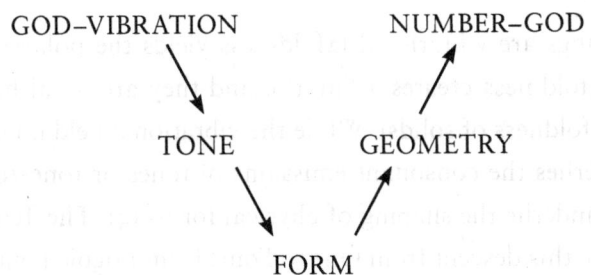

Numbers spell out the formative processes of geometries induced by harmonic vibrational interplay. They are the means by which the mind grasps and contemplates the cosmogonic operations. It may also well be that our mathematics offers the hierarchies of kosmocrators objective insights into the world making operations of their own vibrations and harmonies. We are, Steiner suggests, the religion of the gods:

> People on Earth have all kinds of religions, but all these religions have one thing in common: that man looks up to higher beings, the gods, and worships them. But these higher beings, the gods, also have a religion; they, too, look up to something in awe and reverence. What is this religion of the gods? What is it that the gods revere? It is man. Man is the religion of the gods.[27]

In the end, the "mystery" of numbers and what made them an object of reverence in the ancient world lies in the cosmological *operations* they represent. They are operative principles—particular fractionings of One into harmonic wholes and perfectly balanced geometric units. The first numbers correspond to the phases and types of deployment of One (into two-ness, three-ness, etc.) and the "figurations" that result.

- 1 speaks to the original oneness;
- 2 to division, dualization and polarization—reflected geometrically in the line;
- 3 to triangulation and integrative unity—reflected geometrically in two-dimensional shapes;
- 4 to squaring, corporealization and physicalization, the building of a quaternary world of "solids" out of the four elements—reflected geometrically in three-dimensional structures;

---

27 Reported in Kovaks, *The Spiritual Background to Christian Festivals*, pp. 72–73.

5	to the quintessentializing of the 4fold into a new One, a "soul compounding" of above and below, of threefold and fourfold—reflected geometrically in the pentagonal faces of 12fold self-organizing structures;
6	to the whole gesture and formula of creation, as we shall see below. The Greeks regarded 6 as the number of *Kosmos*;
7	to the delineating and bundling of the archetypal form (octave);
8	to the octave replication of the fundamental "one" in the seed that founds the next septenate (as reflected in the graphism of 8);
12	to the wholistic differentiation of One, which is the self-organizing principle of living forms.

These fractionings are vibrational (2fold-ness yields the polarized field) before becoming tonal (3fold-ness creates a "unit"), and they are tonal before becoming organic wholes (4-foldness of solids). While the vibrational field (2) of the polarized breath of life underlies the consonant emissions of tones or tone-zones, tones and tone intervals (3) underlie the shaping of physical forms (4). The Tetractys captures in one simple figure this descent from One to Four: from Logoic Point to dual waves to threefold tones to fourfold world.

As the singer knows not the geometries s/he creates in the ethers nor the aspirations s/he generates in the souls, the elementals at work in such geometries know not the art they enact, and the gods know not the intricate ways in which their love vibration and penetrating tones arrange stones and crystals, manufacture flowers, turn matter into bones. They do not know how they generate the thought-current that will fertilize a soul petal, or how their tonal pleroma blends so perfectly to shape a beating human heart. Geo-metry and physics, as their name indicate, are of the Earth. Humanity arises to deploy and reveal the creative powers of the gods.

A Mystery-Number in Quantum Electrodynamics—137

The reverence for numbers may seem like a thing of the past and their study an esoteric lore, but a peek in contemporary physics reveals this is not quite so. Most non-physicists will likely be surprised to find that a number-enigma has been occupying the mind of prominent theoretical physicists of our time and elicited a similar reverence. This number is 137—more specifically, 1/137. The fascination arose from the discovery, made in the 1916 by the German physicist Arnold Sommerfeld, of what is referred to as the "fine structure constant"—a constant that would prove to be a key parameter in quantum electrodynamics. Discovered as "Alpha = 0.00729," it was later noted by English astrophysicist Eddington to be equivalent to 1/137 (1/137 = 0.00729), which led to the common reference to 137.

VII. Universality of the Harmonic Model of Creation

Richard Feynman, the 1965 Nobel Prize for Physics, who revolutionized the field of quantum mechanics and formulated the theory of quantum electrodynamics, frequently referred to this number with a mixture of awe and frustration, famously exclaiming:

> It's one of the greatest damn mysteries of physics: a magic number that comes to us with no understanding by man. Immediately you would like to know where this number for a coupling comes from: is it related to pi or perhaps to the base of natural logarithms? Nobody knows. You might say the "hand of God" wrote that number, and "we don't know how He pushed his pencil. We know what kind of a dance to do experimentally to measure this number very accurately, but we don't know what kind of dance to do on the computer to make this number come out, without putting it in secretly![28]

Max Born made it a pronouncement: 1/137 is a law of nature. Should the value of alpha differ, the universe would degenerate.

Wolfgang Pauli, a former student of Sommerfeld, was also deeply intrigued by what he came to regard as a fundamental numerical value, "the theoretical interpretation [of which] is one of the most important unsolved problems of atomic physics." He was quoted as saying that, if the Lord allowed him to ask anything he wanted, his first question would be: "Why 1/137?" Learning from his friend Gershom Sholem, a prominent scholar of Jewish mysticism, that 137 was the number, in Gematria, of the word *kabbalah*" further added to his sense of the numinous value of the constant. When he was hospitalized during what would be the last days of his life, he noticed his room number—137—and remarked gloomily to a friend, "I'm never getting out of here alive." He died ten days later in room 137.[29]

So what is the fine structure constant? Commonly denoted α (the Greek letter *alpha*), it is a dimensionless constant with several different physical interpretations.

The first physical interpretation of alpha by Sommerfeld was as the ratio of the velocity of the electron in the first circular orbit of Bohr's atom[30] to the speed of light in vacuum. The average velocity of the electron is 2,200 meters per second—which equals 1/137 of the speed of light (299,792,458 meters per second).

---

28 Feynman, *The Strange Theory of Light and Matter.* p. 129.

29 Miller, *Deciphering the Cosmic Number,* p. 269.

30 In atomic physics, the Bohr model, introduced by Niels Bohr and Ernest Rutherford in 1913, depicts the atom as a small, positively charged nucleus surrounded by electrons traveling in circular orbits around the nucleus—similar to the structure of the Solar System, but with attraction provided by electrostatic forces rather than gravity (Wikipedia).

Experimentally, the fine structure constant shows up in spectroscopy in the fine spacing between the spectral lines of an atom, revealed when it is exposed to light. When the atom is exposed to light, photons are absorbed or emitted by the electrons, causing their jump between energy levels. The frequencies of the photons involved determine the set of spectral lines observed. Suffice to say for our context that every atomic element produces a unique set of spectral lines, which set is regarded as its "fingerprint." The fine structure refers to the splitting of the main lines, which results from the interaction between the orbital motion of an electron and its quantum mechanical "spin." Alpha is the number that sets the scale of the splitting of spectral lines.

With the development of quantum electrodynamics, the significance of alpha broadened from a spectroscopic phenomenon to a general "coupling constant," understood to characterize the strength of the electromagnetic interaction between electrically charged elementary particles (e.g., electron, proton, muon) and electromagnetic waves/photons (like light or X-rays). All these concepts, abstruse no doubt for the non-physicist, will gather meaning as we go.

The major formula for the fine structure constant is $\alpha = e^2/4\pi\varepsilon_0 \hbar c = 0.0072973525664$. It involves three fundamental constants of nature: e, the electric charge of the electron; c, the speed of light; and, $\hbar$ the reduced Planck's constant. The number for these constants depends on the system of units used for the dimensions involved. For example, the speed of light is 299,792,458 meters per second or 671,000,000 miles per hour. By contrast, when one calculates the constant $\alpha = e^2/4\pi\varepsilon_0 \hbar c$, all the dimensions cancel out. The result is a *dimensionless* number, whose value is 0.007297351, no matter what system of units is being used.

Other formulations exist for alpha, involving other physical measurements. What physicists have continued searching for, however, is a theoretical explanation for this constant. It has been measured experimentally, it has been derived from complex calculations, it has shown its central significance, but where does it really come from, and what does it point to?

※

In the context of our cosmological journey, the fundamental character of the fine structure constant is uniquely compelling, in that the intrinsic unity of the universe calls for intelligible relations between its great fundamentals, in this case alpha and the cosmological numbers we have been contemplating—first and foremost, Phi. Could there exist a relation between alpha and Phi?

From the Phi-based perspective that this expedition has led us to as to a core viewpoint on the universe—the lens of its very "maker"—it would be hard not to

VII. Universality of the Harmonic Model of Creation

see in 137 a close approximation of the familiar "golden angle" we saw at work in the phyllotactic arrangement of leaves around a tree, of petals around a flower, and wonder if this connection might not point to a hidden passageway between this "α" of quantum electrodynamics and the great cosmological constant of proportion (Phi/φ), thus extending another bridge between modern physics and ancient harmonic cosmology.

137°5 and 222°5 are the angular values of the golden secting of a circle of 360°. Similar to how the musical fifth (third harmonic) divides the octave into fifth and fourth, the golden section Phi divides a circle into two angles: 222°5 and 137°5. The term *golden angle* is usually reserved for 137°5. 222°5 is its "complement,"[31] and the geometric equivalent of the fifth in the context of the octave-like wholeness of the circle.

If one spends some time turning around the value of alpha with Phi in mind, it does not take long to find that the constant 0.00729 may also be obtained (approximately) by dividing Phi by 222.5. A simple formula involving Phi might thus account for the birth of this constant, with a margin small enough to consider the possibility that the same cosmological constancy might be at work, approached from different ends. Indeed,

$$\phi\,(1.618) : 222.5 = 0.00727$$
$$\text{Which is equivalent to } \phi : (360:\phi) = 0.00727$$
$$\text{And also to } \phi^2 : 360° = 0.00727$$

The Phi value will probably resonate in the minds of scientists with a significance on a par with the importance they confer to the fine structure constant. Phi is a ubiquitous and universal value present in the physics of life. It is also one that, like alpha, does not depend on physical measurements. It is dimensionless.

However, two facts will raise eyebrows. At least initially. A first objection will be that the exact value is not 0.00727 but 0.00729. From the viewpoint of classical physics, such a discrepancy would suffice to invalidate our investigation. It can only be pointed out that a number of examples in the course of our exploration of the complex yet ordered relations of nature have shown a fluid combination of constancy and inexactitude. Such a mixture, which modern science regards as anathema, is one that nature practices unabashedly in its creations, and that a systems approach to science cannot avoid.

In *The systems view of life*, Capra points out how the emergent new paradigm (systemic versus mechanistic) is necessitating that science accepts and embraces the

---

31   Technically, if the sum of two angles is 90°, they are referred to as complementary angles; if 180°, supplementary angles; if 360°, explementary angles. 222°5 and 137°5 are thus "explementary angles."

reality of approximate knowledge. Such a paradigm shift—from a focus on quantitative measurements of parts in a machine like world mechanism to a focus on relation between parts, and between part and whole—entails indeed a coming into prominence of the factor of adaptation among parts and to the whole. Underlying the slight distortions of archetypal patterns and "constants" by local and transitory influences is this sovereign principle of adaptation. "What makes it possible to turn the systems approach into a proper science," Capra writes, "is the discovery that there is approximate knowledge. This insight is crucial to all of contemporary science."[32] If one hopes to penetrate the phenomena of nature, one cannot risk losing sight of the whole picture by insisting too stringently on exact measurements. One might simply consider how, despite the undeniable constancy of the human form, no one looks exactly like anyone else. While nature unfurls her plants, organisms, and bodies according to the divine proportion, which it deploys further into complex harmonic compositions, she allows many minor deviations, approximations, and disproportions. Nonetheless, the fundamental values preside and remain.

A second objection, and a major one, is the *dimension* of angularity that 222° and 360° introduce to formulate a value prized for its fundamental dimensionlessness. Pondering this objection brings up two worthwhile observations.

The first is that the shape and spin of atoms, the orbiting of electrons, as well as the presence of the constant Pi/π in many equations of quantum mechanics, tacitly acknowledge the spherical "dimensionality" of the space–time framework of particles, organic forms and cosmic systems. Pi, the constant ratio of radius to circumference, addresses the rapport between distance of center-nucleus to periphery on one hand (radius), and orbital circularity of electrons on the other (circumference). Though dimensionless in terms of physical measurements, the fine structure constant is inscribed in the fundamental sphericality of spatial forms and dynamic cyclicity of time (frequency/wavelengths). While all the system-units involved in its formula (meters, seconds, Joule-seconds, etc.) do indeed cancel each other out, so that it can be regarded as dimensionless, the constant is nonetheless "in-formed" by the geometry of earthly shapes and dynamics. Cycles of time, as well as circles of rotation and revolutions in space fundamentally inform the "particularization" of Life and Light into particles of mass-energy. The relation between part and whole, smaller cycles and greater cycles, is precisely what Pi and the number 360 speak to.

The second observation is that the "degree" unit of measurement introduced is a singularly universal one, rather than a cultural one. Universally, since ancient times, the circumference of the circle has been expressed as 360°. 360° is not an arbitrary number, nor is degree a convention of the same order as meters or coulombs. It is

---

[32] Capra and Luisi, *The Systems View of Life*, p. 82.

VII. Universality of the Harmonic Model of Creation

a measure of circle and cycle based on the human experience of time and space: the cycles of Earth, Sun, and Moon. Approximately 365 day/night cycles bring the Sun back to its initial alignment to the stars (one year cycle)—an approximate frequency of 360:1 (Earth rotations per "sun revolution"). Divisible by 12 and 30, 360 embraces into one (year) cycle of time and one (zodiacal) circle of space the three cosmic cycles that structure human existence: 1 revolution of the Earth around the Sun, involving 12+ months of Sun-Moon cycles, each involving 30¨ (27.3 to 29.5) days of Moon-Earth cycles. On the clock, 360 divided by 12 gives us the 60 minutes cycle. 360 is our global unit of time–space and a fundamental cosmological number. The degree is an "earthly" unit of arc (space) based on earthly cycles of time—the angular section of the circle of cosmic space defined by the daily rotational cycle of the Earth in its revolution around the Sun.

Clearly, 360 is not exactly 365 days, and 29 days not exactly 30. 360 speaks to an intelligence of the whole that embraces the three cycles as accurately *and* inclusively as possible—as an *indivisible whole* (year and cycle) that is *also divisible* by each of the three cycles involved—approximately. At work in this global measure of time and space is systemic intelligence.

We saw earlier on in this journey that 360 cycles/second is also the fundamental frequency (360 hertz) and range of the simplest chromatic scale of whole numbers (360–720). 360 speaks at once to a whole cycle of days, and to an octave of 12 whole numbers-tones. Like the scale, the prominent number 360 arises from natural laws that have been integrated into global sets of whole numbers—it does not come from a local system of measurement. The Sun cycle is analogous to the scale, the Sun-Moon cycle analogous to a semitone of the year.

The fundamental phenomena associated with 360 explain its central presence in the earliest known cosmologies—most famously in this hymn from the Rig Veda:

> Twelve spokes, one wheel, navels three
> Who can comprehend this?
> On it are placed together three hundred and sixty like pegs;
> They shake not in the least. (Dirghatamas, *Rig Veda,* Hymn 1.164.48)

In the formula for the fine structure constant: $\alpha = e^2/4\pi\varepsilon_0 \hbar c = 0.00729$, alpha is expressed as an abstract ratio between quantities (electric charge, speed of light and quantum of action or Planck's constant). What the formula Phi:222 would offer, if proven to be more than a coincidence, is a different access to the constant—a geometric access. In that, it would help us gain insight into the constant as a ratio or interval in an actual phenomenon of nature and, using Feynman's words, comprehend indeed "…what kind of dance to do…to make this number come out, without putting it in secretly!"

While the above considerations about 360 and Pi do not eliminate the gap between this geometric formulation of the constant and its "dimensionless" physics equivalent, they establish a potential framework of cosmological approach to the "cosmic number" alpha—be it only as a hypothesis for now.

In α = ϕ : 222 and its equivalent ϕ : (360:ϕ), we may actually "see" a potential configuring—"fractaling"—operation evocative of the "hand of God" and the "pencil pushing" gesture of Creation that Feynman "divines" behind the mysterious constant. What we see is the first curve of a golden spiral reminiscent of the descent of Phi through successive fractals of itself—which we pondered earlier in connection with the diagram represented in figure 75.

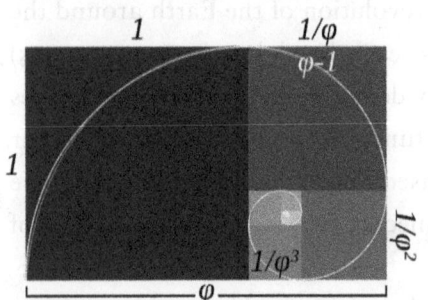

Fig. 75. Phi-based logarithmic spiral

The diagram shows how ϕ divided by ϕ (=1) divided by ϕ (=1/ϕ) unfurls a swirling succession of smaller self-same square-units, which form the framework of the golden spiral. If we substitute 360 for 1—as the fundamental "one" time–space circle of the universe as we experience it—and divide 360 by ϕ, we obtain the first angular golden sectioning of the circle: 360:ϕ = 222.5 (as substitute for 1/ϕ). If we then replace the successive fractaling steps by the successive divisions of the previous angle by ϕ, we generate a Fibonacci like series of partial "arcs" of Phi stepping the golden mean down within the 360° circle. Beginning with 360: ϕ = 222.5, then 222.5: ϕ = 137.5, the descending series reads: "360/222.5, 222.5/137.5, 137.5/84.99, 84/52, 52/32, 32/20, 20/12, 12:7.66, 7.6:4.73, 4:73:2.93, 2:93:1.81, 1.81:1.12—which can be formulated as: 360:ϕ, 360:ϕ², 360:ϕ³, 360:ϕ⁴, 360:ϕ⁵ ... 360:ϕ¹²."

To better understand the process at work, it may be helpful to create a visual imagination of the substitution of 360° for "1" by bending and turning the square-based diagram above into a circle (as shown in fig. 76). If we then imaginatively divide the circular diagram obtained by the golden angle 222°, divide the resulting angle by Phi, then the next angle again by Phi and so on, we obtain a vortex like version of the initial diagram.[33] The squared geometric design turns into a vortical one involving Phi's golden sectioning of the circle into partial spiraling curves.

---

33   The reason we are using the 222° angle rather than 137° is that we are looking at the involutionary movement of descent of the light of life into forms (the "creational picture"). While flowers ascend along the 137° golden angle (clockwise), light descends into "matter" along the 222° angularity (counterclockwise). This is similar to the musical concomitant ascent from fundamental/ descent from higher octave by a fifth, which results in the 5th/4th interval, where the polarized currents of Same and Other conjoin their complementary purposes, Shiva–Shakti-like, in the conception of a tone.

VII. Universality of the Harmonic Model of Creation

The next illustration (fig. 77) shows how this vortex mirrors the forms of nature. The imaginative circular diagram suggests what takes place in the unfolding of leaves and petals—the golden "fractaling" of wholeness into partial yet whole units of life. To follow the unfolding of the leaves along the 222° angle on the figure, one simply goes from 1 to 2 to 3 ... to 11 ... counterclockwise.

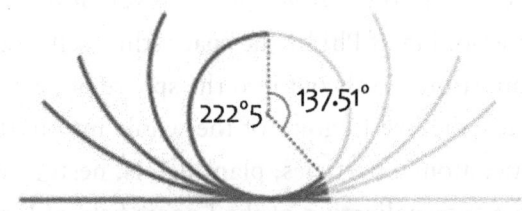

Fig. 76. Golden angle of a circle

Fig. 77. The Aonium exhibits 3 spirals winding in one direction, 2 in the other. The angle between leaves 2 and 3 and the angle between leaves 5 and 6 are both very close to 137.5°. (figs. 76 &77: www.science.smith.edu/phyllo/About/fibogolden.html)

The sequence of angular fractions spelled out above speaks from a different perspective to what we discovered earlier: the Phi-based logarithmic series of nature. Here we see this golden series formulated in term of the 360° cosmological circle. Does it not evoke rather strikingly Feynman's musings about the fine structure constant—as the possible "base of natural logarithms"?

With α = ϕ: (360°/ϕ), as displayed in the imaginatively "rounded out" diagram, we begin to see how the fine structure constant might speak to the fundamental "fractaling" of Phi—the logos-vehicle of the triune oneness of cosmos—into partial wholes—within the existential circle-cycle of space and time. What this formulation of alpha potentially exposes is the fractioning of Phi by its dual angular equivalent within the 360° circle of time–space: 222° and 137°.

Such a "fractioning of the divine proportion" would be a meaningless abstraction were it not for the realization, central to this entire journey, that Phi (the golden mean) is the vessel of our universe's Being, the carrier of the Logos. Proportion (and its essence: harmony and love) is both its vestment and seed-pattern of manifestation. As the fundamental unit-ratio of harmony, the divine proportion is the "mean" by which life wraps itself in forms of matter. The fractioning of Phi is the archetypal "fractaling" at work in nature's dynamics of unfolding. It is no longer unreasonable to consider that a similar pattern would govern the primordial "unfolding" (or atomic "folding-in") of matter, whose wave nature was advanced by De Broglie a century ago. In Max Born's view, alpha "has the most fundamental consequences for the structure of matter in general."

What we actually fathom through the combined pictures-diagrams displayed above is the descent of Phi into the *time–space* vortex of dimensions and particles. What we see is the fractioning of Phi by the space–time self-organization of the universe. The divine proportion spirals out into the sphere of nested spheres that is our universe, "fractalizing" the live Logoity of the whole into partial units of life that are relatively wholistic: atomic particles, plant petals, nested systems. What we see is the unique gestalt of externalization of the Logoic field of Light[34] into particulate forms of mass-energy in space and time—the form building process of life.

As the golden angle is a constant in the unfolding of flowers, trees and spiraling galaxies, so would 0.00729, 1/137, or ϕ/222 point to a fine structure constant in the atom's formation—a quantum stepping down of Light into particle.

The number 137 evokes also the angle associated with the rainbow, which, as a phenomenon, suggests as well an angular "diffraction" (in a broad sense) of the light into denser, slower forms of itself—from particles in electrodynamics, to colors in the rainbow. When looked at on a simple diagram, the primary angle of the rainbow appears to be 42° and is often presented as such. However, a more attentive grasp of the process of refraction and reflection involved shows that the supplementary degree—138°—is the primary angle. Could it be that a similar perfect "fit" of angular harmony between sunlight, reflective particle of water, and perceiving eye as "produces" a rainbow would be occurring between light (photon), electron, and spectroscope to reveal the spectral lines of the atom and their fine structure constant? Could this unique angular "fitting together" of light, mass/water, and eye preside more fundamentally to the descent of light into atomic units of mass/energy?

There may be a scientific no man's land between optical and electrodynamic phenomena, but should this stop us from pondering the analogy between the fractaling geometry of Phi into micro-units via the golden angle (222°5/137°5), the reduction

---

34 The term *Light* is capitalized to suggest the ontological depth of the quintessential field we are referring to, a field that transcends the existential world of dual particles, elements, forms—as well as physical light. Though Light and light are not separate but essentially one, human perception as it operates at this point of collective evolution tends to materialize the light we see, hence de-vitalize the Life abiding in the light.

in speed from photon (light) to electron by a fraction of 1/137, and the unique rainbow effect created by the refraction of the light in drops of water viewed from a 138° angle? Even though the phenomena seem so completely different, are they really? Since "index of refraction" designates the difference between the speed of light in vacuum and the speed of light refracted in a medium, it may not be unreasonable to conceive of 1/137 as the index of "refraction"/densification of the quantum electromagnetic field into an atomic "drop."

That all this be mere coincidence is naturally not out of the question. No matter how much the coherent, symphonic logic of life seems to speak in the deeply suggestive similarity between the golden spiraling of Sun life into forms (the primary divergence angle of petals) and the fine spacing constant between light and particle (photon and electron), without a solid passageway between the dimensionless constant of physics and what might be its geometric translation in the Phi-based value, the "music" stops here.

At this point, I was reconciled with the possibility that these intimations about alpha as a portal between the physics of the atom and the fundamental geometries of nature might remain unconfirmed. All I could hope was that future research would eventually validate the perspectives introduced here by uncovering an actual, measurable link between the physics formula of alpha and its inspiring but unverified Phi-based formula. The significance of such a link would be to make of the fine structure constant a "two-sided fundamental," a fundamental common to quantum physics and harmonic cosmology, hence a channel of exchanges between the two, and a potential seed of unified approach between analytical intelligence of parts and cosmological intelligence of the whole—materialistic science and spiritual science.

These are the questions I was to be left with: where might a Phi-based rapport between light and particle reveal itself from the quantum physics side? In what form might the "harmonic" splitting of the primordial quantum field into nucleus and orbitals, proton and electron, which our whole journey led to hypothesize, show up in the minute measurements and calculations of physicists? Could it appear in terms of energies? Wavelengths? Speeds? The change in speed, which was the first interpretation by Sommerfeld of the value of alpha, certainly pointed to the quantum leap from field to particle, but the calculated value was not associated with an understanding of process that could illumine its underpinnings. Except for the proximity in number-value ($\alpha = 0.00727 = \phi:222$), there was no actual connection between the calculations of physics leading to alpha and the Phi process we were attempting to fathom in the formation and structure of the atom.

This is when the remarkable insights of an Indian research scientist from the Czech Republic, Raji Heyrovska, appeared on my horizon and provided the missing bridge between the Phi-picture and the complex physics of alpha. What Raji Heyrovska first discovered and demonstrated, thereby answering the second question just mentioned, is that the radius of the Bohr atom is divided into two sections by the golden ratio. She explained that this golden secting arises in atomic dimensions due to the electrostatic forces between negative and positive charges (electron and proton)—an understanding that orchestrates the harmonic view of dualization/ polarization as the primordial phenomenon of manifestation. Dualization into attracting proton and attracted electron, centripetal and centrifugal force, would indeed be an essential feature of "particulation" of the field into atoms—and expected to involve the golden mean.

"It was a surprise," she writes, "to find for the first time that the Bohr radius is divided into two unique sections at the point of electrical neutrality, which is the Golden point. The Golden ratio, which manifests itself in many spontaneous creations of Nature was thus found to originate right in the core of atoms. The 'golden ratio' is in the core of the atom."[35]

This discovery led her to the general finding that the golden section is a geometrical constant in atomic and ionic radii, bond lengths and bond angles. Furthermore, she used experimental data on wavelengths of hydrogen spectral lines to evaluate, among other constants such as Rydberg constant and Sommerfeld's relativity factor, the fine structure constant. All proved to be simple functions of the golden ratio.

The connection Raji Heyrovska makes between the fine structure constant and the golden ratio—α and Phi—is the gem we were looking for. A brief look at her demonstration may prove enjoyable to the reader, despite the apparently forbidding intricacies involved. Although beyond the scope of the present work, these intricacies are precious for the magical passageway they reveal to the two-facetedness of alpha—mathematical and geometric—and the deeply unifying implications of this twofold expression of the constant.

A few steps across two or three concepts of atomic physics will suffice to peek at the simple yet radical move that leads from the dimensionless formulation of alpha by physics to its Phi-based equivalent—indeed—based on angular measurements. Not very different from watching a gymnast spin around herself at an angle with great speed, or following the mysterious maneuver downward-up and inside-out that allows Dante and Virgil to pass from hell to Paradise, one barely sees the outside-in spin around of perspective that catapults Raji Heyrovska, and, with her,

---

35 *International Journal of Science,* Research Article vol. 2, May 2013.

VII. Universality of the Harmonic Model of Creation

our grasp of alpha—"to the other side"—from the abstract land of physics to the visual geometry of life.

The passageway she uncovers is based on a particular formulation of the fine structure constant, which also speaks, though in a different way than Sommerfeld's ratio of speeds, to a rapport between light and particle—in other words, electromagnetic field and particle. According to this other formula, alpha equals the ratio of the so-called "Compton wavelength" to the "de Broglie wavelength." De Broglie's wavelength equals the circumference of the Bohr circle, and the Compton wavelength equals the wavelength of a photon whose energy is the same as the mass of that particle. What the ratio between the two provides is a rapport, in terms of wavelengths, between the whole matrix of the light field (represented by its translation into the shorter Compton wavelength) and the particulate condition of the electron (represented by the wavelength of the circumference) when both are calculated to carry an equal energy. This rapport equals alpha.

Raji Heyrovska' ingenious insight was to see that the Compton wavelength amounts to the length of a small arc on the Bohr circle, and to determine that this small arc subtends an angle of 2.627° at the center of the Bohr circle. Noticing the closeness of 2.627 to 2.618 (ϕ^2), hence the closeness of this new value for α (2.627/360) to the ratio 2.618/360 (ϕ^2/360) was the second insight that turned the fine structure constant "outside-in"—from a constant of physical measurements to a geometry-of-life constant—and allowed it to appear in its phenomenological vesture: α = ϕ^2/360. So the formulation we proposed initially for alpha, based on number resonances with harmonic cosmology—ϕ^2/360—proves to be, indeed and in fact, an expression of the fine structure constant of quantum electrodynamics.

Raji Heyrovska's masterful move was to translate length into arc—in a movement not unlike our earlier imaginative modification of the rectangular diagram into a circular one. She also provided an explanation for the small difference, 2.6270° − 2.618° = 0.009°, suggesting it could be due to the Sommerfeld's relativity correction factor.

✪ A fundamental index of the passage from field to particle—whether in terms of slowing down in speed, densification-"refraction" into particles, or polarization into charge/mass, nucleus, and orbiting electrons—the constant alpha is thus directly related to the "involuting" of the Logoic field (Phi) into atomic units by way of the golden angle. Alpha's new formula, ϕ: 222, or ϕ: 360/ϕ, reveals in the constant the geometric key—of angular fractioning—to the vortical, logarithmic involvement of the triune Light of Life into particles, the primordial step of "particulation" of the world matrix of living Light. The fine structure constant speaks to a fundamental rapport/ratio between nondual and dual worlds. It is indeed a "coupling constant"—between whole ("quantum field") and particle (electron). ✪

※

These findings bring us back to the perspective broached earlier. They confirm that the fine structure constant harbors the fundamental geometry of involution-torsion of the ether-matrix of time–space into "units" that replicate its essential threefoldness—its Phi-based "logoity." Particles and forms (be they elementary, biological, human or cosmic) emerge from the harmonious, "golden interplay" between the necessity of wholeness (circle) and the necessity of particularization (secting).

The constant alpha captures the slowing down (by a ratio of 1/137) of the quintessential field of Light as it "particulates" itself into partial "units" by spiraling in and around the 360° cycle-circle framework of space–time-forms through Fibonacci like nodal steps of angular ratios (222/137, 137/84, etc... 7/12). It shows how, faithful to the cosmological process, Phi is "made two" to adapt or "fit" (the Greek meaning of *harmony*) its nondual, transphysical value to the dual spherical vorticity of forms, in order to externalize itself in these forms. Mirroring the primary dualization of Being—into Same and Other, spirit and matter—the triune "Phi field" splits and polarizes itself into complementary pairs of energy/mass, positively charged proton and negatively charged electron—ultimately self and world—by fractaling its vehicle, the golden mean, into Fibonacci-like angle ratios equivalents.

The fine structure constant may thus be regarded as a fundamental of world-manifestation—the singular constant of "refraction" of Light underlying the integrated symmetries and inside/outside dualities of nature's particles and systems. It may be regarded as the logo-rhythmic "refraction index" of world-manifestation.

Measuring up to the physicists' intuition of its significance, the "alpha" of quantum electrodynamics represents nothing less than a third version of the "Logoic" fundamental of the universe. To the divine proportion of geometry and the third harmonic of music, it adds the "coupling" constant of physics—a "divine" index of refraction of the spirit field into particles and forms of matter.

When Feynman mused about alpha as the possible "base of natural logarithms," he was fathoming in the enigmatic constant of quantum electrodynamics the physical profile of the logoic soul of nature.

※

As we proceed further along the descent-externalization of Phi into particles and forms, we find that a sequence of 12 golden angular sectioning of the circle coincides approximately with the completion of 7 full circles: 222° x 12 : 360° = 7.4. In other words, it takes 12 golden steps for the descending energy of life to spiral to a

VII. Universality of the Harmonic Model of Creation

point that also marks the rounding out of a circle. This first convergence of the two cycles—of 222° golden secting and 360° circling—is the first "octave" of the spiral-and-circle pair. It combines a replication of the "origin" (same) with an unfolding of partial harmonics (2/3) in time and space.

The spiraling deployment of the divine proportion ("the Logoic unit") along the golden angular sectioning of the circle is clearly analogous to the successive division of fifth by fifth by fifth, which generates the chromatic scale in music. Similarly, we saw that 12 fifths approximate 7 octaves. The understanding we gained from harmonic laws leads to hypothesize that the octave-like "finiteness" of the atomic particle is likely defined by the "when-and-where" (time-and-space) of approximate convergence between the requirements of separation (7 circles of dimensions) and the requirements of unity (12 angles of golden mediation). This convergence of a duplicating and tri-uniting of One-ness is analogous to the meta-octave of 7 octaves-and-12 fifths, which defined the archetypal scale of musical tones. Having taken 12 golden steps around and down the circle of space time, "Phi" (the vessel of Logoity) touches an octave "unison" with its origin at the seventh circle: it (approximately) closes in on itself to form a whole "atomic enclosure." A "whole unit." The so-called 7 electron "shells" of Bohr's atom mirror these 7 octaves of the fundamental. So do the 7 spirals of Leadbeater's and Babbitt's atoms. It must be noted that we are looking at archetypal patterns. Not all plant-formations will exhibit a 7/12fold structure, yet this is the fundamental harmonic scaffolding of life's forms.

As we saw the Phi-logoity of the Logos emerge in every flower and meet us in every tone, we now see, through the lens of alpha, the same logo-rhythmic process at work in the dynamic formation and "structuring" of the atom out of the quantum field. Given the understanding we reached earlier that Phi is the geometry of the creative love principle of the universe, the formulation of alpha as Phi:222 offers a quasi-visual grasp of the angular descent of the Phi-trinity of the "Light of the world" (the One) into the duality and 7 dimensional orbit-levels of time–space existence—similar to the descent of the "Logos" we followed, with Plato, into 12fold zodiacal circle and 7fold planetary system/platonic solids.

Plotting a golden spiral on a polar graph involving 36 radial lines at 10° increments and 8 concentric circles (fig. 78) shows the perfect inscription of one full angle spiral reaching its "octave" (its departure point on the circle) at the 7th circle via crossing points that are whole angular sections of each circle.[36] Not only does this simple diagram evoke the cosmological deployment of 7 planetary orbits within the 12fold zodiac, or the 7 day week cycle in the 12 months cycle-circle of a year, but it reveals, "plotted" within these circles and cycles, the majestic sweep of One golden

---

36 Illustration from Drunvalo Melchizedek, *The Ancient Secret of the Flower of Life*, p. 218.

spiral—the sweep of Logoic manifestation into the seven days and dimensions of creation. It suggests the profound relation between the dynamics of the cosmological clock and the geometry—or *"cosmo-metry"*—of the golden spiral.

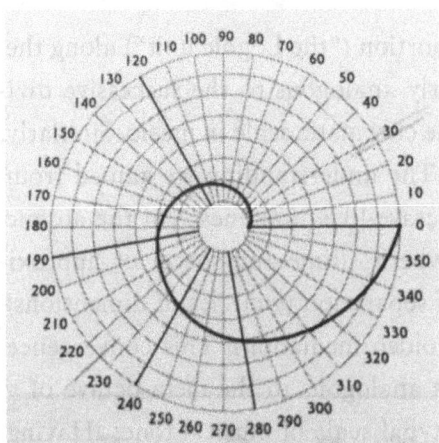

Fig. 78. *Golden mean spiral plotted on a polar graph*

Extrapolating imaginatively, we can see how two golden spirals (one clockwise, the other counterclockwise) issuing from the same point create the heart shape discernible in the forms of nature (leaf, petal, human heart, peacock's tail etc.).

This incursion into atomic physics sheds an interesting light on the great fundamentals of Pi and Phi. As already mentioned, Pi, the ratio of radius to circumference, is present in many constants and equations of quantum physics. What this presence points to is a common operation that physics practices for the purpose of streamlining complex calculations: it converts cycles into linear "radians" via a division by "2Pi."[37]

The sphericity of life, the wholeness of circle and periodicity of cycle is thus surreptitiously done away with in order to obtain quantitative data that can be conveniently added and subtracted, and, in Feynman's words, "put in secretly" in calculations. In the process, any inkling of the cosmological phenomenon they were part of is effectively dismissed and lost track of. Sectioned off the dynamic system-process they are involved in, the gestures of nature disappear, flattened as devitalized segments on the table of human measurements. With the disappearing of cycles, the unit engaged (be it a tone or a standing wave) disintegrates to quantum smithereens. With the disappearing of circumference, the center vanishes to infinity. All that remains are equations supported by an abstract ratio (Pi) useful in the realm of quantity, but disconnected from the relational reality of the soul of life.

This relationality is what Phi safeguards, because it is what Phi is: the dynamic relation of harmony between part and whole on which everything in the created

---

[37] For instance, Planck's constant (h) involves frequency values (i.e., cycles per second, units of space divided by units of time). The "reduced Planck's constant" (denoted ℏ, and pronounced "h-bar"), which is part of the basic formula of α, is a division of h by 2Pi. What the introduction of 2Pi in the Planck constant does, for the purpose of facilitating calculations, is to replace cycles per second (hertz) by radians per second. Time is thus measured in terms of spatial distances, rather than cycles. Linearity is surreptitiously substituted for cycles of time and arcs of space, and sight is lost of the reality of whole units of time–space, and of the ratio and relation of part to whole.

VII. Universality of the Harmonic Model of Creation

universe rests. It is also what Phi has the unique power to restore. Phi is admittedly more difficult to flatten into linear measurements than Pi. More difficult to capture into abstractions, it maintains its phenomenological integrity by limiting its appearances to the discrete hints of its unique angular path and Fibonacci couplings.

While Pi, the ratio of radius to circumference, is an essential interface between the dynamic geometry of life and physical calculations, Phi and the secting by Phi *is* the geometry of life. What the Phi-based formula of alpha does is restore the spiraling profile of nature's holomovement in the filigree of the measurements of atomic physics and quantum electrodynamics.

The inscription of the golden spiral in the polar graph shows how Phi designs the organic way of nature between center and periphery. Complementing Pi's quantitative ratio of radius to circumference, Phi is the qualitative rapport that conducts the back and forth flow of life between center and circumference—spirit and matter. Being at every point a mediation of centrifugal and centripetal motion, the golden spiraling of Phi reveals the "real" (dynamic, space–time) spiraling "radius" of the universe. Phi emerges thus as the soul partner of Pi. Pi leads us to Phi. To restore the gesture of Phi as the fundamental gestalt of our universe is the royal way to kindle a spiritual science of its unitive ground—the relation between part and whole.

Viewed from the cosmological perspective of the present journey, the geometric formula of alpha as Phi:222 awaits as a discrete light-filled opening in the sky of physics—an opening unto *Physis* (nature). Through the opened eye of alpha, Physis gazes at physics, awaiting for physics to catch a glimpse of what it is most after: a unitive view of reality.

What modern physics can fathom, should it be willing to shift perspective through the revolving gate of alpha, is the process of manifestation–incarnation of the light of Life in the circle-cycle of forms in time and space. As there is no finer interval than alpha in the atomic spectrum, there is no smaller angle than the (dual) golden one to embed the wholeness of life into particles and forms.

This theme of incarnation of the archetypal pattern of Life is, remarkably, how Pauli himself interpreted the fine structure constant when it appeared in recurring symbols of "split lines" or "doublets" in his dreams. In a letter to Jung, with whom he corresponded for many years, he shared a dream he had in 1948. In the dream, he recounts, his first "physics teacher [Arnold Sommerfeld] appears, and change in the splitting of the ground state of the hydrogen atom is fundamental." Pauli elaborates, drawing inspiration from—interestingly—Plato's cosmology:

The process is very similar to the one in Plato's Timaeus. The initial stage is a dyadic archetype whose proton corresponds to the "same one" and whose electron corresponds to the other one. Through "reflection of the unconscious," a quaternity is produced. The metal plate, as a symbol of the feminine–indestructible and the physis, corresponds to the physically "divisible" of the Timaeus; the tones, as fleeting–spiritual, correspond to the male principle and the "indivisible." The "Self" that appears here in the form of the physics teacher states that the physis carries permanently with it the image (eidolon) of the tones (*eides*), so that there is a consubstantial unity (*homo-usia*) of both....

The splitting, [he concludes]... *is a symbol of the incarnation of the archetype,* which also accounts for the numinous character of this symbol.[38]

Another memorable dream he had in 1953 about spectral lines elicited further insights into what these "doublets" might mean psychologically.[39] Drawing a parallel with the "division into two" at the moment of birth when a child becomes an independent entity, he interpreted the splitting of spectral lines to mean "the beginning of an assimilation of an unconscious content into consciousness"—the birth into consciousness of an unconscious content.

Clearly, the perspective we are bringing to bear on the constant is not entirely foreign to quantum physicists. Moreover, it is striking to see how Pauli arrives at the very interpretation of the nature of the fine structure constant suggested by its formulation as Phi:222. Phi is indeed the supreme "Tone," which "incarnates" in the atomic unit of physics/*"physis"* through a duplicating process that involves a golden spin (222°/137°) of the whole into a part. Shimmering in the fine structure of the "idolized" atom below is an image of its "Same" above—its "Idea"—and the invitation to recognize the "consubstantial unity of both."

From an opaque formula, quite abstruse to the non-physicist, the fine structure constant becomes an evocative, quasi-pictorial seal of the harmonic-geometric embedding of Phi, the Logos-carrying golden proportion, into "particles" of itself.

Similar to how the third harmonic, the golden fifth, initiates the differentiation of an octave into 7/12 tones, or how the golden mean opens the way for life to descend into 7/12fold forms of matter, and how the 138° rainbow angle opens a perception of the diffraction of light into 7/12 colors, the 222°/137° golden angle unfurls the logarithmic nesting of Light into 7/12fold elementary particles—12 being implied in the number of golden steps that lead to the finite 7fold.

Sommerfeld was first to "hear" it:

---

38   Pauli, *Atom and Archetype,* pp. 102–111 (italics added).
39   Miller, *Deciphering the Cosmic Number,* pp. 180–181

VII. Universality of the Harmonic Model of Creation

The language of the spectra is a true music of the spheres within the atom.

In the end, the fine structure constant of atomic physics illumines throughout the fabric of matter the stunning fact of Alice Bailey's statement, "The principle of buddhi, of cosmic love, is in a mysterious way the principle found at the heart of every atom." As love abides in harmony, and harmony abides in proportion, Buddhi—Love—abides indeed at the heart of the atom.

What moves in the logarithmic spiraling of atomic formation, involving itself in forms of "matter" by means of the logos–proportion Phi, is the Logos. Divine proportion *is* the fundamental pattern of world beginning, the *arche*type of manifestation.

Quite literally stated by John:

> "*En arche en o logos*" (in the "*arche*" is the logos)
> In the beginning is the logos.[40]

NUMBERS AND THE FUNDAMENTAL FORM OF LIFE

> *There is music in the spacing of the spheres,*
> *There is geometry in the humming of the strings.* —PYTHAGORAS

Returning to our contemplation of numbers, we can see how the transitional "figures" they represent between the supra-physical source of cosmos and its physical systems, between heavenly tones and earthly formations, between Gods and their names, pertain to an intermediate realm analogous to that of geometry between patterns of nature and numbers-gods, organic forms and mathematical concepts.

By an obscure phenomenon of resonance between realms, a parallel transition, in this case from gesture of nature to language of mind, occurred at this point of the present journey, sweeping up to a next stage of abstraction one of its central vistas. While pondering how the gods "formulate" themselves through numbers, which inform the geometric morphologies of nature, it so happened that the dynamic form we identified earlier as the archetype of organic systems—the dual vortex or torus—suddenly dropped its swirling gesture and appeared, naked and still, as a simple geometric structure.

When this geometric profile of the spherical vortex, admittedly quite obvious retrospectively, broke into the light of day, it brought with it the sense of an ultimate vista, one that heralded the last step of this journey—for the simple reason

---

[40] John 1: "In the beginning was the logos [Greek: *En Arche en o logos*], and the logos was with God, and the logos was God.... All things were made through him, and without him was not anything made that was made. In him was life, and the life was the light of men. The light (*phos*) shines in the darkness and the darkness has not overcome it."

that it bore the features of some of the oldest symbols found across the globe. The unexpected superimposition of very ancient symbols on what seemed to be a relatively new model of reality (the torus) had the jolting effect of confirming its validity, while it also galvanized the old symbols with a youthful dynamism and a warm iconic glow. This geometric form was the 6-pointed star and its three-dimensional version: the star-tetrahedron. It was hinting at well-known derivatives such as the flower of life or the swastika.

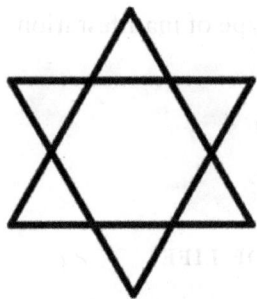

Fig. 79. Six-pointed star

The reader may be familiar with these symbols. They have been the object of much attention from researchers and spiritual scientists, while being part of the esoteric lore of sacred geometers, alchemists, and Freemasons across the ages. Of great significance in our context is the fact that they are regarded as the fundamental patterns of creation. Like the proverbial light at the end of the tunnel, this unexpected break into a landscape of age old symbols and contemporary dialogues released the singular journey of this work into a larger cultural sphere of exchanges between ancient wisdom and newer sciences.

The first figure to signal itself as a geometric blueprint of the dual spherical vortex, the 6-pointed star is also its simplest counterpart (fig. 79). The two interpenetrating triangles can be readily perceived as a stylized version of the intertwining vortices of the dual vortex: involutionary, centrifugal vortex of spirit, and evolutionary, centripetal vortex of matter.

The 6-pointed star captures in a geometric pattern the dynamic process displayed in the torus. It offers a stylized rendering of how, from the interplay of involutive spirit and evolving matter, form is born—a third, a hexagon. Remarkably, Alice Bailey refers to the 6-pointed star as the "Star of Creation," in perfect concord with what we have found to be the process of world formation and the shape of the archetypal unit of "form." The ancient Stanzas of Dzyan echo:

> When he commences work... he passes like lightning through the fiery clouds... takes three, and five, and seven strides through the seven regions above and the seven below... calls the innumerable sparks, and joins them, and forms the germs of wheels. He places them in the six directions of space and one in the middle—the central wheel. He traces spiral lines to unite the sixth to the seventh—the crown.[41]

In Hinduism, the hexagram is revered as the harmonious embrace and mystical union of Shiva and Shakti, which engenders their sixfold progeny. The notion of creation associated with the 6-pointed star also evokes the 7-day model of Genesis, and

---

41 Blavatsky, *Secret Doctrine*, vol. 1, stanza 5.

VII. Universality of the Harmonic Model of Creation

its harmonic-like pattern of separation–union of Heaven and Earth vortices, which produces the kingdoms of the world. 6 is traditionally revered as the sum of the first three numbers, and its association with creation reflects once more the notion that all that is needed to create the world are the first three numbers: 1, 2, 3.

The quintessential "form" of the human being, the causal body of the soul, referred to, symbolically, as the temple of Solomon, is traditionally represented as a six-pointed star called Solomon's Seal, also known as the star of David. The archetype of form—dual vortex and 6-pointed star—harbors the mediating reality born of the conjugation of the dual movement of spirit's manifestation and matter's organization: the soul. "No number is more suited to the soul than the hexad, says Iamblichus, because it contains the numbers that define the musical third, fifth, and octave, and the soul is functionally harmonic by nature, that is, it mediates between the pairs of opposites."[42]

Fig. 80. The heart Chakra, showing the key cosmological numbers: 1, 2, 3, 7, 12

It is only logical that the hexagram should also be central in symbols associated with the heart chakra, the replica in the physical–etheric body of the causal "heart" on the mental plane. The ancient Hindu representation of the heart chakra (fig. 80) does capture in a marvelous seal our harmonic approach to the mysteries of the "central place"—the soul milieu of world beginnings and soul ending: It is one, it is 2fold, 3fold, 7fold and 12fold. The 6 angles-sides of the hexagonal "third" through which the 1 (the primordial "one") externalizes itself reflect the 7fold. The 12 petals of the heart center circumscribe the whole.

Most powerfully associated with Judaism, the six-pointed star also plays a prominent role in Hinduism, Islam, and Buddhism. In the Koran, the hexagram is referred to as the Seal of Solomon and figures prominently in religious architecture. In Tibetan Buddhism, the six-pointed star is called the "origin of phenomenon," a perfect echo of the "star of creation" or the divine intercourse of Shiva and Shakti. In old versions of the Bardo Thodol, the "Tibetan book of the Dead," it appears in conjunction with the swastika.

Painted on the outside walls of nearly every Tibetan Buddhist temple in Tibet and India, the Bhavacakra or Wheel of Becoming (fig. 81), a symbolic representation of the "wheel of life," displays the same fundamentals of creation: The outer rim of the wheel is divided into 12 sections, which encircle the 7 worlds (6+1 at the

---

42 Iamblichus, *Theology of Arithmetic*.

center) associated with the limbs and mouth of Mara. We then note how the three-in-one center of the wheel is replicated in its own middle layer as the trinity of bird (spirit), pig (matter), and snake (mediating, harmonizing wave)—1, 3, 7, 12.[43]

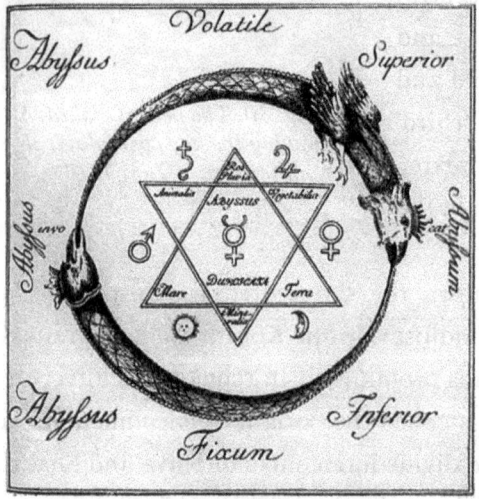

Fig. 81 (above): Bhutanese Painting of the bhavacakra; Fig. 82 (below): alchemical Solomon Seal

In Rosicrucian and Hermetic literature, the hexagram represents a fundamental energetic pattern relating the seven original planets of astrology, as well as the corresponding seven metals of alchemy (fig. 82). It suggests how the Great Work of creation-evolution lies in the balancing of polarities (Sun–Moon, Mars–Venus, Jupiter–Saturn), facilitated by the linking planet Mercury at the center. As its central place suggests, Mercury is the archetypal son of the two (triangles). As Hermes "Trismegistos," he is the transdual, inter-mediating "intelligence" of intercourse between polarities, by which the Hermetic work of evolution-harmonization proceeds. The overall balancing of above and below, involution and evolution, is replicated in the 3fold balance of etheric (Sun–Moon), astral (Mars–Venus), and mental (Jupiter–Saturn) pairs of opposites. So the 1, 2, 3 of cosmological beginnings is replicated—three times. Altogether, four 3fold processes amount to another all-encompassing 12fold.

Three-dimensionally, the six-pointed star is a dual tetrahedron (fig. 86, 2, page 493). The tetrahedron is the geometric outline of the vortex, and the dual tetrahedron is the geometric figure of the spherical vortex, the torus. This 3-dimensional 6-pointed star, or Star Tetrahedron, is also referred to as the *Merkaba*. In Hebrew, *Merkabah* means the

---

43 The traditional interpretation regards the central three as the three poisons of ignorance, attachment and aversion. These would indeed be the core trinity of samsara.

VII. Universality of the Harmonic Model of Creation

throne-chariot of God, which points indeed to the inner reality of the torus: the archetypal form-vehicle of the Logos–cosmos.

The Merkaba is composed of two counter-rotating fields of light—two opposing tetrahedrons spinning in opposite directions. *Mer* means Light. *Ka* means Spirit. *Ba* means Body. When we realize that light is the light of consciousness, the soul, surging from the interplay of Heaven and Earth or the polarity of Same and Other, the word *Merkaba* becomes a perfect designation for what this journey has led us to identify as the 3fold–7fold–12fold quantum of creation, the quantum of spirit-soul-body existentiality.

The notion that the electromagnetic field around our body is shaped like a star tetrahedron has been gaining momentum among spiritual scientists. It converges with the view that the dynamics of the energy field sustaining every living system is toroidal.

We heard, at the beginning of this adventure, how the Vedas sing of the chariot of the Sun in similar terms of numbers-notes: 1, 3, 7, 12. The Sun is the son-"calf" of the Cow goddess, the matrix of the solar system, and the 7/12 solar system is the macrocosmic template of the "chariot" of the human soul. Deployed across the mythopoetic sky of humankind, as far back as 7000 BC, is the central pattern of creation:

> Twelve spokes, one wheel, navels three [hubs]
> Who can comprehend this?
> On it are placed together three hundred and sixty like pegs;
> They shake not in the least.
>
> A seven-named horse does draw this three-naved wheel
> Ageless and irresistible as well,
> Which props all worlds.
>
> Seven steeds draw the seven-wheeled chariot,
> Wherein are placed the sacred notes seven…
>
> Who saw first this structured one when born?
> What time was he born by the unstructured?
>
> Wise poets have spun a seven-stranded tale,
> Around this heavenly calf, the Sun.[44]

The 6-pointed star is intimately related to the "flower of life," an ancient geometrical pattern composed of evenly spaced overlapping circles, arranged in a flower-like design with a sixfold symmetry like a hexagon. A universal symbol found in old manuscripts, art and architecture around the globe, it has been hailed as the key geometry of the universe, containing *all the patterns of Creation*. The Temple of Osiris at Abydos, in Egypt, harbors the oldest known examples of the Flower of Life.

---

44 *Rig Veda*, 1.164.48 and 1.164.1-5.

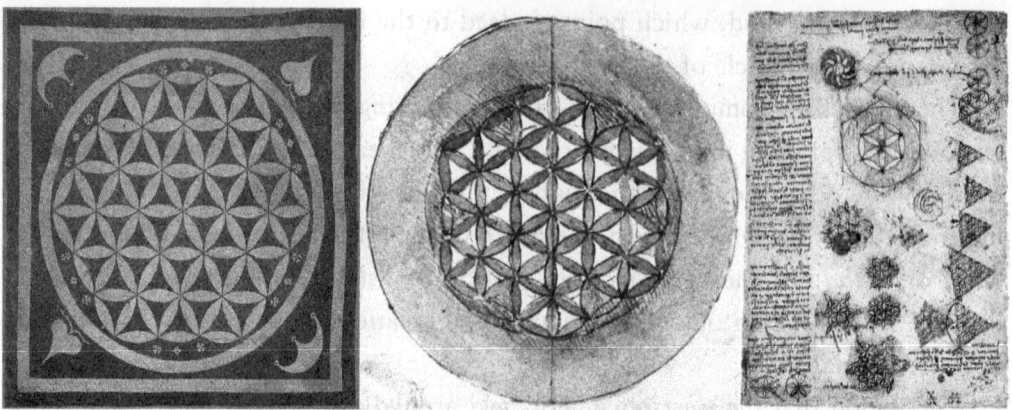

Fig. 83 (left) Mosaic of the flower of life at Ephesus (Miriam I CC BY SA 4.0)
Fig. 84: Leonardo's drawings: the flower of life (center); the seed of life (right)

At least over 6,000 years old, they may date as far back as 10,500 BC—if not earlier. The most common form of the "Flower of Life" is a hexagonal pattern made up of nineteen complete circles and thirty-six partial circular arcs, enclosed by a large circle. The center of each circle is on the circumference of six surrounding circles of the same diameter (fig. 83).

Leonardo da Vinci studied the shape and mathematical properties of the Flower of Life. He also drew what is referred to as the Seed of Life, along with geometric figures representing the Platonic solids, a sphere, a torus, and a flower-like spiraling vortex (fig. 84). The "Seed of Life" (fig. 84, right, and fig. 85) is composed of seven circles arranged in a sixfold symmetry. It acts as a basic component of the Flower of Life's design and has been regarded by some as the symbol of the 7 days of creation. The central circle is the seventh day of rest, as well as the archetypal day of world beginning.

Not only do the Seed and Flower of Life contain the geometry of the six-pointed star, but they also infuse the geometric lines of the star with the wave-like flow of its formative dynamics, intimating the phenomenon of harmony that generates it:

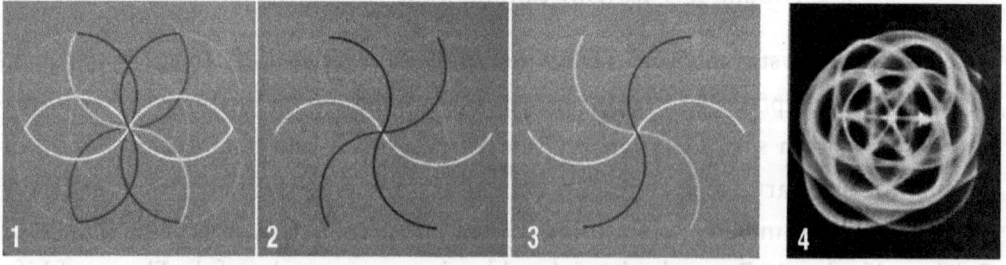

Fig. 85. (1–3) The seed of life; (4) Buckminster Fuller's "seed of life"

VII. Universality of the Harmonic Model of Creation

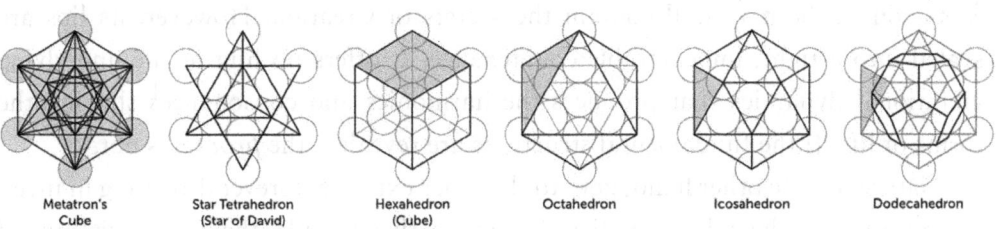

Fig. 86. *The five platonic solids are inscribed in the Flower of Life*

Intermediary between geometry and harmony, between lines and spheres, the Flower of Life offers a stylized representation of the spiraling dinergy at the core of nature. Its petals correspond to the paths of six sets of waves, which profile the interweaving currents that generate particles and forms. In suggesting the formative process at work in the spiraling of vegetables, fruit, and flowers, the Flower of Life invites the mind into the flow of life at the source of forms in a way that its geometric equivalent simply cannot.

What Buckminster Fuller's experiments brought to light, most strikingly illustrated in what reads now as a "seed of life" (fig. 85–4) are the harmonic laws at the source of all forms of life. Unsurprisingly, the geometries that carry out the transition between the formless harmony of the world soul and the archetypal forms of nature, namely the five primary solids, are all inscribed in the flower of life (fig. 86).

As the figure also shows, the Flower of Life contains the geometric basis of Metatron's Cube, one of the most sacred of ancient geometrical symbols. A three-dimensional structure of twelve spheres surrounding a central one, it was regarded to contain all of the geometric patterns of creation.

Metatron's cube is associated with the Biblical figure of Enoch, a 7th-generation descendant of Adam and Noah's great grandfather. At the end of a 365-year long life on Earth, he was brought through the 7 stages of heaven without dying. Upon meeting God, he turned into a pillar of flame and became the supreme angel Metatron, the only angel allowed to sit in the presence of God. He received the secrets of creation and is said to hold the scroll of genesis. He is often represented holding "Metatron's cube"—confirming the cosmogonical secrets sealed in the geometric symbol we are contemplating. Combined with the 12 spheres, the recurrent symbolism of 7 highlighted in the narrative and reflected in the central 7fold of the Cube, resonates with our entire context, pointing to Metatron's Cube as the symbol par excellence of the 3/7/12 archetypal template of living systems and beings.

The Kabbalah's Tree of life can also be derived from the Flower of Life; so can the tube torus be produced by ratcheting the Seed of Life.

Metatron's cube may well contain the secrets of Creation. However, its lips are sealed into a tight, impenetrable countenance. It offers no hint at the underlying vibrational dynamics that produces the harmonics and consonances shaping the forms of life. If the fundamental structures are revealed, the *process* is not.

Nature, on the other hand, goes to the other extreme, forever drowning in inventive adornments the naked simplicity of her creative formula, the precise stitching of her every nervure, the vesica of every leaf, and the unerring steerings of her golden compass. Her shape-shifting gestures hide the majestic logarithmic profile, flashing with newness yet ever the same, of the invisible Logos who steers through her motions the myriad wheels of his chariot.

Humankind's "quintessential" place in evolution invests it, individually and as a whole, with the challenge to unfold *out of its self* the Logoic countenance—to integrate the forces and dimensions of its being, to steer its "free wheels" to the mediating curve and tempered consonance of the heart of the world, to master the mute secret and command of metatron to engineer harmony and generate love. Proportionality of thought, feeling and action emits harmony and fosters love, the Logoic life of its cosmic host and being. What takes the individual and the collective to oneness is a continuous pressing for proportion, analogous to the golden course of the Logos in cosmos.

In a note written during the days of Nazi occupation in France, Simone Weil echoes Doczi in pointing to the mysterious power of limitation that guides the process of proportional becoming:

> Brute force is not sovereign in this world.... What is sovereign is limit.... Every visible and palpable force is subject to an invisible limit, which it shall never cross. In the sea, a wave mounts higher and higher, but at a certain point...it is arrested and forced to descend.... That is truth which bites at our hearts every time we are penetrated by the beauty of the world. That is the truth which bursts forth in matchless accents of joy in the beautiful and pure parts of the Old Testament, in Greece among the Pythagoreans and all the sages, in China, with Lao Tzu, in the Hindu scriptures, in Egyptian remains.[45]

---

45   Weil, *The Need for Roots*, pp. 285–87, 293.

EPILOGUE

> *We shall not cease from exploration*
> *And the end of all our exploring*
> *Will be to arrive where we started*
> *And know the place for the first time.*
> —T. S. Eliot, *Four Quartets*

Once the end touched the beginning, and the singularity of this expedition reached a landscape of universal symbols that reverberated its sightings, it became clear that a circle had rounded itself whole, and an abstract question had fleshed out its white bones with the vibrant folds of nature and cosmos. As the metaphysical wisdom of old was stepping down a path of encounter and conversation with the novel germinations of science—as if to draft the *"uni-versing"* flow of a toroidal spiritual–scientific model—the subjective quest of the book was being met by an objective periphery already mirrored in its content.

Could it be that human thinking, as it forms coherent views of the universe, operates in the archetypal way of the flow form assumed by the mushroom, the embryo, the heart, and the brain—all of which demonstrate the forming of a spherical vortex? Divine in the creations of nature, human in the creations of thought, a forward impulse eventually embodies itself by turning back on itself—"one and the same curling round on itself of the world."[1]

Is it not noble logic indeed that the creative gesture of the human mind would aim at begetting a likeness of the creative gesture of the divine mind in nature? And that a "reflection" of the universe should indwell the soundness of a coherent, "living" thought—whether the thought of a poet or scientist or the marvel hovering in a synchronicity?

As the Logos involves itself into the objective diversity of cosmos while also turning back on itself to evolve in partial, subjective impulses towards unity-universality, the creative subjectivity of thinking pulses out of the human heart, whose longing to become, create, and know is bound to be met by a periphery of resonant realities, facts, or universal symbols—and turned into "a blessed god, conversing with itself."

---

1 Purce, *The Mystic Spiral*, p. 8.

Both scientist and poet throw their vibrant lines of inquiries and intuitive inklings into the ethers. Spun of pregnant data and equations for one, of stirred feelings for the other, the lines of human aspirations long to catch and be caught by a golden ray of universality that will swirl them into a vortical sphere, a self-animated "ball" of vivid significance—a whole, live "realization."

William Blake saw it as the essence of human destiny to surge with the ray of Logoity aglow in our "selves" and let its golden source wind it into a "Heaven's Gate."

> *I give you the end of the golden string,*
> *Only wind it into a ball,*
> *It will lead you in a Heaven's Gate*
> *Built in Jerusalem's wall.*

As Steiner or Teilhard intimate in converging insights, the radial, outgoing impulse at work in everything alive is implicitly engaged, unknowingly and unwittingly, with the "tangential" or spherical "act of Being" of the cosmic whole, ever ready to round out the forward impulse of the part and wind it into a new and unique living whole, which is the offspring of their intercourse.

Is it not highly reasonable to envision how the flow-form of a full thought or coherent system of thought longs to imitate that of its god-made instruments—the brain, the heart—and so offer a hologram of the universe? Is it not a sound, symphonic truth to say that the subjective thinking activity we project into the vibrant space of the universe is meant to turn into an *"imago mundi"* in which the world finds itself reflected, thereby rediscovering itself in its own reflection?

Such is the ancients' idea about the harmonic universe. So is Einstein's equation, $E = MC^2$, a stellar example of universal thought—a mathematical toroidal flow formula, born of the harmonic blend of a human impulse to understand with the divine longing of the universe to involve itself in the partial waveforms of elements and thoughts, and wind unitive "balls" of consciousness, charged with a live kernel of cosmic intelligence eager to swirl the collective mind onward.

"Come and see!" says the Zohar. "Thought is the origin of all that is."[2] The universe is a thought, and thought is the golden mean handed to humankind to create songs and equations, theories and symphonies, religions and sciences, myths and cultures that recreate the universe and offer it to itself and to us all—donned in human garments woven of unitive cognition.

❋

What came to light in the aftermath of this entire journey is that the resulting book was, however unintendedly, a next installment in the quest that produced the

---

2   *Zohar*, I, 246b.

first—*The twelvefold archetype of the human soul*—and that the perhaps unseemly, but intriguing shape of this first book, as it glided of its own movement from subjective into objective realms, turning its stage inside out from philosophy to history as our tracking of the soul compelled it, was that of a vortex ring, albeit a poorly proportioned one. It then emerged as simple evidence that the title for this second book could very well be: *The threefold archetype of the soul of the world.*

So much hovers, unfathomed, in the trekking of a book: A mysterious destination clears the path to be trodden, but remains elusive; vistas are partly overcast; voices echo in the distance, but muffled by "the eternal silence of the infinite spaces." Yet, the hovering steadfastly beacons, blowing its pregnant breath and luminous intensity on the obtuse borders of the mind.

Does this not epitomize what every species experiences at the periphery of its constitution—as the breath of its approaching Logos? It is a compelling and moving recognition indeed, when we contemplate the kingdoms of nature around us, to discern how a next and upcoming world dimension draws near each of them, and how the world-being patiently and gradually assembles itself, kingdom after kingdom, by integrating, one by one, its seven dimensions.

Drawing near the mineral kingdom, yet external to its units and "elements," is the dimension of "system formation," the seven/twelvefold organization of a life form, which the next evolutionary step of the universe, the plant kingdom, will begin to integrate. In the mineral, organic wholeness still lies outside the bounds of each atomic element per se, but it hovers above and around them all as a dominant feature of the kingdom as a whole.

Drawing near the plant kingdom, its mineral constituents and its etheric life, is the next whole dimension of astral sentiency. Autonomous, "systemic"–organic life has been integrated. Still external to the plant is the sentient networking of information associated with a nervous system. Yet it draws near. As ethno-pharmacologist Dennis McKenna points out, mycelium root networks, which can be miles in diameter, are thinking neural networks. Peter Wohlleben, the author of *The Hidden Life of Trees* and a life long forester, makes a convincing case, based on his observations, that trees communicate and the forest is a social network. He develops the notion of the "wood-wide web," through which nutrition and signals are exchanged among trees.[3]

The animal kingdom internalizes this next dimension in the form of an "information" system: an instinctual-emotional (astral) organism, mediated by its own nervous system, *animated* by desire and fear. This internal neural network organizes the new, *animal* life form, a form equipped with the faculty of autonomous

---

3   The notion was first introduced by Dr. Simard in the journal *Nature*.

movement and the capacity to preserve itself and reach out for what it needs. What remains external to the animal is the self-soul principle. While elements have a group organism, and plants a group nervous system, animals have a "group soul." In contrast to the sense of psychospiritual uniqueness we have in any encounter with a human being, the sense we have about the "individuality" of cats and dogs, lions or eagles, is, generally said, *as a species*—which does not negate the "personality" traits and inklings of "selfhood" any companion to a cat or a dog will note in his animal friend.

The human kingdom heralds the integration of the dimension of "self." Language, self-awareness, and reflectiveness signal the incorporation, in this new life form and kingdom, of the quintessence of Logoity. The internalized principle of Egoity or I AM-ness reorganizes from within the entire countenance and organism of a "living being." The new being is shaped by an etheric body, animated by an astral body and minded by an Egoic principle. He "stands" between Heaven and Earth, self-aware and essentially motivated by a creative soul potential and a monadic spark of Logoity.

So, the question naturally arises: What next dimension of the world being might be drawing near humankind? What hovers above our disorders and wars, economic throes and woes, cultural and social developments? The dim intuition of a similar hovering arrested Teilhard de Chardin as he gazed at the live frontline of World War I:

> As I turned back to take a last look at that sacred line, the warm and living line of the front, in a flash of half-formed intuition I thought I could see it taking on the form of some higher, very noble, entity: I felt that it was coalescing as I watched it, but that it would have called for a more perfect mind that mine to grasp it and understand it. I thought, then, of those fantastically vast cataclysms which, aeons ago, had only animals to witness them. And it seemed to me, at that moment, that confronted with this great thing that was coming into being, I was like a dumb beast whose soul is awakening and can perceive groups of correlated realities and yet cannot see what it is that holds together the single reality they represent.[4]

The coalescing "form" Teilhard perceived above the world camps pitted against each other by nationalistic urges, may have been the "noble entity" that began to step down, after the end of World War II, into a set of international organizations intent on grounding one facet or another of a new world order, reflective of a global self awakened by global suffering: Peace in unity, international justice, world bank, world health etc.... Though far from constituting a world organism, such global

---

[4] De Chardin; quoted in Henri de Lubac, *The Eternal Feminine*, p. 114.

versions of major sectors of society speak nonetheless to the dawn of global thinking and global will in our world history—the dawn of the "noble self" and form of humankind.

The "noosphere" anticipated by Teilhard, the intuition of which germinated in his mind precisely during the war, has become a tangible reality with the advent of a world wide vehicle of information and communication—the world wide web—the brain basis of a global mind. With the world wide web, humankind has assembled the mechanism, *logistics* and "logiciels" (French for software) of a global brain sphere of information sharing.

Tesla voiced the same intuition with predictive precision in a 1926 interview:

> When wireless is perfectly applied, the whole Earth will be converted into a huge brain.... We shall be able to communicate with one another instantly, irrespective of distance. Not only this, but through television and telephony we shall see and hear one another as perfectly as though we were face to face, despite intervening distances of thousands of miles; and the instruments through which we shall be able to do this will be amazingly simple compared with our present telephone. A man will be able to carry one in his vest pocket.

Reminiscent of the neural network of plants, which anticipated the animal or *animated* kingdom, the world wide web vehiculates a rhizomatic proliferation of intelligent activity. A purely neutral and neural technological field, it is *technically* global and not animated, as of yet, by a self-organizing principle—a brain circuitry without an indwelling consciousness at the center. A pure carrier of information, the sophisticated vehicle engineered for the intelligent activity and exchanges of the world is like a body without a soul. However soulful its particular components/sites might be, the field as a whole is of the order of a marvelously constructed, automated air traffic system of information sharing. The *"logiciels"* do not host a Logos, or only to the very limited extent that they are "logic" based information "wares," hard and soft automatons programmed to execute sequential operations that make up their virtual connective fabric.

The fragmented contents and cacophonic tones of the web are operating outside any concern for harmony. However, the feeling of global interconnectedness that accompanies the internet phenomenon does hint at the soul dimension, as if humans had managed to engineer a mechanical, technological version of the unitive milieu of the soul, and constructed below something that mimicked the above. As planes mimic birds, and cars animals, the cold, metallic cases of global Internet connections harbor a linking machinery that mirrors, albeit as a mindless simulacrum, the subtle bonding milieu of mind and soul. A tool of and for interconnections, it has, as of yet, very partial, fragmented and incoherent repercussions on global concerns and global will.

Nonetheless, the noble logic of cosmos suggests that, as the transition from plant to animal kingdom illustrates, this neural network, newly internalized by humankind, anticipates and announces a next dimension and stage of our kingdom—a stage of collective soul actualization. As of yet unfathomed, yet deeply logical (*cosmo-logical*), is the eventual animation or ensouling of this neuro-mental global system by a global "self" in need of a mental apparatus. However blind we necessarily are, as the dumb beasts Teilhard felt a sudden kinship with, to the gestures of our own evolution, we may sense a higher entity, who is the "I" and soul of humankind, hovering above our world wide web of exchanges. However unaware we are of the collective purposes that transcend yet guide what we do, these deeper purposes are at work in us, and we may catch their reflection in our creations, such as this technological feat, which, while transmitting all kinds of destructive and distorted informations, is the potential *"noospheric"* body of the soul centered selfhood humankind is to awaken to. Unbeknownst to the computer engineer and programmer contributing to it, the noble logic of the universe points to this technological achievement as a stepping stone provided by humanity to its descending global "self."

Teilhard gave the name "Christosphere" to the next evolutionary sphere of humankind. To give this "kingdom-to-come" the name of the archetypal bearer, for our age, of the Logoic love principle of our universe makes deep logical sense. Other names such as kingdom of god, White Brotherhood, Grail community, Spiritual Hierarchy, Akashic paradigm (Laszlo[5]) carry the same intuition of an upcoming soul-centered stage and spirited dimension of humankind. The insight introduced by Alice Bailey that the spiritual hierarchy is currently engaged in a process of externalization–incarnation points to the complementary polarity of this heartening of the noosphere. So does Steiner's visionary prescience about the "return of the Christ in the etheric"—that is, in the etherized, Christed soul-forces of humankind.

Clearly, the economic and sociopolitical life of humanity and its relation to the environing kingdoms are still far from coalescing, as the world continues to awaken to itself. International egos and shadows are getting to know each other,

---

5   Laszlo and Dennis, *Dawn of the Akashic Age*: "The new Akashic paradigm recognizes that the coherence of the whole is a precondition of the functioning of the parts... The right way to act and to be is not merely to enhance our own, individual coherence, but to contribute to the coherence of the systems that sustain our life. This means safeguarding our coherence with our fellows in a community, in a state and nation, in a culture, and in the living worlds as a whole" (p. 198).

and the global tension between self-centered forces and good will energies is at a uniquely intense point.

Nonetheless, the supranational structures that have appeared since the world wars signal the formation of vehicles of coordinated activity toned and tuned—partly unconsciously (the internet), partly subconsciously (economic market), partly consciously (sociocultural and ecological organizations)—by the drawing near of our global self, and a growing capacity, on the part of humankind, to orient itself toward the global good and its golden rule of right relationships.

This next step of our universe's spiral is to emerge from the gradual integration of a common feeling, viewpoint, and commitment to global care, wisdom, and goodwill—an active surrender to the Logos of the Earth and a decisive internalization of the soul principle. "For the observers of the Future," Teilhard stated during a conference in the early 1930s, "the greatest event will be the sudden appearance of a collective humane conscience."

To venture closer to this unitive sphere drawing nearer to our collective mind and to glimpse its golden viewpoint will have been in the end the purpose of this journey.

What draws near all along is the God—the universe's Being. Gradually and measuredly, the stature and splendor of the world being reveals itself in the forms of space and eons of time, until his heart quickens in a human likeness—a humankind in which he begins to stand, open his eyes and speak his glory. "Men do not sufficiently realize," urges Bergson, "that their future is in their own hands. Theirs is the responsibility for deciding if they want merely to live, or intend to make the extra effort required for fulfilling, even on their refractory planet, the essential function of the universe, which is a machine for the making of gods."[6]

---

6   Bergson, *The Two Sources of Morality and Religion*.

ILLUSTRATION CREDITS

Public Domain: Figs. 3, 4, 5, 8, 11, 15, 17, 18, 22, 25, 42, 53.2, 59.1, 69, 70, 82, 84

Grateful acknowledgment to Wikimedia Commons for the following illustrations:

Fig. 13 , CC BY SA-3.0 , Just plain Bill; Fig. 14, 15, 16, 34, 37, 41;

Fig. 39, CC BY SA-3.0 : Cyp; Fig. 40, CCO 1.0, Jahobr; Fig. 43, CC BY-SA 2.0, Astronomy at Stevens; Fig. 46.1; Fig. 50, CC BY SA-3.0, Chris 73; Fig. 56, 58.3;

Fig. 58.7, photo by u/SunriseLand, www.reddit.com

Fig. 59.2 CC BY SA-2.0, Jean-Pierre Dalbera; Fig. 66, by Ramanarayanadatta astir; Fig. 67.1, 73; Fig. 81, CC BY SA-3.0, Stephen Shephard; Fig. 83, Miriam I CC BY SA 4.0; Fig. 86. Fig. 64, license by vectorstock.com.

Fig. 60, reproduced from *Cymatics: A Study of Wave Phenomena and Vibration* by Hans Jenny, © 2001; used by permission from Macromedia Publishing, Eliot, ME (www.cymaticsource.com).

Thanks also to Shambhala Publications for permission to reproduce illustrations and selected text from *The Power of Limits* by Gyorgy Doczi, © 2005. Reprinted by arrangement with The Permissions Company, LLC on behalf of Shambhala Publications, Boulder, CO (www.shambhala.com).

Thanks to Thames & Hudson for permission to reprint illustrations and text from *Sacred Geometry: Philosophy and Practice* by Robert Lawlor, © 1982. Reprinted by kind permission of Thames & Hudson, London.

BIBLIOGRAPHY

Agassiz, Louis. *Essay on Classification*. (1857); http://www.spirasolaris.ca/sbb4d2bs.html.

Arnould, *Sous le voile du cosmos*. Paris: Albin Michel, 2015.

Attar, Farid-ud-Din. *Le Memorial des saints*. Paris: Éditions du Seuil, 1976.

———. *Muslim Saints and Mystics*. Abingdon on Thames, UK: Routledge, 1966.

Aurobindo, Sri. *Rebirth and Karma*. Wilmot, WI: Lotus Light Publications, 1952.

Azuonye, Chukwuma. *Dogon*. Rosen Publishing Group, 1996. Penguin Books, 1990.

Babbitt, Edwin Dwight. *The Principles of Light and Color*. Whitefish, MT: Kessinger, 1942.

Bailey, Alice. *Esoteric Astrology*. New York: Lucis, 1989.

———. *Initiation Human and Solar*, New York: Lucis, 1922.

———. *Labors of Hercules*. New York: Lucis, 1974.

———. *Rays and Initiations*. New York: Lucis, 1972.

———. *Serving Humanity*. New York: Lucis, 1977.

———. *A Treatise on Cosmic Fire*. New York: Lucis, 1989.

———. *A Treatise on White Magic*. New York: Lucis, 1934.

Barnstone, Willis (ed.). *Apocryphal Acts of John: The Other Bible*. New York: HarperCollins, 1984.

Bergeron, Marie-Ina. *La Chine et Teilhard: Parole d'homme*. Ligugé, France: Aubin Imprimeur, 1976.

Bergson, Henri. *The Two Sources of Morality and Religion*. New York: Henry Holt, 1932.

Besant, Annie, and Charles Leadbeater. *Occult Chemistry: Clairvoyant Observations on the Chemical Elements*. London: Theosophical Publishing, 1919.

Blavatsky, Helena Petrovna. *The Secret Doctrine*, 2 vols. Pasadena, CA: Theosophical University Press, 1988.

Bohm, David. "On the Intuitive Understanding of Nonlocality as Implied by Quantum Theory." *Foundations of Physics*, vol 5. 1975.

———. *Wholeness and the Implicate Order*, London: Routledge, 1980.

Brisson, Luc, and F. Walter Meyerstein. *Inventing the Universe*. Albany, NY: SUNY Press, 1995.

Bock, Emil. *Moses. From the Mysteries of Egypt to the Judges of Israel*. Floris Books, 1986.

Burger, Bruce. *Esoteric Anatomy*. Berkeley, CA: North Atlantic Books, 1998.

Campbell, Joseph (ed.). *The Portable Jung*. New York: Viking, 1971.

Capra, Fritjof. *The Tao of Physics*. Boulder, CO: Shambhala 1975.

Capra, Fritjof, and Pier Luigi Luisi. *The Systems View of Life: A Unifying Vision*. Cambridge, UK: Cambridge University, 2014.

Cook, Theodore Andrea. *The Curves of Life*, New York: Dover, 1978.

Corbin, Henri. *Alchimie comme art hieratique*. Paris: L'Herne, 1986.

———. *Temple et contemplation*. Paris: Flammarion, 1980.

Cousto, Hans. *The Cosmic Octave*. Mendocina, CA: Life Rhythm, 1988.

Currivan, June. *The Wave: A Life Changing Journey into the Heart and Mind of the Cosmos*. Winchester UK: O Books, 2005.

D'Olivet, Antoine Fabre. *La musique expliquée comme science et comme art et considérée dans ses rapports analogiques avec les mystères religieux, la mythologie et l'histoire de la terre*. Œuvre posthume. Paris: Hachette Livre-BNF (1896 ed.), 2012.

Dante Alighieri. *The Vision: Or, Hell, Purgatory, and Paradise*. New York, 1868.

Dantzig, Tobias. *Number: the Language of Science.* New York: Pi, 2005.

Davis, T. Anthony, and Rudolf Altevogt. "Golden Mean of the Human Body" (see https://www.fq.math.ca/Scanned/17-4/davis-a.pdf).

Deleuze, Gilles. *A Thousand Plateaus: Capitalism and Schizophrenia.* Minneapolis: Univ. of Minnnesota, 1987.

Doczi, Gyorgy. *The Power of Limits: Proportional Harmonies in Nature, Art, and Architecture.* Boulder, CO: Shambhala, 1981.

Edinger, Edward. *The Mysterium Lectures: A Journey through C. G. Jung's Mysterium Coniunctionis.* Inner City Books, 1995.

Edmonson, Amy C. *A Fuller Explanation: The Synergetic Geometry of R. Buckminster Fuller.* Boston: Design Science Collection, 1987.

Einstein, Albert. *Autobiographical Notes.* Chicago: Open Court, 1949.

Einstein, Albert, and Max Born. *The Born-Einstein Letters.* New York: Walker, 1971.

Einstein, Albert, and Leopold Infeld. *The Evolution of Physics: From Early Concepts to Relativity and Quanta.* New York: Simon and Schuster, 1942.

Fideler, David. *Jesus-Christ, Sun of God. Ancient Cosmology and Early Christian Symbolism.* Wheaton, IL: Quest Books, 1993.

Fechner, Gustav. *Elemente der Psychophysik,* Leipzig, 1860.

———. *The Little Book of Life after Death.* Newburyport, MA: Red Wheel/Weiser, 2005.

Fernando, Diana. *Alchemy: An Illustrated A to Z.* New York: Sterling, 1998.

Feuerstein, Georg. *Tantra: Path of Ecstasy.* Boulder, CO: Shambhala, 1998.

Feuerstein, Georg, Subhash Kak, David Frawley. *In Search of the Cradle of Civilization: New Light on Ancient India.* Wheaton, IL: Quest Books, 1995.

Feynman, Richard. *QED: The Strange Theory of Light and Matter.* Princeton, NJ: Princeton University, 2014.

Ghyka, Matila. *The Geometry of Art and Life.* New York: Dover, 1977.

———. *The Golden Number: Pythagorean Rites and Rhythms in the Development of Western Civilization.* Rochester VT, Inner Traditions, 2016.

———. *Le Nombre d'or.* Paris: Éditions Gallimard, 1931.

Godwin, Joscelyn. *Harmonies of Heaven and Earth: Mysticism in Music from Antiquity to the Avant-Garde.* Rochester, VT: Inner Traditions, 1987.

———. *The Harmony of the Spheres: A Sourcebook of the Pythagorean Tradition in Music.* Rochester, VT: Inner Traditions, 1993.

———. *The Mystery of the Seven Vowels.* Grand Rapids, MI: Phanes, 1991.

Goethe, Johann Wolfgang von. *Dramatic Works of Goethe.* London, 1851.

———. *Maxims and Reflections.* Los Angeles: Enhanced Media, 2017.

———. *The Metamorphosis of Plants.* Cambridge: MIT, 2009.

Goswami, Amit. *The Self-aware Universe: How Consciousness Creates the Material World.* New York: Penguin, 1995.

Greene, Brian. *The Fabric of Cosmos: Space, Time, and the Texture of Reality.* New York: Vintage, 2004.

———. *The Elegant Universe: Superstrings, Hidden Dimensions, and the Quest for the Ultimate Theory.* New York: Norton, 2010.

Griaule, Marcel, *Conversations with Ogotemmeli: An Introduction to Dogon Religious Ideas.* Oxford, UK: Oxford University, 1965.

Hall, Manly P. *The Secret Teachings of All Ages.* New York: Tarcher Perigee, 2003.

Hallett, *Studies in Ray Correspondences.* Adyar pamphlet no. 142.

Hamel, Peter Michael. *Through Music to the Self: How to Appreciate and Experience Anew.* Boulder, CO: Shambhala, 1979.

Hauschka, Rudolf. *The Nature of Substance: Spirit and Matter.* Forest Row, UK: Rudolf Steiner Press, 2003.

Heisenberg, Werner. *Physics and Philosophy: The Revolution in Modern Science.* Northampton, UK: John Dickens, 1958.

———. *Physics and Beyond: Encounters and Conversations.* New York: Harper and Row, 1971.

Hiley, Basil, and Peat, F. David (eds.). *Quantum Implications: Essays in Honour of David Bohm.* London: Routledge, 1991.

Hoeller, Stephan A. *The Gnostic Jung and the Seven Sermons of the Dead: And the Sermons to the Dead.* New York: Theosophical Publishing House, 1982.

Huseman, Armin. *The Harmony of the Human Body: Musical Principles in Human Physiology.* Edinburgh: Floris Books, 1994.

Iamblichus, *The life of Pythagoras* (tr. T. Taylor). Rochester, VT: Inner Traditions, 1986.

Jambet, Christian. *Se rendre immortel suivi du "Traité de la resurrection" par Mollâ Sadrâ Shîrâzî.* Chicago: Fata Morgana, 2000.

James, Jamie. *The Music of the Spheres: Music, Science, and the Natural Order of the Universe.* New York: Grove Press, 1993.

James, William. *A Pluralistic Universe.* New York: Longmans, Green, 1909.

Jeans, James. *The Mysterious Universe.* Cambridge, UK: Cambridge University Press, 1930.

———. *Science and Music.* New York: Dover, 1968.

Jenny, Hans. *Cymatics: A Study of Wave Phenomena and Vibration.* Eliot, ME: Macromedia, 2001.

Jho, Chakradhar. *History and Sources of Law in Ancient India.* New Delhi: Ashish Publishing House, 1986.

Jung, Carl Gustav. *The Archetypes and the Collective Unconscious.* Princeton, NJ: Princeton University, 1969.

———. *Memories, Dreams, and Reflections.* New York: Vintage 1989.

———. *Mysterium conjunctionis,* Princeton, NJ: Princeton University, 1977.

———. *Psychology and Alchemy* (2nd ed.). Princeton, NJ: Princeton University, 1980.

———. *The Psychology of Kundalini Yoga: Notes of the Seminar Given in 1952.* Princeton, NJ: Princeton University, 1999.

———. *The Structure and Dynamics of the Psyche.* Princeton, NJ: Princeton University, 1960.

———. *Synchronicity: An Acausal Connecting Principle.* Princeton, NJ: Princeton University, 1960.

———. *Two Essays on Analytical Psychology.* Princeton, NJ: Princeton University, 1966.

Kepler, Johannes. *Harmonies of the World* (*Harmonices mundi,* tr. C. G. Wallis). http://sacred-texts.com/astro/how/index.htm.

Koestler, Arthur. *The Sleepwalkers: A History of Man's Changing Vision of the Universe.* New York: Penguin Classics, 2017.

Kovaks, Charles. *The Spiritual Background to Christian Festivals.* Edinburgh: Floris Books, 2007.

Koyré, Alexandre. *The Astronomical Revolution: Copernicus – Kepler – Borelli.* Paris: Hermann, 1961 (English: Ithaca, NY: Cornell University, 1973).

Kramer, Samuel Noah. *The Sumerians: Their History, Culture, and Character.* Chicago: University of Chicago, 1963.

Kükelhaus, Hugo. *Urzahl und Gebärde: Grundzüge eines kommenden Massbewusstseins.* Baar, Germany: Klett & Balmer, 1980.

Lamborn Wilson, Peter. *Angels.* London: Thames and Hudson, 1980.

Laszlo, Ervin. *Science and the Akashic Field: An Integral Theory of Everything.* Rochester, VT: Inner Traditions, 2004.

Laszlo, Ervin, and Kingsley L. Dennis. *Dawn of the Akashic Age: New Consciousness, Quantum Resonance, and the Future of the World.* Rochester, VT: Inner Traditions, 2013.

Lawlor, Robert. *Sacred Geometry: Philosophy and Practice.* London: Thames and Hudson, 1982.

Le Mee, Jean. *Ad Quadratum Construction and Study of the Regular Polyhedra.* http://www.gatewaycoalition.org/files/millennium_sphere/products/AdQuadratum.pdf

Leadbeater, Charles. *The Chakras.* Pasadena, CA: The Theosophical Company, 1987.

Leadbeater, Charles, and Annie Besant, *Occult Chemistry: Clairvoyant Observations on the Chemical Elements.* http://www.gutenberg.net, 2005.

Leviton, Richard. "Ley Lines and the Meaning of Adam." http://www.bibliotecapleyades.net/ciencia/antigravityworldgrid/ciencia_antigravityworldgrid06.htm.

Madsen, Jon. *New Testament: A Rendering.* Edinburgh: Floris Books, 1994.

Mayer, Julius Robert. *Die organische Bewegung im Zusammenhang mit dem Stoffwechsel* (The organic motion in its relation to metabolism). Germany: Nabu, 2011.

McClain, Ernest G. "Musical Theory and Ancient Cosmology"; in *The World and I*, Feb. 1994. http://www.ukraine.com/forums/religion/2412-musical-theory-ancient-cosmology.htm; also http://www.oocities.org/blackinkal/MusicalTheory.html.

———. *The Myth of Invariance: The Origin of the Gods, Mathematics, and Music, from the Rig Veda to Plato.* Lake Worth, FL: Nicholas Hays, 1976.

———. "A New Look at Plato's Timaeus"; in *Music and Man*, vol. 1, no 4, 1975.

———. *The Pythagorean Plato: Prelude to the Song Itself.* Newburyport, MA: Weiser, 1984.

Melchizedek, Drunvalo. *The Ancient Secret of the Flower of Life.* Flagstaff, AZ: Light Technology, 1990.

Mendeleev, Dmitrii Ivanovich. *Principles of Chemistry*, vol. 2. Andesite, 2015.

Miller, Arthur I. *Deciphering the Cosmic Number: The Strange Friendship of Wolfgang Pauli and Carl Jung.* New York: Norton, 2009.

Morya, Master. *Agni Yoga.* New York: Agni Yoga Society, 1928. See http://www.agniyoga.org/ay_en/Agni-Yoga.php.

Nambirajan, Srinivasan. *The Mystic Citadel of 22 Srutis Music.* http://www.22sruti.com, 2006.

Nichol, Lee (ed.). *The Essential David Bohm.* New York: Routledge, 2003.

Oshawa, Georges. *The Art of Peace.* Chico CA: Oshawa Macrobiotic Foundation, 1990.

Pauli, Wolfgang. *Atom and Archetype: The Pauli/Jung Letters 1932–1958.* Princeton NJ: Princeton University, 2001.

Plato. *The Timaeus* (tr. B. Jowett). https://www.ellopos.net/elpenor/greek-texts/ancient greece/plato/ plato-timaeus.asp.

Plotinus. *The Essential Plotinus* (E. O'Brien, ed.). New York: Mentor books, 1964.

Plutarch, *Plutarch on Plato's Procreation of the Soul in Timaeus,* sect. 24. See https://books.google.com/books?id=fBpFDwAAQBAJ&pg=PA24&source=gbs_toc_r&cad=3#v=onepage&q&f=false.

Popper, Karl. *Quantum Theory and the Schism in Physics: From The Postscript to the Logic of Scientific Discovery.* New York: Routledge, 1982.

Proclus. *Commentary on the Timaeus of Plato.* Cambridge, UK: Cambridge University, 2007.

Purce, Jill. *The Mystic Spiral: Journey of the Soul.* New York: Thames and Hudson, 1980.

Radhakrishnan, Sarvepall. *The Principal Upanishads.* Indus/Harper India, 1994.

Richardson, John Atkins. "Speculations on Dogon Iconography," in *African Arts*, vol. 11, no. 1, pp. 52–57 (Oct. 1977).

Richter, Jean Paul. *The Notebooks of Leonardo da Vinci,* vol. 1. New York: Dover, 1970.

Rohr, Richard. *The Divine Dance: The Trinity and your Transformation.* London: Society for Promoting Christian Knowledge, 2016.

Rudhyar, Dane. *The Magic of Tone and the Art of Music.* Boulder, CO: Shambhala, 1982.

Ruff, Willie. *A Call to Assembly: The Autobiography of a Musical Storyteller.* New York: Penguin, 1991.

Sagan, Carl. *Cosmos.* New York: Random House, 1980.

Schneider, Michael S. *A Beginner's Guide to Constructing the Universe: Mathematical Archetypes of Nature, Art, and Science.* New York: HarperPerennial, 1995.

Schwaller de Lubicz, R. A. *Du symbole et de la symbolique.* Paris: Éditions Dervy, 1978; English, *Symbol and the Symbolic: Ancient Egypt, Science, and the Evolution of Consciousness.* Rochester, VT: Inner Traditions, 1981.

Bibliography

Schwenk, Theodor. *Sensitive Chaos: The Creation of Flowing Forms in Water and Air.* Forest Row, UK: Rudolf Steiner Press, 1965.

Schrödinger, Erwin. *Mind and Matter.* Cambridge, UK: Cambridge University, 1958.

Srinivasan, Nambirajan, *The Mystic Citadel of 22 Srutis Music.* 2006. http://www.22sruti.com.

Steiner, Rudolf. "The Alphabet." Spring Valley, NY: Mercury Press, 1982.

———. *Anthroposophical Leading Thoughts: Anthroposophy as a Path of Knowledge: The Michael Mystery.* Forest Row, UK: Rudolf Steiner Press, 1973.

———. *Art History as a Reflection of Inner Spiritual Impulses.* Great Barrington, MA: SteinerBooks, 2016.

———. *Balance in Teaching.* Great Barrington, MA: Anthroposophic Press, 2007.

———. *Goethe's Worldview,* Spring Valley, NY: Mercury Press, 1984.

———. "The Human Heart." Spring Valley, NY: Mercury Press, 1985.

———. *The Inner Nature of Man and Life between Death and Rebirth.* Forest Row, UK: Rudolf Steiner Press, 2013.

———. *Interdisciplinary Astronomy: Third Scientific Course.* Hudson, NY: SteinerBooks, 2019.

———. *Karmic Relationships: Esoteric Studies,* vol. 3. Forest Row, UK: Rudolf Steiner Press, 2002.

———. *Karmic Relationships: Esoteric Studies,* vol. 5. Forest Row, UK: Rudolf Steiner Press, 1984.

———. *Man in the Light of Occultism: Theosophy, and Philosophy.* Blauvelt, NY: Garber, 1989.

———. *Occult History.* London: Rudolf Steiner Press, 1982.

———. *Occult Signs and Symbols.* New York: Anthroposophic Press, 1972.

———. *Original Impulses for the Science of the Spirit.* Lower Beechmont, Australia: Completion Press, 2001.

———. *The Philosophy of Spiritual Activity.* Great Barrington, MA: SteinerBooks, 2007.

———. *Reading the Pictures of the Apocalypse.* Hudson, NY: Anthroposophic Press, 1993.

———. *Spirit as Sculptor of the Human Organism.* Forest Row, UK: Rudolf Steiner Press, 2014.

———. *Supersensible Knowledge.* Hudson, NY: Anthroposophic Press, 1987.

———. *True and False Paths in Spiritual Investigation.* Forest Row, UK: Rudolf Steiner Press, 1969.

Steiner, Rudolf, and Édouard Schuré. *The East in the Light of the West/Children of Lucifer: A Drama.* Blauvelt, NY: Spiritual Science Library, 1986.

Talbot, Michael. *The Holographic Universe.* New York: Harpercollins, 1991.

Tame, David. *The Secret Power of Music: The Transformation of Self and Society through Musical Energy.* Rochester, VT: Destiny Books, 1984.

Tansley, David. *Subtle Body: Essence and Shadow.* London: Thames and Hudson, 1984.

Taylor, Richard. *Chaos, Fractals, Nature: A New Look at Jackson Pollock.* Eugene, OR: Fractals Research, 2006.

Teilhard de Chardin, Pierre. *Activation of Energy* (tr. R. Hague). New York: Harcourt Brace Jovanovich, 1978.

———. *The Heart of Matter.* Harcourt Brace Jovanovich, 1978.

———. *Human Energy.* Harcourt Brace Jovanovich, 1971.

———. *Hymn of the Universe.* New York: Harpercollins, 1969.

———. *Let Me Explain.* New York: Collins, 1970.

———. *The Phenomenon of Man.* (tr. J. Huxley). New York: Harper and Row, 1966.

———. *Science and Christ.* New York: Harper and Row, 1965.

———. *Writings in Time of War.* New York: Collins, 1968.

Theon of Smyrna. *Mathematics Useful for Understanding Plato: Or, Pythagorean Arithmetic, Music, Astronomy, Spiritual Disciplines.* San Diego: Lawlor, 1979.

Thiessen, Del. *Bittersweet Destiny: The Stormy Evolution of Human Behavior.* New York: Routledge, 2017.

Trinh Xuan Thuan and Mathieu Ricard. *The Quantum and the Lotus: A Journey to the Frontiers Where Science and Buddhism Meet* (tr. I. Monk). New York: Crown, 2001.

Twiss, Sumner B. *Explorations In Global Ethics: Comparative Religious Ethics and Interreligious Dialogue.* New York: Routledge, 2000.

Van der Wal, Jaap. "The incarnating embryo—The embryo in Us," in *Foundations of Morphodynamics in Osteopathy: An Integrative Approach to Cranium, Nervous System, and Emotions.* Torsteń Liem and Guus van der Bie (eds.). Pencaitland, UK: Handspring, 2017.

Vreede, Elisabeth. *Le Ciel des dieux: Lettres sur l'astronomie.* Picardie, France: Triades Éditions, 1954 (English: *Astronomy and Spiritual Science: The Astronomical Letters of Elisabeth Vreede.* Great Barrington, MA: SteinerBooks, 2007).

Ward, Keith. *Pascal's Fire: Scientific Faith and Religious Understanding.* London: Oneworld Publications, 2006.

Watts, Alan. *The Two Hands of God.* Brooklyn, NY: George Brazilier, 1963.

Weil, Simone. *The Need for Roots* (tr. A. Wills). New York: Putnam, 1952.

West, John A. *The Serpent in the Sky : The High Wisdom of Ancient Egypt.* Wheaton, IL: Quest Books, 1993.

Wilber, Ken. *Quantum Questions: Mystical Writings of the World's Great Physicists,* Boulder, CO: Shambhala, 2001.

Wilhelm, Richard (trans.). *The Secret of the Golden Flower: A Chinese Book of Life.* New York: Harcourt, 1962.

Woodroffe, John (Arthur Avalon). *The Serpent Power.* New York: Dover, 1974.

———. *The Garland of Letters.* Madras: Ganesh, 1971.

———. *Shakti and Shakta.* New York: Dover, 1978.

Young, Arthur. *The Reflexive Universe* (rev. ed.). Cambria, CA: Anodos Foundation, 1999.

———. *Nested Time: An Astrological Biography.* Cambria, CA: Anodos Foundation, 2004.

INDEX OF TOPICS, NAMES, AND NUMBERS

A

Abel, 359
addition, 68,
and multiplication, 174–86, 248, 336, 359, 381
    and subtraction in chain of being, 189
    in golden spiral, 217–18
Akhenaton 128–29
Al Hallaj, Mansur, 159–60, 168, 403, 414, 449
alchemy,
    opus of, 15
    psychology of, 397, 406–07, 409–22
alpha, *See fine structure constant*
Amoli, Haydar , 433
analogy, analogia, 137, 358
angle, 268
    astrological aspects as, 438
    equiangular spiral, 171
    golden angle, 222, 473, 476–82, 486
    in golden spiral, 174
    Maraldi angle, 93–94, 380
    musical intervals as, 273
Anu, 263, 454–55, 458
arithmetic, allegory of, 67–68
Artaud, Antonin, 448
Aspect, Alain, 315
Aristotle, 79, 83, 122, 129
Arthur, King, 434
astrology, 17–19
    as cosmo-psychology, 422–440
    as 3fold, 320, 362
    as 7fold, 145, 362, 367
    bridge between cosmos and self, 20–21, 23, 164, 198
    in China, 115
    Jung on, 257
    Kepler on, 86
    Plato, 437
atom, 98–101, 109–110, 331, 346
    Blavatsky & Bailey, 191, 362

Bohr, 339
Einstein, 340
golden ratio in, 480, 483
heart of is love, 149, 487
Leadbeater & Besant on, 240–242, 263
music of the spheres in, 486
Planck, 336
Platonic solid shape of, 239
spectral lines of, 471–472, 478
torus-shaped ether, 263
versus cell, 185, 386
See also Bohr, Anu
Aurobindo, 391, 414, 503
axiom of Maria, 418

B

Babbitt, Edwin, 166, 263–266, 362, 390–91, 448, 483
Bailey, Alice, 11,
    love principle, 148–9, 158, 487
    music of creation, 162, 399
    number 144, 404–05
    on atom, 362
    on sun, soul, zodiac, 13–21, 52, 122, 262, 283, 432–34
    seven, 188
Besant, Annie, & Charles Leadbeater, 170, 245
    atomic elements, 98–100, 237–41, 263, 269–270, 367, 407, 448
Bhagavad Gita, 233, 265, 288–89, 292, 294
Big Bang, 2, 4, 119, 206, 262, 293–295, 298–301, 305, 319, 344, 352, 369–370, 448, 455, 464
bindu, 298, 302–309, 320, 326
biosphere,
    in Teilhard de Chardin, 167, 169, 204, 341
Blavatsky, Helena Petrovna, 26, 187, 190–92, 357, 488
    atom, 362

dodecahedron, 118
logos as Christos, 159
law of motion in nature, 165
man as seven-leaved plant, 147, 360, 375
on selfhood, 51, 126, 191
Bode law, 80, 88-89, 238, 254
Bohm, David, 133, 292, 297, 318, 351-54
Bohr, Niels, 291-92, 311
    Bohr's atom, 339, 471, 480-83
Born, Max, 471, 478
Brahma, breath of, 293, 294, 297-302, 304-09, 345-46, 351, 367, 372, 379

C

caduceus, 240-41, 244, 265-66, 391, 415
caelum, 415-17
Cain, *See Abel*
cancer, *See capricorn*
Capra, Fritjof, 297, 308, 310, 331, 474
capricorn, cancer
    solstices of, 283-84
carbon, 98, 107, 110, 111
causal body, *See soul*
chakra, 10, 11, 14
    and planets, 142, 427
    as quantas of being, 342
    crown, 306-7
    causal body, zodiac, solar system as, 17, 18, 71, 121, 122
    clairvoyant description, 244-45
    diagram of, 266
    harmonic setting of, 139-142, 147-48, 375
    heart, 16, 25, 26, 29, 84-85, 148, 154, 206, 281, 400, 403
    Jung on, 420
    petals add to 12 x 12, 404-05
    sevenfold set, 137, 166, 188, 208-9, 246, 401
    symbolic representation, 489
chi, *See X*
China, Chinese, 11, 114-16
Chladni, Ernst, 143, 378
chlorophyll, 110-12, 120, 242, 361

Christ
    as archetype of selfhood, 431
    as logos, 130, 156, 158-59, 219
    as mediator, 28, 288, 324
    birth of principle, 404
    crucifixion, 74, 107, 160-61, 226, 324
    resurrection, 220, 402, 432
    Teilhard on, 168-70, 334
Christmas, *See Capricorn*
circle of fifths, 46-48, 51, 53, 66
    archetype of mediation, 206
    and zodiac, 435-36
    blossom of third harmonic, 385
    musical dodecahedron, 104-05
clef, G-clef, 31, 226-27
color, 137-8
    Goethe, 144-47
    Goethe/Newton, 393, 447
comma,
    and yearly cycle, 42, 102, 224
    as archetypal gap, 207, 445, 447
    definition, 39
consonance
    as symphonia, 366
    and astrological aspects, 438-39
    degrees & types of, 225, 273, 343, 377, 390
    five elements as, 240, 249, 379
    forms as bundles of, 374, 379, 392-93
    inner, 404, 407
    ontological and cosmological, 71, 441, 446, 467
    perfect, 154-155, 274
    pneumata vs. somata, 366
    Pythagorean laws of, 34, 336
    resonance vs, 272, 365-368, 372
Copernicus, 80-81, 86, 93, 166
Crookes, Sir William,
    periodic tables, 98-99, 269-270
crop circles, 371
crystal, 89, 98-100, 102, 105-6, 109-10
    seven types, 241
crystallization, 143-44, 170, 188, 370
cymatics, 143, 268-69, 273

D

d'Olivet, Antoine Fabre, 31, 381, 384
Dante
    turn in Divine Comedy, 209, 258, 278–79, 288, 480
    vision of the rose, 124, 153–54,
Death, life after, 180
De Broglie, Louis, 328, 331, 478, 481
Deleuze, Gilles, 448
dimensions, world–dimensions,
    See levels of being
dinergy, dual motion, 351, 493
    in cosmos, Taoism, EMF, 236–37, 345–46, 349, 396
DNA, 108–110, 250–51, 276–77, 348,
    in A. Young's model, 364
Doczi, György, 136, 138, 247–48, 260, 275, 409, 463
dodecahedron, 77, 79, 81–82, 93–94, 97–98
    and circle of fifths, 102
    and DNA, 109, 251
    and golden ratio, 104–05, 380
    and human body, 135
    and Sumerian pantheon, 454–55
    archetype of self–organization, 112–14
    as quintessential element, 437
    foundation stone, 154
    in cymatics, 269–70
    secret character of, 446
    universe as, 119–20, 122
Dogon, 442–60
duplication, dualization
    and mediation, 78, 205, 232, 233
    and threefolding in science, 317–20
    as self-sameness, 225, 271
    cell duplication, 276
    in Dogonic cosmogony, 444–45
    in Hindu cosmology, 300–01
    in Jung, 359
    in Plato, 76
    in Taoism, 307
    octave as symbol of, 40, 45, 207, 372, 374
    of world-soul in Plato, 72–75
Dzyan, stanzas of, 147, 357, 444, 488

E

Edinger, Edward, 415–17
egg, cosmic egg,
    and Big Bang, 455
    Blavatsky on, 444
    Dogons', 443–48
    human auric, 265
    mandorla as, 27–30
Ego/Egoic lotus, *See soul*
Einstein, Albert, 6, 102, 195, 292
    and Blavatsky, 292, 357
    and music, 336, 339–40, 356
    $E=MC^2$ as trinity, 309–10, 315, 316–17, 324–27, 332–33, 349
    energy emits "wholes", 330–32
    relativity, 200, 290, 299
    speed of light, 322, 23
electromagnetism, electromagnetic field
    and ether, 379–80, 391
    and quantum physics, 321–32, 337–38, 342
    bidirectionality, 345–46, 349, 396
    earth grid, 117–18
    in ancient teachings, 302–23
    Maxwell, 296, 321–27
element, elements
    and chakras, 245, 401
    and harmonics, 241, 379, 389
    and platonic solids, 76–80, 90–92, 97, 269–70, 273, 367–68
    and senses, 193
    atomic, 237
    Chinese, 117
    clairvoyant investigation of, 98–101, 106, 241
    fifth, quintessence of, 78–9, 104–45, 136, 357, 359, 380
    four, 136, 436
    Hindu, 113, 385
    intimacy with, 160, 403, 414, 466
    logarithmic system, 363
    marriage of, 256
    of life, 114
    organization, 109–11, 184–85, 497
    periodic table, 99, 238, 363, 389

spectral lines, 472
elemental
    levels of being, 49, 90, 107, 112, 189, 433
embryology, 275–76, 396
Enoch, 493
ether, 114, 126, 236, 380, 386,
    Aristotle's, 79
    hair of Shiva, 113, 385
    Hindu, 113–14, 379
etheric body, 14, 16, 128
    and zodiac, 17, 121–22, 127
    Tesla on, 357

F

Fechner, Gustav, 176–183, 203–05, 211, 250
Feuerstein, Georg, 318–19, 401–03, 412
Feynman, Richard, 471, 476–77, 484
Fibonacci, 220–26, 242, 274, 463, 476, 482
    gap with golden mean, 224
Fideler, David, 32–34, 137, 202, 467–68
field, primordial wavefield, 45, 119, 134, 186, 206, 293–97, 300–03, 446, 448
    from field to form, 270–72, 365, 367–70, 372, 386, 478–83
    in physics, 321–23, 325–29, 331, 335–37, 344–58
    of resonance, 239–40, 246–49, 251, 379
        chakra as, 244
    planets as harmonic, 257–58, 438
    quantum, 242, 308, 310
    unified, 316–18, 321, 422
    See also electromagnetic field
fifth
    as dominant, 148, 271, 387
    and love, 154, 159, 288
    and Venus, 50, 105
    element vs principle, 387
    fourth and, 32–35, 64, 66, 145, 210, 358–59
    human being as, 153, 279
    in Kepler's third law, 82
    identity with golden mean, 102–03, 108, 224–27, 473
    integration of whole & part, 367
    interval of mediation, 31–35, 37–41, 489
        genesis, 44, 53–54

principle of life, 106, 112
principle of mind & soul, 48–54, 90, 105–56, 126, 191–92, 200
    source of sevenfold, 381–2
    Sumerian Enki, 454, 458
    See also circle of fifths & element, fifth
fine structure constant, alpha, 470–87
fourth, *See fifth*
Fuller, R. Buckminster, & Hans Jenny, 143, 265, 268–72, 388, 492–23
fullerenes, 107

G

galaxy,
    black hole center, 263
    Milky Way, 253
    toroid form, 265–66, 280
    12-sided structure, 120
Galileo, 86–87
gematria, 468–68, 471
Ghyka, Matila, 105–07, 135, 187, 366
gnomon, 213–20
Godwin, Joscelyn, 89, 139, 199, 377–78, 436, 438
Goethe, Johann Wolfgang von, 8, 108, 110
    and Newton, 393
    on colors, 144–47
    on nature, 152, 229
    plant, 184
gold as, 408
golden mean, section, ratio, divine proportion, Phi
    adaptability, 222–25
    and physics, 320, 370, 476–86
    archetype of fractality, 219
    as heartbeat of cosmos, 230–34
    as logos, 130, 215, 226–28, 233–34
    as number, 103
    as phi, 96
    as quintessence, 386–88
    as trinity, 227
    as vessel of love, 148–61, 288
    definition of, 58–60
    encryption of the world-soul, 228, 361
    in body, 134–6

in DNA, 251
in phyllotaxis, 254–55
in psychology, 410–12, 416
in solar system, 82, 254–55
in Sumerian pantheon, 458
in Tarot, 462
in torus, 274
logoic number, 223
ratio of reason, 223, 259
seed of platonic solids, 97, 104–08
thinking as, 127, 201–02
uniqueness, 75, 97
golden spiral, *See spiral*
golden angle, *See angle*
   See also fifth and Fibonacci
Goswami, Amit, 311–13, 319, 371
Griaule, Marcel, 442–43
Grof, Stanislav and Christina, 422
guna, 244, 304–05

H

harmonics, definition, 34–37
   and Platonic solids, 96
   astrology as science of, 428, 435
   vowels and, 139
   See also chakra, elements, field, vowel
harmonization
   creation & evolution as, 91, 125, 157–58, 196, 223, 229, 490
   power of three, 62
   psychology of, 397, 401, 406–07, 430, 439–40
harmony
   and tone, 377
   definition, 31
   formula, 32–34
heart
   and individuation, 403–04
   center, 10, 16, 25, 63, 84–85, 148–49, 400, 413, 489
   dual vortex, 263–66, 268, 277–80
   midnight as, 282–6, 288
   Rudolf Steiner on, 25–26
   zodiac as, 13, 17–18

Heisenberg, Werner, 1, 79, 205, 291, 329, 343, 356, 370
helix, 237, 247, 250–51, 258, 343, 347–48
Hercules, 161, 404
Hermes, 157, 490
hexagram, 488–490
Heyrovska, Raji, 480–81
Hindemith, Paul, 371
holomovement, 244, 248, 343–46, 351, 353, 485
horoscope, 20, 426, 429
   Jung on, 257

I

I Ching, 116, 307, 375
Iamblichus, 31, 32, 174, 358, 455, 464–66, 489
ida, pingala, sushumna, 244, 246, 266, 304, 306, 391, 401, 450
individuation
   completion, 420, 431, 440
   notion of, 23, 152, 190, 399,
   path of, 404–411, 413, 417–20, 422, 424
initiation,
   Dante, 154
   fourth, 258, 431
   gate of Capricorn, 283
   Hercules, 161
   human path, 186, 219, 284–87
   Jung, 421 fn.
   Orpheus, 466–67
   Plato, Pythagoras, 30–32, 56–57, 160, 446, 455, 464–65
   12 world-initiators, 433
integer, 32, 37, 70
   harmonics as, 248
   in torus knots, 274
   reciprocals, 73
   smallest octave, 33, 94–95, 149, 337
   versus proportion, 223, 225
interval (musical), 32–42, 51–54, 63–72
   define platonic forms, 93
   fourth and fifth, 210
   of duplication and mediation, 75
   See also ratio, harmonics

J

Jaap, van der Wal, 275–6
Jenny, Hans, *See Fuller, R. Buckminster*
Job, 76, 159, 230
Jung, Carl, 22–24, 53, 73, 151, 359,
    404–21, 431, 462, 485
  astrology, 23, 257, 424
  love as kosmogonos, 149
  on yoga, 420
  psyche as organism, 15
  synchronicity, 313–15

K

kingdom
  animal, 123
  human, 123, 201, 215–16, 218, 386–88,
    390, 400, 425, 498
  mineral, 107, 237
  molecular, 364
  of nature, 14, 50
  of soul, 158, 189, 334, 360
  partials of Logoic tone, 192, 209, 245,
    342–43, 391–92
  plant, 110–12, 185, 222, 242, 361, 386,
    388, 390
  seven, 147–48, 189, 369
  sevenfold, 375
  unfolding of, 497–500
Kepler, Johannes, 49, 51–52, 80–92, 125, 128,
    134, 183, 225, 253, 371, 376–77, 379,
    428, 438
  third law, 82
kundalini, 30, 246, 265, 307, 401, 420

L

Lambda, Plato's, 61–62, 64, 67, 72–74, 383–84,
    456
language, 31, 62, 184, 215, 229–30, 313, 368,
    387, 461, 468, 486
Laszlo, Ervin, 296–97, 328, 369–70, 397, 500
Lawlor, Robert, 64, 67, 96–7, 111, 119, 123,
    212–14, 231, 249–50, 305, 344, 350,
    374, 408
Le Mee, Jean, 92–96, 165, 380
Leadbeater, Charles, *See Annie Besant*
Lemaitre, Georges, 293, 305, 448, 455
letter
  and numbers, 467–68
  Hebrew letters, 461
levels of being
  notion of, 11, 14, 26, 48–50, 52, 138, 141,
    184–189, 245–46
  theosophical map of, 12
Leviton, Richard, 114, 117–18
life
  versus Life, 76 fn
light,
  as heart of the story, 333
  speed of, 309, 316–7, 322–6, 337,
    471–72, 475, 479
  versus Light, 478 fn,
logarithm, logarithmic
logarithmicity, 176–82, 184–87
  and harmony, 205–08, 213
  as threefold principle, 199
  blend of outward/ inward process, 194
  chain/scale of being, 187–92
  co-arising of world and soul, 209–12
  dynamic of creation, 199, 200
  growth, 213–14, 218
  in physics, 331–2, 336, 339, 343–44, 350–
    51, 354, 356, 363, 477, 482, 486–87
  in sense organs, 195–97
  in standing wave, 247–49
  in phi, 230–31, 233–34
  of mind, 201–03
  Phi, archetype of, 223, 228, 233
  scale, 177–8
  sense perception is, 176–83
  series/progression, 175–76, 179
  spiral, 172–75, 236–38
logoic number, *See golden mean*
logos
  as bindu, 305–06
  as fifth, 210
  and logarithmicity, 174, 178–79, 184, 187,
    200
  as Phi, 228–34, 387, 445, 477, 486
  Christ as, 130, 324
  cosmos and, 181, 195, 199, 201, 210–20,
    267, 419

Greek notion of, 60
Kepler, 83, 85
logic of, 186, 196
proportion, 137, 156, 168, 487
solar, 18, 74, 107–08, 149
triune, 121, 227
Word, 157–162

M

mandorla, 27–31, 43–45, 243, 265, 374
Maxwell, James Clerk, 296, 321–22, 327–30, 336
McClain, Ernest G., 32, 42, 67, 72–73, 94–95, 178, 337–38, 458–59
mean
    musical, 33–34, 36, 40, 63–66
    See also golden mean & tritone
Melchizedek, Drunvalo, 265, 276, 483
Mendeleev, Dmitri, 100, 237, 239, 362–63
metal, 187, 363–64, 419, 490
Metatron, 493–94
mineral, 107, 109, 112, 153, 237–38, 241, 258, 386
    See also kingdom
Mitchell, Edgar, 293
Mithras–Phanes, 28–30, 44, 243, 265, 284, 431, 446, 468
monad, 14–15, 18, 31, 42, 49, 122, 126, 161, 189–92, 230, 334, 397, 402, 407
monotheism, 129–30
Moseley, Henry, 363
Moses, 128–29, 434
multiplication, *See addition*
music,
    etymology, 384
    Indian, 449–64
    of spheres, 487, 365, 441, 486
    of the Lord of the world, 162

N

Newton, Isaac, 83, 86–88, 102, 128, 157, 291, 296, 407
    and Goethe, 144, 147, 393

O

octave, 40
    law of, 372–82
    living system as, 271–72, 367
one, *See 1, 2, 3 below*
Oppenheimer, Robert, 292, 325
Orpheus, 403, 465–67

P

Palmer Hall, Manley, 42, 120, 138, 141, 397, 467
particle
    elementary, 2, 4–6, 79, 132–33, 263, 270, 296–97, 307–21, 326–33, 336–40, 343–58, 367–70
    from light to, 478–83
    12 particles, 362
Pauli, Wolfgang, 412, 471, 485–86
perception, 123–24, 139, 175–85, 192–94
periodic table, *See elements*
Phanes, *See Mithras*
phenomenon,
    nature of, 228, 310–13
    of creation, 30, 121, 144, 155, 332
    origin of, 489
Phi, *See golden mean*
pi, 484–85
Planck, Max, 323, 329–40, 351, 354–56, 472
plant, *See kingdom*
Platonic solids,
    in Plato, 77, 79
    in Kepler, 81–82, 84
    See also golden mean, atom, harmonics, element
Pleiades, 76, 106, 159, 165, 253
Plotinus, 155, 179, 293
Prana, 127, 243–44, 297, 391
proportion,
    in being, 467, 494
    in quantum physics, 330–32
    keys of, 101, 106, 155, 249, 298
    musical, 31–34
    principle of, 58–60, 75–76, 78
    pulse of love principle, 275

See also logos
Pythagoras, 32–4, 38, 42–6, 67–69, 81fn, 256, 366, 397–98, 465–68, 487

Q

quark, 320, 355
quintessence, 49, 51, 53, 126, 136, 141, 147, 192
   alchemy, 406–08

R

rainbow
   and musical scale, 138
   angle of, 478, 486
ratios, major, 225, 273, 366
   See also interval
rationality, 201–05, 222–23, 230, 425, 440
relativity, Einstein's theory of, 290, 299, 316–17, 321, 327
resonance, *See* consonance
resurrection, 30. 108, 219–20, 284, 287, 324–25, 402
Revelation, Book of, 11, 360, 405
Ricard, Matthieu, 313, 296, 299
Rig Veda, 10–1, 17fn, 94, 121–22, 243, 259, 338, 358, 360, 475, 491
rose
   fragrance of, 123, 249
   of heaven, 124, 153–54, 258, 403
   Venus cycle as, 50
Rudhyar, Dane, 39–41, 54, 188, 376, 378

S

self-sameness, *See* duplication
scale
   chromatic, how obtained, 38–41, 46–47, 66, 69, 96, 104, 178, 483
   diatonic, 32–33, 38, 48, 137, 142
   diatonic & chromatic, 10, 137, 208–09, 393
   dominant in, 221
   in Babylonian times, 459
   in experiments, 268–74
   names of seven notes, 376
   rationale for, 381–82
   with smallest integers, 94–95, 337, 475

Schrödinger, Erwin, 291, 311, 315, 327–28, 331
Schwaller de Lubicz, R. A. 202–03, 216, 274, 363
Schwenck, Theodor, 143, 251–53, 264, 395
seven, *See* 7
Shakti / Shiva, 40,
   in hexagram, 488
   polarization of Brahma, 302–07, 337, 345, 347, 444
   reunion of, 412
Shirazi, Molla Sadra, 190, 192, 335
shruti, sruti, 64, 450–57, 461
Solomon, 87, 150–51, 154, 489–90
Sommerfeld, Arnold, 470–71, 479–82, 485–86
   atom, 486
   music of the spheres in atom, 486
soul, Ego, Egoic lotus, causal body
   and proportion, 183
   as fifth principle, *See* fifth
   as mediation, 25–31, 199–200
   as milieu of evolution, 125–26
   as self-awareness, 60
   co-arising of cosmos and, 210
   conjugation of body and, 194
   corporealization of, 74–78
   definition & structure, 13–16
   geometric vessel of, 148
   harmony as vessel of, 53, 55, 78, 106, 148
   in plant world, 111
   incarnating/excarnating, 257–58,
   involution/evolution, 163, 171
   musical translation of, 40–47
   world-soul, composition of, 57–71
   See also kingdom of, twelvefold
space and time
   and mind, 54, 230
   are one, 360
   combination of in standing waves, 36
   Einstein unified, 296, 299–300, 316–17
   fundamental numbers in, 95
   intertwining of, 163–65, 213–14, 264, 347
   limit to partitioning, 335
   matrix of cosmos, 121–22, 309

octave as whole of, 373
    partition of, 9–11
Speed of light, *See light*
spiral
    and human being, 166
    etymology, 235, 446
    golden spiral, 172–74, 176, 211–13, 216–18
    in cosmos & nature, 166–70
    in Dogonic egg, 446
    in sense organs, 195–97
    logarithmic, 172–76, 186
    mediates, 194
    motion of life, 165,
Steiner, Rudolf
    auric egg, 265
    Dante, 144
    earth radiance, 288, 324
    fifth, 153
    Greek art, 131
    heart, 25–26
    midnight hour, 284–85
    night & sleep, 281–82
    music, 338, 365
    radial / spherical, 396, 496
    science, 392
    sound, 64
    spiral, 166
    sun, 254
    thinking, 127, 128
    4 & 3, 147
    7 & 12, 393–95
    12 world-initiators, 433–35
stone, circles of, 118
    alchemical, 124, 406–08, 415, 417, 435
    foundation, 154
Stone Randolph, 127, 166, 236, 245, 305, 346, 349–50, 391
string theory, 294, 300, 308, 354–56
Sumer, 449, 453–64
synchronicity, 202, 313–15, 412–13

T

Tao, Taoist
    dual spiraling, 345
    logarithmic spiral, 236
    union, 412, 414
    1, 2, 3, 300, 307
tarot, 29–30, 461–62
technology, 256 in, 447–48
Teilhard de Chardin, Pierre
    birth of reflection, 201, 341–42
    on love, 148, 168, 334, 402
    on matter/spirit, 109, 169–70, 302
    radial & tangential, 395–96, 496
    spiral evolution, 167–71, 204, 498–501
    union differentiates, 152
Tesla, 289, 294, 357, 499
tetractys, 397–99, 415–19, 470, 366
three, *See 1, 2, 3 below*
time, *See space*
torus, *See vortex / dual vortex*
Trinh Xuan Thuan, 313, 296, 299
trinity
    alchemical, 407
    and logos, 130, 375, 483
    and nuclear bomb, 325
    and quaternary, 147, 359, 381–82, 427
    as flow, 267
    astrological, 429
    Christian, 114
    Egyptian, 307
    Hindu, 246, 304
    human, 85, 395
    in Einstein's equation, 309, 319
    in science, 301
    of harmony, 64–65
    Planck's, 332–34
    Plato's, 74
    quantum of spirit soul body, 392
    Roublev, 227
    second trinity (3,7,12), 208, 211–12, 215, 303, 383, 385
    Taoist, 62, 114, 300, 307, 427
    Tibetan, 490
tritone, 34, 149–50, 454
two, *See 1, 2, 3*

U

unison, octave, 224, 374, 380
   experience of, 401–02, 408, 413–14, 417–18, 453
   note D as, 382
   plant as, 375
universe
   as Kosmos, 31
   as living system, 4
   as man, 14
   as monochord, 42
   as torus, 261–63
   as uni-flow, 71, 283
   etymology, 237
   "our," 18
   7-levelled, 48, 12
Unus mundus, 412–22

V

valency, 98–99, 239, 269
Venus, fifth essence, 49–52, 368
   cycle of, 105, 255
vesica piscis, 27–28, 40, 43
vir unus, 415, 417–18
Vishnu, 122, 347–48, 460
vortex
   and consonance, 272–3
   and particle physics, 344, 352, 477–8
   archetypal organ, 252–53
   celestial, 141
   chakras as, 244, 256
   creation as, 237, 348
   dual, spherical vortex, 259, 261–66, 322, 462
   evolution as, 169, 258, 335
   geometry of, 488–92
   human being as, 277–88
   platonic solids dynamic, 269–70
vowels, harmonics & chakras, 139–40
   seven, 141
Vulcan, 255

W

wave, sound-, 36
   standing wave, 37–38, 43–44, 247–48
   standing wave patterns, 143, 252
   structures, *See dual vortex*
Wilber, Ken, 187–88, 190
Woodroffe, John (aka Arthur Avalon), 113, 127, 244, 294, 341
World-soul, composition of, 57–71
   as external/internal whole, 101, 206
   as third & child, 102
   dodecahedron embodies, 119–20
   splitting of, 160–61

Y

yin (yang), 114, 117, 236, 244, 307, 345
   See also trinity
yoga, 246, 251, 400–06
   Jung and, 419
Young, Arthur,
   "fiat" of beginning, 331, 333, 335
   nature's seven folds, 145, 364
   on astrology, 20
   torus, 263, 274

Z

zodiac,
   and primal tones, 115, 435–37
   and solar system as dual vortex, 258, 262, 280–81, 286, 385
   and soul, 9–10, 18, 30–31, 125, 395, 404, 431, 434
   as heart chakra, 13, 16–17
   etheric matrix, 121–22, 127–28, 133–34
   etymology, 70, 119
   in Vedas, 113, 122
   time–space wheel, 29
   twelve entities, 433
zygote, 226, 275–76

Numbers

1, power of 1,2,3, 61–62, 397
2, See *dualization, duplication*
3, see *mediation and trinity*
1, 2, 3
    as world matrix, 96, 436
    meaning of, 184, 244, 366, 383
    Plato on, 208, 287
    powers of, 60–66, 383
7, seven, sevenfold, 44
    in human plant, 375
    in torus, 273–74
    law of, 376–83
    See also *chakra, kingdom, vowel, color, scale diatonic, octave, crystal, fifth*
12, twelve, twelvefold,
    circling of the square, 226
    housing of being (life) in time and space, 112, 460
    See also *scale chromatic, soul, zodiac, circle of fifths, mediation*
7/12
    body/soul, 208, 246, 385
    bundling of in scale, 39–40, 436
    in astrology, 425, 437
    in Dogonic cosmology, 447
    in embryology, 275–77
    in forms of world/self, 205–07
    in Metatron, 493
    in physics of particles, 486
    in Plato, *see lambda*
    in Rig Veda, 491
    integration of in torus, 268, 274
    planets/zodiac system 74–75
    space–time, 163
    twin birth, 44–45
    within the musical fourth, 359
3, 7, 12
    quantum of spirit soul body, 392
    See also *trinity, second*
22, 446–62
137, 222, 470–82, 486
144, 221, 404–05
222, 473–86
256,
    in Dogonic cosmogony, 443, 445–46, 449
    in music, 338
    in Plato, 68–69
    in technology, 447
360,
    in alpha, 473–83
    in degrees, days, scale, 94–96
    Veda, 121–22
432, 95, 338
720, 94–96, 475

Numbers

1, power of 1, 2, 3, 61–62, 307
2, see duplication, duplication
3, see mediation and trinity
1, 2, 3
  as world matrix, 96, 436
  meaning of, 183, 242–306, 383
  Plato on, 208, 258
  powers of, 60–65, 183
7, seven, sevenfold, 44
  in human plant, 373
  in lotus, 273–74
  law of, 376–83
  See also chakra, kingdom, vowel, color, scale diatonic, octave, crystal, fifth
12, twelve, twelvefold,
  circling of the square, 226
  housing of being (life) in time and space, 112, 360
  See also scale chromatic, soul, zodiac, circle of fifths, meditation

7, 12, 3
  body/soul, 208, 240, 385
  bundling of in scale, 39–40, 436
  in astrology, 424, 437
  in Dogonic cosmology, 443
  in embryology, 29–32
  in forms of world/self, 205–07
  in Metatron, 191
  in physics of particles, 486
  in Plato, see lambda
  in Rig Veda, 491
  integration of in torus, 268, 174
  planets/zodiac system, 74–75
  space-time, 181
  twin birth, 44–45
  within the musical hybrid, 570

3, 7, 12
  quantum of spirit-soul-body, 394
  See also trinity, second
  22, 446–612
  137, 222, 470–85, 556
  141, 212, 407–05
  222, 473–86
  256,
  in Dogonic cosmogony, 443, 445–49, 449
  in music, 358
  in Plato, 68–69
  in technology, 447
  360,
  in alpha, 473–83
  in degrees, days, scale, 94–99
  Veda, 121–22
  432, 95, 339
  720, 94–96, 475